Sixth Edition

Basic
Technical
Writing

Herman M. Weisman
Fellow, Society for Technical Communication

Merrill, an imprint of
Macmillan Publishing Company • New York

Maxwell Macmillan Canada
Toronto

Maxwell Macmillan International
New York • Oxford • Singapore • Sydney

Editor: Stephen Helba
Developmental Editor: Monica Ohlinger
Production Editor: Sheryl Glicker Langner
Art Coordinator: Vincent A. Smith
Cover Designer: Robert Vega
Production Buyer: Patricia A. Tonneman

This book was set in Palatino by Waldman Graphics, Inc. and was printed and bound by Book Press, Inc., a Quebecor America Book Group Company. The cover was printed by Phoenix Color Corp.

Macmillan Publishing Company
866 Third Avenue
New York, NY 10022

Macmillan Publishing Company is part of the
Maxwell Communication Group of Companies.

Maxwell Macmillan Canada, Inc.
1200 Eglington Avenue East, Suite 200
Don Mills, Ontario M3C 3N1

Library of Congress Cataloging-in-Publication Data
Weisman, Herman M.
    Basic technical writing / Herman M. Weisman. — 6th ed.
       p.   cm.
    Includes bibliographical references and index.
    ISBN 0-675-21256-1
    1. Technical writing.   I. Title.
    T11.W43   1992
    808'.0666—dc20   91-13124
                                                                                    CIP

Printing: 1 2 3 4 5 6 7 8 9   Year: 2 3 4 5

*To Margaret, Harlan, Lise, Abbi, Sally, Stanley, Mark,
and to Jason, Sara, Daniel, Nathan, and Jennifer*

Preface

A preface by tradition is supposed to give an author the opportunity to talk directly to his or her readers to tell them why his or her book will, at worst, not harm them, and, at best, increase their wisdom, add wit and charm to their character and, perhaps, brighten, if not their futures, their hopes. I ought not make such claims because I cannot help but agree with Francis Bacon (1561–1626), a great writer of prefaces who candidly admitted that prefaces are a great waste of time, and despite their attempts at modesty, were a showy display of boasts and excuses. Yet I, shamelessly, will continue the venerable custom (without boasts or excuses), because an author is obligated to explain his or her rationale so readers—students and instructors—can better understand and use the textbook.

The purpose of *Basic Technical Writing* is to provide classroom learning experiences in the art and craft of writing about technical matters. The word *technical* describes the specialized, sometimes complex, type of work certain professionals do. Inherent in being a professional is the capability to communicate effectively. What about those of you who believe you were born without such talent? It's true some persons have more innate ability to use words than others; but *no one* is innately unable to communicate. Such an attitude is a cop-out. The essence of effective writing is clear thinking. Do any of you feel you can't think?

Of course you can think! And of course you have the innate ability to use your mind to learn both the principles and the art of communication. With that ability you can master the craft of technical writing. This book is based on the premise that an understanding of how the process of technical communication operates and the problems inherent in the process, will enable you, *with practice,* to communicate effectively. *Basic Technical Writing* aims to help you think more clearly so that you can use your innate abilities to organize your messages and to communicate them with maximum clarity and effect. You can attain that capability by grasping the principles presented in the text and by applying the principles in the practice contained in the Discussion Problems and Assignments at the end of each chapter. For the past thirty and more years that *Basic Technical Writing* has been in print, students and professionals like yourself have found this to be so.

Introspecting about the changes in the technical writing field since the first edition appeared, I recognize the remarkable role of the computer. There is little doubt computer technology has made the technical writing activity more efficient and cost-effective. But what about the *process* of technical writing and the principles the process employs? Has computer technology advanced the technical writing process itself? The technical writing activity is experiencing changes similar to the revolution in the automation of manufacturing procedures of the late '70s and early '80s.

Technical writing, however, is not a factory activity. While its end-products may be automated by way of desktop publishing or by other new presentation technologies such as videotext, technical writing is an intellectual *process* for conveying thoughts and ideas by means of a natural language; its objective is to convey *meaning* about a technical subject. The operative word in the professional pursuit of technical writing is *writing*, not technical word manipulation nor technical word processing.

Given the pervasiveness of the computer in today's technical writing activity, how can or should textbooks for courses in technical writing prepare students? The answer is linked more or less to contemporary trends in the teaching of the subject.

Some observers have classified technical writing texts into two categories: product and process. The term *product* refers to a focus concentrating on techniques and models of a finished document (product) of writing. *Process* focusses on the prewriting activities and the various thought processes involved in completing the writing task.

Basic Technical Writing combines the two approaches. Technical writing is both an art and a craft. The process approach is necessary to the art of creating and/or developing information. Craft is involved in the algorithms or techniques for designing and executing the information that was generated into a document. That approach was expressed in the first edition and carried through the sixth. Important segments of the text deal with prewriting processes, with emphasis on scientific method, reasoning, operation of communication, language, and meaning. With this background, students can better absorb and practice the craft—algorithms and techniques—of technical writing.

The poet, the playwright, the novelist must master their art and their craft before they can give form and *meaning* to what they want to convey. So must the conveyer of technical information. If technical writers are to reach the minds of other persons, they must use language pertinently symbolic of their thoughts and use it with precision, clarity, and grace. The technical writer explains facts, expounds theory and principles, describes and analyzes concepts, objects, processes, events, and data for the purpose of informing and sometimes persuading the recipients of the information.

The sixth edition of *Basic Technical Writing* has been updated, reorganized, and enlarged. (Also revised: I have been influenced by Mark Twain's observation that the difference between the right word and nearly the right word is the difference between lightning and the lightning bug.)

Part One provides backgrounds for understanding the communication process and prewriting avenues, as well as immediate hands-on learning experience for writing about technical subjects. Part Two gives students a full grasp of fundamentals and algorithms of the technical writing craft and an understanding and use of word processing technology. Part Three takes the class through a step by step program of instruction on the methods for developing a written report. It comprehensively covers the phases of research and information gathering—including the new information technology—data analysis, methods for organizing the report material, and the writing process. The chapter on graphic presentation shows students how visuals can better organize, confirm, and underscore data and their interrelationships. Part Four provides instruction on shorter technical forms, proposals, oral reports, and writing technical and scientific articles.

Part Five, a Reference Guide to Grammar, Punctuation, Style, and Usage has been greatly expanded. Extensive exercises to test student skills in language use and grammar are also included. Lastly, I have updated the Bibliography to cover the full range of activities under the technical writing umbrella. Because computer technology may be unfamiliar to some students, I have provided a Glossary for the field.

To aid student learning, each chapter begins with a Chapter Objective, followed by identification of the chapter's Focus. Each chapter ends with a brief paragraph of review and reiteration of the chapter's focal points. Chapter exercises have been revised, updated, and reorganized into two categories: Class Discussion (to permit clarification of fundamentals, principles, and writing craft) and Assignments. A few chapters have team assignments. Some chapters have suggested additional readings. Each chapter has a summary at its end. More critical sections of some chapters may have end-of-section summaries to emphasize the principles or techniques involved.

Basic Technical Writing is comprehensive. Not all of its material may be telescoped into the time limitations of an academic course or fit the particulars of every institution. Instructors tell me they can comfortably and confidently select sequences and portions from *Basic Technical Writing* to cover those aspects more appropriate to their individual needs.

I am indebted to the many students who have used the book and contributed to it. To the many colleagues, teachers, and professionals who have been users of the book and who have, by their comments and suggestions, enabled me to strengthen the treatments at many points, I wish to express publicly my thanks. Also, let me acknowledge in alphabetical order the following people who have provided me with help in this edition: William Booth, *The Washington Post*; Sally Browne, Electronic Industries Association; Dr. Joe Caponio, National

Technical Information Service; Dr. Lise J. Duran; Paula A. Hollingsworth, Epi Tech; Captain Eric H. Petersen, Air Force ROTC Division, Maxwell Air Force Base; Susan Porter-Kuchey, Group 1 Software, Inc.; William C. Stolgitis, Executive Director, Society for Technical Communication; Abbi J. Williams; Harlan F. Weisman, M.D.; Sally Weisman; and Mary R. Wise, STSC, Inc.

I owe a debt of appreciation to the following colleagues who reviewed the manuscript of the sixth edition, *Basic Technical Writing*: Albert Blankenship, Central Missouri State University; Myna D. Frestedt, DeVry Institute of Technology; Jeanne Graham, Indiana Vocational Technical College; David W. Porter, Southern University; Robert Rule, Griffin College; and Ruth M. Walsh, University of South Florida.

I would also like to acknowledge the following editors of Merrill, an imprint of Macmillan Publishing, who gave me incalculable help and encouragement: Stephen Helba, Executive Editor; Monica Ohlinger, Developmental Editor; Sheryl Langner, Production Editor; and Vince Smith, Art Coordinator.

Finally, I must acknowledge my gratitude to the authors and copyright holders for permission to reprint their material and for sharing their insights with the readers of *Basic Technical Writing*. These permissions are inherent and recorded within the text.

Now a word to you, my readers: If you find that this book does increase your wisdom, sharpen your wit, magnify your charms, and brighten your hopes and futures, that is merely incidental. That potential was within *you* all the time.

Herman M. Weisman

Contents

3
The Process of Communication   95

PART TWO
Modality and Media   115

4
Technical English, Technical Style   117

PART FIVE
Short Technical Writing Forms   435

16
Short Reports   437

17
How to Prepare Proposals—From Concept to Document   453

PART ONE
Background

1

Technical Writing—How We Transfer Factual Information and Knowledge

Chapter Objective

Provide a background and understanding of the process of technical writing—what it is, what it does, and who does it.

Chapter Focus

- History of technical writing
- Definition of technical writing
- Who its practitioners are and what they do
- Problems in factual communication

What is technical writing? Though a long-established and important professional activity, **technical writing** as a term is not found in present dictionaries. The words *technical* and *writing* are defined separately. Webster's *New World Dictionary* defines technical as "having to do with the practical, industrial, or mechanical arts or the applied sciences." *Writing* is defined as "the occupation of the writer . . . the practice of composition." We might combine these two definitions to define technical writing as "writing about science and technology." People in the profession might want to restate this as "Technical writing is a specialized field of communication whose purpose is to convey technical and scientific information and ideas accurately and efficiently."

History of Technical Writing

Technical writing is the term used for a very old and very common communication process. Human-like creatures have inhabited this planet for well over a million years. For most of the time, they lived like other animals. It is only in the last 25,000 years that we have seen civilization and progress. The invention of writing has contributed immeasurably to this advance. Writing gave us the ability to record and communicate our experiences and knowledge. Each generation thereafter did not have to begin all over again. The recordings of predecessors became an available source of information and ideas that stimulated further thought and progress. It took us over 1 million years to arrive at the agricultural revolution; only 25,000 years to get to the industrial revolution; a mere 150 years to come to the atomic edge of the space age; and but 1 year to bring azidothymidine (AZT) from idea, to research stage, to commercial production and sale.[1] The process of technical writing has given continuity to human efforts and has accelerated the tempo for increasing knowledge and progress.

Technical writing can be traced to prehistoric cave paintings in France and Spain that illustrate primitive man's techniques for hunting buffalo. It has more direct origins in the earliest cuneiform inscriptions of the Akkadians and Babylonians. Modern science had its beginnings in Babylonia. The ancient Babylonians accomplished much in astronomy and mathematics; they were the first to divide the year into 360 days and the circle into 360 degrees. They were also our earliest agricultural scientists. We know this from their technical writing, which has survived in the form of clay tablets. The New York Metropolitan Museum, for example, has a clay tablet from about 2000 B.C. It is a technical manual, giving instructions for making beer.

The Chinese invented paper in the second century A.D.; but the Egyptians had invented papyrus, a paperlike substance, at least 700 years earlier. We have remains written on papyrus of Egyptian technical writing in the fields of medicine and mathematics that date to about 500 B.C. Among the more prolific technical writers were the ancient Greeks. Their writings on mathematics, the

---

[1]AZT is used to control the effects of AIDS (autoimmune deficiency syndrome).

physical sciences, biology, and psychology are more than a historical curiosity. Present-day mathematics, physics, and medicine show the influence of Euclid, Archimedes, and Hippocrates. Aristotle's writings on physics, astronomy, biology, and psychology were used as texts far into the fifteenth century. Writers today can learn much about principles of writing from Aristotle's *Poetics* and *Rhetoric*.

An important area in technical writing is the instruction book, or manual. Modern instruction books originated in the sixteenth century when the first manual on military weapons was written. Early instruction books covered the complete range of weapons for a branch of the service. Until about 100 years ago, an artillery man learned "the book." A single book covered everything his branch of the service was supposed to know—types of weapons, how to use them, how to clean them, elements of drill, and tactics. The infantry had its own book. The first technical manual for a specific weapon was written in 1856 by Commander John Dahlgren for a new naval gun. Soon after, the military produced other specific instruction books for such weapons as the Gatling machine gun.

From the middle of the nineteenth century until after World War I, military weapons were comparatively simple. If instruction manuals were inadequate, the gun crew usually had time to learn about the weapon through trial and error. World War II's advanced weapons, with their elaborate electrical and hydraulic systems, made instruction books more necessary.

World War II brought a tremendous speed-up in research and technology. The country needed a quick and efficient way to explain new scientific devices and weapons to the nonscientists and soldiers who were going to use them. How could these recruits be taught quickly to operate and maintain their equipment? The answer was instruction manuals. Every piece of equipment needed an instruction manual, and the field of technical writing grew up almost overnight.

The computer and other advanced technologies have today replaced defense systems as the primary impetus for research, development, and production. Today's society more than ever needs technical writers to explain how to use the new systems, and consumer products and services, spawned by recent advances in agriculture, biology, chemistry, computer sciences, engineering, and physics. The work of technical writers crowds the pages of periodicals, publications, and instruction manuals, and fills training videos, CDs, and cassettes. There isn't a camcorder, microwave oven, VCR, automobile repair manual, or computer software disk that is not accompanied by instructions written by a technical writer. More than 100,000 writers are engaged in the computer field alone.

What Technical Writing Is

The term *technical writing* covers a large territory. In some organizations and companies, the designation *technical writer* is given to those persons who write instruction manuals, procedure guidelines, reports, specifications, and proposals. Other companies might call them information developers or learning products specialists. In the computer industry, they might be labeled software development writers, documentation writers, or even systems analysts. Whatever they are called, the writers do similar work. They write reports and/or instruction manuals, as well as magazine articles and press releases that explain their organization's scientific and technological developments. In this latter activity, the technical writer is doing much the same work that the journalist does, but the journalist does not call herself or himself a technical writer. Newspaper and magazine writers who concentrate on science and medicine call themselves science writers or medical writers, but their work is similar to that of technical writers working for Westinghouse or Ciba-Geigy Corporation who translate technical material into terms stockholders will understand when reading their company's annual report.

Not only the physical and life sciences but also the social sciences and the humanities require the services of technical writers. For example, a discussion of suprasegmental phonemes or metathesis in linguistics or an explanation of the vagaries of the business cycle or the economics of technological obsolescence can become very "technical," as can the problem and demands involved in constructing and explaining a sociometric diagram of an individual's interaction with other group members. Demography offers a variety of technical aspects involving complex statistical data and mathematical formulas, such as cohort analysis and racial typology. In psychology, the physiology of stimulus-response and of memory; the techniques of learning to read faster; and analyses of processes like association, cognition, and information processing pose problems of explanation, interpretation, and simplification. Such problems are similar in kind to those facing the writer in explaining scientific processes like laser radiation or genetic replication.

Some people like to distinguish between the writing done by scientists and engineers for their colleagues and the writing done by technical writers who serve as intermediaries—interpreters or chroniclers of scientific and technological matters researched, developed, or created by another person. However, though there may be a distinction in the source of the writing and in the purpose for which it is written, the principles of the writing process are the same, whether the reporter is the developer of a new high-tech electronic device, a technical writer producing an instruction manual for it, or a journalist telling about it in the Sunday supplement of a newspaper. This text is concerned with the principles common to the requirements for technical writing, no matter who does the writing or by what means it is being disseminated. This is said even though I recognize that the information manipulation capabilities of computers are increasing to such an extent that some observers believe that such developments as hypertext and desktop publishing (see Chapter 5) have the potential of converting authorship into an on-line editorial craft. The computer with its capacity to embellish is only a writing tool: The principles of rhetoric, composition, and creativity will still apply.

What Technical Writing Does—Factual Communication of Experience

Let us begin with first principles. Molière, the great French comic playwright of the seventeenth century, wrote an amusing play about a rich shopkeeper who wanted to attain culture and become a gentleman. This Monsieur Jourdain hired a pedagogue to teach him the social graces.

"Teach me to talk in prose and poetry!" he ordered.

"But, sir," replied the teacher, "You do talk in prose."

"I do?" asked the incredulous Monsieur Jourdain.

"Yes, you do. You have been talking in prose all your life."

"I have? Fancy that!" exclaimed the delighted shopkeeper. "I have been talking in *prose* all my life!"

Many of us might be similarly astonished when we discover that we are scientists on occasion. Part of human experience is to observe and to induce or deduce valuable conclusions, just as the professional scientist does; for *science*, as the dictionary defines it, is concerned with observation and classification of facts for the discovery of general truths and the operation of "natural" laws.

Let me illustrate this with a personal example. One day, a number of years ago, I took my family for a drive in the country. We drove past a farm where cattle were grazing. "See the cow, Daddy!" my then two-year-old daughter, Lise, exclaimed. "Cow is hungry." She observed the cows, saw them browsing, and induced from her own experience the generalization that the cows were eating because they were hungry. Her comment was a basic example of the scientific principle of induction based on observation and past experience. Of course, Lise's reasoned inference was based on a limited observation, but it had the same probability of being correct as some accepted scientific "laws" based on a similar proportion of observations; for example, metal expands when heated, or altitude affects the boiling point of water.

To discover a fact (I am using the scientific definition of *fact*: a verifiable observation) we need not peer through a microscope nor trace a pattern on an oscilloscope. To reach a general truth, it is not necessary to collect and examine a whole notebook of data. Science is the knowledge of the physical world we see, hear, touch, taste, and smell.[2] In other words, science is learning from experience. Technical writing is the factual recording of that experience or knowledge for the purpose of disseminating it.

---

[2]According to psychologists, perception and apprehension of one's environment come about in a number of ways. The human organism has developed a variety of receptor organs of amazing sensitivity and complexity with which it receives and interprets "messages" from its environment. While it is traditional and logical to divide sensations on the basis of five senses, psychologists say that there are at least ten distinct sensory systems, and a typical sensory response involves a "global" response of the organism to a message of the environment which can almost never be limited to the response of just one of the traditional five senses. Perception is a complex process involving the sensations specific to a sensory modality interacting with past experience and other ongoing sensations from other systems.

"Do you mean," you might ask, "that if I record whatever I perceive through my five senses I am doing technical writing?" In a very limited way, yes—provided you are accurate both in your observation and in your writing. "What does that mean?" you might ask.

First, let me quickly state that most single-sentence definitions are over-simplified (as you might infer from footnote 2), but they can serve a useful purpose. For the moment, I need to provide you with a background for understanding the process of technical writing. Many elements within the process are common, but some have aspects which are specialized, not easily apparent, and often difficult.

Now, to return to your question: let me explain what I mean by *accurate observation*. Take, for example, the familiar story of the six blind men and the elephant. These blind men had often heard of elephants but had never "seen" one. One morning an elephant was driven down the road where they stood. When they were told that the elephant was before them, they asked the driver to let them touch it. They thought that by touching it they could learn just what kind of animal an elephant was.

The first blind man felt the elephant's side and called it a wall; the second grabbed the animal's tusk and labeled it a spear; the third touched the trunk and believed the beast to be a snake; the fourth grasped one of the elephant's legs and judged the animal to be like a tree; the fifth, a tall man, took hold of the animal's ear and was certain the elephant was like a fan; the sixth seized the tail and was convinced the elephant looked like a rope.

The oversimplified moral here is that even people who have eyes and all their senses sometimes do not observe carefully. But as we think about the story, we realize something more significant: *No matter what we see and report, there is always something left out in our reporting.* When what we omit, for whatever reason, is more significant than what we include, our observations are inaccurate and distorted.

Problems in Factual Communication

The accurate recording of what we perceive through our senses is complicated further by our means of recording—language. As you know, language is a body of *words and methods of combining words* to communicate our experiences, feelings, and thoughts to others. People use artificial symbols to represent things and ideas. Complications follow because words are not actually things or ideas but mere representations of them. This artificiality is the "occupational hazard" of language. We may intend a word to convey a certain meaning but the reader or listener, because he has a different background, experience, and personal makeup, receives a different meaning.

This hazard can be explained with a simple illustration. Ask a teenage lover of hard rock to listen to and describe an aria from the opera *Pelleas et Melisande* by Claude Debussy. Then ask an opera buff who is appreciative of the vague, dreamy tonal quality of Debussy to listen to and describe the recording of "Janie's Got a Gun," by the hard rock group Aerosmith. What the hard rock lover and the opera buff describe would be impressions created by a combination of present sensations and past experiences. Neither description would be objective.

In this distinction between subjective and objective lies the crux of the matter. The subjective lies within our minds and is colored by our personality and experience; the objective is independent of our minds and has its source in external things. This important distinction corresponds to the difference between literature and technical writing. Literature is an interpretive record of human progress and is based on imaginative and emotional experiences rather than the factual record of human achievements. Technical writing, however, is not history. (History is a record of our past; it subjectively interprets and offers value judgments about the past.) Literature is concerned mainly with our thoughts, feelings, and reactions to experiences. Its purpose is to give us insight. Technical writing concerns itself solely with factual information; its language appeals neither to the emotions nor to the imagination, but to the intellect. Its words are exact and precise, and its primary purpose is to inform. Its information is the activity and progress of science and technology. Technical writing is the term given to this type of informative reporting because much of the writing deals with highly technical matters. Technical writers report technical and scientific progress in books, reports, magazines, and professional and trade publications for the expansion and utilization of knowledge.

Chapter in Brief

To give an understanding of the process of technical writing, we have defined the term, examined its history, considered its practices and practitioners, and identified its major objective—the efficient communication of factual experience.

Chapter Focal Points

- History of the practice of technical writing
- Practitioners of technical writing
- The practice of technical writing

Questions for Discussion

1. Technical writing falls into the larger category of factual writing. *Factual writing* concerns itself with verifiable observations and with reasoning based on facts. It adheres to the disciplines of accuracy, order, and precision. *Literature* is concerned with the imaginative and subjective world as the writer sees and interprets it. Poetry, drama, and fiction belong to literature. In the category of factual writing are news stories, biography, history, informational articles, reports, essays, reviews, and technical writing. In what category would you place the following? Briefly explain why.

   a. The lead, front-page news item in today's newspaper

   b. The lead editorial in the newspaper

  c. A biography of Madame Eve Curie

  d. The annual report of the IBM Corporation

  e. "The Chambered Nautilus," by Oliver Wendell Holmes

  f. *The Origin of Species*, by Charles R. Darwin

  g. *On the Nature of Things*, by Lucretius

  h. An instruction sheet for assembling a tricycle

  i. *The Guaranteed Goof-Proof Healthy Microwave Cookbook*, by Margie "The Microwhiz" Kreschollek

  j. *Basic Technical Writing* by Herman M. Weisman

2. Would the following descriptions defining a rhyme scheme qualify as technical writing? If yes, why? If not, why not? The *terza rima* is a three-line stanza form borrowed from the Italian poets. The rhyme scheme is aba, bcb, cdc, and so forth. One rhyme sound is used for the first and third lines of each stanza and a new rhyme introduced for the second line; this new rhyme, in turn, is used for the first and third lines of the subsequent stanza. The meter is usually iambic pentameter.

3. a. Who reads and uses technical writing? Name at least four types of users and comment on how they might benefit from the process.

  b. Now consider a homeowner who opens a large crate and begins to assemble the various components of a power lawn mower with the aid of an enclosed instruction manual; a maintenance worker reading turbine specifications while searching for the sputter in the engine roar of a 747 aircraft as it warms up on the tarmac; a mother who reads a pamphlet on nephritis because a consulting physician has tentatively diagnosed her son's ailment as being that; a local councilwoman examining a communication objecting to fluoridation of the town's water supply. These four people are reading explanations of a particular fact or condition. What generalizations about technical writing can you make from these examples?

4. Can you write objectively, factually, and scientifically using the first person singular? Give reasons for your answer.

5. Which of the following three statements is/are accurate? Which is most accurate?

Water freezes at 32 degrees Fahrenheit.

The freezing point of H_2O is 0° Celsius.

At sea level, pure fresh water starts to freeze when the temperature falls to 32 degrees Fahrenheit.

6. If words cannot mean the same thing to two people, how is communication possible, especially in technical writing? After discussing this question in class, refer to Chapter 3, which examines the communication process in detail.

Assignments

1. Find and bring to class examples of technical writing from yesterday's newspaper, last Sunday's newspaper supplement, a general reader's magazine, and a professional journal. Explain briefly why you believe these items to be examples of technical writing. Is it their subject matter? Purpose? Tone? Style? Vocabulary? Use of graphics?

2. Examine the want ads in the Sunday edition of a large metropolitan newspaper. Openings for technical writers may be listed under several categories. What are these categories? What qualifications are employers looking for? What do the ads tell you about the level of professional skills technical writers need to have?

Suggested Further Reading

1. Examine current issues of any of the following periodicals:

   Technical Communication, Journal of the Society for Technical Communication

   Technical Writing Teacher

   Technical Writing and Communication

   IEEE Transactions on Professional Communication

2. Moran, Michael G., and Debra Journet, eds. *Research in Technical Communication, A Bibliographic Sourcebook.* Westport, CT: Greenwood Press, 1985.

Technical Correspondence

2

Chapter Objective

Provide understanding of the principles and techniques of technical correspondence and provide proficiency in its various types, with emphasis on job applications.

Chapter Focus

- Role of correspondence in science, technology, and industry
- Psychology of correspondence
- Prewriting
- Format considerations
- Employment letters
- Inquiries and responses.
- Quotation letters
- Claim letters
- Letters of instruction
- Computer's role in correspondence
- Facsimile (FAX)

Correspondence is the basic communication instrument in business and industry. Much of the activity of science and technology is conducted through letters; many technical reports are written in letter form. It is not only technical administrators or their subordinates but also the people in the laboratory who daily must use correspondence to accomplish their work. Even the basic science researcher in the relative isolation of the laboratory is called upon to write letters. He sends out inquiries and requests, answers inquiries and requests, orders equipment, sends acknowledgments, writes letters of instruction, and on certain occasions, writes sales letters and letters of adjustment. When he wants to change jobs, he will write an application letter.

Because the technical worker's successful activity depends very much on social interaction, it would be well to examine the mechanics of correspondence, the form constituting a major portion of such interaction. Moreover, many of the principles of letter writing apply to report writing. While letters are intended for a single reader and reports for a wide range of readers, letters, like reports or any piece of organized information, receive the classical structure of a beginning (which indicates the purpose), a middle (which elaborates on or develops the purpose), and an ending (which completes the purpose). Neither the report nor the business letter is written for the pleasure of the writer or reader. It is intended for some practical objective. The distinguishing dissimilarity between the report and the business letter is the definite intrusion of personal elements in the letter. Modern business letters are reader centered. Their style is based on the premise that a business letter communicates an attitude as well as a message. Reports, on the other hand, though directed toward a specific audience, are impersonal and objective. The letter tends to establish a personal relationship between the writer and the reader. The stress in the report is on fact; in the letter, it is on rapport.

The intention of this chapter is not to replace a text or handbook on business correspondence but to offer fundamentals, principles, and techniques of modern business correspondence that will be useful to the technical person. A list of some of the better texts on business correspondence will be found in the references at the end of the book.

Psychology of Correspondence

The modern business writer is much concerned with the reader. The "You Psychology" plays an important role in letter composition. "Put yourself in your reader's shoes" is the maxim. When you compose a letter, therefore, remember you are writing to a specific reader who is a human being. The reader of your letter is interested only in how your message will benefit *him*. He will not buy your product or service merely because you want his business. He will not hire you merely because you want to work for him. He will not accept delay in the delivery of your product only because you are having procurement problems. He will not buy for cash just because you ask him to. If your letter is to appeal to him, it must be constructed in terms of benefit to him. You must convince him that, by hiring you, he will get financial returns. You must show him that the material you want to use in the product he ordered is worth his waiting for. You must prove that his buying for cash is the best thing for him. Visualize your reader; tailor and personalize your letter specifically for him. Begin your letter with something which will be of interest to him. For example, do not say, "In order to help us simplify our problems in processing orders, we require customers to include our work order number in their correspondence to us." Would it not be better to say, "So that your order may be promptly serviced, please include our assigned work order number in your correspondence"?

Write simply, write naturally. Make your letter sound as if you were talking directly to your reader. Business English is not pompous English; it is a clear and friendly language. Stilted, formal letters build a fence between the writer and the reader. Compare "Please be assured, kind sir, of our continued esteem and constant desire to be of service in any capacity" with "We are always glad to help in any way we can."

Be sure that the general appearance of your letter creates a favorable impression. We like certain people because their appearance impressed us favorably the first time we met them. The same principle holds for letters. If you receive a letter that is neatly and evenly typed; well centered; correctly punctuated; and free from typing, spelling, and grammatical errors; your impression of the writer is likely to be a favorable one. Your letters can be one of the best public relations and advertising mediums because they reveal the quality of the service which can be expected from the writer.

Prewriting—Determining Your Purpose

No one can write a good letter without being exactly sure what he or she is after. Good letters are based on planning. Before starting to dictate or write, ask yourself, What do I want this letter to do? What action do I want the reader to take? What impression do I want to leave with the reader? If you have to answer a letter, read carefully the letter you have received; underline questions or statements to be answered; jot down comments in the margin. In composing a reply, it is often helpful to look through past correspondence. When you have gained all the background and facts, ask yourself, What is the most important fact to the reader? Usually the most important fact should be dealt with first; let the rest follow in logical sequence.

Organizing Your Letter—Role of the Paragraph

Set a number of paragraphs—one for each main thought or fact—before you start. This will force you to order your thoughts and prevent confused ramblings. Make use of 1-2-3 or a-b-c lists wherever possible to streamline your message. This will help to clarify and emphasize.

Use the first paragraph to tell your reader what the letter is all about. Link it with any previous correspondence, but do not repeat the subject of the other person's letter as a preliminary to your reply. The result will be something trite and clumsy like, "In reply to your letter of April 25 requesting. . . ." Neither is the participial type of opening, like "Regarding your letter of April 25," any better. A simple "Thank you for your letter of April 25" or "Thank you for your quotation request of April 25" is effective. Business people do not read letters for pleasure. They want letters to be brief and specifically to the point. Begin your letter directly. For example, "Production tells me we can now promise delivery of the 48 items of your order No. P-1465 on August 18, or a day or two sooner." Or "Here is some information about our multiplexers, which we are very glad to send you."

Try to see the closing paragraph before you begin; keep moving toward it when you set down your opening sentences. Use the last paragraph to make it easier for your reader to take the action you want taken, or use it to build a favorable attitude when no action is needed. When your message is completed, stop. For example, "Please send us the completed forms by May 15. We'll do the rest." Or "If you have a special measurements problem, our engineers might be able to help you. Just call us, and we'll be glad to send someone to see you."

Keep sentences and paragraphs short. Business letter experts recommend a sentence averaging twenty words; but vary sentence length. Opening and closing paragraphs normally should be brief. The longest paragraphs usually come in the middle of the letter. Paragraphs keep thoughts together that belong together. They enable the reader to get your meaning as easily and as certainly as possible.

Read your letter carefully before you sign it. Once it goes out, *you*, the signer, are responsible for any errors or confusion.

Format of the Letter

While content is certainly more important than format or style, all of us recognize that style and format are to a letter what dress and appearance are to an individual. None of us would appear for a job or interview sloppily dressed or covered with mud. Similarly, when we send a letter, it speaks for us in a business situation. A single error may nullify an otherwise well-written letter. The receiver of the letter can evaluate the message only by the letter's total impression. If even a minor aspect suggests carelessness, slovenliness, or inaccuracy, the reader loses confidence in the more fundamental worth of the message. Therefore, an examination of format and style is important.

Most organizations and companies use one of three format styles in their letters—the **block**, the **semiblock**, and the **simplified** style. Examples are shown in Figures 2.1, 2.2, and 2.3.

HILL AND KNOWLTON, INC.

Public Relations Counsel
150 East Forty-second Street
New York, N.Y. 10017

February 1, 19__

Professor Herman M. Weisman
1801 Richmond Road
Westport, MD 43214

Dear Professor Weisman:

I am sorry that it has taken so long to reply to your letter of November 27, requesting information about forms used in business correspondence.

Hill and Knowlton, Inc., has issued a <u>Secretarial Guide to Style</u> as a guide to the forms preferred when writing letters, memoranda, and reports. As can be noted from some of the samples given at the back of the Guide, indentation is preferred, but block form, which I am using, is also acceptable. Single space for letters and memoranda is also preferred, but length will sometimes dictate style.

I am sorry that we are not able to provide specimens of letters, memos, and reports, but feel that the material contained in the Guide will be helpful. As indicated in the introduction, the booklet is a supplement to the <u>Complete Secretary's Handbook</u> and the <u>Correspondence Handbook,</u> but does not conform in all instances to these aids.

I hope that we have been able to supply some of the answers you are seeking. We shall be happy to assist you if you wish additional information.

Sincerely yours,

Thelma T. Scrivens
Administrative Assistant
Education Department

TTS:lt
enclosure

Figure 2.1
The Block Style

THE FORD MOTOR COMPANY

The American Road
Dearborn, Michigan
June 12, 19__

Dr. Herman Weisman
1801 Richmond Road
Westport, MD 43214

Dear Dr. Weisman:

Thank you for the opportunity to help you prepare your students for what the business world will require in the field of report and letter writing. The importance of this field cannot be emphasized too much, for the ability to write good reports and letters is a basic requirement for business success.

All of the Ford Motor Company training courses on writing stress three things:

1. Be brief—Businesspersons do not read letters and reports for pleasure, so make them short and to the point.

2. Be specific—Do not make the reader interpret your letter.

3. Be conversational—Stilted, formal letters build a barrier between the writer and the reader.

The type format preferred by our company is described in the enclosed copies of our standards on internal and external communications. Also enclosed are copies of letters and reports written by our employees.

If there is any additional information you might desire, please contact us.

Very truly yours,

FORD MOTOR COMPANY

W.J. Gough, Jr., Supervisor
Offices Services Section
Administrative Services Department
Finance Staff

WJG:rn
Enclosures

Figure 2.2
The Semiblock Style

HANSON PRINTING MACHINERY

2222 22 Street S.E.   Washington, D.C. 20020

10 July 19__

Dr. Herman M. Weisman
1801 Richmond Road
Westport, MD 43214

Enclosed is the Handy Type Index and Price List you recently requested
from American Type Founders.

We are area distributors for American Type Founders and stock
practically all of the currently popular faces in the Washington, D.C.,
area. You may be assured that prompt attention will be given your order
when it is received.

We will be pleased to furnish any further information concerning
printing equipment or supplies that you may need. Please give us the
opportunity of serving you.

W. Wayne Gilbert

WWG:mje
Enclosure

Figure 2.3
The Simplified Style

Some organizations may use all three styles or combinations of them. Many federal agencies have decreed the use of the simplified style to increase the productivity of their typists and word processor operators. A good many companies have style manuals for letters and memoranda issued by personnel of their companies in order to achieve uniformity and excellence. If the organization or company for which you work has such a manual, follow its recommendations.

The semiblock differs from the block format only in that the paragraphs in the body of the letter are indented. The block and simplified block formats are used in about 80 percent of all types of business letters. The simplified letter is very efficient in that it eliminates the need of indenting and tabulating by the typist, but some readers are disturbed by its unbalanced appearance and its omission of the salutation and complimentary close. Management consultant firms have a tendency to use this form because it gives the appearance of efficiency. Many advertising firms also use it because of the breezy appearance. The simplified letter form will, on occasion, replace the missing salutation with a subject line, and frequently also omits the dictator's and typist's initials.

Mechanical Details

Stationery

Stationery should be of a good quality of unruled $8\frac{1}{2}'' \times 11''$ bond paper, preferably a 20-lb. weight composed of 25% rag/cotton stock, usually white. Company or organization preprinted letterhead should be used when appropriate. For continuation pages, bond of the same size, quality, and color as the letterhead should be used. Some companies also use shorter letterhead—$8\frac{1}{2}'' \times 7''$. This size is used for letters of one or two very short paragraphs. Tissue sheets are used for carbon copies. Letterhead bond of the appropriate quality is used for photocopies, although some businesses prefer simply to copy the letterhead along with the communication.

The Envelope

The first thing the recipient notices about your letter is its envelope. The envelope deserves as much care as the letter it contains. Assure its accuracy by checking it with the address printed on the letterhead of the company to which you are writing. The elements of neatness and attractiveness are just as important in typing the envelope as in typing the letter. There are two standard sizes of envelopes in use today—the $9\frac{1}{2}'' \times 4\frac{1}{8}''$ and the $6\frac{1}{2}'' \times 3\frac{5}{8}''$. The larger size is in greater favor because of the greater ease in folding the letter sheet.

Addresses on envelopes should be typed in block style, double-spaced if the address is in three lines and single-spaced if the address contains more than three lines. Postal authorities require that the zip code be used. Foreign countries are typed in capitals. Use no abbreviations or punctuation for the end of the line. The person's name, title, and the name of the branch or department should precede the street address, city, and state. *Special delivery* or *Registered* should be typed in capitals several spaces above the address. If an attention line is used in the letter, it should be typed on the envelope in the lower left-hand corner. If a *Personal* notation is desired on the envelope, it should appear in the same position as the attention line.

Mr. Frank J. Hill
Senior Vice President
Hill and Langer Company
93 Mill Pond Road
Dobbs Ferry, New York 10522

Personal

The Sawyer Corporation
62 Broadway
New York, New York 10004

Attention: Personnel Department

Framing a Letter on the Page

A typed letter should be so placed on the page that the white margins serve as the frame around it. Practice and convention call for more white space at the bottom than at the top unless the length of the letter and the size of the letterhead make this impossible. The side margins should be approximately equal and usually should be no wider than the bottom margin. Side margins are usually from one to two inches, depending on the length of the letter. The bottom margin should never be less than one inch below the last typed word. The right-hand margin should be as even as possible. The body of the letter proper should be placed partly above and partly below the center of the page. A short letter makes a better appearance if it is more than half above the center of the page. The placement of the letter on the page, frequently referred to as centering, can be achieved by starting the first line of the inside address at a chosen depth to give the most pleasing appearance of white space above and below the letter. The shorter the message, the lower the inside address is placed.

The Heading

When preprinted letterhead is not used, the writer includes a heading to help identify the source of the letter. The heading includes the address, but not the name, of the writer. The street address is on one line and the city, postal zone, and state on another. The date follows the city and state line. The heading is placed at the right side of the page. Here are some examples of headings:

1303 Springfield Drive	87 Cherry Creek Lane
Ft. Collins, Colorado 80521	Minneapolis, Minnesota, 55421
November 8, 19__	March 29, 19__

605 West 112th Street
New York, N.Y. 10026
December 21, 19__

The usual top margin is 1½". If the letter is short, the top margin will, of course, be longer. The placement of the heading should be planned so that the last line ends at the right-hand margin of the letter.

The Date Line

The purpose of the date line is to record the date the letter is written and signed. When preprinted letterhead stationery is used, it is the first item typed on the page. If ordinary bond paper is used, in practice, the date line has become the last element of the heading. (Note the examples above.) With letterhead stationery, the date line can be centered or placed on the right or the left depending upon the style of layout. Sometimes the design of the letterhead will suggest one position or the other. When the date line is placed on the right, it should end flush with the right-hand margin of the letter. Its usual placement is about four spaces below the last line of the letterhead. If the letter is very short, the date line should be dropped down to give a better balance to the page. The month should be spelled out in full and no period is used after the year.

The Inside Address

The purpose of the inside address is to identify the receiver of the letter by giving the complete name and address of the person or organization to whom the letter is being sent. It is usually placed four to six lines below the date line. No punctuation is used at the end of any line in the address. Write out *street* or *avenue*, as well as the name of the city and the state. Use a courtesy title, such as Mr., Mrs., Miss, Ms., or M. (when the gender is unclear or you have only initials to work with), when addressing an individual. When a title follows the name of an addressee, it is written on the same line, except where it might be unusually long; then the title is placed on the next line:

Mr. Thomas B. Morse	Ms. Abbi J. Weisman
1620 Dakota Avenue	1303 Springfield Drive
Cincinnati, Ohio 45229	Fort Collins, Colorado 80521

Mr. Jason Duran, President
General Computers, Inc.
43 Madison Avenue
Milwaukee, Wisconsin 53204

Dr. Sylvester Tarkington
Assistant Director, Engineering
True Ohm Resistor Corporation
Suite 2400
Pennsylvania Building
Philadelphia, Pennsylvania
19102

The Attention Line and Subject Line

The attention line is intended to direct the letter to the person or department especially concerned. An attention line provides a less personal way of addressing an individual than placing his name at the head of the inside address. It is losing favor, however, since current practice is to address letters to individuals. If a subject line is to be used, it should be typed below or in place of the attention line.

The J.B. McConnell Company
1812 Atlantic Avenue
Brooklyn, New York 11233

Attention: Mr. Joseph Rich, Comptroller
Subject: Payment for Crystal Diodes

Gentlemen:

The attention line is placed two spaces below the last line of the inside address; it should not be in capitals, nor is it usually indented. Practice varies on how it appears. No end punctuation marks are used. Here is an example:

General Motors Corporation
General Motors Technical Center
P.O. Box 117, North Penn Station
Detroit, Michigan 41202

Attention: Mr. Raymond O. Darling
Educational Relations Section

The purpose of the subject line is immediate communication of the topic of the letter, thus expediting its objective. Also, the subject line is an aid to filing. It should be placed conspicuously, though practice varies on exact placement. Some company correspondence manuals use the word *Subject* or *Reference* or the abbreviation for reference, RE. Others omit these. The prevailing use of the subject line in government and Defense Department correspondence has stimulated its wide use. Also, the National Office Management Association encourages its use. The subject line may be placed on the same line with the salutation or two line spaces below the last line of the inside address or so placed that it ends at the right-hand margin of the letter. Here are some examples:

Avco Manufacturing Corporation
420 Lexington Avenue
New York, New York 10017

Subject: Our Purchase Order No. T1052

Avco Manufacturing Corporation
420 Lexington Avenue
New York, New York 10017

Gentlemen: Our Purchase Order No. T1052

Avco Manufacturing Corporation
420 Lexington Avenue
New York, New York 10017

Our Purchase Order No. T1052

Gentlemen:

The first letters of important words in the subject line should be capitalized. Underscoring may or may not be used depending upon personal reference.

The Salutation

The salutation originated as a form of greeting. The simplified letter form omits it entirely as old-fashioned and superfluous. It is placed two line spaces below the inside address or six spaces if an attention or a subject line is used. Use a colon after the salutation. There is an increasing tendency to use the open pattern of punctuation. An open pattern omits commas and other punctuation marks unless meaning would be misinterpreted by the omission. A closed punctuation pattern makes liberal use of commas and other punctuation marks, placing one wherever tradition or grammatical structure allows. In the open punctuation pattern, the colon is omitted in the salutation and the comma omitted after the complimentary close. In business correspondence, if the writer and addressee have a friendly relationship, a comma often replaces the colon after the salutation. Conventionally, the word *Dear* precedes the person addressed unless a company or group is the recipient of the letter:

Dear Mr. Smith: (Preferred) *or*
Dear Sir: (closed)
Dear Bill   (open)

Dear Dr. Jones: (Preferred) *or*
Dear Sir: (closed)
Dear Dr. Jones   (open)

Dear Mrs. Smith: (Preferred) *or*
Dear Madam: (closed)
Dear Mrs. Smith   (open)
Dear Ms. Smith: (but never Dear Madam when Ms. is used)

Gentlemen: is the correct form in addressing a group of men, and *Ladies*: in addressing more than two women. When two persons are addressed, use both names:

Dear Dr. Jones and Mr. Smith:
Dear Mrs. Brewer and Ms. Handley:

"Gentlemen" has been the predominant way of addressing a company or organization. More recently there has been the recognition that women as well as men make up a company or organization. Consequently, more and more letters addressed to a group use the salutation, Dear Ladies/Gentlemen.

American Psychological Association
1200 17th Street, N.W.
Washington, D.C. 20036

Dear Ladies/Gentlemen:

Chevy Chase Savings & Loan Inc.
8401 Connecticut Avenue
Chevy Chase, MD 20815

Dear Ladies/Gentlemen:

The Body of the Letter

Begin the body of the letter two line spaces below the salutation. Whether the body is indented, of course, depends upon the style used. Single-space letters of average length or longer. Short letters of five lines or less may be double-spaced. Always double-space between paragraphs in single-spaced letters. Quoted matter of three or more lines is indented at least five spaces from both margins. Very short letters that are double-spaced should use the semiblock form to make paragraphs stand out as separate units. Double-spaced material does not receive extra space between paragraphs.

The Complimentary Close

By convention, the purpose of the complimentary close is to express farewell at the end of the letter. In a simplified style, the complimentary close is frequently omitted, but conventionally, the complimentary close should be typed two lines below the last line of the text. It should start slightly to the right of the center of the page, but it should never extend beyond the right margin of the letter. The comma is used at its end, unless the colon was omitted after the salutation. In that case, omit the comma after the complimentary close. Only the first word of the complimentary close is capitalized.

Yours truly, Very truly yours, Sincerely, and *Sincerely yours,* are the proper and conventional complimentary closes most frequently used. *Yours cordially* implies a special friendship; *Respectfully yours* implies that the person addressed is the writer's social or business superior.

The Signature

When the sender's name and title are printed in the letterhead, they are frequently omitted after the signature. Otherwise, the name and the title are typed three to five spaces directly below the complimentary close. First letters of each word are capitalized; no end punctuation is used. The letter is not official or complete until it is signed. Only after the dictator or writer has signed the letter does he become responsible for its contents. In routine letters, a secretary may, at times, sign the name of the official sender. In such a case, the secretary or stenographer signing the sender's name adds his or her initials. Sometimes her initials are preceded by the word *per*. Salutation and complimentary close must match in tone.

Yours truly, — (closed)
Stanley J. Duran
Stanley J. Duran, C.P.A., J.D.
Controller

Sincerely yours — (open)
Charlotte Billings
Charlotte Billings
President

Respectfully, — (closed)
Tom Swift
Tom Swift
Assistant Engineer

Cordially — (open)
Marilyn Troupe
Marilyn Troupe, Ph.D.
Chairman, Sociology

If a company, instead of the writer, is to be legally responsible for the letter, the company name should appear above the signature. The use of company letterhead does not absolve the writer; so if you want to protect yourself against legal involvement, type the company name in capitals a double space below the complimentary close; then leave space for your signature before your typed name:

Your truly, — (closed)
STERLING PRODUCTS
James Joyce
James Joyce
Design Engineer

Sincerely yours — (open)
AERO SPACE, INC.
Theodora Blum
Theodora Blum
Technical Editor

Reference Information

In order to identify the dictator or official sender and the typist, the dictator's initials followed by the stenographer's are typed two line spaces below the signature, flush with the left-hand margin. The dictator's or sender's initials are conventionally typed in caps, with a colon or a slant (/) and the typist's initials following, for example, HMW:mhb or HMW/mhb.

Enclosures

If there are enclosures, type *Encl.* one line below the reference initials. If there is more than one enclosure, type *2 Encls.* or *3 Encls.*, as the case may be. If the enclosures are significant, list their identifications:

> 2 Encls. (1 Contract no. N2021)
>   (2 Proposals No. 42C)

Copy Notations

When a letter requires copies for other than the addressee, the designation *c:* for copy or *cc:* for courtesy copy should appear in the lower left-hand corner of the copies. Under certain conditions, the dictator may wish to have the addressee informed of the distribution of copies, in which case, the notation should also appear on the original. In other instances, the sender may not wish the addressee to know that a copy was sent to a certain individual; then the letters *bc* (standing for "blind copy") are penned or typed under the *c* notation. The copy notation appears two lines below the enclosure data.

The Postscript

The postscript, years ago frowned upon, is being used more and more in business correspondence. It is used to add extra emphasis to some particular item or to give additional information. It is placed two line spaces below the referenced information, flush with the left-hand margin. The initials *P.S.* are sometimes used but are frequently considered unnecessary, and may be omitted.

Second and Succeeding Pages

When a letter carries over to succeeding pages, use bond paper that matches the color and quality of the first page. Some firms have preprinted second-page stationery. Always start at least one inch from the top of the page. The name of the addressee should appear at the left margin. The page number, preceded and followed by a hyphen, should be centered, and the date should be at the right—forming the right-hand margin. Sometimes these elements are single-spaced, one below the other, on the left margin. The second page should carry at least three lines, although five lines are preferable. Never divide a word at the end of a page.

New Hampshire Electronics   -2-   July 6, 19__

or

New Hampshire Electronics
Page 2
July 6, 19__

The Memorandum

Purpose and Format

The **memorandum** or **memo** has been borrowed from the practice of military correspondence. It is the most common form of written communication in business and industry. Formerly, its use had been restricted to interoffice, interdepartmental, or interorganizational communications. However, it is now being circulated with greater frequency out of the originating organization.

The memorandum format is similar to that of the letter, but its tone is impersonal. Most organizations have printed memo forms for companywide communications. The format has been highly conventionalized, although details will vary from one organization to another. There may be a preprinted heading identifying the company and department originating the memo. Instead of the inside address, the following three lines are used:

TO:

FROM:

SUBJECT:

A date line is placed usually in the upper right-hand corner. A project or file number may also be included. In place of the complimentary close, there may be a signature of the writer, or the writer may sign or put her initials following her name in the "From" line. If the memorandum is a long communication, it may be organized into a number of sections and subsections with headings. Memorandums (*memoranda* is also used as the plural form) are frequently the most convenient mechanism to convey reporting information. (Figure 2.4).

The purpose of a memo is to circulate information, to request others to take care of certain work, to report on what occurred in a meeting or on a trip, to keep members of an organization posted on new policies, to report on an activity or situation, etc.

The primary purpose of the memo is to save time for both the reader and the writer. Amenities and courtesy are sacrificed for conciseness. Format and

GENERAL DYNAMICS CORPORATION

INTRA-CORPORATION COMMUNICATION

TO: Mr. F. R. Crane   New York
FROM: John Doe   Date: December 18, 19___
SUBJECT: Executive Orders Manual

Here is your copy of the General Dynamics Executive Orders Manual
containing the formal instruction, policy statements, announcements, and
other advices of the President.

The Orders in the Executive Orders Manual have superseded all previous
Executive Orders and include all Orders currently in effect. The Manual
will be kept current by the Office of the Vice President—Organization.

Your receiving a personal copy of the Executive Orders Manual carries
with it responsibility for complying with the instructions and policies
in the Manual and for insuring compliance by all personnel under your
direction.

Enclosure   J.D.

(a)

OFFICE MEMORANDUM   Date: November 27, 19___

To: PUBLIC RELATIONS STAFF

From: J. K. ABBOTT

       We have arranged through Glen Cross to view the video cassettes of
the first three presentations in the special program on "Orientation for
some of the Company's non-insurance personnel." These tapes will be run
on Monday, Tuesday, and Wednesday, December 2, 3, and 4 at 4 P.M. in our
conference room. Anyone interested is invited.

Subjects of the first 3 talks are:

1. Monday—"Organization and History of the Company"—Mr. Meares
2. Tuesday—"The Ordinary Insurance Operations Organization and
   Actuaries Place in the Company"—Mr. Phillips
3. Wednesday—"Our Product, Ordinary Insurance, and Annuity"—
   Mr. Ryan

(b)

Figure 2.4
Sample Memorandums

language are directed to move the message along. Memos are frequently written under the pressure of time, and the writer is required to analyze a message situation quickly and to formulate it succinctly. The writer must reduce the subject—no matter how complex—to its substance in a terse, single statement in the "Subject" line. The reader is given direct, concise information and facts, with conclusions and recommendations (as appropriate) to provide clear but ample background to arrive at a proper decision and necessary action. Consider the following attempt at a memorandum as an illustration of a typical situation calling for this medium.

TO   : Mr. Stanley J. Duran, President

FROM   : Leslie S. Bruback, Administrative Assistant

SUBJECT   : Trip Report on the American Management Association
                (AMA) Conference on Employee Personnel Problems

Due to very bad weather my plane was delayed leaving our airport and it arrived more than two hours and thirty-three minutes late to New York. Instead of landing as scheduled at Kennedy International Airport, it was rerouted to Newark, N.J. Fog and other foul-ups delayed us at least another two hours and fifteen minutes.

This brought me to the Waldorf Astoria at 11:00 PM, five hours beyond the six o'clock deadline for my hotel reservation. Just as I feared, all the rooms were gone and I had to settle for a second rate hotel six blocks from the Waldorf where the conference was being held. It rained the three days of the conference, adding to the fun of getting to and from the meetings.

Despite these frustrations and inconveniences, the three days were not a washout, due mostly to a session on the third day conducted by Dr. Dennis Nagel, a consultant in industrial psychology. But I am getting ahead of myself.

I shall detail all the sessions even though all of them except for Dr. Nagel's were pretty much warmed over, old soup.

The first session on Monday—(thank God for the coffee served through the excellent arrangements of the AMA—It knows how to run meetings!)—was on "Things to Watch for in Interviewing Salaried Staff," etc., etc., etc.

Young Mr. Bruback is reliving the trying three days of his trip to New York to attend a conference. Stanely J. Duran, though a kind and sympathetic boss, cares little about the inconveniences his assistant suffered while broadening his background at a conference and cares less to have them recorded formally. All Duran wants to know is: What did you learn that we might want to consider applying at our company? He wants information, short and sweet. Of course, he will not hesitate to let Leslie S. Bruback know this. A little red in the face, Bruback tries again. He is a bright young fellow, quick to learn. His next effort Mr. Duran wants shared with his division managers. That memo might read as follows:

TO   :   Division Managers

FROM   :   Leslie S. Bruback, Administrative Assistant to the President

SUBJECT :   What Makes Good Employee Morale

     I have just returned from an American Management Association Conference in New York. A session I found extremely valuable was on Improving Employee Morale, conducted by Dr. Dennis Nagel, a well-known industrial psychologist. Mr. Duran has suggested that I pass on to you a summary of Dr. Nagel's research and recommendations for improving employee morale.

Factors Important to Morale
1. Economic security
2. Interest in work
3. Opportunity for advancement
4. Appreciation
5. Company and management
6. Intrinsic aspects of job assignment
7. Wages
8. Supervision
9. Social aspects of job
10. Working conditions
11. Communication
12. Hours
13. Ease
14. Benefits

Please note that wages are halfway down the list. We should not jump to the conclusion that money is unimportant to an employee. Once his or her basic needs are satisfied through adequate pay, Dr. Nagel says, other non-monetary factors of an employee's job take on an ever-increasing significance.

Recommendations for Improving Employee Morale
     Dr. Nagel said that there are no simple ground rules. Every situation is unique—no two employees nor two companies are identical. However, psychological research on morale and employee attitudes has indicated the following recommendations:

1. Tell and *show* your employees that you are interested in their ideas on how conditions might be improved.
2. Treat your employees as individuals, never deal with them as impersonal ciphers in a working unit.
3. Improve your general understanding of human behavior.
4. Accept the fact that others might not see things as you do.
5. Respect differences of opinion.

6. Insofar as possible, give explanations for management actions.
7. Provide information and guidance on matters affecting employee's security.
8. Make reasonable efforts to keep jobs interesting.
9. Encourage promotion from within.
10. Express appreciation publicly for jobs well done.
11. Offer criticism privately in the form of constructive suggestions for improvement.
12. Train supervisors to think about people involved insofar as practical, rather than just work.
13. Keep employees up-to-date on all business matters affecting them, and quell rumors with correct information.
14. Be fair.

Mr. Duran suggests we be prepared to discuss the implementation of these recommendations at the next Management Circle Meeting on Monday, October 17.

Employment Letters—How to Apply for a Job

For a period of time, I was associated with an electronics company. One of my duties was personnel administration and recruitment of technical and professional personnel. During that period, I had the opportunity to read applications from several hundred people seeking positions with our company. I also interviewed a substantial number of applicants. I suppose within this group was a good cross-section of young people looking for their first job, as well as those wishing to change from one position to another. The applications fell into four general categories.

The first included those so poorly written and/or so carelessly and sloppily done that I threw them into the wastebasket at once on the assumption that the person was so careless that he could not possibly be a first-rate, professional employee.

The second group contained resumes and cover letters that did not give sufficient information for me to form any judgment of the writer's prospects. Because our company was public relations conscious, these were acknowledged in polite terms but not filed.

The third included unsolicited applications written by people with fairly good backgrounds, training, and experience, but because no vacancy existed or seemed likely to occur for their particular backgrounds and capabilities, there was no reason at present for our further correspondence. However, these letters were acknowledged individually; the reply would indicate that we were impressed by the applicant's background and that we would keep the inquiry on file. If we had information about suitable vacancies at other companies, this was mentioned.

Finally, a few of the letters and resumes presented a clear picture of an individual who seemed capable, personable, well trained, and experienced for the opening, and seemed to have the personal qualifications most organizations want in their staff. Such persons were invited for a personal interview. When the interview confirmed first impressions, the individual was further interviewed by department heads in whose section or division his or her talents and experience would be appropriate. If the interviews further confirmed earlier impressions, a job offer ensued.

The situation I have just described is the typical one in the employment market. Many job applicants, no matter how capable or experienced, are so obsessed by their own personal situation that their approach in applying for the job is unfortunate. In business and industry, the resume plays a prominent role in helping to place individuals in appropriate openings. It might be well to picture the setting into which an application arrives. For purposes of illustration, let us imagine that you are an employer. Let us say you have an opening for a very responsible position that pays a competitive salary. You are anxious to find a qualified person to fill this vacancy. You place an ad in the Sunday edition of the city newspaper because you want it to receive a wide reading.

Response to the ad brings you more than 100 envelopes bearing applications. Although you are delighted to get so many, you are also overwhelmed by the problem of how to select the right person for the position. As you look at the envelopes, you notice one that is lavender and scented. That is in such unbusinesslike form you file it promptly into the circular file (wastebasket). You notice another envelope that has misspelled both your name and street address. There is another with the remains of the applicant's breakfast on it. "Well," you tell yourself cheerfully, "I wouldn't want anyone so careless as these two filling this position," and you file both envelopes unopened in the circular file. As you continue to leaf through the pile, you find several more you can dispose of quickly, but about 100 still remain. So you roll up your sleeves and start reading.

As you read the cover letters, each sounds like the previous one; the resumes also seem like clones: bunched text, their details running full up and down and across the page—words, words, dizzying words. You almost want to give up, but the need to fill the position urges you on.

After reading about ten letters, you find one that catches your interest; you read with eagerness. The first few sentences are not only appropriate but lead immediately into qualifications that have merit. Moreover, the writer is able to describe herself, her experience, and her background so that she appears to be an individual, apart from the others. "This person has character; she has ability; she has the background and the experience. I want to see this person!" you tell yourself, and you put her application carefully aside.

You return to read; for the most part you go no further than the first paragraph of the cover letter and give a perfunctory glance at the resume. You are looking for applications like the one that caught your interest previously; ones that make their writers come alive. After almost two hours, you have four piles of applications: those to be thrown away; those to be acknowledged politely; those to be acknowledged with more care and filed or entered into your computer for possible later openings; and those—about seven or eight—from candidates you want to interview.

From this hypothetical but typical situation, we can see that the reception your application will receive depends on these things:

1. Its initial impression, its appearance. If the cover letter and resume are neat and well-framed on the page, they favorably color the mood in which they will be read.
2. The first few sentences of the cover letter. If they are interesting (not bizarre) and lead the reader to the applicant's qualifications, the prospective employer will continue to read. The employer has a good idea of what he is looking for. The writer should emphasize at once her qualifications that she believes to be the most important to the opening. (An applicant's qualifications include, of course, experience, education, and personal matters.)
3. Establishing one's individuality. An employer becomes dizzy reading several scores of applications. One applicant, even though qualified, begins to sound like another. Hence, the writer needs to set herself apart from other applicants. She needs to make herself remembered. This, of course, is not easy, but we shall soon see how it can be done.

Job Strategy

So much for acquainting you with the job market environment. Let us now go to the beginning of the process. For the graduating student entering the job market, finding a job involves the following steps:

1. Self-appraisal (identifying salable qualifications and personal strengths),
2. Preparing a dossier,
3. Researching job possibilities,
4. Developing a resume,
5. Writing a cover letter/executive briefing,
6. Being interviewed for a position,
7. Writing a post-interview letter, and
8. Writing a letter of acceptance or declination of the job offer.

Self-Appraisal

You are the *product* you will want to sell to a prospective employer. Sales persons who are successful know their product. So before you prepare your sales instrument—the resume and its cover letter—you need to know yourself: you need to analyze and inventory your strengths, weaknesses, skills, experience, education, aptitudes, and goals. How do you do that? You ask yourself some questions and write down the answers honestly, after giving careful thought to each question. Consider such questions as the following:

What can I do well?

Can I speak well? Write? Draw? Socialize? Have I won any awards?

Do I speak or write any foreign languages?

Do I have military experience? Is any aspect of that experience appropriate to employment?

What skills have I acquired from my schooling? From jobs held? From hobbies?

What technical instruments, machine tools can I operate? How skillfully?

How well do I get along with others? Am I a leader? A follower? Can I supervise others?

What important courses have I taken?

How good am I working under stress?

What do I want out of life? What do I want from the job I am seeking? Security? Money? Power? Prestige? Excitement? Travel? Something else?

What is my goal? What is my career objective?

Your answers become an analytical inventory that will help you construct your job application material (resume and cover letter) and help you in a job interview.

Preparing a Dossier

Your college/university placement office is a valuable resource. It provides counseling, arranges on-campus job interviews, provides information on job openings, and helps you to establish a personal file, or **dossier**, and serves as its caretaker. The dossier contains information that substantiates your resume and cover letter, and contains other supporting documents such as transcripts of your college grades, listings of job experience, copies of awards and honors earned, unsolicited letters of praise, and other information that may benefit you or be of interest to a prospective employer. Copies of your dossier or some appropriate elements of it will be sent to a prospective employer at your specific request.

Researching Job Possibilities

Jobs don't come to you; you have to search for openings. Here are some suggested places to begin looking:

The Help Wanted section in newspapers, particularly the Sunday edition of a metropolitan paper.

Your college placement office. It posts announcements of openings and arranges for job interviews with recruiters visiting your campus.

The office of your department major. It also posts job opportunities. Check with your faculty for placement suggestions.

The professional journal(s) of your field. They carry national listings of openings.

Regional and national professional and trade association meetings. Such meetings almost always include recruiting sessions for members (commonly called "slave markets").

State and local employment offices.

The Office of Personnel Management of the U.S. Federal Government. Write to ask for information on opportunities in your field.

Your reference librarian. He or she can help you locate reference sources of information such as directories of corporations, government agencies, and professional organizations, and guides to other business information.

Job Analysis

Many persons, even the experienced and capable, approach an application for a position from the point of view of their own eagerness for it. They forget that a prospective employer has his own viewpoint and needs; so, unfortunately, the letters the applicants write express only their eagerness for this position, how much they would like the position, and how they feel the position would be right for them. Despite a great show of sincerity on the applicant's part, the employer cannot get interested. He is looking for a person who knows the employer's requirements and desires and who can show how the applicant's experience, training, and personal qualifications can be of benefit.

The first step—one that begins before the application is written—is to analyze the available position. This analysis must be based on the information describing the opening. This, of course, is found in ads or announcements revealing the availability of the opening. A good device is to itemize the position's qualifications listed in the announcement. Then, in a column alongside, list an inventory of your own qualifications, background, training, experience, and personal characteristics that would be appropriate for the opening. This type of inventory helps you understand, through comparative analysis, the type of qualifications and their importance to the employer.[1] It also allows you to look objectively at your own qualifications in light of what the prospective employer is looking for. If you do not have what the employer wants, there is no point in applying. Save yourself the time, energy, and emotional investment. It is true, sometimes, that your qualifications may not be exactly what the employer is looking for, but you may have some of the qualifications plus additional ones that would be attractive to the prospective employer. There is no formula to gauge just how many qualifications you need to make it appropriate for you to apply. This depends on your analysis and knowledge of the requirements of the type of position being advertised.

---

[1]Later, if appropriate, this comparative inventory can be converted into an executive briefing (see pages 60–61).

If the ad reads this way:

> SOFTWARE ENGINEER WANTED
> with five years experience
> in defense industry in ADA/VAX software systems,
> software configurations management, and
> IBM MVS/VM systems programming.

there is no point in applying if you have just graduated from engineering school and lack the necessary experience in the defense or intelligence communities.

Yet, it is also true that employers are looking for certain personal qualities, which no amount of schooling and actual work experience can provide. There have been many surveys to identify the reasons why employees lose their jobs. Incompetency is very low on the list. Very high on the list are personal factors— matters such as absenteeism, belligerency, hypochondria, alcoholism, drug addiction, inability to work with others, and so forth. Personnel managers, therefore, scrutinize applications very carefully for clues on prospective employees' instability, health and emotional problems, and personality difficulties. Employers are always on the lookout for prospective employees who will be responsible, who have stability, initiative, and characteristics of personal growth. Employers want not only individuals who have the personal knack of getting along well with others but also those who may have leadership and managerial abilities. Demonstration of such factors is eagerly searched for in resumes.

In your inventory, note carefully those qualities that would be attractive to employers. Granted, all of us have a sincere belief in our own abilities and that we are popular and capable of working productively with others. However, that mere assertion is not enough. What is required is evidence in the form of proven accomplishments. If you have been elected to an office in your fraternity, sorority, or campus organization, you have demonstrated some leadership abilities. Recall any experience in extracurricular matters that offered challenges. Indicate instances of productive accomplishments.

Developing a Resume

In current practice, applications for positions have two parts. Part one is a cover letter, or in some instances an executive briefing; part two is the resume, which is written before the cover letter and is the more important element.

The resume is a factual, succinct statement that highlights your qualifications. If this piece of technical writing is mechanically flawed, or is too long, too detailed, too difficult to follow, or too sketchy, it will be discarded. A manager does not read resumes for fun or sport. The resume evolved as an employment instrument to save a manager's time. Most managers spend less than sixty seconds skimming an application. The resume must grab attention immediately and speak clearly, succinctly, and persuasively of your value as a potential employee.

Types of Resumes

Professional resume mavens say there are three approaches to presenting your credentials to a prospective employer:

Chronological

Functional

Chrono-Functional (A combination of the above two)

The **chronological** resume lists and describes experience and education sequentially, beginning with the current or most recent and moving backward in time. For most students, education should be placed before employment experience, since education may be their most important qualification. Do not list your high school. Indicate your major field of study and the date or expected date of graduation. Do not list your courses unless some are significant to the position you are applying for. Honors, awards, professional affiliations, extra-curricular activities, and offices held should also be indicated.

The **functional** resume focuses on experience and capabilities rather than on the time or context in which they were acquired. It is the approach to take when you have a specific career goal and you want to zero in on a specific job. This approach can help to convince the employer you have the necessary experience, abilities, and aptitude for the opening.

The **chrono-functional** resume combines the previous two approaches. It is an apt resume tool for the person who is on a career track with a strong performance record. It contains a career summary either at the beginning or end of the resume, spotlighting professional skills, personal traits, and accomplishments, and aims toward future career growth. The chrono-functional resume starts with functional skills and experience relevant to job/career goals without reference to employers. This is followed by a chronological employment history, working backward, listing company names, dates, titles, duties, and responsibilities. A statement on education is then listed.

Preparing the Resume

Ideally, resumes, no matter which style is followed, should not be longer than two pages typed on one side. Note the attractive, framed appearance of the resumes in Figures 2.6, 2.8, and 2.10–2.13. Use paper of a good quality, 20- to 25-lb. bond of white or slightly off-white stock. Printing your resume on colored or pastel paper is poor form. The type should be black and clear. Avoid fancy or unusual type fonts, although you should take advantage of word processors to provide italics, bold, and capitalization for emphasis.

Basic Resume Ingredients

A resume is like a recipe. The ingredients of two entrées may be essentially the same, but what determines gourmet taste is the method of preparation, and the special touches by which the ingredients are combined to produce a unique flavor. Some resume ingredients are universal, some may sometimes be included, and there are special touches that can be added to pique the taste (interest) for particular situations.

The basic ingredients of a resume are:

Contact information (your name, address, phone number)

Career/Job objective statement

Experience

Education (include any special training or courses, and list any academic honors or awards)

Personal information (optional)

References (do not list but indicate availability)

Professional honors, awards, publications, patents, affiliations

Military service (if training and experience are appropriate to career/job)

Summary statement (used in chrono-functional resumes)

Let's examine these ingredients and see how they are combined to contribute to an effective resume.

Contact information

Use your name as the heading, *not* the word *Resume.* Center and type it all in capitals at least one inch from the top. Give your complete address; do not abbreviate any part of it, except the state element. Include your zip code and your telephone number; include the area code.

DANIEL M. WILLIAMS
12171 Carriage Square Court
Silver Spring, MD 20906
301-555-5980

Unless your current employer knows of your job search, do not include your business telephone in your resume. If appropriate, list the number in your cover letter with a caveat.

Career/Job objective statement

A career objective relates to your life work. A job objective relates to a specific occupational process; they are seldom interchangeable. Your career objective can change with the exigencies of time; a job objective relates to the opportunity of the moment as influenced by your qualifications. Young persons starting out on a professional career might delay indicating a career objective until they have attained enough work experience to be confident of the direction they want their careers to take.

A statement of a job objective provides a thematic conception to your resume. It helps the resume reader to focus on your qualifications as they relate to the opening to be filled, and provides a retrievable index term under which to file your resume. Further, a job objective statement indicates to a prospective employer your present level of competence and a willingness to work for advancement in the company. Job objectives are preliminary to career objectives.

An effective articulation of a career objective requires brainstorming, self-analysis, and self-evaluation. You have often faced the question, "What do you want to be when you grow up?" Now is the time to ask yourself, "What kind of a job do I want? What am I qualified to do? What skills, talents, aptitudes do I have? What experience? What skills and experience do I need to obtain the kind of work that will lead me to the career I want? On the practical level, what kind of job am I qualified for that will send me on the road I want to go?" After careful thought and self-analysis, compose a job objective statement as precisely as you can that can lead to a career objective. Do not say

> Objective: Chance for advancement in the computer industry. (This is too general and wishy-washy.)

Instead, say

> Objective: To obtain a responsible and challenging position as a programmer where my initiative, education, and work experience with IDMS and COBOL would enable growth into advanced system design.

Experience and Education

In a chronological resume format the placement order of the next two categories, *Experience* and *Education,* varies. The applicant who has been out of college a number of years and whose experience has been extensive places the experience section before that of education. If you are an entry-level applicant, one just graduating, list your education before the experience entries. Use reverse chronology in listing experience and education.

Under *Experience,* your entries should tell not merely the title of the job but should also specify specific duties. Accomplishments should be listed. Recent graduates should include summer and part-time jobs. If you worked your way through college, fully or partially, you should so indicate. If you have computer literacy or proficiency, provide the information. Technical knowledge, especially expertise in computers, is critical in today's world.

Under *Education,* begin with the latest degree and work back, listing degrees, diplomas, and special training beyond high school. List college courses relevant to the opening. Note laboratory work, field work, and independent study. Indicate any accomplishments, honors, awards, and extracurricular activities.

Personal Information

In accordance with federal employment laws enacted to prevent discrimination, you are not required to give information about your sex, race, national origin, religion, or marital status. Nonetheless, plain sense should tell you what personal items to include that would add to your qualifications for the job. For example, if you are applying for a position requiring travel, by indicating you are single, you inform the employer that you do not have family obligations that might prevent you from traveling. So, include personal information that enhances your qualifications.

You should list affiliations with professional societies; that indicates a dedication to your profession. Omit references to religious, political, or potentially controversial affiliations. Your resume should reflect your professional, not personal, life. If you have publication credits, list them; these demonstrate creativity and extended effort.

References

There are differences among resume experts about the appropriateness of listing references in a resume. Some believe interviewers are not interested in checking references before they meet and develop an interest in an applicant. Also, according to the Federal *Fair Credit and Reporting Act,* employers are forbidden to check references without your permission. Further, listing references takes up valuable resume space. Most employers, these experts say, assume references are available for the asking. On the other hand, some resume experts feel that an employer may recognize a name on your list and thus pay special attention to your name and qualifications. I recommend using the simple statement: "References available on request."

Professional Affiliations, Honors, Awards, Publications, Patents

List, if any, your affiliations with professional associations and societies related to your field; such memberships show dedication to your career. Employers know that memberships can be helpful and important to networking. Include honors and awards received; they bespeak achievements and will impress a reader. Publications and patents manifest creative thought. They tell the prospective employer that you, the candidate, invest effort above the call of accepted professionalism. Publications carry much weight in professions and industries where literary visibility is important. Patents are a positive factor in technological and manufacturing fields. List publications and patents, if you have any, at the end of your resume as follows:

PUBLICATIONS

Jones, Mary, "A Program for Making DOS User-Friendly in the Russian Language," *Personal Computing,* October, 1991.

PATENTS

Non-Carcinogenic Briquettes for Outdoor Grills. U.S. Patent 00,000,000, September 20, 1990.

Matters to Be Excluded from the Resume

Some information does not belong in the resume; its inclusion detracts and likely will fail to get you an interview. Among such matters are:

Written testimonials. They are absolutely in poor form (and no one believes them).

Salary expectations. This subject is best left to the interview.

References to salary (past and present). Too high or too low will knock you out of the running.

Photographs. Only if you are seeking a job in the entertainment field.

Exaggerated qualifications (or lies). Never, *never* lie about yourself or exaggerate your skills. Don't claim degrees or qualifications you don't have. Many employers have become skeptical of the data supplied in resumes and make it a point to verify a resume's claims. Stigmas of deceit can follow you for the rest of your life.

Preferences for work schedule, days off, overtime. The resume is not a negotiating instrument.

Writing the Resume

Review what we have discussed about the resume. Now consider your background. Which organizational style is appropriate for you—chronological, functional, or chrono-functional? Examine the examples of each type in Figures 2.10, 2.11, and 2.12. The chronological approach seems a simpler form and can be effective. Try it first, because it provides the basis for developing either of the other more difficult patterns. Now, examine the layout and page design of the resumes in Figures 2.6–2.13 and use them as a model. You will have to tinker a little to fit your own situation.

Try a rough draft. Following the layout of the models, enter the Contact Information. Then compose your Objective statement. Express it in as few words as possible, but in words that will bring a clear picture to the reader's mind. Keep the statement short, limiting it to one or two sentences.

If you have little work experience relevant to the position you are seeking, your next element will be Education. The sequence of data is degree, institution, its location (city and state), date graduated, course of study, thesis and supervisor if relevant, important courses if appropriate to job opening or job objective, and honors and awards, if any. Data in an entry are usually separated by commas. Periods are frequently used following dates. A period is used after a completed entry.

Under Experience, the sequence of data is date of employment, name of company, location (city and state), job title, and duties. Do not use complete sentences; use action verbs, such as those in the following list:

> achieved, administered, advanced, analyzed, assembled, assisted, authored, built, chaired, coached, completed, computed,

conducted, contributed, coordinated, created, decreased, demonstrated, designed, determined, developed, edited, engineered, established, evaluated, expedited, formulated, generated, guided, implemented, increased, initiated, invented, launched, led, marketed, monitored, operated, organized, performed, persuaded, planned, prepared, presented, processed, produced, programmed, provided, published, recommended, reduced, researched, restructured, retrieved, reviewed, scanned, screened, simulated, solved, specified, standardized, stimulated, streamlined, supervised, systematized, trained, trimmed, upgraded, wrote.

In the Personal section, list activities that could support your candidacy, for example the extent of your computer literacy or experience (if not previously covered), special skills, for example: speaks, reads, and writes Spanish fluently; captain, University Golf Team; served on community's Big Brother organization, junior and senior years.

Under References, indicate they are available on request or available from the university placement office.

As you examine the resumes reproduced in this chapter, you will note that sentences are truncated yet understandable. The pronoun, *I*, or the name of the applicant does not appear in the text lines; that saves space and allows you to "toot your own horn" without seeming boastful. There is an instance in which you can inject the personal *I*, which can give a human appeal to your resume. That instance is the inclusion of a *Personal Statement* at the end of your resume, as for example:

> I am competitive and frequently seek new approaches to solve problems, have excellent interpersonal skills, and am capable of working with and for people at all levels. I am able to plan, develop, organize, and implement ideas into viable projects.

A rule of thumb: the length of a resume is one page for every ten years of experience. For the entry-level candidate's resume, the length should be limited to one page. However, don't worry about length while writing your draft. Worry may hamper you. Usually two and one-half pages of double-spaced handwriting makes one page of typescript. After you have typed your resume and find it is longer than a single page, review it to see what can be deleted. Ask yourself: Can I cut any superfluous words? Where have I repeated myself? If in doubt, cut. Leave nothing but facts and action verbs.

Appearance of the final resume is important—so important that a worthy candidate's document can be sent to the trash can unread if it contains typos, erasures, and misspellings, or if it does not conform to current practice in style and substance.

Type your own resume if you are a competent typist, or better yet, use a word processor if you have access to one. Preparing your resumes and cover letters with a word processor offers many conveniences and advantages.[2] You can easily

- correct typos, misspellings, and punctuation errors,
- delete or insert words, phrases, and sentences without need for retyping the page,
- move items to more strategic locations on the page,
- change format of headings and their type style,
- update your resume as your situation changes,
- tailor your resume and cover letter to specific openings, and
- save different versions on diskette for quick retrieval.

Once your application material is stored, you have a permanent, easily recalled, attractive resume for use or revision as your situation requires.

After you are completely satisfied with the content and appearance of your resume, you need copies. Never send carbon, dot matrix, or messy reproduction copies. If you have the use of a letter quality printer, you can run any number of attractive, clean copies. If not, a photocopying machine is your best bet; or, for a modest cost you can have your resume reproduced by a job shop that offers photocopy reproduction or offset printed copies. The neat, professional-looking copies are worth the cost.

I suggest you avoid resume preparing services. While their final result appears attractive, their text material has a "canned" character. Employers often recognize the source by its style and might conclude you are unable to communicate or organize data on your own.

Writing a Cover Letter

Now that you have completed your resume, your next challenge is its **cover letter**. The resume on its own may or may not lead you to the position you seek. The job market, especially for professional openings, is so crowded that employers receive numerous resumes as impressive as yours. How can you stand out? The cover letter or executive briefing can be the answer. (I shall discuss the executive briefing later.) By writing an effective cover letter or executive briefing, you can persuade an employer to grant you an interview.

The cover letter of the resume is a sales letter, constructed to attract attention and to create interest on the part of the employer. It should speak with conviction about the applicant's "merchandise," and it should stimulate the recipient to the desired action.

To attract attention, the letter should have an interesting opening tailored to the specific position being applied for; to create desire, the letter should describe the applicant's major qualifications in a way that will induce the reader to examine the resume. Descriptive language is not enough; the tone must be

---

[2]Word processing is covered in detail in Chapter 5.

sincere, and the letter must convince the reader that the applicant has the desired qualifications.

Few people are hired on the basis of a letter and resume alone. The purpose of this application material is to create interest and stimulate the reader to invite the writer to appear for an interview.

How to Attract Favorable Attention

The very first sentence or two of a cover letter are critical. Good opening sentences are unhackneyed, simple, and direct. There are a number of ways to begin a letter of application to interest the reader.

Original Openings

We have seen in our earlier hypothetical situation that a majority of the letters begin with the commonplace, "In reply to your Sunday's ad," or "Having read your ad in the morning *Post*," or "Regarding the ad you had in the recent issue of . . . ." Effective beginnings reflect the *you* attitude; they are simple, direct, distinctive but not odd or extravagant. Here are some examples of effective novel beginnings:

> The April 14 issue of *Science* had a feature story on IBM that convinced me that mathematicians employed by IBM have enviable opportunities for professional growth and personal advancement. As a graduating mathematical statistician, I believe I could do no better than to start my mathematical career under one of the training programs you are advertising. Won't you consider my qualifications?

> Your notice in a recent issue of *Computerworld* on developments in transaction processing caught my eye and your company name caught my attention. This letter is to introduce me and to explore any need you may have in VAX/VMS architecture.

> I am a young engineer eager to secure a position where I can earn my keep and, at the same time, learn how to be a better chemical engineer. I understand from Mr. Ben Holloway, of your Synthetics Department, that you are expecting to add a Junior Engineer to your staff. May I have a moment to acquaint you with my background and qualifications for this position?

In your effort to be original, be careful that you do not become bizarre. No one can deny that the following two openings are original and unhackneyed, but they defeat their intended purpose by their oddness:

> Stop! Go no further until you have read my letter and resume from beginning to end. I feel sure you will find me to be the best qualified metallurgist for your position in the quality control department.

> Symbiosis is the biological term used to indicate cooperation between two or more parties toward a mutual goal. Your company has an outstanding training program for young executives. I feel I am qualified to meet your expectations, so, let us get together.

Name Beginnings

If you have learned of the availability of a position through a person whom the prospective employer knows, or whose name or title will command the respect of the prospective employer, then beginning the letter with the name of that individual can be very effective. Here are some examples of name or reference beginnings:

> Professor J.K. Wagner, Department Head, Forest Recreation and Wildlife Conservation at Colorado State University, called me into his office this morning to see if I were interested in the permanent ranger position described in the notice you sent him.

> The Placement Office of our university has informed me that my qualifications might be of interest to you for the Management Trainee Position in your Overseas Service Department.

> Dr. James C. Donovan, Vice President of Engineering Research in your Denver office, suggested this afternoon that I write you concerning an opening in your Chicago office. He said, "Ted, because I know you have an unusual five-year educational combination of engineering and management, I think you have just the right background, and, of course, the ability to make yourself quite useful to the Industrial Engineering Department in our Chicago office. Would you be interested in dropping them a note at once?" I was happy to hear of this opportunity and am acquainting you with my interest and background.

Summary Beginnings

Another effective way to begin is to present immediately a summary of the most significant qualifications for the opening. Here are some examples:

> Since graduating from Wharton School of Business three years ago, I have served as Assistant Budget Officer in Keely and Company, with the principal duties of preparing budget requirements, undertaking statistical studies, making comparison of methods, costs, and results with those of other refining companies.

> My four years' work in the field servicing of electronic testing equipment, as well as a degree in Electrical Engineering, make me confident that I can qualify for the position you have open in your Quality Control Department.

Question Beginnings

Beginning a letter with a question can be effective, especially when one is applying for an unadvertised position. Here is an example:

> Do your plans for the future include graduating Electrical Engineers with practical experience in model shop and breadboard design? If so, please consider my qualifications.

Starting a cover letter with a question equally can be effective for an advertised opening.

> Wouldn't that Mechanical Engineer you advertised for in the Sunday *Post* be more valuable to your department if he had as much as three and a half years' experience as maintenance technician in the Navy?

How to Describe Your Qualifications

The letter's beginning has the purpose of capturing the reader's attention. You must now create an interest in your qualifications. Entry-level applicants have four major qualifications: education, experience, personal qualities, and personal history. Just which of these four elements are emphasized in your letter will depend, of course, on your analysis of the requirements of the position. The principle is to write first and most about the qualification which your analysis deems most important to the position. If the ad reads: "Wanted—experienced design engineer," obviously the employer will want to know first and most about your experience as a design engineer. While the employer will be interested to know that you were graduated *magna cum laude,* that fact may not be as important as four years' experience designing certain kinds of instrumentation. If you have experience which is appropriate to the requirements of the opening, that is the most important qualification. Therefore, discuss it first and most fully.

If your strongest point has been your education, then that is what you need to start with. If your personal qualities are your strongest point, then show that you can develop or are capable of gaining the kind of experience necessary to do the job and that you are able to grow into more responsible positions within the company. After you have presented your strongest point, take up other qualifications you may have that are important to the position. Details should be specific. Give the name of the firm where you have acquired experience, the kind of position held, and the duties connected with the position. If you accomplished anything outstanding in the job, tell about it.

General and ineffective: "For three years, I served as a technician."

Specific: "For three years, I served in the model shop of the research and development department of the Radio Corporation of America in Camden, N.J. My work consisted of making breadboards, fabricating prototype models from rough engineering sketches of development engineers. This required not only expert use of a soldering iron, but also machine shop tools and equipment."

In discussing education, focus attention on those subjects directly or closely related to the position being applied for. For example, a student applying for a job with an agricultural machinery manufacturing company might indicate his educational qualifications as follows:

> My studies concentrated in the area of machine design, including such courses as kinematics of machines, dynamics of machinery, computer aided

design, machine analysis, agricultural power, and agricultural machinery. Some other pertinent courses include statics, dynamics, strength of materials, fluid mechanics, computer graphics, physics, chemistry, and robotics. In my Special Problems Course in Agricultural Machinery, I designed and built a corn conveyer feeding machine presently being used on our Agriculture Hog Farm.

By its substantiation of qualifications, the central portion of the application letter creates an interest in, and a desire to interview the applicant.

Securing Action

The final paragraph of the application letter has the duty of persuading the reader to grant the applicant an interview. Mention should be made, either in the last portion of the letter or in the middle portion, that a resume with complete information on the applicant is attached. Give as much care to the ending of your letter as you did the beginning. Don't be hackneyed or weak and don't be impertinently brash or timorous:

> If you think I can qualify for your position, I'd be grateful if you gave me a chance [negative and timorous].
>
> I suppose it's only fair to tell you, you better call me before the end of this week, because I have several fine job offers I am considering and I would like to compare yours with the others before I reach my decision [brash and impertinent].
>
> Trusting you will grant me an interview at your convenience. I shall await your call [hackneyed].

To obtain the action you want—securing an interview—calls for a more effective way to close your letter. Be direct yet circumspect. Here are some examples:

> I believe my qualifications are appropriate for the position you advertised. I would be happy to provide details beyond those outlined in my attached resume. May I have an interview at your convenience? Won't you write me at my home address or call me at 287-3873 between the hours of 5:00 and 9:00 P.M.?
>
> I shall be in Chicago from December 22 to January 2. Would it be convenient for me to come in to tell you more about myself and to learn more about the position you have available?
>
> May I have an interview at your convenience? I have enclosed a self-addressed postal card for you to let me know when I might stop by to better acquaint you with my qualifications.
>
> Whether you have an opening at present or not, I would appreciate the opportunity to meet with you so that you may better judge my qualifications for possible employment with your organization. Toward that end may I call you on the 15th for a mutually convenient meeting?

Model cover letters and resumes appear in Figures 2.5–2.13.

698 South Grant
Fort Collins, CO 80521
January 6, 19__

Mr. John Koldchuk
Division Manager
Johnson Technology Corporation
Florence, CO 81226

Dear Mr. Koldchuk:

Do your plans for this new year include graduating electronic engineers with a varied hands-on experience? If so, please consider my qualifications.

My curiosity in the "magic" world of electronics prompted me to enlist in the navy where the opportunity to become an electronics technician was available. I was granted this opportunity and also the chance as an instructor to pass the knowledge I acquired on to others. I learned to operate and teach others about airborne equipment, as well. I was able to make first-class petty officer during my four-year enlistment.

The experience in the navy was but a mere acquaintance with electronics and an invitation to learn more. With an assist from the GI Bill, I enrolled in the College of Engineering at Colorado State University, where I took such courses as Circuit Analysis, Computer Systems, AC and DC Machinery, Robotics, and Human Engineering.

To meet the financial requirements in acquiring my degree, it was necessary to supplement my military service benefits with parttime and summer employment. Besides financial assistance, these jobs offered acquaintance with, and some practical experience in, the agriculture and petroleum fields; these jobs also helped increase my sense of responsibility, gave me confidence, and taught me to appreciate my fellow workers.

The knowledge, training, and experience I've acquired thus far have been mainly for my own satisfaction. I now want to put them to wider use. If my background and experience is of interest to Johnson Technology Corporation, I would appreciate an interview. You can contact me at the address and phone number listed on my attached resume. I look forward to hearing from you.

Sincerely yours,

Victor P. Norris

Figure 2.5
Sample Cover Letter

VICTOR P. NORRIS
689 South Grant
Fort Collins, CO 80521
303-555-1367

EMPLOYMENT OBJECTIVE: To serve as an electronic test instrument design engineer with an opportunity to continue professional growth.

EDUCATION
B.S.E.E., Colorado State University, Fort Collins, June, 19__
    Important Courses: Electronic Circuits, Circuit Analysis, Computer Systems, Network Analysis, Robotics, Engineering Economics, Psychology, Technical Writing, Feedback Systems.

EXPERIENCE
September, 19__ to present. Part-time Handyman, Physical Plant, Colorado State University. Electrical repairs; designed automated feeders for large animals at Agriculture Station; carpentry.

Summers, 19__, 19__, 19__. Rustabout Pan American Petroleum Corporation., Midwest, Wyoming. Pulled rods and lines and helped maintain and repair oil well equipment and flow lines.

MILITARY EXPERIENCE
1985–1989. U.S. Navy, Great Lakes Naval Training Center, Aviation Electronics Technician. Norfolk Naval Air Station, AT1 rating; served as radar instructor.

PERSONAL
    Single; Excellent health; expert photographer

REFERENCES
    Available on request.

Figure 2.6
Sample Resume

728 West Laurel
Fort Collins, Colorado
April 9, 19__

Mr. Thomas Welk
Supervisor
Mesa Verde National Park
Colorado 81330

Dear Mr. Welk:

Professor James Moss, department head of Forest Recreation and Wildlife Conservation at Colorado State University, called me into his office this morning to see if I were interested in the permanent ranger position described in the notice you sent to him.

The saying "opportunity knocks once" immediately flashed through my mind. To find this opportunity open at this point in my career was a pleasant surprise.

Why? I have two reasons: (1) I have long had a hobby and interest in Indian lore and archaeology, which would be of tremendous value in this type of position and (2) I have trained in a Forest Recreation major for National Park work, hoping to eventually secure a position where study and interpretation of Indian culture are carried on.

Professor Wagner stressed the background needed for this position and the education necessary to qualify. My background has included various jobs where public relations were required. Helping manage a swimming pool one summer gave me invaluable practice in controlling groups of people and in gaining their confidence and interest. At Yellowstone National Park on busy weekends I helped guide tours to points of interest giving interpretive talks along the trail. From my hobby of Indian lore I have gained a knowledge of various Indian cultures that directly relate to Mesa Verde's Indian culture.

I will be in Mancos, Colorado, from May 22–29 doing research for the University on increased tourist visits to this community resulting from your location near the park. I will be happy to call at your convenience for a personal interview. If this is not convenient for you, I can be reached at my Fort Collins address after May 31.

Sincerely,

Robert D. Weeks

Figure 2.7
Sample Application Letter

ROBERT D. WEEKS
728 West Laurel
Fort Collins, CO 80521
Telephone: 303-555-3825

OBJECTIVE: A ranger position in a U.S. national park where study and interpretation of Indian culture are carried on.

SUMMARY OF QUALIFICATIONS
Varied experience with the National Park Service and the U.S. Forest Service; degree in Forest Recreation.

EDUCATION
B.S., Forest Recreation and Wildlife Conservation, Colorado State University, June, 199_.
Major courses: Principles of Wildland Recreation, National Park Management, Field Recreation Studies and Management, Ten-week Forestry Summer Camp, Principles of Wildland Management, Wildlife and Forestry Ecology.

EXPERIENCE
Summer, 199_, Grand Canyon National Park, AZ, Fire lookout and fire suppression.
Summer, 199_, Yellowstone National Park, WY, Tour guide.
Summer, 199_, Forest Field Summer Camp, Pingree Park.
Summer, 199_, Skyline Acres, Denver, Swimming pool Manager and Lifeguard.

PERSONAL
Single. Member, National Society of Park and Recreation; served as delegate, Western Clubs Convention, 199_ and 199_.

SPECIAL INTERESTS
Indian lore, mountain climbing, skiing.

Figure 2.8
Sample Resume

203 West Lake Street
Fort Collins, Colorado 80521
December 1, 19__

Mr. E. A. Ferris
Administrator of Training
Combustion Engineering, Inc.
Prospect Hill Road
Windsor, Connecticut 06095

Dear Mr. Ferris:

Colorado State University has a steam heating system for its buildings on campus. One of the steam generators is a Combustion Engineering boiler erected in 1950. Since 1950 this boiler has performed satisfactorily with a minimum of maintenance. Modifications were recently completed for modernization of the heating plant and your boiler appears ready for another 45 years of service.

The performance of this boiler and the reputation of your company have prompted me to write you. If your company has an opening for a graduating mechanical engineer with previous mechanical experience, please consider my qualifications.

After attending college for one year, I joined the army to gain practical experience and maturity. While in the army, I attended school at Ft. Belvoir, Virginia, for the maintenance of engineering equipment. After equipment repair school, I was stationed in Germany where I had 2½ years of practical experience in the field repair of heavy equipment. My tour of duty convinced me that I had been right in choosing mechanical engineering for a career, so I returned to Colorado State University to finish my studies for a B.S.M.E. degree.

To help pay college expenses, I supplemented my scholarship and savings by summer work and part-time work during the school year. Last summer I worked as a junior engineering at Boeing Airplane Company. My job involved an efficiency and cost analysis of both steam and hot water heating systems, evaluation of economy suggestions and inspection of building repairs, and maintenance.

If my qualifications as outlined in the enclosed resume are of interest, I would appreciate an interview. I can be reached at the address and telephone number shown on my resume.

Sincerely,

John B. Dunne

Figure 2.9
Sample Student Employment Letter

JOHN B. DUNNE
203 West Lake Street
Fort Collins, CO 80521
303-555-1473

CAREER OBJECTIVE: A mechanical engineering position in which I can use my education, experience, and enthusiasm for challenges to advance to a supervisory position with an engineering equipment manufacturing company.

EXPERIENCE:
Summer, 199_, Junior Engineer, Plant Services Section, Boeing Airplane Company, responsible for study to determine efficiency and cost analysis of steam heating system of factory, engineering, and administration buildings.
Summer, 199_, Mechanic, Colorado State University, maintenance of motor pool trucks, tractors, bulldozers, also repaired automatic machine tools.
1984–1989, U.S. Army Corps of Engineers. Specialist Second Class E-5, repair and maintenance of heavy equipment.
Summer, 198_, Laborer, Gardner Construction Company, Colorado. Worked directly from blueprints, duties included tying steel rods for concrete walls.

PROFESSIONAL AFFILIATIONS:
Member of A.S.M.E. student section; Honors: Sigma Tau, honorary engineering fraternity, and member of Phi Beta Phi.

SPECIAL INTERESTS:
Computers, mechanical repairs, skiing, reading, travel.

REFERENCES:
Available on request.

Figure 2.10
Example of a Chronological Resume

JOHN B. DUNNE
203 West Lake Street
Fort Collins, CO 80521
303-555-1473

CAREER OBJECTIVE: A mechanical engineering position in which I can use my education, experience, and enthusiasm for challenges to advance to a supervisory position with an engineering equipment manufacturing company.

SUMMARY: A B.S.M.E. degree with emphasis on computer-aided machine design, plus eight years of both hands-on and supervisory experience, maintaining and repairing engineering equipment from bulldozers to automatic screw machines.

MAINTENANCE and REPAIRS
> Crawler tractors, cranes, grader ditchers, bulldozers, bucket loaders, computer-controlled machine tools, and automatic screw machines.

RESEARCH
> Conducted efficiency and cost analysis study of heating systems of factory, engineering, and administration buildings at Boeing Airplane Company

MILITARY
> U.S. Army Corps of Engineers, 4-year enlistment, with 2-1/2 years in Germany, responsible for field repair of heavy equipment.

WORK EXPERIENCE
> Boeing Airplane Company, Junior engineer, Summer, 199_
> Colorado State University, equipment mechanic, part-time and summers, 199_–199_.

EDUCATION
> B.S.M.E., Colorado State University, 199_.
> Certificate, Engineering Equipment, Corps of Engineering Service School, Fort Belvoir, VA., 198_.

REFERENCES
> Available on request.

Figure 2.11
Functional Resume

JOHN B. DUNNE
203 West Lake Street
Fort Collins, CO 80521
303-555-1473

OBJECTIVE

Employment with an engineering equipment
manufacturing company in a position with supervisory
potential.

SUMMARY

B.S.M.E. degree with emphasis on computerized machine
design, plus eight years of both hands-on and supervisory
experience, maintaining and repairing engineering
equipment from bulldozers to automatic screw machines.

MAINTENANCE and REPAIRS

Crawler tractors, cranes, grader ditchers, bulldozers,
bucket loaders, computer-controlled machine tools,
and automatic screw machines.

RESEARCH

Conducted efficiency and cost analysis study of
heating systems of factory, engineering, and
administration buildings at Boeing Airplane Company

MILITARY

Corps of Engineers, 4-year enlistment, 2-1/2 years in
Germany, responsible for field repair of heavy
equipment.

Figure 2.12
Example of Chrono-Functional Resume

Page 2 of 2   John B. Dunne   303-555-1473

EXPERIENCE
Summer, 199_, Junior Engineer, Plant Services Section,
    Boeing Airplane Company. Responsible for research
    study to determine efficiency and cost analysis of
    steam heating systems for factory, engineering, and
    administration buildings.
Summer, 199_, Mechanic, Colorado State University,
    maintenance of motor pool trucks, tractors,
    bulldozers. Operated and repaired heavy equipment
    and computerized machine tools.
199_–199_, U.S. Army Corps of Engineers, specialist second
    class E-5, field repair and maintenance of heavy
    equipment.
Summer, 199_, Laborer, Gardner Construction Company,
    Colorado. Worked directly from blueprints, duties
    included tying reinforcing steel rods for concrete
    walls.

PROFESSIONAL AFFILIATIONS
    Member of A.S.M.E. student section; Sigma Tau
    Honorary Engineering Fraternity, and Phi Beta Phi.

REFERENCES
    Available on request.

Figure 2.12 (continued)

SALLY H. WEISMAN

2589 Valley Parkway
Malvern, Pennsylvania 19355
215-555-6123

EXPERIENCE

Sinai Hospital of Baltimore, Baltimore, Maryland

1989–90	Productivity Management Director Implemented a computerized cost quality management system to analyze and monitor quality, efficiency, and productivity of hospital departments. Instituted policies and procedures to assure highest level of patient services with optimal resource utilization and maximal level of reimbursement.
1988–89	Management Information Systems Clinical Systems Coordinator Developed and managed a computerized clinical patient care system. Supervised conversion and maintenance of the new system to ensure optimal performance. Analyzed work flows and developed procedures for increased productivity. Coordinated and supervised user training and education.
1986–88	Management Information Systems Financial Systems Coordinator Designed and implemented a computerized financial system. Supervised training of financial personnel in the use of the system. Served as liaison between Finance Departments (Patient Accounting, General Accounting, Reimbursement, Budgeting, and Payroll) and Management Information Systems.
1983–86	Management Engineering Management Engineer Evaluated procedures, policies and practices in hospital departments. Provided recommendations and devised procedures to improve work flow, productivity, and overall operational efficiency. Developed new systems for operational organization and work flow.

Figure 2.13
Sample Resume of a Professional

EXPERIENCE (continued)

Mount Sinai Medical Center, New York, New York

1980–83   Medical Records
Admission, Discharge and Transfer Unit Coordinator
Established and supervised a computerized information system
to provide accurate admission and discharge data. Coordinated
communication between the Departments of Nursing, Admitting
and Medical Records.

EDUCATION

University of Baltimore, Baltimore, MD
M.B.A. expected in 1991

Hunter College, City University of New York, New York, NY
M.S. in Health Science, 1982

State University of New York, Buffalo, Buffalo, NY
B.A. in Psychology/Communications, 1978
Magna Cum Laude, Phi Beta Kappa

PROFESSIONAL SOCIETIES

American Hospital Association

Health Care Information and Management Systems Society

Maryland Society of Health Care Information Systems

REFERENCES

Available on request

Figure 2.13 (continued)

Writing an Executive Briefing

The **executive briefing** has been developed by the employment recruiting field to replace the cover letter, particularly for a specific opening. It is based on the belief that the initial resume reviewer (a member of the personnel department) has only a vague understanding of the job opening. Cover letters, it was reasoned, have become stereotyped, so the executive briefing was developed as a streamlined, effective instrument to call attention to an applicant's qualifications for a specific opening. It is actually an abstract of an applicant's resume, highlighting in "sound-bite" terms specific items of experience, competencies, and background as these relate to the opening. An aid to developing the executive briefing is to make use of the comparative inventory discussed on page 36.

The executive briefing is less appropriate for the person just beginning a career, but it is becoming an effective device by which an experienced professional can call attention to the complete resume. Figure 2.14 is an example of an executive briefing.

Guidelines for Preparing Resumes

1. Don't use the word *Resume*. Begin with your name, centered, as the heading. (You want the contact information to stand out.) If you use a word processor, boldface the letters in your name. Center your address and telephone number below your name. Some resume writers place their address and telephone number on the same line in order to save space.

2. Your Career Objective statement follows below the contact information or heading. The words *Career Objective, Employment Objective,* or *Objective* may be centered, with the statement across the page below it, or the objective statement may be placed in a prominent heading style at the left margin. Be consistent in your style when formatting headings. Refer to the sample resumes in this chapter.

3. If you decide to use a Summary statement (usually used in a functional or in a chrono-functional resume) place it to follow the Career or Employment Objective statement. (In some functional resumes it replaces the Employment Objective statement.)

4. The chronological resume is most appropriate for the entry-level, less experienced applicant. Use reverse chronology in your listings; that is, begin with your current or most recent experience and work backward in time.

5. For recent graduates, your education usually is your best qualification, so list it before the section on Experience. Again use reverse chronology in listing the items.

6. The usual sequence of data within each entry under Education is: degree, institution, location of institution (city and state), date graduated, course of study, grade-point average (optional), honors received, list of major courses significant to the opening or objective.

7. Sequence of data for each job experience is: job title, name of company, address (city and state), notation if part-time, dates worked, (sometimes

<u>EXECUTIVE BRIEFING</u>
for a senior programmer analyst
as advertised in <u>Computerworld</u>

Nathan M. Duran
118 Meadow Vista Way
Encinitas, CA 92024
619-555-4361

     To help you evaluate the attached resume and conserve your time, I have prepared this executive briefing. It lists your advertised needs on the left and my experience and skills on the right. My resume will give you further details.

<u>Job Title:</u>	<u>My Current Title:</u>
Senior Programmer Analyst	Senior Program Analyst
Required Experience:	Relevant Experience:
8 yrs. Programmer Analyst	8 yrs. as a program Analyst
of which at least 3 were	of which 5 yrs. as a senior program
as a senior analyst	analyst, in addition, 4 yrs. as a
	programmer with 2 yrs. as a project
	leader
<u>Hardware:</u>	<u>Hardware:</u>
Prime 850,750,550	DEC VAX 11/70, System 10, PDP-11/45
IBM360	PDP-8, Prime 850,750,550, IBM360,
	IBM3083
<u>Software:</u>	<u>Software:</u>
DOS, Primos, TAPR	DOS, PRIMOS, TAPR, Datatrieve,
Datatrieve	VMS(RMS), TOPS-10, TSX-11M,
	Apple Plus II
<u>Languages:</u>	<u>Languages:</u>
COBOL, FORTRAN,	COBOL, Info/Basic, MACRO-11, DCL,
Info/Basic, PASCAL	GEOMAP, RPG, FORTRAN, Easycoder,
	PASCAL, IBM370 Assembly, Apple
	Assembly

Figure 2.14
Example of an Executive Briefing

dates are placed columnarly on the left), responsibilities and duties, and achievements if any. Be specific in describing duties; use action verbs, but don't clutter up your resume with too much data. Include only what reflects well on you. Avoid using personal pronouns. (Instead of saying, "I was responsible for maintaining UNISYS equipment," state, "maintained UNISYS equipment.")

8. Data in an entry are usually separated by commas. Some resumes use periods after dates. A colon follows after the item "duties." Items listed under duties are separated by commas. The completed entry is followed by a period. Be logical and consistent in the punctuation you use.

9. Don't include physical details (such as height, weight, health, gender). Use your date of birth, not your age. Marital status may be omitted. If you have computer literacy or proficiency, indicate it.

10. Do not list references, but indicate they are available on request.

11. Don't include so much data that your resume gives the appearance of clutter and runs three or four pages. The employer will not read it. A well-organized resume with carefully selected information need not run more than one page for an entry-level applicant, at most two.

12. An attractive, neat resume invites reading. Proofread it carefully. There must be *no* typographical errors, misspellings, or lapses in grammar. Do not bind the resume in a folder. Clip the cover letter to it; if your resume runs two pages, staple them together. The paper for both cover letter and resume should be a good quality white bond no less than 16 lbs, preferably 20-lb, 25% cotton/rag stock. A good quality business-size envelope should be used. Of course, the typing on the envelope should be neat, attractive, and accurate.

The Job Interview

The purpose of a job interview is to verify the impressions and credentials your application material provide the prospective employer, and to give a basis for judging whether you are the best among several candidates for the opening to be filled. The employer wants to see firsthand how you look, act, and react.

The prospect of an interview can be an anxiety-ridden experience, but it need not be a terrorizing one. It is normal for the prospect to give you butterflies in the stomach, but don't be frightened by it. Remember you have already passed a major hurdle—you have been invited for an interview! It is an opportunity for you to directly exhibit your qualifications.

The secret of success is preparation. Learn and research more information than you had before applying about the company/organization—details such as its products, services, its history, achievements and successes, its role/status in the industry/field, its prospects, branch locations, and its community activities. If time permits, request copies of company literature and annual reports. Research for information in financial publications such as *Dun and Bradstreet* and *Standard and Poor's Corporation Records*, and appropriate trade and professional periodicals. Your faculty is also an excellent source.

After you have all the useful information that is practical for you to obtain, give yourself time to analyze and relate it to your situation. With careful preparation, you will be competent and confident in your interview. To attain that type of security, consider the following factors:

You were not invited to be humiliated.

You are being given the opportunity to prove yourself: to perform one-on-one, so that you can be evaluated on how you behave and manage yourself in a stressful situation, how you think on your feet, how you speak, and how you interact.

With that in mind, here are some matters an employer may ask or say to you during your interview:

Tell me about yourself (in one or two minutes).

Why do you want to work here?

What sort of job are you looking for and why?

What do you know about our company/organization?

What qualifications do you have for the position?

What is your educational background/training? What did you learn? How does/ did your course of study prepare you for this opening? What extracurricular activities did you have? What benefits did you derive from them?

What do you consider your strongest qualifications?

What do you see your greatest shortcomings to be?

What are you hoping to do in the future? (Career objective.)

Where would you like to be in ten years?

What is your secret ambition?

What makes you happy? What makes you unhappy?

You, in turn, should be prepared to ask some questions of your own:

What specifically does the position entail?

In what department/section is the work performed?

How does the department/section fit into the scheme of things in the company/ organization's objectives/activities?

Who will the selected candidate be working with and for?

What responsibilities/opportunities are there for the selected candidate and the department in which he/she works?

What is the stability of the position and of the company?

Consider carefully answers to those questions an employer may ask you. Consult with your faculty, knowledgeable friends, and family. They may have questions and topics not listed here. Write out the answers, but don't memorize. Extemporize so that you are spontaneous in your responses during the interview. Practice the answers and rehearse the interview with a friend.

Bring a copy of your resume with you to the interview. The employer will have a copy of it, but it will help you to elaborate on any matter the interviewer may ask. If there are new additions to the resume's information, it is a good idea for you to present them at the interview. For example: I received my degree two weeks ago. Or, I have received notice that I am graduating *magna cum laude.*

If a silence occurs during the interview, don't feel obliged to talk, unless you have something pertinent to say. It is the employer's ball game; let the interviewer make the next move. Perhaps, the interviewer wants to gauge your reaction to the silence. Don't ramble on just to talk. Successful interviews last about an hour. When the interviewer wants the meeting to end, he or she will so indicate, usually by rising and saying something to the effect that the company/organization was pleased to meet you and that you will be informed about a decision after all interviews are completed. At the end of the interview, express tactfully but enthusiastically your interest in the position and thank the interviewer for the opportunity to present your qualifications. It is appropriate before leaving to inquire when you might expect to hear the decision regarding your application.

Writing a Post-interview Letter

The application cover letter is a sales letter. Good salespeople know a sale frequently depends upon repeated efforts and follow-ups. Follow-up letters, after an interview, create a favorable impression upon a prospective employer. People admire determination and persistence. After you have been granted an interview, express your appreciation and reemphasize those particular qualifications which you have determined during the course of the interview are important to the prospective employer. If you are able to present additional information or additional qualifications to enhance your application, it is well to do so in your follow-up letter. The follow-up letter should be neither too long nor too short and should be timed, in most cases, to reach your prospective employer the day following the interview (Figures 2.15 and 2.16).

Acceptance and Refusal Letters

When you receive an offer for a job, whether by telephone or letter, it is advisable, and polite, to write a letter accepting or declining the position. If you are informed by phone and are accepting the offer, it is appropriate to request a written confirmation and to respond with a formal letter of acceptance. The offer of employment can be considered a contract; the letter formalizes the terms of your employment. Your response should reiterate the terms you are accepting. Your letter, though formal and brief, should express your pleasure at the prospect of working for the company. Figure 2.17 is a sample acceptance letter.

112 40th Street
Apartment 720
Baltimore, Maryland 21210

Bernard Faulkner, M.D.
President
Sinai Hospital of Baltimore
2401 West Belvedere Avenue
Baltimore, Maryland 21215

Dear Dr. Faulkner:

Thank you for the opportunity to meet with you last week. I thoroughly enjoyed the time you spent acquainting me with Sinai Hospital. It was kind of you to arrange my meeting with Mr. Benson and Ms. Berkowitz.

I am excited about the prospect of continuing my career in hospital administration at Sinai hospital. I am confident that I can contribute to the success and effectiveness of the management operations of your institution. I am looking forward to this opportunity.

Sincerely yours,

Sally H. Weisman

Figure 2.15
Sample Employment Follow-Up Letter

Dear Mr. Fisher:

I appreciate the time you spent with me yesterday discussing the sales engineering position you are seeking to fill. Your description of some of the problems the position entails offers a challenge I would like to tackle. My work as a Junior Applications Engineer for Moore Hydraulics Company during the summer of my senior year at Engineering College clinched my resolve that Sales Engineering would be my life's pursuit. It also gave me the kind of invaluable experience needed for engineering sales by working with customers on problems that needed modification of off-the-shelf Moore engineering products.

The products of Digital Equipment Corporation, Mr. Fisher, were pointed out to me early in my laboratory courses by my professors as among the finest in the industry. Because a salesman must believe in the products he sells, I would be proud to have an association with your organization. The problems you mentioned in our conversation yesterday aroused my enthusiastic interest. I would very much like to help you meet them and look forward to doing so.

Sincerely yours,

Figure 2.16
A Follow-Up Letter[3]

---

[3]In this and most subsequent sample letters, some headings, dates and signatures are omitted for purposes of saving space.

Dear Mr. Beachem:

    I am happy to accept the position of Technical Writer at Interactive Software Company at a yearly salary of $28600.

    I will telephone Ms. Stearns in your Personnel Office for instructions on the reporting date, physical examination, and employee orientation.

    I am looking forward to a fruitful career with Interactive Software Company.

                                     Sincerely,

Figure 2.17
A Sample Acceptance Letter

Every now and then you may face the enviable problem of receiving more than one job offer and will have to turn an offer down. After due but prompt consideration, write a letter of refusal. Since an employer has invested time and effort considering your qualifications and interviewing you, you are obligated to explain your reason for not accepting the offer. Do not begin your letter with a blunt refusal. Lead into it with sincere, complimentary remarks about the interview and the position that was offered. Be candid but cordial, and always leave a door open for future possibilities. Sometimes, the job you accepted does not turn out the way you had hoped and you may wish to explore the job you turned down. Figure 2.18 is a sample of an effective refusal letter.

Company Replies to Job Applications

The sample letters in Figures 2.19a and 2.19b illustrate typical replies to job applications. The policy of most firms with well-organized placement and personnel departments is to offer the courtesy of a letter explaining an unfavorable employment decision (see second letter, Figure 2.19b). The graciousness expressed takes the sting out of the refusal.

    If you should receive a rejection letter like the example in Figure 2.19b from Collins Radio Company and are still interested in working for the company to which you applied, it would be good strategy to send a reply similar to the one in Figure 2.20.

Dear Mr. Ellis:

I am gratified by your offer of employment as a junior mycologist in your clinical laboratory. I have discussed it with my major professor at State University, Dr. Edna Sapolsky. She tells me I should be flattered by it and I am. Dr. Sapolsky believes that I would find no better opportunity to pursue microbiology than in your laboratory. Unfortunately, while your offer has given me a great deal of happiness, it has also presented me with a conflict. You see, I had already decided to pursue studies in aquatic ecology before your letter arrived, and had been considering a similar offer from a laboratory in Cambridge, Massachusetts.

After further discussion of my career aspirations with both Dr. Sapolsky and my parents, I have come to the decision that my interest in aquatic ecology can best be developed further by Ph.D. study. The offer from the other laboratory permits me to continue part-time graduate study at company expense. The Massachusetts Institute of Technology has accepted my application for this study. Under these circumstances, so advantageous to my aspirations, I have decided, not without mixed emotions, to accept the offer from the Cambridge laboratory.

I do appreciate the wonderful opportunity you extended to me. I shall always recall the pleasant experience I had during my interview visit to your laboratories. I know it would have been a joy to have been part of your laboratory organization.

Sincerely yours,

Figure 2.18
A Sample Refusal Letter

COLLINS RADIO COMPANY
CEDAR RAPIDS, IOWA 52403
March 5, 19__

Mr. F. R. Billings
3030 North 10th Street
Milwaukee, Wisconsin 43206

Dear Mr. Billings:

It was a pleasure to meet you when we were on your campus recently interviewing prospective graduates concerning an association with Collins Radio Company.

Our company has a number of openings which, we believe, offer an opportunity for the career you seek. We would like to have you come to Cedar Rapids at our expense to get better acquainted with Collins Radio Company by seeing our headquarter's facilities and meeting some of the members of our staff. It will fit our schedule better if you are able to plan your visit any Monday through Friday between March 18 and April 15. However, please arrange to come whenever it is most convenient for you.

You will need one full day here from 8:30 a.m. to 5:00 p.m. to cover the program we have planned. Select whatever mode of transportation is best for you. Cedar Rapids is serviced by United Air Lines and the Milwaukee and Northwestern Railroads, although you may prefer to drive. We sincerely hope your wife will be able to accompany you to Cedar Rapids as her considerations are most important in the selection of your future home.

When you have reviewed available transportation schedules and have selected a convenient date for your visit, please advise me. If timing is a problem, call me collect at (319)555-2661, Extension 505. We will be happy to reserve hotel accommodations for you as soon as we know your plans. We will confirm completed arrangements. Since we will be having a number of candidates in during this period, it will be helpful if you will suggest an alternate date or two, as we are able to accommodate only a limited number each day.

An application blank is enclosed for you to complete and return to my attention, along with a copy of your college transcript. If your transcript does not include courses completed recently, please attach a list of these, as well as the courses you will complete before being graduated.

We are looking forward to your visit to Cedar Rapids and to discussing the part you and Collins Radio Company may play in the bright future of the electronics industry.

Very truly yours,

J. J. Field, Assistant Director
Treasurer's Division

JJF:DT

(a)

Figure 2.19
Sample Replies to a Job Application

COLLINS RADIO COMPANY
CEDAR RAPIDS, IOWA 52403
May 14, 19___

Mr. James C. Mitchell
3520 East Jackson Boulevard
Lansing, Michigan 48906

Dear Mr. Mitchell:

Miss Williams has forwarded your completed application and your letter so that we may consider you for a possible association in our Controller's Division. We sincerely appreciate your interest in our Company.

We should like to congratulate you on your efforts to acquire a specialized Accounting education. You will find it an exceptionally fine background for any field of endeavor in which you engage.

The Accounting system of Collins Radio Company is necessarily somewhat complex because we are engaged in both government and commercial enterprise. Your application indicated you have completed approximately nine semester hours of Accounting work with LaSalle Extension and Drake University. At the present time, we are using, almost exclusively, Accounting majors in our Controller's Division, which has the responsibility for the Company's accounting program. Because of this fact, we think it unfair to you to consider your application at this time. Most of your competition in that division would have from two to three times as many hours of specialized accounting background as you have, in addition to considerable exposure in the area of economic theory. Most likely your progress here with that type of competition would be seriously limited.

May we suggest that you continue your Accounting studies and perhaps get in touch with us when you have completed that curriculum. Your application would seem to indicate you have the ambition and aggressiveness which we like to find in Collins' employees.

Thank you for your interest in Collins Radio Company. May we extend our best wishes for your continued success in your Accounting studies.

Very truly yours,

J. J. Field, Assistant Director
Treasurer's Division

JJF:DT

(b)

Figure 2.19 (continued)

Dear Mr. Field:

Thank you for writing me to explain why my application for employment with your company cannot be favorably considered at this time. Although I regret that there is nothing appropriate available for my background, I hope that with additional study and experience, which your letter has stimulated me to pursue, there will be an opportunity for which I may qualify in the future. I shall look forward to your examining my improved qualifications at that time.

Sincerely yours,

Figure 2.20
A Sample Reply to a Rejection Letter

How to Write Inquiries

Letters of inquiry are probably the most common type of correspondence in business and industry. This letter seeks information or advice on many matters, technical and otherwise, from another person who is able to furnish it. Sometimes the inquiry may offer potential or direct profit to the person or company addressed. Frequently, the person addressed has nothing to gain and much time and energy to lose. Most organizations will answer all letters of inquiry as a matter of good public relations. Whether they send the inquirer the information requested frequently depends on the writer's ability to formulate his inquiry. A properly formulated inquiry will make the receiver want to answer it and makes the job of answering easier. A poorly written request may go unanswered or may receive an answer of little value.

A well-formulated letter of inquiry will have the following organizational pattern:

1. The opening paragraph should be a clear statement of the purpose of the letter. It should define for the reader the information desired or the problem involved: what is wanted, who wants it, why it is wanted.
2. The second paragraph should lead into the inquiry details. It should be specific and arranged in such a way as to make the answer as easy as pos-

sible. A good technique is to state specific questions in tabulated form. However, the request should be reasonable. The writer should not expect a busy person to take several hours to answer a long, involved questionnaire. Many concerns will not answer inquiries from people who are not customers, unless they know why the writer wants the information. This is especially true if information requested is of a proprietary nature and would be of benefit to a competitor.

3. The final paragraph should contain an expression of appreciation with a tactful suggestion of action. The letter might conclude with a statement that the writer would return a similar favor or service.

The letter of inquiry falls into two categories: the solicited letter and the unsolicited letter. The solicited letter is written in response to an advertisement inviting the reader to write for further information about a certain product or service. The unsolicited letter is written when the writer takes the initiative for making his request or asking for information or advice. Figures 2.21 and 2.22 provide examples of solicited and unsolicited letters.

Department B
Venard Ultrasonic Corporation
Herricks Road
Mineola, N.Y. 11501

Ladies/Gentlemen:

Please send me some further information on the features and costs of the Venard line of Ultrasonic equipment, which you recently advertised in Electronics magazine.

Sincerely yours,

(a)

Figure 2.21
Examples of Solicited Letters of Inquiry

Sales Department
Bakelite Company
30 E. 42nd Street
Room 308
New York, N.Y. 10036

Ladies/Gentlemen:

I have noticed your advertisement in the September 10 issue of the <u>Wall Street Journal</u>. I would appreciate your sending me a booklet, "Products and Processes."

In addition, I am particularly interested in epoxies—and their application in the construction of dies and molds. I would like to know the physical characteristics of epoxies, such as bearing load limits, heat resistance, and wear resistance. Perhaps a representative of your company might call on us to discuss our possible uses of epoxies.

                                        Yours truly,

(b)

Figure 2.21 (continued)

PRD Electronics
202 Tillery Street
Brooklyn, New York 11201

Ladies/Gentlemen:

At the recent IEEE convention, I noticed an interesting photograph in your exhibit of an engineer calibrating micrometer frequency meters.

Can you supply me with an 8 × 10 glossy print, along with some rough ideas and notes, about what is actually being done in that operation? We would like to use it in the "New Production Techniques" section of our <u>Electronics Production</u> magazine.

                                        Yours truly,

(a)

Figure 2.22
Examples of Unsolicited Letters of Inquiry

Public Relations Department
The Ford Motor Company
Dearborn, Michigan 48126

Ladies/Gentlemen:

Will you give us the benefit of your experience? We, as a university, want
to prepare our students going into business and industry to be able to
write effective letters, memoranda, and reports. To make our course
work as practical as possible, we would like to know what business and
industry require in these areas. Will you give us your ideas?

Do you have a preference for, or a prejudice against, any of the forms
presently used in business correspondence—block, semiblock, complete
block, the National Office Management's Simplified Letter Form, or the
hanging indentation form? Why? Is the form used universally
throughout your company?

We have another favor to ask. Examples are often the best teacher. May
we ask for specimens of your letters, memos, and reports that may be
helpful to train our students along the line you suggest?

We hope you can help us in preparing our students to meet the
communications problems they will find in business and industry.

Sincerely yours,

(b)

Figure 2.22 (continued)

759 Cherrygate Lane
Rosemont, PA 19010
August 6, 199_

Security Chemical Company
Lima, Ohio 45802

Ladies/Gentlemen:

Would you please have your Agricultural Division recommend
appropriate fumigants for my gladiolus corms while they are in winter
storage and a soil sterilant for the control of gladiolus thrips? Please let
me know where these materials may be obtained locally.

Won't you send me a copy of your garden catalogue?

Sincerely yours,

Sara Rachel Weisman

(c)

Figure 2.22 (continued)

Answers to Inquiries

Replies should begin with a friendly statement indicating that the request has
been granted, or granted to the extent possible. Then there should follow the
complete and exact information that was desired by the requester, including
whatever explanatory data might be helpful. If part of the information wanted
cannot be provided, this fact should be indicated next and accompanied with
an expression of regret and an explanation of the reasons why the complete
information cannot be given. Additional material which might be of value to
the requester is also included. Finally, the reply could end with the courteous
offer to provide any further information that might be wanted that is possible
for the writer to provide. Figures 2.23 and 2.24 show examples of answers to
inquiries.

The Quotation Letter

The quotation letter is a reply to an inquiry about prices. Complete description
of the product is usually given in an accompanying bulletin. Figures 2.25 and
2.26 show examples of quotation letters.

Mr. John Billings
Purchasing Agent
Johnston Laboratories
Stockertown, Pa. 18083

Dear Mr. Billings:

Your letter of September 8, requesting the minimum shipping quantity and price per pound on bulk shipment of Dioctyl Phthalete has been referred to this office for reply.

We are pleased to quote our current price schedule for Plexol Plasticizer DOP, which is our product trade name for this material:

Plexol Plasticizer DOP: in tank car lots, $1.00 per lb. in tank truck lots (4,000 gal. minimum), $1.35 per lb. in compartment tank truck lots (1,000–4,000 gals.)

Delivered to your plant at Stockertown, Pa.

We appreciate your interest in Dioctyl Phthalete, and if we can be of any further assistance, please communicate with us. Our district sales office, located at 12 N. 6th St., Philadelphia, Pa., is convenient to you. The telephone is 628-5555.

Sincerely yours,

Figure 2.23
An Answer to an Inquiry Letter

THE CUMBERLAND COMPANY
PAPER MANUFACTURERS

Cable Address
Cumberland Boston

Cumberland Mills
Central Mill
Copsecook Mill

80 Broad Street, Boston, Massachusetts 02401

30 December, 19___

Professor Herman M. Weisman
1801 Richmond Road
Westport, Maryland 43214

SUBJECT: Correspondence and letter-writing practices

As you see, Professor Weisman—
we rate as an ardent supporter of the simplified form of letter-writing—so-called.

In fact, we're credited with being the originator of the practice. We've been using the simplified system of letter-writing, and successfully, for at least 50 years. Here's why:

There's no reason in the world for using the endearing term in the salutation part of a letter. In our opinion, the simplified form of correspondence automatically eliminates this practice, and the communication becomes "alive" immediately. That's what we call the fast start in letter-writing. Of course, there are many other reasons, too numerous to mention here in favor of the practice. But for the benefit of your students, perhaps the easiest way for us to demonstrate our technique is this: We'll arrange to send you, automatically, copies of typical correspondence covering a period of two weeks. Through this means, you can determine by actual experience, instead of theory, how we operate in progressing the practice of simplified letter-writing.

Incidentally, perhaps we should tip you off in advance that most of our correspondence is with people engaged in the advertising profession. And may we further add that the simplified form of letter-writing evidently appeals to them. We've never had even the semblance of a squawk from our customers to the effect that the technique is "flip" or undesirable.

Please believe, sir, we welcome this opportunity to be of some service to you. Here's wishing you much continued success in your worthy educational endeavors. With heartiest of season's greetings to you for the New Year, we are, as always,

Promotionally optimistic,

Randy Raymond

RR/emb
Advertising Department

Figure 2.24
An Answer to an Inquiry Letter

The Oakland Company
St. Paul, Minnesota 55102

Attention: Mr. Elmer Peterson, Purchasing Agent

Gentlemen:

Thank you for the opportunity of quoting on our Road Runner Deluxe 40-Channel CB Radio Model RR2922.

Model RR2922	$209.95
6% sales tax	12.60
Total	$222.55

The attached specification sheet describes this extremely popular model offering what we believe is the greatest value in CB radios on the market today. This model features bright LED channel display, Delta tune, and RF gain. It has a switchable automatic noise limiter, noise blanking, and illuminated S/RF/SWR meter, with an antenna warning indicator. It is compact in design and weighs only 7 lbs. All our CB's carry a one-year guarantee. Terms are 30 days net, and we have a Road Runner Deluxe 40-channel CB Radio ready for immediate delivery to you.

Be sure to call us if you have further questions.

Sincerely yours,

John Doaks, Sales Dept.

Figure 2.25
A Sample Quotation Letter

Mr. John Kane
Chicago Paint Company
1217 Marshall Street
Chicago, Illinois 60623

Dear Mr. Kane:

Thank you for your letter of September 12, requesting a sample and price quotation on our Triethanelamine. We are sending you, without charge, 8 ounces of Triethanelamine for your evaluation.

Our current price schedules on this material are as follows:

In less than carload or less than truckload lots
      (in 55 gallon drums) . . . . $2.58 per lb.

On less than 55-gallon drum lots
      In 1 gallon containers . . . . $3.70 per lb.
      In 5 gallon containers . . . . $2.99 per lb.

Shipments less than 55 gallon drums are F.O.B., South Charleston, West Virginia; 55-gallon drums or less are F.O.B. delivery point of rail carrier (when shipped by rail carrier) or delivered to your plant (when shipped by truck) at Chicago. Shipments are made in nonreturnable 1, 5, and 55 gallon containers holding 9, 45, and 500 lbs., net, respectively, for which no extra charge is made.

This opportunity to be of service is appreciated. If we can be of further assistance, please let us know. Or, if more convenient, our Chicago District Sales office is located at 930 South Michigan Avenue in Chicago.

                     Yours very truly,

                     George Smith, Sales Manager

Figure 2.26
A Sample Quotation Letter

How to Write Claim Letters

As long as the human equation operates in business and industry, mistakes will be made, claims will be filed, and adjustments will have to be made. No matter what the cause, the claim or complaint needs to be expressed calmly, courteously, and objectively. Facts must be stated positively and truthfully. No matter how tempted the writer is and how justified he may be, any impatience, sarcasm, or discourtesy must be avoided. Vituperative or sarcastic language throws obstacles into the proceedings and makes adjustments more difficult and almost certainly will cause further delay in solving the problem.

The following suggestions are a practical guide in writing a letter of claim:

1. Explain what the problem is or what has gone wrong. Give necessary details for identifying the faulty product or service. Include dates of order shipment and arrival or nonarrival. Specify breakage or the kind and extent of damage. Give model number, sizes, colors, and whatever information is necessary to enable the reader to check into the matter. It is proper to include a statement of inconvenience or loss which has resulted from the cause of complaint.
2. Motivate the reader for the desired action by appealing to his sense of fair play or pride.
3. Include a statement of what adjustment you would consider to be fair. Figures 2.27a and 2.27b are two examples of claim letters.

The Adjustment Letter

How adjustments are met is usually determined by company policy. Most companies grant adjustments whenever a claim seems justified. Claims are usually decided on their individual merits. However, every complaint or claim, no matter how trivial, is answered courteously and promptly. Where a claim is granted, the adjustment letter has the following structure:

1. The writer is thanked for calling attention to the difficulty or problem.
2. The problem may be reviewed and explained. Alibis are considered poor form; whenever possible and expedient, whatever caused the difficulty is dealt with frankly.
3. The writer grants the adjustment, emphasizing a sincere desire to maintain good relations with the customer, and finally, appreciation for the customer's business is expressed.

When an adjustment is refused, the letter begins, similarly, on a positive note. The writer is thanked for calling attention to the difficulty. The situation is reviewed; particularly, the facts surrounding the claim are examined from the point of view of the decision.

The adjustment is refused with an explanation. The writer needs to show the reader that she understands the reader's problem. She must also, with friendly candor, make sure that the reader will understand the writer's situation. Examine Figures 2.28 and 2.29 (pp. 83–84) for some samples of adjustment letters.

Mr. A. H. Brown
Contracts Administrator
Homes Electronics
14 Lincoln Way
Lexington, Mass. 02173

Subject: Our Purchase Order No. BDLA83760

Dear Mr. Brown:

Our project engineer, Mr. T. H. Sherman, has informed me that despite an earlier agreement, your company has decided not to provide the feature of easy means and convenient interchangeability between crystal diode and bolometer in the instrumentation you are developing for us.

Please refer to our Purchase Order No. BDLA83760, which stipulates that the instrument under development will confirm to specifications.

I am sure this misunderstanding can be cleared up in time for delivery to be made in accordance with the required dates.

                                        Sincerely yours,

                                        R. H. Fanwell
                                        Assistant Purchasing Agent

(a)

Figure 2.27
Sample Claim Letters

Knight Laboratories
2330 Eastern Avenue
Los Angeles, California 90022

Gentlemen:

Our last order included 1 × 100 Deserol Ampules. The order was delivered
but the Deserol drug was not included within the packaged order, which
arrived by United Parcel. Since your shipping document listed seven boxes
of Deserol and this item was also included in your invoice attached with the
shipment, we want to let you know of this discrepancy. What is more
important, we are badly in need of this drug for filling our prescriptions.

Sincerely yours,

Barry Nelson

(b)

Figure 2.27 (continued)

The Letter of Instruction

Technical personnel, because of their specialized knowledge, are frequently
called upon to instruct others who have a need for their specialized knowledge
to perform a particular task. Letters of instruction are a convenient instrument
toward this end. If the matter is a complex one, requiring diverse contributions
by a number of people with a number of devices, tools, and equipment, the
letter may be more conveniently structured as a report. Frequently, however,
matters can be related most conveniently in the form of a letter or a memo.

Instructions are mastered more readily, are better remembered, and carried out more intelligently if the person being instructed knows the reasons for
the procedures indicated. The letter of instruction, therefore, usually begins
with a covering explanatory statement, which provides the background for the
writing of the letter. The language of the letter is structured in accordance with
the background, experience, and intelligence of the person for whom the instructions are intended.

The imperative mood is used for giving specific instructions. The imperative is the form of verb used in stating commands or strong requests. The
subject of the verb is not expressed but understood. (The rest of this paragraph,
for the sake of illustration, is structured in the imperative mood.) Be careful in

Mr. R. H. Fanwell
Assistant Purchasing Agent
King Electronics
1231 Jay Street
Brooklyn, N.Y. 11222

Subject: Your Purchase Order NO. BDLA31767

Dear Mr. Fanwell:

Your letter regarding the above purchase order has been reviewed with our Technical Personnel who have been working with your engineers on this instrumentation. Unknown to us, work requirements for this Purchase Order were left to informal understandings between some of our technical personnel and your engineers.

Our Contracts Department interpreted the work requirements in accordance with the written specifications of the purchase order and, accordingly, ruled that incorporation of a means for convenient and easy interchangeability between the crystal diode and bolometer was not called for. The interpretation was based on the note in purchase order BDLA31767, which reads: "Items 1a, 1b, and 1c to be in accordance with Proposal No. 1401 of May 24, 1980." Nowhere within Proposal No. 1401 is the requirement of interchangeability of crystal diode and bolometer mentioned.

Because of the informal understanding between our technical people and your project engineer, we will accomplish without any further cost, the interchangeability requested. Delivery of the low-powered unit will be made within the next two-week period. We would appreciate notification of acceptability of that aspect of the work as soon as you have tested the delivered unit.

Sincerely yours,

Figure 2.28
Sample Adjustment Letter

Mr. Barry Nelson
City Drugs
12 University Avenue
Denver, Colorado 80206

Dear Mr. Nelson:

Thank you for calling our attention to the omission in your order of the Deserol drug. We did not intend to include it with your order. Listing it on the Bill of Lading and on the invoice was a mistake.

Deserol is packaged one ampule per box, seven boxes to a shipping container. Since it is kept under refrigeration, we normally ship it by air parcel post, special delivery, rather than with regular orders. Because of this, and because we were uncertain of the amount you desired, we did not include it in your shipment. If you will send us your order indicating how many Deserol ampules you would prefer, we would be glad to make immediate air shipment.

We appreciate your interest in our products. Please call on us if we can be of any further assistance.

Very truly yours,

Figure 2.29
Sample Adjustment Letter

using the imperative that you do not appear unduly brusque or imperious. This is especially advisable in dealings with clients or persons outside your organization. The imperative mood is used for the specific directions required, so be sure the instructions are in parallel grammatical construction; be sure, also, that your instructions are complete and that all words and statements are specific and clear. Whenever the reader is to use his or her own judgment, so indicate. Be sure there is logical order in the presentation of the steps of your directions. When instructions can be arranged in successive or chronological order, organize them in accordance with the demands of the situation so that they ensue as a planned sequence of activity. Write instructions concisely and precisely; avoid roundabout expressions, vague and unnecessary words and generalities. Use such devices as numbering, underlining, and indention of headings because these help clarify the indicated procedures. Include the time or date by which the action must be accomplished. See Figures 2.30 and 2.31 for examples of letters of instruction.

Computer Technology

Computer technology has entered the technical writer's work place. Automating the writing process is discussed in Chapter 5. Three topics related to technical correspondence, however are covered here.

Mr. James Mitchell
19 Cherry Creek Lane
Jericho, New York 11753

Dear Mr.Mitchell:

We are sorry that our salesman did not explain that all Hollywood beds are delivered unassembled. I am sure you understand that it would be most awkward and inconvenient to ship, in protected cartons, a bed completely assembled.

Such bulk would add to storage and transportation costs and would be reflected in the price we would have to charge you. Actually, assembling one of our Hollywood beds is a very simple process that can be done in four very easy steps. Won't you try them? We are sure you will find our instructions very easy to follow:

1. Place the two sections of the frame in a parallel position on the floor, so that both headboard plates are at the same end. Swing out the cross rails, until they are at right angles (90°) to the side rails.
2. Slip tension clamps on cross rails to your left, wing nuts facing out. <u>Do not tighten the clamps</u>. Overlap right and left cross rails so that left cross rail is on the <u>bottom</u> at the head end and on <u>top</u> at the foot end. Slide tension clamps to center of overlap of cross rails. <u>Do not tighten clamps</u>.
3. Measure the width of the boxspring across the bottom. Adjust the width of the bed frame to the same measurement. Tighten both clamps securely with the wing nut. Insert casters, glides, or rollers, whichever you have. Exert pressure downward on the frame until they click in.
4. Place boxspring and mattress on the frame, which should hold the boxspring securely. If not secure, loosen clamps, adjust frame for snug fit, and tighten clamps.

Thus you will have assembled the frame of your comfortable, but sturdy, Hollywood bed in not more than 15 minutes. I am attaching, also, for your information, drawings illustrating each step.

Sincerely yours,

Figure 2.30
Sample Letter of Instruction

Mr. G. W. Schmidt
Field Engineer
Colorado Sugar Corp.
Berthoud, Colorado 80513

SUBJECT: <u>Instructions for recording soil and drainage data</u>

Dear Mr. Schmidt:

The data you collect on a soil and drainage investigation go to the Central Design office. The men who use your data have probably not seen the investigation site. In order that your data be complete and consistent, it is desirable that you follow a standardized procedure for recording such data.

Use Form D16, a coy of which is included. This form is available from the Engineering Office. Divide the recording data into three main parts:

A.   Generalized site description
B.   Specific soil and material description
C.   Symbolic profile representation

A.   <u>Generalized site description</u>

Use the spaces at the top of this form to record:

1.   Area—the particular land development project.
2.   Pit Number—the assigned number.
3.   Photo Number—if the area is covered by an aerial photograph survey, note the photograph number and the scale of the survey. Example: 17–30, 1:8100 (Photos are filed by number under the survey scale).
4.   Ground water: Record the depth of the ground water. If ground water was not present at the time of the survey, write "none."
5.   Date: Date is important in conjunction with water table fluctuations.
6.   Surveyor: Write your last name.
7.   Site description: Note all conditions at the particular site. Land use, crop condition, if cultivated, natural vegetation, slope, salinity, and alkalinity.

B.   <u>Specific soil and material description</u>

1.   Indicate any changes in the soil profile by drawing a horizontal line across the form using the depth marks as guides.
2.   Write out the soil texture for each horizon. Abbreviations can cause confusion.

Figure 2.31
Sample Letter of Instruction

3. Indicate the structure of each horizon since structure is a guide to permeability.
4. Estimate the permeability of each horizon using standardized permeability classes.
5. Write out the moisture condition in each horizon.
6. Indicate any molting by standard nomenclature. Molting, when present, gives an indication of post ground water conditions. Example: Even when ground water was not encountered at the time of the survey, dark, rust-brown molting near the surface might indicate drainage problems sometime in the past.
7. If free water occurs in the pit, indicate by writing "Water Table" at the point of the Free Water Surface. Note the nature of the water table, true or perched.
8. Indicate any samplings of soil materials by blocking out the depth marks, "marks on the form corresponding to the depth of the sample."
9. Indicate water samples, by writing "Water Sample" in the supersaturated zone.

C. Symbolic profile representation

1. Use the space on the left side of the form, D16, to show the complete soil profile in symbolic form.
2. Use standard soil and material symbols.

If you follow this method of recording field data, you can be sure your data will be complete and clearly understood by those using it.

Sincerely yours,

Figure 2.31 (continued)

Electronic Mail

Electronic mail, sometimes referred to as **E-mail**, permits direct transmission of letters and memos electronically to a recipient's computer. It offers the speed and convenience of a telephone call, plus the advantage of a permanent copy of the message. Though it costs more than a postage stamp, it is less costly than overnight air express mail. E-mail must be part of a dedicated system; it can be sent short distances between offices or very long distances across the world by satellite. The message can be read on the terminal, stored in the computer file, or printed at any time.

To use electronic mail, your computer must be linked to a system or service. You access the system or service by proper code identification. You key your message to your destination, which must have a linkage to your system or service. The message is received and placed in the recipient's file space (mail box).

There are two general types of E-mail:

1. An inhouse network, which connects computers or terminals within an organization, and
2. An external network, connecting computers at different locations, even around the world.

Electronic Bulletin Boards

The **electronic bulletin board** is the equivalent of the conventional bulletin board with the exception that its system enables a linkage of enlarged readerships from great distances. Its purpose is to exchange information. Key to the electronic bulletin board is the system operator (or sysop), who oversees the activity. Connection to the bulletin board is made by a **modem**, a device that converts binary digital data to audio tones suitable for transmission over telephone lines and vice versa. Users can post messages for others to see; read messages posted by others; communicate with the system operator to ask questions, report problems, make suggestions, share programs, or obtain programs.

Electronic bulletin boards are often operated by individuals interested in computers, and who wish to communicate with others with a similar interest to share information and software programs. Companies and organizations have also started electronic bulletin boards as a service to clients and members. Some businesses use them for customers to place orders, ask questions, and have problems solved. Electronic bulletin boards are a useful resource for technical writers in practicing their profession.

Facsimile (FAX)

In some ways similar to electronic mail, **facsimile (FAX)** is a technology using telephone links to provide a hard copy duplicate of an original document. The image of the document is scanned at the point of transmission, reconstructed at the receiving station by another FAX machine, and duplicated as hard copy.

Facsimile machines send an exact replica of a page—text, illustration, and even halftones—over telephone lines. The FAX uses a built-in scanner that "reads" 200 dots of printed matter in every square inch. As the scanner passes over letters (including handwriting) or illustrations on a page, it reads an "on" signal for a dark dot and an "off" signal for white space. At the receiving end, a high resolution printer translates the on/off signals to dot/no dot instructions and presents a picture of the page.

Advances in technology now permit personal computers to serve as FAX machines. A FAX modem, known as a fax board, allows both sending and receiving messages by way of a user's personal computer. Computer-generated outgoing FAXes are more readable and better appearing than messages sent by regular FAX machines, because the image does not have to go through the degradation of the FAX machine scanner. Improved confidentiality is another advantage; messages enter the computer's disk instead of the FAX machine tray.

A new device has come on the market that combines a FAX, a copier, an image scanner, and a document printer and is designed to tie together these now-separate operations. Using this machine, a technical writer by pushing a button on his or her computer can order up one or several hundred high-quality bound copies of a large report or send a FAX of the document to another computer user. In addition, the machine can electronically scan paper documents for storage into a computer, and that document later can be manipulated on the computer screen for other needs. Price consideration at present make this machine affordable only to large companies and organizations.

Chapter in Brief

In this chapter we examined the role correspondence plays in science and industry. We looked into the psychology of correspondence, examined the conventions and formats of letters, and how to organize and compose them. We concentrated on correspondence for the employment situation, particularly the resume and its cover letter. Other types of technical correspondence were examined—letters of inquiry, complaint, quotation, and responses to each, as well as letters of instruction. We ended the chapter by discussing aspects relating to correspondence utilizing computer technology—electronic mail, the electronic bulletin board, and facsimile.

Chapter Focal Points

- Psychology of correspondence
- Pre-writing—planning the message
- Format conventions
- The Memorandum

- How to apply for a job
  The Resume; its types
  The Cover letter
  The Executive briefing
- Claim and adjustment letters
- Letter of instruction
- Electronic mail
- Electronic bulletin board
- Facsimile

Questions for Discussion

1. Many technical personnel have secretaries type their letters. Do you think a "Federal case" has been made in this chapter on correct format, layout, and typing considerations? Defend your answer.

2. Should a letter dealing with technical matters be concerned with the "You Psychology"? Before you answer, recall that technical writing is objective and impersonal. Defend your answer.

3. One of the largest irrigation systems in Colorado has an opening for the position of assistant supervisor. The applicant must have a degree in irrigation, agricultural or civil engineering, and must be between the ages of 25 and 35. Business ability is necessary. An applicant has written the following letter to the Fort Lyon Canal Company, Box 176, La Junta, Colorado:

704 S. College Ave.
Fort Collins, Colorado 80521
October 5, 199___

Fort Lyon Canal Co.
Post Office Box 176
La Junta, Colorado 81050

Gentlemen:

A coyote can live very well in a zoo, but he is not happy there. He longs to trade the security of his stuffy zoo quarters for his home on the prairie where he can battle the elements for himself and get away from dense population. I am like the coyote in many respects. I grew up in the Arkansas Valley and learned to love it. Cities, heavy traffic, and design rooms do not appeal to me. I am home in the Arkansas Valley working to try to increase its productivity. That is why I want to be assistant supervisor of your company.

I grew up on irrigated farms under the Fort Lyon and farmed there before entering the Air Force. As you know, my father farmed in the Arkansas Valley for 54 years, most of which was under the Fort Lyon. I am well acquainted with the problems of your company and of the farmers, a quality which would not only be useful in making decisions, but also in gaining the confidence of the farmers.

I will receive a Bachelor of Science degree from Colorado State University next month in Agricultural Engineering, irrigation option. My technical electives were taken in irrigation, and most of my nontechnical electives were taken in the fields of business and economics.

I would like to have an interview on any Saturday this month. If this would be impossible, I can appear at your convenience.

Sincerely yours,

How do you think the prospective employer will react to the opening? If you deem an unfavorable reaction, how would you rewrite the opening? Does the applicant meet the specifications called for in the ad? What specific information is lacking? Can this be supplied in the resume? Structure a resume supplementing the letter. What merits do you find in the letter? What elements would you improve upon?

4. What is your reaction to the following letter?

Mr. J. R. Barnes, CLU
General Manager
New York Life Insurance Company
1740 Broadway, Suite B-304
Denver, Colorado 80202

Dear Mr. Barnes:

I want to hitch my wagon to a star . . . not any star, but the New York Life star of the insurance galaxy.

I can help put your district even higher into the insurance sales atmosphere than it is now . . . if I can only be given the opportunity.

I will graduate from Colorado State University in June, 19__, having completed my degree in the business administration field of study. I have no outstanding achievements that could be cited from my college records, but I have what I believe to be most essential to help you reach even higher sales records . . . the sincere desire to make a name for myself and your district by selling more insurance my first year than any previous salesman.

I believe I am qualified to do this with your able training assistance in the insurance sales field. I have previously worked in the district sales category for Curtis Publishing Company, Wearever Aluminum, Norge, Frigidaire, Maytag, Zenith and RCA on commission sales. Other sales work included advertising copy sales work for my hometown paper in high school and college paper at Kansas State College all in the period from age 16 until the present time.

I like meeting people, making friends, doing things for these people, and most of all I like to make money. My motives are not entirely selfish because I don't have to say that to make this money in the insurance field one has to work hard . . . right? This I am ready to do.

If you will glance up the left margin of this letter you can see that every paragraph starts with "I" . . . this is because I am thinking of what I can do for you and New York Life, if given the opportunity. May I have a personal interview with you at your convenience for furthering this discussion? My home telephone number is 555-5713, in case you are in town and wish to call.

                                        Sincerely yours,

What attitude is prevalent in the above letter? What kind of person do you think would write such a letter? Do you think it would impress a sales manager? How would you revise this letter?

5. What is your reaction to the following letter? Why? How would you revise it?

                                        Room 115 Green Hall
                                        Ft. Collins, Colorado 80521
                                        May 13, 199___

Personnel Manager
King's Clothing
Lakeside Shopping Center
Westminster, Colorado 80030

Dear Sir or Madam:

Beautiful women! What do these words bring to your mind? Are the images tall, slender, immaculate, and graceful models? Do you see them sitting in a pseudo-swing, posing in a horse and buggy, or simply stepping from a fashion runway?

Without a doubt they will be lovely and a pleasure to look upon. I would like to be able to make these visions come alive for you and your customers. Fashion shows are the media through which I would work. My experience in this field was acquired by modeling for an American dress chain store and in fashion shows of Job's Daughters and high school. I love to organize and carry through brainstorms using lovely models and select clothing such as you carry.

But even the best fashion show will not succeed without commentary that complements and emphasizes hidden features that make an outfit complete but are not obvious at a first glance. Journalistic training on the Denver Post, my high school newspaper, and *CSU Collegian* provide experience in composing proper comments and extensive college speech training make appearing before the public a pleasure.

Being 25, female, and in excellent health further qualify me to work with you to advance interest and sales through fashion shows. May I come in and talk with you about your position available as a fashion show director? I shall be happy to call you at 2:00 Monday for an appointment.

Enclosed are three letters of recommendation you might be interested in referring to when considering me for employment. Also included is a recent photograph.

Thank you for your consideration.

Very truly yours,

4 Enclosures

Assignments

1. For training purposes, build up a collection of letters. Obtain them from friends and relatives in business, industry, and from various types of organizations. Study and analyze these letters in accordance with the various principles propounded in this chapter. In this analysis, write down for each letter the writer's purpose. If there are several purposes, note if they are organized from an overall objective; if there is more than one purpose, do they create confusion? Rewrite such letters. Are there letters in which the writer does not seem to be successful in establishing his purpose? How would you revise those letters?

2. Study the want ads in the classified section of a major newspaper or the listing of positions available in a professional magazine of your field. Write a two-part application letter for a permanent position on your graduation from college. Try at least three different types of opening paragraphs—original, summary, name beginning. What effect does each type of opening have on the rest of the letter?

3. Now compose a letter for an unsolicited opening—that is, write to a company for whom you wish to work although you do not know that it has a position available. How much of the previous letter (problem 2) can you use? Should your resume or data sheet be slanted for each job you apply for, or does it remain constant? Why?

4. An influential friend of your family, or a relative, who commands respect in the field in which you seek employment has recommended you for a position. Apply for it by letter.

5. Write follow-up letters following interviews for positions in problems 2, 3, and 4. Write job acceptance letters for problems 3 and 4.

6. You have not been offered a job in the situation mentioned in problem 4. Write a follow-up letter.

7. Write letters of acceptance for problems 3 and 4.

8. Write two letters:

   a. A letter to a company, a government agency, or an educational institution in which you ask for information about a product, a piece of equipment, a bulletin, or a program of study.

   b. A letter giving the information requested in 8*a*.

9. a. Write a letter to a manufacturer ordering six separate catalog items; for the sixth item, request certain changes from the catalog specifications. Ask for your order to be acknowledged and the acknowledgment to include a price list.

b. Write the acknowledgment to the order in *9a*; indicate two alternatives to the special revision of the catalog item. Explain the advantages and disadvantages of each.

10. Write two letters:

   a. A letter giving instructions to a group of workmen under your supervision.

   b. Write a letter report in response to the letter reporting the results after the instructions have been carried out.

11. Recall a product or service you have had that was unsatisfactory. Write a letter to the company concerned telling of your lack of satisfaction and asking for a refund or adjustment.

12. Write two replies:

   a. One granting the adjustment.

   b. The other refusing the adjustment.

13. Write a letter to a consulting engineer describing a technical problem; ask him whether he is interested in solving the problem for you and what his fee would be.

14. a. Write a letter to an eminent scientist inviting her to be the speaker at the annual banquet for your student scientific activity.

   b. Write a letter the scientist might write accepting the engagement.

   c. Write a letter, after the banquet, thanking the scientist for helping to make your student affair a·success.

15. Write a letter requesting information about price, efficiency, cost of operation, and maintenance of a piece of equipment for your college laboratory.

16. Reply to problem 15 above.

The Process of Communication

3

Chapter Objective

Provide an understanding of the process of communication and the roles language and semantics play in the process.

Chapter Focus

- Nature of language
- Nature of communication
- How the process works
- Nature of semantics
- Nature of meaning
- Communication—Bridging the gap between words and thought

*E*fficient communication of scientific thought or technical information— whether conveyed by a technical writer, a computer systems analyst, a scientist, or an engineer—does not just happen. Effective technical writing is a craft, never an accident. The poet, playwright, and novelist must master their art before they can give form and meaning to what they have to say. So must the conveyers of technical information. If they are to reach the minds of other people, they must use language with precision, clarity, and grace. In technical writing, one explains facts, expounds theory and principles, and analyzes concepts, objects, processes, events, and data.

The purpose of this text is to teach you how to communicate technical information effectively. To begin, you need to know how language and semantics affect communication, and how their various elements operate.

Origin and Nature of Language

Although linguists and anthropologists disagree as to how language originated, there was a time when language started, just as there was a time when human beings first used fire. In *An Introduction to Philosophy*, Sinclair claims that language originated as and remains "symbolic pointing." He says the following:

> Speech is simply gesture made audible. . . . Speech is an extremely complex system of more or less standardized and conventionalized noises, and writing is an even more highly standardized and conventionalized system of visible marks upon a surface, but in principle speech and writing are as much gesture as is pointing with the finger.
>
> The origin of language appears to have been roughly as follows: Our remotest human ancestors when they attempted to draw the attention of others of their kind to anything in particular (or when they behaved in a way which did in fact draw attention to it, intention or no intention) pointed and gesticulated in its direction. . . . The gestures and the noises together resulted in drawing attention to the object or situation in question. . . . Later, after thousands upon thousands of years, conventionalized marks upon surfaces came to be employed to represent the conventionalized noises.
>
> Consider an example. When a biologist writes and you read, that the hind leg of a horse is homologous with your own leg . . . what is happening? What he does is to arrange the printer's black marks on paper so cunningly that you, sitting in your chair by the fire, have your attention drawn to certain matters which he could otherwise have drawn your attention to only by leading you up to a horse and pointing with his hands. . . .
>
> If the biologist in our example were unskilled in pointing, you might have difficulty in understanding what he was driving at when he pointed. To be effective, he needs some natural aptitude and a good deal of acquired skill gained by long practice. If he were unskilled in handling the English language, you might similarly have difficulty in understanding what he was

driving at when he wrote. . . . Since man through countless generations has conventionalized and developed the system of symbolic pointing which we call language, we can generally find suitable words to direct other people's attention to what we wish, provided that it is the commonplace things of life. . . .

On the other hand when we wish to direct their attention and our own to less plain and obvious matters, we fall into difficulty, because we have to use for that purpose words which were originally developed and used for other and more matter-of-fact purposes, i.e., we have to use words as metaphors. The parallelism between words to write with and fingers to point with is very close. [7:113–19][1]

Anthropologists say that language brought the birth of civilization. Whether the Neanderthal was anthropoid or human depends less on cranial capacity, upright posture, or even use of tools and fire than on whether or not he or she communicated with other fellows. Aldous Huxley (3:13) observed that human behavior as we know it became possible only with the establishment of relatively stable systems of relationships between things and events on the one hand and things, events, and words on the other. Behavior is not human in societies where no such relationship has been established, that is to say, where there is no language. Only language enables people to build up a social heritage of accumulated skill, knowledge, and wisdom and allows them to profit by the experience of past generations. Our mastery over reality, both technical and social, depends on our knowledge of how to use words. This is true of craftsmanship within a primitive community or in our own highly technological society. Knowing the name of a "gizmo" is frequently the direct result of knowing how to use a gizmo. The right words to describe the activity of a craft or trade or the right words to describe techniques and abilities assume meaning to the extent that the person becomes the master of them or acquires the ability to carry out the proper action. To put an ability, craft, or technique into communicable words provides a basis for cooperative action. Civilization depends on cooperation; cooperation depends on communication.

The Nature of Communication

The word *communication* comes from the Latin *communis,* meaning common. When we communicate, we are trying to establish a *commonness* with someone. Dictionaries define the process as "the giving and receiving of information, signals, or messages by talk, gestures, writing, signals." Communication in-

---

[1]The first number in the brackets refers to a reference listed in sequence in the back of the chapter. If other numbers follow a colon after the reference number, they refer to appropriate pages in the cited reference (see pages 379–380, Chapter 14).

volves the sharing of experience with others. While I want to stress the *human symbolizing* activity of language (which is the basis of human thought), I want you also to be aware of the following two points:

1. Human experience allows communication without recourse to formal language, through signs, signals, symbols, and so forth.
2. Other living beings—animals and insects, for example—also communicate.

Nonverbal Communication

Two lanterns in old North Church were a sign to Paul Revere to get on his horse. A runny nose, watery eyes, hoarseness, and a cough communicate to a parent that a child should stay in bed. A horn honking behind you as the traffic light turns green tells you to get your car moving. Blasts on a ship's whistle in a fog warn other vessels of its presence. A school bell tells the teacher and the students that classes are ready to begin or that classes are over. Thus, people communicate or receive communication from natural phenomena in their own experiences and from devised conventional signs such as gestures, facial expressions, pictures, codes, flags, lights, bells, signs, bar codes, alphanumerics, and so forth.

Animals communicate. A bee returning from flowers with a load of nectar performs a dance that tells the other bees in the hive where to get nectar. The length of the dance indicates the number of flowers there are and the number of workers needed to transport the nectar. If the bee dances for only a few seconds, a few bees will go out; if for two minutes, twenty or more may go. Long dances indicate a very rich food supply. The direction of the dance indicates the direction for the bees to go (2:36–40).

Animals know and read signs. Overcast skies tell birds it will rain. Thirsty cattle know that low, thick foliage means water. Dogs communicate their wants by barking, whining, scratching, and running. Wild geese in a V flight formation coordinate and communicate their next movement to each other to maintain their V formation throughout the flight.

Biologists tell us that through chemical, optical, auditory, tactile, electrical, and other signals there are communications not only between organisms of the same species and between organisms of different species, but also that there are communications between cells of the same organism and between parts of the same cell. Without inter- and intra-cellular communications, life would not be possible.

Language, however, is the presumably unique human characteristic that has enabled us to rise above the insects, birds, and other animals. As Lee has observed:

> A bird builds a nest. A man builds an engine. Each uses time, effort, and materials, but soon their purposes have been served. A new generation comes. The new bird builds again and the next nest is not unlike it. With the new man, however, come new possibilities, new searchings, new ways

of looking, new experimenting, so that when the engine is built, it is built anew. To the work of the past something is added from the present. For with the symbol-using class of life, the power to achieve is reinforced by past achievement. . . . [4:15]

Human Communication—How It Works

Let us examine the great human achievement that has made civilization and progress possible. The diagram in Figure 3.1 will help us understand the process.[2] We can borrow terminology from a radio or telephone circuit, because the theoretical concepts of radio and telephone communications are based on the human communication system.

There are three basic elements in human communication:

1. The source, or sender
2. The message, symbols, or signal
3. The destination, or receiver

The **source** may be a person speaking, writing, keyboarding, drawing, or gesturing. It may be a communication organization such as a magazine, book publisher, radio or television station, motion picture studio, or computer network. The **message** may be in the form of writing, Braille, printed characters, diagrams, pictures, sound waves, electrical impulses, dot patterns on a screen, a gesture or a grimace, lights, bells, flags, smoke signals, or any form of signal capable of meaningful interpretation. The **destination** may be a person listening, watching, reading, or perceiving by any of the senses.

What happens when the source tries to establish commonness of experience with the intended receiver? The answer is diagramed in Figure 3.2. First, the source (person) encodes the message by encoding the information, feeling, thought, or idea he or she wants to share into a form that can be transmitted. *Encoding* is the thought or formulation process that takes place in the mind of the source for translation into a signal, message, or symbol for sending. The message or signal is sent in the form of gestures, grimaces, speech, handwritten or keyboarded words, drawn diagrams, pictures, or other graphic illustrations,

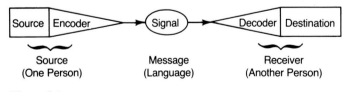

Figure 3.1

---

[2]Figures 3.1, 3.2, 3.4, and 3.5 are based on diagrams in Schramm's excellent article cited in reference 6.

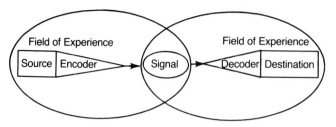

Figure 3.2
Commonness of Experience Is Requisite for Communication

printed means, or radio pulses. The message is received and decoded by the perceptual system of the individual receiving the message. The receiver can translate the message only within the framework of personal experience and knowledge.

The source can encode and the destination can decode only in terms of the experience each has had. If the circles have a large area in common, communication is easy. If the circles do not meet—if there is no common experience—communication is impossible.

Plainly, the communication system can be no better than its poorest link. In engineering terms, there may be noise (a disturbance that does not represent any part of the message), filtering (abstraction or abridgment), or distortion (perversion of meaning) at any stage in the communication process. If the person sending the message does not have adequate or clear information, the message is not formulated or encoded adequately, accurately, or effectively into transmittable signs. If these signs are not transmitted accurately enough to the desired receiver, or if the message is not decoded in a way that corresponds to the encoding, or if the person receiving the message is unable to handle the decoded message so as to produce the desired response, then there is a breakdown in the communication system.

Look at Figures 3.1 and 3.2 and steal a glance at Figure 3.6. Let us say an electronics engineer, assisted by a technician, is designing a new instrument. The engineer is testing a circuit. If it works, it will solve an important problem in the design of the instrument. At a critical moment the engineer tells the assistant to connect a lead to a scope. The assistant asks, "How?" "Use an alligator clip," the engineer answers. "Quick!" The technician connects the alligator clip to the lead of the instrument and then to the terminal of the scope. "Cool!" exclaims the engineer, and watches the trace of the circuit on the scope.

While it is difficult to imagine an electronics engineer unable to communicate to an aide what an alligator clip is, let us suppose that the engineer, because of incompetence, emotional pressure, or difficulty of speech, is unable to formulate the proper message about the alligator clip to the technician. Or, suppose that the engineer's helper is inexperienced, or has an auditory or other perceptual difficulty, or does not speak or understand the engineer's language. Noise, filtering, and distortion would then occur at the signal encoding or decoding elements in the communication process. The source can encode and the destination can decode only in terms of the experience each has had. If the circles in Figure 3.2 have a large area in common, communication is easy. If the circles do not meet, there is no common experience, and communication is impossible. The most important factor about a system like that in our diagram is that the receiver and sender must be in tune. Although this is clear in the case of a radio transmitter and receiver, it is more difficult to understand when it means that the human receiver must be able to understand the human sender.

If we have never learned Russian, we can neither encode nor decode in that language. If we have had no experience with an alligator clip, unless the message includes a proper description, explanation, and background, we might think an alligator clip is a device that clips the claws of an alligator. *The source must always be aware of the experiential background of the receiver* so that he or she can encode the message in such a way that the destination will be in tune with the message and so that the receiver can relate it to the personal experience most like that of the source.

In actuality, the communication system can get much more complicated than the preceding two simple diagrams indicate.

Figure 3.3 represents a situation that frequently occurs before the intended report reader (destination) receives a report about a problem in which she is interested. The source might be the engineer working on the words used to formulate the message about the problem. The first receiver in this case might be a tape in a recorder into which the engineer dictates the message. The tape becomes the first relay. A stenographer/word processor operator (the next transmitter) transcribes the message into a rough draft. The transmitting agent is the rough draft. A technical writer may be the receiver. Reading the draft, she is also its destination. After she has polished up the message or report with a "blue pencil," she becomes relay 2. She (the transmitter) sends her polished version (transmitting agent) to a word processor operator (receiver or destination) who acts upon it. The newly revised copy is relayed by the operator to a supervisor. The supervisor (relay 4) goes over the polished version and, in turn, amends it. This version is reprocessed and a new chain of communication may begin, with editors and decision-makers entering the chain. The final version may undergo desktop publishing procedures; it may be sent in disk form to the recipient's computer system; or it may be faxed by telephone links. Other communication chains may be started at each iteration of the report's preparation process. Noise, filtering, or distortion may take place so that the final message or report may have elements of incomprehensibility, inaccuracy, or distortion in some or all of its elements. Breakdowns in the message may occur at any of the links in communication chain. Conversely, the report may be better for all the collaboration it received en route. (See also Figure 3.9.)

Figure 3.3 shows the links in the communication chain and points out the importance of watching for any weaknesses in the chain. Each person in the communication process is both an encoder and decoder; each receives and transmits. Communicators get feedback of their own messages by listening to their own voices as they talk or by reading their own messages as they write (Figures 3.4 and 3.5). We are thus able to correct our mispronunciations or catch mistakes

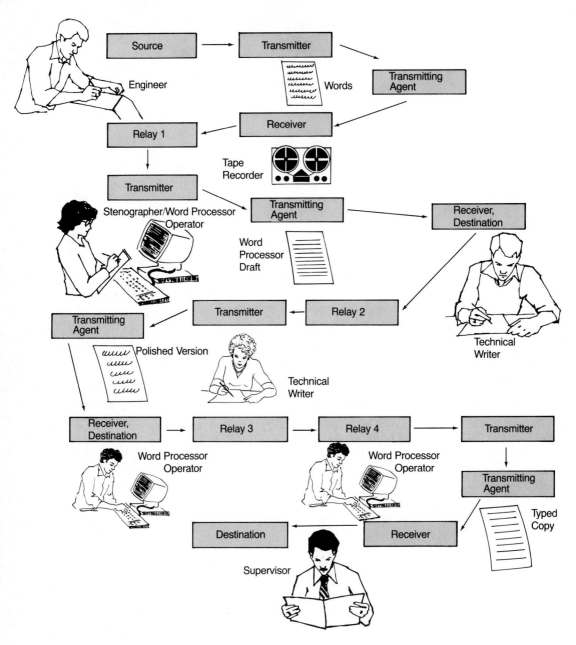

Figure 3.3
An Example of How the Communication Chain Works

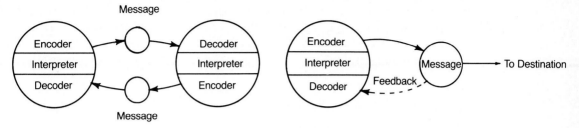

Figure 3.4   *Figure 3.5*

in our writing or add information as we recognize omissions. The feedback process enables us to catch weaknesses in the communication link as they occur.

Semantics

Words as Symbols of Reality

Semanticists remind us that words and language are not the things they represent. Semanticists say we live in two worlds and we should not confuse those two worlds. Irving Lee called them "a world of words" and "a world of not-words."

> If a word is not what it represents, then whatever you might say about anything will not be it. If in doubt, you might try eating the word *steak* when hungry, or wearing the word *coat* when cold. In short, the universe of discourse is not the universe of our direct experience. [4:16]

Semanticists also caution us about **symbols**. Symbols are not the things they are supposed to represent. The picture of Uncle Sam, although it represents the idea of the United States, is not the United States. *Au* is the chemical symbol for gold. It may do a proper job in a lab notebook, but it will not crown a tooth or replace currency.

Symbols have a physical effect in human experience. They are able to cause sensations within us. They permit us to deal with absent things as if they were present. The use of symbols allows our minds to handle dollars, pounds of apples, oceans of water, and billions of stars without having these things physically present. Humans are able to do this because we have developed an elaborate substitute system for experience and thought in signs, symbols, signals, words, and language. Unfortunately, reactions to a particular symbol or word are not the same for one person as they are for another. For example, Uncle Sam may represent one thing to an American, another thing to an Iraqi, and a third thing to a Britisher. For good or bad, words and symbols have acquired emotional associations (Figure 3.11).

We can manipulate words and symbols independently of what they represent. Semanticists point out that the structure of our language does not always correspond to the structure of the world. To make their point, semanticists often refer to a simile of a map. A road map that puts St. Louis north and west of Denver would not be much use if you were driving from New York to St. Louis. Language is a map; unless it corresponds to the territory, the driver is in trouble.

Process of Abstraction—The Problem of Using Words as Symbols

All words and symbols are an abstraction of reality; that is, they are removed from life, as the definition of the word signifies. When you and I use words, we use them as arbitrary agreed-upon symbols to represent reality. For example, the word *chair* is an arbitrary symbol for many different types of furniture on which people sit. If I wanted or needed to be more specific in a situation, I might say "easy chair." To be even more specific, I might say "the brown, overstuffed easy chair my wife Margaret purchased at Macy's Department Store, which I use every night to sit on and relax in after I get home from work." The word "chair" symbolizes, or is an abstraction of, the particular chair to which I refer. It can also refer to any number of chairs existing in reality or capable of existing in the minds of any number of people. The word *chair* is more abstract than my particular chair. While it is more specific than the word *furniture,* it still might be considered a term of higher order abstraction. Semanticists have constructed what they call *abstraction ladders.* Abstraction ladders are graphic devices to help us understand how to make words more accurate maps of the fact territories they represent.

Here is a typical abstraction ladder that I have borrowed from Chase. To get at the most accurate map in the abstraction ladder, we begin at the top and work downward:

> *Mountains:* What can be said about mountains which applies in all cases? Almost nothing. They are areas raised above other areas on land, under the sea, on the moon. The term is purely relative at this stage: something higher than something.
>
> *Snow-capped Mountains.* Here on a lower rung, we can say a little more. The elevations must be considerable, except in polar regions—at least 15,000 feet in the tropics. The snow forms glaciers which wind down their sides. They are cloud factories, producing severe storms, and they require special techniques for climbing.
>
> *The Swiss Alps.* These are snow-capped mountains about which one can say a good deal. The location can be described, also geology, glacier systems, average elevation, climatic conditions, first ascents, and so on.
>
> *The Matterhorn.* Here we can be even more specific. It is a snow-capped mountain 14,780 feet above the sea, shaped like a sharp wedge, constantly subject to avalanches of rock and ice. It has four faces, four ridges, three glaciers. It was first climbed by the Whymper party in 1865, when four out of seven were killed—and so on. We have dropped down the ladder to a specific space-time event. [1:143–44]

I have illustrated the process of abstraction as it applies to the concrete term *mountain.* A *concrete term* or word designates something that exists in the physical world. A concrete term may be specific; that is, it may designate only one specific object, such as the Matterhorn, Pikes Peak, Mississippi River, the General Motors Corporation, the Macmillan Publishing Company, Albert Einstein, my daughter Abbi, or my son Harlan. It may be general or designate a class of objects, such as mountains, rivers, corporations, scientists, children, or baseball players. An *abstract term* usually designates something that does not exist in the physical world. It applies either to classes of things or to attributes or relationships. It does not have a specific reference against which its meaning can be checked. Examples of abstract terms are culture, Americanism, beauty, art, philosophy, evil, honesty. Some abstract words are more definite. These are based on concepts derived from relationships that our minds associate with elements of our experience. Among such words are mass, force, magnetism, dimensions, absence, profit, rigid, cost, ripe, red, confused, equal.

Abstract words are related to another set of words grammarians call *relative* words. These words name qualities. Their meaning depends on their relation to the experience and/or intention of the person who uses the word. The word *dark,* for example, implies total or partial absence of light. We can say "Brown is a dark color" or "A dark shade of red at sunset" or "It is getting dark." We can say that someone's mood was dark, meaning gloomy; we can say a person is "in the dark" if he does not understand something. We can say that someone's "intentions were dark," meaning her intentions were insidious, obscure, or secret. The word *dark* in any of these examples gives the receiver some difficulty determining the specific meaning the sender intends. **Meaning** depends on a frame of reference or the context in which words are used.

Words representing colors, degrees of warmth, weight, and pressures are also relative words. Technical writers must be precise in their communication; so they must be particularly careful in their choice of terms. When they need to refer to a color, to a weight, or to a pressure, technical writers will describe colors by their chromatic or achromatic scales; they will indicate that an object is not just heavy, but that it weighs exactly 163 pounds, or 1 milligram, as the case might be; they will specify that a force exerts 1000 pounds of pressure per square centimeter or that a pulse has a frequency beat of 100 kilohertz.

Problems in Achieving Meaning

Since words and language were invented to represent reality and to help human existence, endeavor, and progress we might wonder, "Why can't we devise our language to stabilize the meanings of words the way a mathematician has stabilized the quantities of that science and the way a chemist has stabilized the symbols of the elements?" We could avoid so much confusion if words meant *one* thing and one thing *only.* We should not allow a word to mean one thing in one context and another thing in a different context. Language should be perfected so that each word has its own meaning.

Figure 3.6
Simple Idea

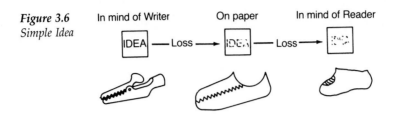

This notion seems ideal, but there are an infinite number of things and experiences in human reality. While each person may have universal experiences, each experience is unique for that person. To structure a language with an infinite number of words for the infinite number of possible experiences would make communication through language impossible. People would be inventing words for each spontaneous situation. As Schiller has pointed out:

> . . . if a word has a *perfectly* fixed meaning, it could be used only once, and never again; it could be applied only to the situation which originally called for it, and which it uniquely fitted. If, the next time it was used, it retained its original meaning, it could not designate the actual situation but would still hark back to its past use, and this would disqualify it for all future use. . . . It is evident that the intellectual strain of continuously inventing new words would be intolerable, and that the chances of being understood would be very small. . . . Thus the price of fixity would be unintelligibility. The impracticability of such a language might not be regarded as its "theoretical" refutation; but after all languages are meant to be used, and human intelligence, at any rate, could make no use of it. [5:56–57]

Urban, in his book *Language and Reality*, describes effectively the character and obligation of language:

> In the animal world the meanings of things are largely instinctive and the signs of these meanings are adherent to the thing signified. In human society, on the contrary, all is mobile, so a language is required which makes it possible to be always passing from what is known to what is yet to be known. There must be a language whose signs—which cannot be infinite in number—are extensible to an infinity of things.

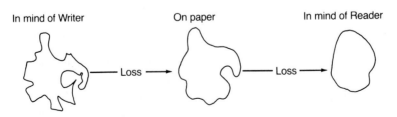

Figure 3.7
Complex Idea

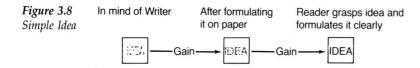

Figure 3.8
Simple Idea

The mobility of signs, the tendency of a sign to transfer itself from one object to another is characteristic of language. This phenomenon of transference forms the central fact of linguistic meaning.

Every meaning-situation . . . has two . . . components, the sign and the thing signified—that which means and that which is meant. Meaning in its most elementary form is then, this relation. . . . [8:107–15]

Urban further points out that there is no semantic meaning situation that does not involve a speaker (source) and a hearer (destination) as necessary components. Words are signs (signals), but they are expressive signs. As such, they imply communication, either overt or latent [8:115–16].

Some diagrams will help us understand what occurs when a source successfully or unsuccessfully communicates a message to a receiver. Figure 3.6 represents the communication of a simple idea. By substituting the alligator clip for the square, you can clearly see the extent of loss that can occur in a communication situation. Figure 3.7 represents the extent of loss in the communication of a complex idea.

If we reverse the diagrams, we can see how the process of encoding and sending the signal to the receiver—the communication process—can help formulate an effective message that the destination receives and in the decoding process helps to clarify. Figures 3.8 and 3.9 represent a situation where the source may begin with only a vague idea. While the idea is encoded or formulated, it begins to take shape. The sender conveys the message in a formulated concept or in words and sentences that the receiver decodes according to her experience. The receiver may act upon the message in such a way that when she receives it, it is decoded in a clear, definite statement. The idea or concept,

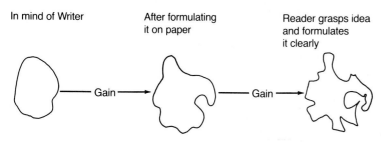

Figure 3.9
Complex Idea

as the receiver had decoded it by her own thought process, becomes complete, clear, concise, and meaningful. The receiver, using her experience and associations stimulated by the signals she has received, organizes the signals through her cerebral process and makes the proper sense dictated by the requirements of the message situation. This has been illustrated operationally by Figure 3.9. *This gain process is, of course, highly unusual.* The diagrams help us visualize what frequently occurs in the revision stages of writing. The writer may begin with a blurred idea in a first or second draft. As he or she writes out the idea, it becomes clearer (see also Figure 3.5) so that by the time the third or fourth draft reaches the reader, the idea is totally clear.

How Meaning Is Conveyed

When we speak or write to one another, we transmit our thoughts through physical signals. Properly structured and organized, these signals become messages of orderly selected signs and symbols. The signs, then, embody our messages or thoughts.

Three elements are involved in the conveyance of meaning:

1. A person having thoughts
2. A symbol
3. A referent

These three elements are frequently represented by three corners of a triangle as in Figure 3.10.

The symbols or words have no direct relationship to their referents and can be identified only indirectly with the physical fact they symbolize through the reader's mental processes. Line *AC* of the triangle in Figure 3.10 becomes a direct relationship only after the reader pronounces the word symbols of the fact; the symbolization (the sentence "Acid turns blue litmus paper red.") be-

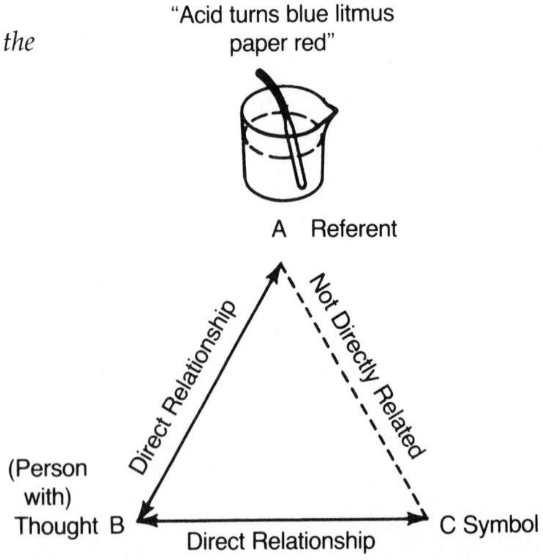

Figure 3.10
Triangle of Meaning Operating in the Mind of the Encoder of a Communication Message

comes identical (means the same thing in the mind of the reader) with the fact it represents.

Words represent facts. A person perceives the fact that acid turns blue litmus paper red. He selects the proper words to form a message that will designate (mean) the fact or thing perceived (the referent). Words, then, refer to facts. Memories of past experience and one's external environment influence proper selection of the symbolic response to a referent by the person encoding and sending the message. The receiver's meaning emerges from the operation indicated in Figure 3.2.

We need to recognize three types of meaning:

1. What the communicator (source) intends to indicate
2. What is suggested to the receiver (destination) by the message
3. The more or less general habit of using a given symbol to indicate a given thing

What a communicator intends and what a receiver understands both depend heavily on meaning 3, which is often influenced by the frame of reference or context in which the words are used.

Bridging the Gap between Words and Thoughts

Language usage is the occupational hazard of technical writers, who must communicate knowledge clearly and precisely. Language is the link between the reality of the outside world and the concepts or thoughts in people's minds. Because everyone's nervous system is different, each person sees reality differently. This can cause problems, as illustrated in Figure 3.11.

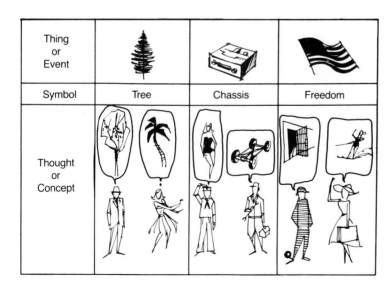

Figure 3.11

Effective technical writing is clear, precise, and accurate. Knowledge of semantics can help the technical writer and the reader.[3] It guides the writer in the correct choice and use of both technical and nontechnical terms, and it guides the reader in the correct interpretation of the symbols and terms used. Knowledge of semantics also makes the technical writer realize that words and terms have emotional associations. Therefore, the writer must avoid placing words and terms in a context that will make them emotionally "loaded." Finally, technical writers must always keep their readers' background in mind so that the terms, words, and other communication symbols they use will be familiar to the reader.

Role of Writing in the Scientific Method

Because technical writing is as much a way of thought as a way of expressing thought, these observations by the linguist Whorf are an appropriate conclusion to our discussion of communication:

> The revolutionary changes that have occurred since 1890 in the world of science—especially in physics but also in chemistry, biology, and the sciences of man—have been due not so much to new facts as to new ways of thinking about facts. . . .
> I say new ways of thinking about facts, but a more nearly accurate statement would say new ways of *talking* about facts. It is this *use of language upon data* that is central to scientific progress . . . we must face the fact that science begins and ends in talk; this is the reverse of anything ignoble. Such words as "analyze, compare, deduce, reason, infer, postulate, theorize, test, demonstrate" mean that, whenever a scientist does something, he talks about this thing that he does. As Leonard Bloomfield has shown, *scientific research begins with a set of sentences which point the way to certain observations and experiments, the results of which do not become fully scientific until they have been turned back into language, yielding again a set of sentences which then become the basis of further exploration into the unknown. . . .* [9.220]

Chapter in Brief

To provide an understanding of the process of communication, we have examined the origin and nature of language and the roles semantics and meaning play in the process. The part meaning plays in bridging the gap between

---

[3]We have had time for only an introduction to semantics and communication theory in this chapter. A deeper and more comprehensive study would be rewarding for you. The references at the end of this chapter and in the Bibliography at the end of this text list some excellent sources. In the area of semantics, I suggest your reading Hayakawa's *Language in Thought and Action* and Lee's *Language Habits in Human Affairs*. Urban's *Language and Reality* is difficult but comprehensive. Ogden and Richards' *The Meaning of Meaning* is outstanding for an examination of meaning. Cherry's *On Human Communication* provides both a mathematical and nonmathematical introduction to communication theory.

thought and words was described. To help you understand the process of communication, we used diagrams to illustrate its various aspects.

Chapter Focal Points

- Human communication
- Nonverbal communication
- The Communication process
- The Communication chain
- Language
- Semantics
- Abstraction ladder
- Meaning

Questions for Discussion

1. The Arabic language is said to have almost 6000 terms to describe camels. In other languages of the world there are no terms for camels at all. What generalization about languages can be made from this? Would you consider the automobile the English vocabulary equivalent to the camel in that respect? Check your answer in an unabridged dictionary. What other terms in the English language might approach the camel in that respect? If you know a foreign language, can you think of a word, idiom, or expression which cannot be adequately translated into English and vice versa? Why should there be such a difficulty? Does this mean that there are things, processes, and concepts that a technical writer can never adequately describe? Justify your answer.

2. When technical writers forget or ignore the fundamentals of the communication process, "noise" or "interference" occurs. Indicate the type of noise that may occur in writing for various levels of readership. One of the best ways for overcoming interference is to use "the principle of feedback." Does the principle of feedback operate for the technical writer? If you say no, explain why not. If you say yes, list several types and explain.

3. Do we know the meaning of a word, a term, or a concept when we know to what the word or concept refers? Explain.

4. This is a class experiment to illustrate the principles of communication: Your instructor will bring to class a picture or an advertisement showing several persons involved in some sort of action in a business or industrial setting. The instructor will ask for three volunteers to be sent out of the classroom for no more than ten minutes. The instructor shows the class the picture, allowing two minutes for viewing. There is to be no discussion about the picture. After two minutes, the instructor removes the picture and asks the class to write a "scenario" of 25–50 words which might explain what the picture is about. The instructor then asks volunteer one to return to the class, and then calls on a class member to read aloud to volunteer one the scenario he or she wrote. The instructor then asks volunteer two to return to the classroom. Volunteer one explains to volunteer two what the picture is about, based on what the

reader read. Volunteer three is then called back into the room. Volunteer two explains the picture as it was explained by volunteer one.

The instructor then shows the picture and asks volunteer three whether adequate representation was provided. Volunteer three must justify the answer given.

This experiment was devised to illustrate the principles of communication, languages, and meaning explained in this chapter. Write a paper between 750–1200 words identifying the principles illustrated by the scenario as it occurred in the class.

Assignments

1. Library research will be necessary for the following exercises:
   a. Within the context of human communication, define:
      1. Signal
      2. Sign
      3. Symbol
      4. Message
   b. Distinguish between *language* and *code*.
   c. Distinguish between *semantics* and *general semantics*.
   d. Distinguish between *phonetics*, *phonemics*, and *graphemics*.

2. Make a list of ten common symbols, as for example: $, *Dollar*; ¶, *paragraph*; %, *percent*; π, *ratio of the circumference of a circle to its diameter*, and denote what each symbol stands for.

3. Read one of the following books and write a review. Include a brief summary of the contents, an indication of the purpose, scope, and value of your understanding of the communication process:

   Borden, George A. *An Introduction to Human-Communication Theory*. Dubuque, Iowa: Wm. C. Brown.

   Chase, Stuart, *Power of Words*. New York: Harcourt, Brace.

   Cherry, Colin. *On Human Communication*. New York: John Wiley and Sons.

   Condon, John C., Jr. *Semantics and Communication*, 3rd ed. New York: Macmillan.

   DeVito, Joseph A. *Communication, Concepts and Processes*. Englewood Cliffs, N.J.: Prentice-Hall.

   Doubleday Pictorial Library of Communication and Language, New York: Doubleday.

   Hayakawa, S. I. *Language in Thought and Action*, New York: Harcourt, Brace, Jovanovich.

   Lee, Irving J. *Language Habits in Human Affairs*, New York: Alfred A. Knopf.

   Walpole, Hugh R. *Semantics: The Nature of Words and Their Meaning*, New York: W.W. Norton.

References

1. Chase, Stuart. *Power of Words*. New York: Harcourt, Brace and Co., 1954.

2. Haldane, J. B. S. "Communication in Biology." In *Studies in Communication*. London: Martin Secker and Warburg, 1955.

3. Huxley, Aldous. *Words and Their Meaning*. Los Angeles: The Ward Ritchie Press, 1940.

4. Lee, Irving J. *Language Habits in Human Affairs*. New York: Harper and Brothers, 1941.

5. Schiller, F. C. S. *Logic for Use*. New York: Harcourt, Brace and Co., 1930.

6. Schramm, Wilbur. "The Nature of Communication Between Humans." In *The Process and Effects of Mass Communication*, rev. ed., edited by Wilbur Schramm and Donald F. Roberts. Urbana, Ill.: University of Illinois Press, 1971.

7. Sinclair, W. A. *An Introduction to Philosophy*. London: Oxford University Press, 1944.

8. Urban, Wilbur M. *Language and Reality*. New York: Macmillan, 1951.

9. Whorf, Benjamin Lee. "Linguistics as an Exact Science." In *Language, Thought and Reality, Selected Writings of Benjamin Whorf*, edited by John B. Carroll. New York: John Wiley and Sons and Massachusetts Institute of Technology Press, 1956.

PART TWO
Modality
and
Media

4

Technical English, Technical Style

Chapter Objective

Provide guidance on the stylistic qualities and elements of technical writing.

Chapter Focus

- Qualities of technical style
  Clarity and precision
  Conciseness and directness
  Objectivity
  Voice
- Elements of technical style
  Sentences
  Paragraphs
- Reader analysis and readability

*T*echnical English is not a substandard form of expression. Technical writing meets the conventional standards of grammar, punctuation, and syntax. Technical style may have its peculiarities—in vocabulary and at times in mechanics—but it is, nevertheless, capable of effective expression and graceful use of language. The purpose of this chapter is to examine the characteristics of good technical style.

Jonathan Swift said that the "proper words in the proper places" make for style. Seneca and Lord Chesterfield said that style is the "dress of thoughts." Expressed simply, **style** is the way a person puts words together into sentences, arranges sentences into paragraphs, and groups paragraphs to make a piece of writing express thoughts clearly. Technical style, then, is the way you write when you deal with a technical or scientific subject.

Technical style can be described by the richness or poverty of vocabulary, by the syllabic lengths of words, by the relative frequency of sentences of various lengths and types, by grammatical structure, by pictorial representation and tabular matter integrated with text. The style varies with the writer and subject matter. Nevertheless, it has certain basic characteristics.

Qualities of Technical Style

By tradition, technical style is plain, impersonal, and factual. Its fundamentals and principles were stated very early in the modern scientific movement by Sprat in his *History of the Royal Society*, 1667. The members of the Society tried

> to reject all of the amplifications, digressions, and swellings of style: to return back to the primitive purity, and shortness, when men delivered so many *things*, almost in an equal number of *words*. They have exacted from all their members a close, naked, natural way of speaking; positive expressions; clear sentences; a native easiness; bringing all things as near the mathematical plainness as they can: and preferring the language of Artizans, Countrymen, and Merchants, before that, of Wits and Scholars. [12:13]

The tradition of plainness and directness has come down to the present time, but the scientist's technical vocabulary has extended far beyond that of "Artizans, Countrymen, and Merchants" and has made plainness of style a difficult achievement. Scientists have discovered so much and named qualities and things in such numbers that the average person has come to identify "big, strange, and unpronounceable words" as the most conspicuous trait of scientific writing.

Technical style is characterized by a calm, restrained tone; an absence of any attempt to arouse emotion; the use of specialized terminology; the use of abbreviations and symbols; and the integrated uses of illustrations, tables, charts, and diagrams to help explanation. Technical writing is characterized by exactness rather than grace or variety of expression. Its main purpose is to be informative and functional rather than entertaining. Thus, the most important qualities of technical style are clarity, precision, conciseness, and objectivity.

Clarity and Precision

Clarity and precision are frequently interdependent. Clarity is achieved when the writer has communicated meaning fully to the reader. Precision occurs when the writer attains exact correspondence between the matter to be communicated and its written expression.

Faults in clarity and precision result when the following occur:

1. The writer is not familiar enough with the subject matter to write about it
2. The writer, though generally familiar with the subject, cannot distinguish the important from the unimportant (The essence of the problem has escaped the writer.)
3. The writer has a thorough mastery of the subject matter but is deficient in communication techniques
4. The writer is unfamiliar with the reader and has not directed the communication to the desired level of audience understanding

Conciseness and Directness

Concise writing saves the reader time and energy because meaning is expressed in the fewest possible words and readability is thereby enhanced. Directness also increases readability; it eliminates circumlocutions (roundabout expressions involving unnecessary words) and awkward inversions. Take this example:

> An important factor to be cognizant of in the relation to proper procedures along the lines necessary is to consider first and foremost in connection with the nature of the experiment that effectuation is dependent on a fully and complete darkened interior enclosure.

The following sentence says the same thing more concisely, directly, and clearly:

> The experiment requires a completely darkened room.

Circumlocutions and awkward inversions so plague technical and scientific writing that many company manuals on report writing invent humorous and exaggerated examples to make the point:

> A 72-gram brown Rhode Island Red country-fresh candled egg was secured and washed free of feathers, etc. Held between thumb and index finger, about three feet more or less from an electric fan (General Electric No. MC-2404, Serial No. JC23023, non-oscillating, rotating on "high" speed at approximately 1052.23 ± 0.02 rpm), the egg was suspended on a string (pendulum) so it arrived at the fan with essentially zero velocity normal to the fan rotation plan. The product adhered strongly to the walls and ceiling and was difficult to recover; however, using putty knives a total of 13 grams was obtained and put in a skillet with 11.2 grams of hickory smoked Armour's Old Style bacon and heated over a low bunsen flame for 7 min. 32 sec. What there was of it was excellent scrambled eggs. [2:8]

A more direct and concise rewrite is this:

> Very good scrambling was produced by throwing an egg into an electric fan. The product was difficult to recover from the walls and ceiling, but the small amount that was recovered made an excellent omelet. [2]

Objectivity

Scientific style is characterized by objectivity. Personal feelings are excluded; attention is concentrated on facts. Furthermore, the use of the passive voice and the third person point of view is a long-established tradition in technical writing. The feeling is that the exclusion of personal pronouns produces a style consistent with objectivity and the use of the passive voice places emphasis on the subject matter. As a result, technical writing is much more impersonal and much drier than it has to be. True, directions for assembling a window fan are quite different from explanations of basic research. It is also true that much "literary" writing is as guilty of mechanical, lifeless expression as is much technical writing. Still, a major complaint is that basic research is reported in such a cut-and-dried manner as to drive all interest and excitement from it. The personalness, directness, and thoroughly human qualities that emerge clearly from the explanations of Harvey, Newton, Huxley, or Watson make for fascinating and comprehensible reading; whereas, much current scientific research is reported in such an impersonal manner as to be as dull and lifeless as a phonebook.

Opinion differs as to whether impersonality in scientific and technical writing requires the third person and passive voice. Writing for the general or less knowledgeable reader requires the frequent use of devices to inveigle interest. The first person point of view and the active voice help to create interest. Informative and functional technical writing can be objective and still include personal pronouns. Many successful technical writers can establish effective rapport with readers by playing an active role in their narration as if they were the principal actors involved. Readers will often allow themselves to be involved in exposition if it is made interesting for them by the actual participation of a human being elaborating on data which could otherwise be as foreign and dry to them as a lobster's tail. This is exactly what Huxley did, and it was Huxley who did more to popularize science than any other person at a time when scientific activity was beginning to expand and was in need of public support.

Two excerpts from Huxley's writing follow. Both exemplify excellent scientific style and effective communication. Each is aimed at a different audience. The first example is aimed at an intelligent, but uneducated and unsophisticated, audience.

The Lobster's Tail

Note the use of the first person pronoun, I. Huxley uses excellent psychology to involve the uninformed reader immediately by asking a

I have before me a lobster. When I examine it, what appears to be the most striking character it presents? Why, I observe that this part which we call the tail of the lobster, is made up of six distinct hard rings and a seventh terminal piece. If I separate one of the middle rings, say the third, I find it carries upon its under surface a pair of limbs or appendages, each of which consists of a stalk and two terminal pieces. So that I can represent a transverse section of the ring and its appendages upon the diagram board in this way.

    If I now take the fourth ring, I find it has the same structure, and so have the fifth and the second; so that, in each of these divisions of the tail, I find parts which correspond with one another, a ring and two appendages and in each appendage a stalk and two end pieces. These corresponding parts are called, in the technical language of anatomy "homologous parts." The ring of the third division is the "homologue" of the ring of the fifth, the appendage of the former is the homologue of the appendage of the latter. And, as each division exhibits corresponding parts in corresponding places, we say that all the divisions are constructed upon the same plan. But let us consider the sixth division. It is similar to, and yet different from, the others. The ring is essentially the same as in the other divisions; but the appendages look at first as if they were very different; and yet when we regard them closely, what do we find? A stalk and two terminal divisions, exactly as in the others, but the stalk is very short and very thick, the two terminal divisions are very broad and flat, and one of them is divided into two pieces. I may say, therefore, that the sixth segment is like the others in plan, but that it is modified in its details. [4:21]

question about the matter he will be explaining: "What do you think is the most characteristic aspect of the lobster?" Its tail, of course. Huxley, speaking in the first person, involves the reader in examining the special physiological components of the lobster's tail. He has the reader examine, as he does in his explanation, the separate parts of the tail. Notice how he explains the technical, anatomical term homologue, an organ or part of the animal (lobster) similar to another part. Huxley maintains interest by involving the reader in the examination and asking questions which the uninformed would ask, and then answering the questions.

    The other example from Huxley's technical writing is an excerpt from a textbook intended for advanced students of zoology. Note that the first person is entirely missing and that most of the sentences are structured in the passive voice. Recall the use of nontechnical terms in the first passage and compare the use of technical terminology in the one below, where the details are more complete, the approach is more formal, and the sentences are much longer.

The Abdomen of the English Crayfish

The body of the crayfish is obviously separable into three regions—the *cephalon* or head, the *thorax,* and the *abdomen.* The last is at once distinguished by the size and the mobility of its segments. And each of its seven movable segments, except the telson, represents a sort of morphological unit, the repetition of which makes up the whole fabric of the body.

    The fifth segment can be studied apart. It constitutes what is called a *metamere*; in which are distinguishable a central part termed the *somite,* and two *appendages.*

    In the exoskeleton of the somites of the abdomen several regions have already been distinguished; and although they constitute one continuous whole, it will be convenient to speak of the *sternum* (Fig. 36, st. XIX), [These are call outs in the drawing of a transverse section through the fifth abdominal somite.] the *tergum* (t. XIX) and the *pleura* (pl XIX), as if they were separate parts, and to distinguish that portion of the sternal region, which lies between the articulation of the appendage and the pleuron, on each side, as the *epimeron* (ep. XIX). Adopting the nomenclature, it may be said of the fifth somite of the abdomen, that it consists of a segment of the exoskeleton,

Here the reader is a college student. Huxley uses some scientific terms without definition. Students of an advanced course in zoology are expected to know what morphological unit, exoskeleton, transverse section, ganglion, extensor muscles, etc., mean. However, note that Huxley will define a term that is new to the student; for example, "the cephalon or head." His explanation of new terms considers the intelligence and knowledge of the reader. For instance, all he needs to do to explain metamere is to say it is the fifth segment of the abdomen of the crayfish (English lobster).

Notice how Huxley's organization of the description of the abdomen fits the principles of the written technical description explained in Chapter 7. In this excerpt, Huxley deals with an anatomical part of the crayfish, the abdomen. He identifies the segments of this anatomical division, tells what each is, its purpose, appearance, shape, and how each interrelates.

divisible into tergum, pleura, epimeron, and sternum, with which two appendages are articulated; that it contains a double ganglion (gn. 12), a section of the flexor (f.m.) and extensor (e.m.) muscles, and of the alimentary (h.g.) and vascular (s.a.a., i.a.a.) systems. [5:141]

When to Use Active and Passive Voices in Technical Writing

In grammar, *voice* is the term used for a form of a verb to show connection between the subject and the verb. The voice of a verb tells the reader whether the subject performs an action (active voice) or receives an action (passive voice). In active voice construction, the subject is the doer of the action or is the condition varied by the verb:

> The canal starts at Whalen Dam.
> Man bites dog.
> Vertebrates learn mazes readily.
> The variable resistor controlled the circuit.

When the subject of a verb receives the action, the verb is in the passive voice:

> The canal at Whalen was built many years ago.
> The dog was bitten by a man.
> Mazes are learned readily by vertebrates.
> The current was controlled by the variable resistor.

The passive voice form is used in all tenses. It usually consists of a form of the verb *to be,* but the past tense of the verbs *to get* and *to become* also is sometimes used to form a passive:

> The component became overheated.
> The traveling wave tube is to be used in the circuit.
> During mitosis, the heterochromatic segments get stained more strongly than the euchromatic regions.

Much technical writing is concerned with the description of work so objective that the reader does not care who did it. The reader is interested solely in the work itself and is not at all interested in the agency or agent involved. The conventional, impersonal passive construction is suitable for this kind of subject. Compare the next two sentences with the two that follow:

> The other end of the tie beam was connected to an anchor pile by a bolt of 2½" diameter, inserted through an opening in the pile.

As the experiment progressed, additional water was added to equal the amount lost from evaporation.

I connected the other end of the tie beam to an anchor pile, using a bolt of 2½″ diameter which I inserted through an opening in the pile.

As I went on with the experiment, I decided to add water equal to the amount I determined was lost through evaporation.

A comparison shows the third person passive construction to be more objective and efficient than the first person active example.

But compare this sentence:

It is desired to ascertain how the success was achieved in increasing the yield of Russian wild rye.

with this one:

We want to know how you increased the yield of Russian wild rye.

and this sentence:

The agglutination is caused by substances analogous to antibodies that are present in the serum.

with this one:

Substances analogous to antibodies present in the serum caused the agglutination.

or the following:

It is believed that the city should increase its reserve water supply.

with this sentence:

We believe the city should increase its reserve water supply.

Active verbs are more lively than passive verbs and call for simpler sentence structure. Usually, therefore, they are more efficient and writers should prefer them. But let us remember that passive verbs are as grammatically correct as active verbs and that there are instances when the passive voice is to be preferred:

1. When the doer of an action is not known to the writer or when the writer does not want to be identified:
   a. In 1947, when this site was selected, the water table was low and pump irrigation was necessary in many areas of the valley.
   b. The decision was made against the employee.
2. When the writer desires to place the emphasis on the action or on the doer at the end:
   a. The mineral is mined in Wyoming.
   b. The research was carried out by the director.

In normal usage, nonetheless, passive constructions are considered weak because they have actionless verbs, invert the natural word order, and require additional phrasing.

Some publications and organizations have a specific policy forbidding the use of *I* or *we.* In writing for such a publication or organization, live with its rules unless you have achieved the position and skill of a Thomas Huxley. Then you can change the rules.

Elements of Style

So far I have been discussing qualities of style. Qualities are overall impressions or characteristics. They result from the writer's typical or individual use of the elements of style: grammatical construction, diction, phrasing, sentence length, and figures of speech. **Sentence structure** and **diction** (choice of words) are two of the major concerns of the technical writer.

Types of Sentence Structure and Their Capacity for Expression

The technical writer should understand the capacity of the sentence for expressing simple and complex relationships. This understanding is valuable to the technical writer in formulating observations, generalizations, and conclusions.

Simple Sentence

For stating an uncomplicated, unqualified observation, a **simple sentence** is used:

> Lavoisier, the great French chemist, named the new gas oxygen.
> Honesty is the best policy.
> Fill the test tube with water.
> The child patted the dog.
> The light is absorbed in the retina by a pigment called visual purple or rhodopsin.

Compound Sentence

The **compound sentence** expresses coordinate ideas in balance or in contrast:

> "And there was evening, and there was morning, one day" (Genesis 1:51—RSV).
> The flow is dark brown and blocky on the surface, but it continues to steam from hot viscous lava beneath.
> "Science is nothing but trained and organized common sense, differing from the latter only as a veteran may differ from a raw recruit; and its methods differ from those of common sense only as far as the guardsmen's cut and thrust differ from the way in which a savage wields his club" (Thomas H. Huxley).

Complex Sentence

In the **complex sentence**, dependent clauses are used to express ideas subordinate to the thought of the main clause. The word *complex* is derived from Latin. It does not mean difficult, but literally "woven into." The dependent clause (an incomplete thought having a subject and predicate that modify or support in some way a word or sentence element of the whole sentence) is woven into the design:

A substitute for wood, which would really take its place, has not been developed.

> *The dependent clause,* which would really take its place, *is not of equal rank to the independent clause, a* substitute for wood has not been developed. *It provides an explanation of the main thought.*

We wondered when you would come.

> When you would come *tells the reader the necessary information about what the main thought,* we wondered, *is concerned.*

Any sentence can be analyzed if you use sufficient patience and don't go into unnecessary detail.

> If you use sufficient patience and don't go into unnecessary detail *explains why* any sentence can be analyzed, *the major thought in the sentence.*

When all else fails, read the directions.

> Read the directions *is the important thought in this sentence, but it is meaningless unless it is explained by the descriptive qualifier,* when all else fails.

If the revised process is to succeed, the first stage may be completed under the careful supervision of the shop personnel, but the second stage must be directed by high-level engineers.

> If the revised process is to succeed *is a dependent clause that explains (modifies or describes) the two independent clauses connected by the coordinating conjunction* but, *which connects the two statements actually in opposition to each other,* the first stage may be completed under the careful supervision of the shop personnel *contrasted with* the second stage must be directed by high-level engineers.

The value of the thought ex-pressed in the independent clause, uniformity in re-port writing undoubtedly has some advantages, *is clarified by the qualifying dependent clause,* but when the uniformity results in dull and difficult reading. *The opposing thought ex-pressed in the independent clause,* I seriously ques-tion its validity, *provides the intended meaning.*

Uniformity in report writing undoubtedly has some advantages, but when the uniformity results in dull and difficult reading, I seriously question its value.

Complex-Compound Sentence

A **complex-compound sentence** contains two or more independent clauses (ex-pressing coordinate ideas in balance or in contrast) and at least one dependent clause expressing a thought(s) subordinate to the main clause it modifies.

Professional writers today use about 50 percent complex sentences, 35 percent simple sentences, and only about 15 percent compound sentences, com-pound-complex sentences, and fragmentary sentences. (A fragmentary sentence actually is not a sentence, but a phrase or dependent clause, expressing an incomplete thought.) Out of twenty sentences, three are compound; seventeen are either simple or complex. The most frequently written sentence in current English is a complex sentence from twelve to thirty-six words long.

Which type of sentence to use depends mainly on the thought to be com-municated. Other factors pertinent to the communication are the purpose of the communication and the reader of the message. Efficient technical writing depends on clarity, precision, conciseness, and objectivity in the composition. For the sake of example and analysis, let us examine the sentence below. At first, it seems to express an uncomplicated thought concisely and clearly.

A heavy object will sink readily.

On closer examination, the thought of the sentence is not as clear as it might be. The meaning of the modifiers *heavy* and *readily* is vague. Heavy and readily are subjective judgments. The sentence, though a short one, does not express a clear, precise, objective thought. Shortness is not the equivalent of conciseness. Conciseness requires succinctness and clarity. Let us try rewriting the sentence to communicate the thought more clearly and efficiently.

An object of great density will sink very quickly.

The new attempt, also a simple sentence, is an improvement. Though the word *density* carries a certain factual significance in the rephrased thought, the adjective *great* reflects subjectivity, as does the intensive *very*. *Quickly,* while having an aspect of subjectivity, is a more appropriate word than *readily* because it suggests an inherent capability for speed of action; whereas, *readily* suggests speed in compliance or response. Let us try again. This time let us aim for adequate expression no matter how many words or clauses are to be used in the sentence.

> A heavy object is defined as one whose density is such that it will not displace its own weight when placed in water, and, therefore, it will sink quickly.

This twenty-nine word complex-compound sentence has achieved objectivity. Its heavy-handed structure is frequently seen in current technical writing. It can be improved:

> An object that does not displace its weight in water will sink quickly.

Both the critical qualities of the weight factor, which causes the object to sink, and the manner of sinking are clearly, objectively, precisely, and concisely expressed in the complex sentence. The complex sentence is effective in technical writing because the dependent clause can be a modifier of exacting precision and clarity.

The Paragraph—The Basic Unit in Writing

Every serious piece of writing has a structural design. Many professional writers develop an outline for that purpose. The outline (which I will discuss in detail in Chapter 12) helps the writer subdivide and arrange the subject matter into related units or topics. In the composition (encoding) process, each topic becomes the structural basis of a paragraph. A **paragraph** is a group of related sentences that a writer presents as an organized unit in development of the subject. Technically, then, the function of the paragraph is to break the text into readable units. It has two principal uses:

1. It holds together thoughts or statements that are closely related;
2. It keeps apart thoughts and statements that belong to different parts of the subject.

The Topic Sentence

A **topic sentence** expresses the main idea developed by a paragraph. Writing a paragraph has been compared to sorting material into labeled baskets: the material is the sentences conveying ideas or information; the label on the material is the topic sentence. The topic sentence can be a generalization that summarizes what the paragraph is about—its central thought—or a statement that tells what the paragraph is to do or indicates the direction the discussion will take. Grammarians call this latter type a **pointer** sentence, since it points or guides the reader in the direction the paragraph is to develop. Often a paragraph with a beginning pointer sentence will have a summarizing topic sentence at its close.

By steering the design of its paragraph, the topic sentence controls the text. It provides:

The first sentence of this paragraph is both a topic and a pointer sentence.

The italicized words provide details—four specifics on how the topic sentence "steers" the design of the paragraph. The indentations and italics provide emphasis and give the text graphic prominence.

The indentation also provides special balance to the material of the paragraph, as well as to supplemental details.

The final sentence summarizes the central idea of the paragraph.

> *unity* (its material is related),
>
> *coherence* (its material is connected),
>
> *emphasis* (its important material is strategically positioned or given graphic prominence), and
>
> *proportion* (its material is balanced and it has harmony in the amount of information it has positioned).

The topic sentence is usually, though not always, placed at the beginning of the paragraph. It may apear at any position in the paragraph, depending on the factors indicated. If you compose your topic sentence with care, you will not only help your reader grasp the meaning of the paragraph, but also help yourself by providing the necessary information that supports and clarifies your intended meaning.

If you are an inexperienced technical writer you would do well to begin your paragraphs with a topic sentence. It is a useful technique that helps set the flow of necessary information on paper. A good procedure for developing paragraphs is to ask yourself two questions:

1. What is the main point of this paragraph?
2. What must I tell my reader to support, explain, clarify, or accept it?

Expository devices such as examples, analogies, analyses, comparisons or contrasts, explications, and details are used to develop the topic sentence.[1]

A few examples can illustrate how the reader's expectations are satisfied by the development of the topic sentence:

> It is a commonplace fact that scientific discoveries are a function of methods used.

The reader's natural question is why is it a commonplace fact? The writer has an obligation to give the reasons in the rest of the paragraph.

> Several occurrences during the experiment confirmed this opinion.

The reader logically expects specific instances to confirm the opinion of the writer.

> "In certain other aspects, especially its spatio-temporal aspects as revealed by the theory of relativity, nature is like a rainbow" [7:3].

The reader is interested in knowing how the analogy applies. The writer (in this case, Sir James Jeans) elaborates an effective explanation of his analogy.

---

[1]Chapter 6 discusses these and other expository devices in detail.

The rattle is the most characteristic feature of the rattlesnake and is one of the most remarkable structures in nature.

The reader expects an explanation to answer the question Why?

Operation of an autopilot may be seen from the diagram in Figure 12.

The rest of the paragraph becomes supplemental to the diagram in the explanation of how an autopilot operates. Graphic illustrations are used with great effect in technical writing to aid the development of the topic sentence of a paragraph. Frequently, they are the most important substantiating details in such development.

What causes wind shear?

A question and its response can be a useful technique for organizing diverse aspects of information. To answer the above question, the various types of turbulence that cause wind shear are identified and elucidated. If the elucidation requires much informational material, each item in the answer may become a paragraph and each separate item becomes the topic sentence of that paragraph.

Test your paragraphs by these criteria:

1. What is the central idea?
2. What details are needed to support or explain the central idea for the reader?
3. Is there anything in the paragraph not related to that idea?
4. Are the sentences organized in a sequence that is sufficiently logical to support or explain the topic sentence clearly?

Structural Paragraphs

There is considerable similarity between the topic sentence of a paragraph and the first paragraph of a section. Both are necessary to summarize, preview, and connect—to orient the reader and offer a proper perspective. Furthermore, closing paragraphs of sections often summarize the contents of that section and show their significance to the whole.

Structural paragraphs are commonly placed between the major heading and the first subheading of the section. Such paragraphs usually point both backward and forward. Their primary purpose is to introduce the subject to be discussed. Like a road map, their function is to help the reader as she moves along from one phase of the presentation to the next. This type of paragraph prods her to look back to where she has been and to look ahead to where she is going. The technique is simple: the writer tells the reader briefly the major points covered, and then indicates what is coming next. The structural paragraph serves as a connecting link for the topics, sections, and chapters of information you present to your reader.

An example of such a paragraph is the one that begins Chapter 6 of Dewey's book, *How We Think*:

Sentence 1 recaps the previous chapters. Sentence 2 gives the major thesis of those chapters. Sentence 3 indicates what we will read next—examples of what was discussed in previous chapters.

We have in previous chapters given an outline account of the nature of reflective thinking. We have stated some reasons why it is necessary to use educational means to secure its development and have considered the intrinsic resources, the difficulties, and ulterior purpose of its educational training—the formation of disciplined logical ability to think. We come now to some descriptions of simple genuine cases of thinking, selected from the class papers of students. [1:91]

A well-organized paper has structural paragraphs that have duties of introduction, transition, review, and summarization. Like headings, they keep the reader informed of the design of the whole composition; in addition, they connect the larger parts of the report. They are an aid to readability and comprehension.

Paragraph Principles Summarized

In writing, the paragraph is the basic unit in the development of a subject. The paragraph's basic developmental unit is the topic sentence. The topic sentence provides unity and coherence, and orders and emphasizes the material of the text. If you are to meet your reader's needs efficiently, you must discuss one thing at a time. Keep things together that *belong* together. Make clear relationships among the ideas of your paragraph with reference words, repetition, connectives and relative words, phrases, and clauses. Indicate important ideas by properly subordinating the unimportant ones. Finally, help your reader make transitions from one idea to the next, one subject to the next, or one section to the next by using linking words, sentences, and paragraphs.

Reader Analysis—How to Write for Your Reader

In every English composition class that you have had, you've been told that you need to pay attention to your reading audience, otherwise your message will not get through. In Chapter 3, we learned how communication works. When we communicate, we are trying to establish a commonness with someone. To communicate, then, a writer needs to establish a common frame of reference with the reader. In the human communication system, as we have seen, there is a *source*, a *message*, and a *receiver*. Inexperienced writers, often unwittingly, are self-centered and more concerned with themselves (the source) and their message than with the recipient of the message. They forget that the message they want to transmit must be encoded (written) within the framework of the experience and outlook of the receiver (reader). To be comprehended and accepted, messages must be structured to meet the needs, desires, interest, and background of the reader.

Your writing experience—like that of most students—has usually been limited to the classroom. Your intended reader has been your instructor, who supposedly knows more about the subject than you do. The objective of your written assignment often is academic. Its purpose is to demonstrate that you can write adequately, if not competently, and coherently about the subject, thus proving you have mastered the assignment. In the real world of professional activity your readership varies. Some readers, like your supervisor, will know a good deal about your subject; some will know nothing at all. But most readers will not know what you know or what your considered judgment is about the subject. If you have done your job well, you will have brought a sharp focus on the problem of interest. That is why your message is or should be important to the reader.

Professionals write to people within and outside their organizations. Although their documents may be directed to a specified reader-client, they may yet have to meet the needs of many groups:

> readers within their own groups, e.g., supervisors and colleagues;
>
> readers within the organization with a specialized need to know, e.g., upper management personnel, the legal office, the public relations office, the advertising office, and so on.
>
> readers outside the organization with a specialized interest in the subject, the most prominent being the client. Such outside readers may lie in the expert class or they may range from persons with little formal education to experts in fields other than the subject matter.

You can see that a document may have to serve the needs of many different kinds of readers. A good example is a report on hazardous waste leaks or spills. The effects of hazardous chemical effluence on a community like Love Canal in Niagara Falls, New York or on Times Beach, Missouri are of interest to a wide variety of readers: all levels of the Environmental Protection Agency (EPA), other concerned federal agencies, the governors, mayors, legislators of the states and communities directly and potentially concerned, the mass media, environmental scientists and professionals in related disciplines, public safety personnel, the organizations or persons said to be responsible for the effluence, the medical profession, public spirited and concerned citizens, the legal profession, members of the community directly and indirectly affected, and many, many others too numerous to list.

To prepare a document relevant to such a diverse universe of needs is not easy—certainly not for the inexperienced writer. Our purpose in this text is to lessen the pitfalls and to make the task easier. (Notice, I said *easier*—not easy.) With the guidance of experience, many writers have learned to meet successfully the needs of such a wide universe of readers. Our purpose in this section is to provide you with the thinking and guidance to analyze your reader requirements and to offer helpful instruction on the means to reach your readers.

Focussing on the Reader

If we stop to think a minute about what is involved in analyzing the readers of our writing, certain obvious things are clear. You can list them as well as I:

Who is the primary reader? (The primary reader will use the writing to act on the information it contains.)

Who is the secondary reader? (These are readers who could be directly affected by the document or by the uses their organization may make of it.)

Who is the immediate reader? (The immediate reader is the formal transmitter of the document or of its information.)

You will certainly list other specifics to help characterize your reader or readers:

What is the reader's job or role?

What is the reader's background, education, technical knowledge, and experience?

What are the reader's chief responsibilities, concerns, needs, and desires?

What does the reader know about the matter or problem the writing deals with?

What are the concerns and requirements of the reader's organization?

What are the political realities of the situation or problem and how can you best meet them?

What personal characteristics does the reader have that affects the comprehension, reception, and acceptance of the document?

What attitudes, preconceptions, biases, or misinformation does the reader bring to the problem?

Who, if anyone, will be offended or pleased by the document?

How will the reader use the document?

- as a guide for immediate action?
- as background for future action?
- as a reference?
- as an archival document?

Such prewriting, planning, and thinking, you'll find, will help because they enable you to:

- identify your reading audience;
- determine the purpose and scope;
- organize your facts and ideas to meet the purpose of your study and needs of your reader(s);
- keep your writing on track from the introduction section, to the body, to the conclusion section.

Your introduction indicates immediately for whom the writing is intended, what the writing addresses, and the reason it was prepared. (See Chapters 12 and 13 for further elaboration.) When a reader is motivated and understands why he is supposed to read, he is more receptive to investing his reading time. To maintain that interest, you will remind him from time to time where your information has taken him and where he will be going. Such transition devices, as we saw in our discussion of structural paragraphs, keep the reader oriented and economize on his attention and time.

Making Writing More Readable

Research has been going on since the early 1920s to find a formula for more readable writing. One of the more prominent **readability** researchers is Rudolf Flesch, who has been a consultant for the Associated Press, various newspapers, and many corporations in an endeavor to make their printed communications, annual reports, and other writing more readable. According to Flesch, writing will have "reading ease" if sentences are short—an average of not more than 19 words each; if sentences have short words—not more than 150 syllables per 100 words; and if the writing has liberal use of words and sentences possessing human interest—that is, liberal use of personal pronouns [3].

Parents who have looked at the primers their children bring home know that a page made up of short, simple sentences—even with personal pronouns—is dreadfully monotonous. Variety in sentence length—change of pace—helps sustain reader interest. How short is a "short" sentence? What does *readable* mean? Shortness and readability are, of course, relative terms. A person's concept of sentence length changes with age, education, and reading experience. One study has shown that sentences written by schoolchildren increase in length from an average of 11.1 words in the fourth grade, to 17.3 in the first grade of high school, and 21.5 in the upper college years [11:180].

A sentence that looks long to the fourth grader may look short to the college junior. A person whose only readings are Andy Capp, tabloid headlines, and TV captions will find a sentence long that looks short to the student of literature.

A story for children has sentences averaging less than fifteen words, whereas sentences for educated adult readers may average between twenty and thirty words. A higher or lower average indicates that a writer should examine his sentences critically, but *it does not necessarily mean* that anything is wrong. Variety, type, construction, and length, on the whole, are more important than the average number of words.

A long sentence is suited to grouping a number of related details clearly and economically. Thus, the writer of a newspaper story answers in her first sentence the six essential questions: Who? What? Where? When? Why? How? Causes and reasons, lists, results, characteristics, and minor details may all be expressed tersely and clearly in long sentences. One of the most effective sentences in the English language is eighty-two words long:

> It is rather for us to be here dedicated to the great task remaining before us—that from these honored dead we take increased devotion to that cause

for which they gave the last full measure of devotion—that we here highly resolve that these dead shall not have died in vain—that this nation, under God, shall have a new birth of freedom—and that government of the people, by the people, for the people, shall not perish from the earth. [Lincoln, *Gettysburg Address*]

Readability formulas may be used sometimes as diagnostic instruments. They are not a quick and easy answer to making your writings more readable. The chief ingredient for readability is within the definition of communication: the establishment of a common interest between writer and reader. Readability formulas will not guarantee comprehension. Consider the following typical sentence from Gertrude Stein's *Four Saints in Three Acts*:

Short longer longer shorter yellow grass Pigeons large pigeons on the shorter longer yellow grass alas pigeons on the grass. [13:533]

Or, the opening paragraph of Franz Kafka's enigmatic novel, *The Metamorphosis*:

As Gregor Samsa awoke one morning from uneasy dreams he found himself transformed in his bed into a gigantic insect. He was lying on his hard, as it were armor-plated, back and when he lifted his head a little he could see his dome-like brown belly divided into stiff arched segments on top of which the bed quilt could hardly keep in position and was about to slide off completely. His numerous legs, which were pitifully thin compared to the rest of his bulk, waved helplessly before his eyes. [8:1432–33]

Both of these excerpts would score higher in readability formulas than Lincoln's *Gettysburg Address* but would prove confusing to many readers. Readability formulas do not measure content, organization, or cogency, since these characteristics cannot be measured quantitatively; nor will readability formulas improve the writer's style in a written communication. But formulas do point out certain qualities in written communication pertaining to comprehension. Therefore, they may be helpful if discriminatingly used.

Research on how learning takes place has contributed significant insights to readability. Certain writing techniques, researchers have found, help the reader to receive and comprehend information. Such techniques include:

Advance organizers,
Overviews,
Inserted questions,
Prompting clues, and
Graphic aids.

Advance Organizers

Headings (as we shall see in Chapter 14) are the best examples of advance organizers. They are road signs telling the reader the text territory to be visited.

Introductory text material also prepares and helps the reader organize and bridge from section to section the elements the document deals with. For example:

> This report will present a comparative cost analysis of high pressure hot water (HPHW) and steam systems as heat transfer media for a four-building complex at Boeing Airplane Company's Transport Division plant at Renton, WA. First, the report will describe the HPHW system in two of the buildings and the steam system servicing the other two structures. Cost for labor, materials, and maintenance will then be examined and analyzed. Conclusions and recommendations based on the cost efficiency of the system most practical for the Renton facilities will complete the report. A time table to implement the recommendations is also provided.

Overviews

Similar to advance organizers, *overviews* give the reader a quick glance at the subject of the writing, plus an indication of its significance. This technique whets the interest of the reader because it reveals at once the importance of the topic. For example:

In an article entitled "Food and Drug Interactions," appearing in the *FDA Consumer,* there is an overview above the article that begins:

> *If you're taking a drug, the food you eat could make it work faster or slower or even prevent it from working at all. Eating certain foods while taking drugs can be dangerous. And some drugs can affect the way your body uses food.* [10]

Overviews, like this one, are usually set off by italics, boldface typography, indentions, or by other graphic embellishments to catch the reader's eye. The purpose of the overview is to grab the reader's attention and arouse interest with the promise of details contained in information to come. The reader will then want to read the entire material. The hints provided in the overview help the reader to follow the details.

Inserted Questions

If you are asked a question, you pause to think whether you know the answer or you become curious to learn the answer. A writer can take advantage of our innate inquisitiveness by devising a question or set of questions that epitomize the topic or any of its aspects. For example, in the article just mentioned, "Food and Drug Interactions," the writer begins with several questions, which immediately grab the reader's attention and lead to the point of the article, that eating certain foods while taking certain drugs can be dangerous:

> Would it occur to you not to swallow a tetracycline capsule with a glass of milk? Or avoid aged cheese and Chianti wine if you are taking a certain medicine to combat hypertension? Or to eat more green leafy vegetables if you are on the pill? Probably not. Yet the effects foods and drugs have on each other can determine whether medications do their job and whether your body gets the nutrients it needs. [10]

Prompting Clues

Giving the reader a clue about the significance of the content to be presented helps the understanding and makes the details easier to follow. The simplest form of clue is typographical: use of italics, boldface, or underlining. Color and illustrations can also be used. Rhetorical devices such as examples, analogies, or comparisons can be effective. They prepare the reader for the substance and particulars because such prompting devices have made the path to comprehension more familiar and comfortable. Note in the example below—how the metaphoric use of "pump," "duplex apartment," and "pipes," as well as italics, enables a lay reader better to understand how the heart works and what happens in a heart attack:

> To understand what happens in a heart attack, we have to think of the heart, a muscle, as a pump (Figure 4.1). Blood from the body enters the right side of the heart. The heart has four chambers, which have special roles in the pumping process. The upper chambers are called the *auricles*; the lower chambers, the *ventricles*. The auricle and ventricle on each side form an independent part of the heart somewhat like rooms in a duplex apartment; in effect they make up a "right heart" and a "left heart." There is no connection for the blood to flow between right and left hearts; they are separate "duplexes." Each pumps its own circuit; the right side sends blood under low pressure into the lungs where it is combined with oxygen. From the lungs, the left ventricle pumps blood under relatively high pressure to the rest of the body, supplying oxygen and nutrients to the tissues. So the heart is like a pump, squeezing and forcing blood throughout the body. The most important part of the heart is the middle muscular layer of the heart wall, the *myocardium*. Like all muscles in the body, the myocardium must have oxygen and nutrients to do its work. Unfortunately, the myocardium cannot use oxygen and nutrients directly from the blood within the chambers of the heart. Nutrients and oxygen are furnished by blood vessels outside the heart.

Figure 4.1
A Normal Heart

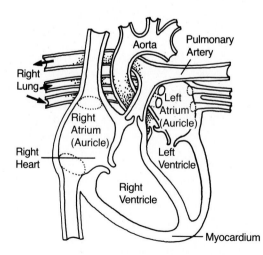

The two most important vessels are the right and left coronary arteries or veins that begin at the base of the *aorta,* the large artery that carries blood from the heart to other parts of the body.

As a person grows older, the coronary arteries are sometimes narrowed by fatty deposits or *plaque.* This process is called *atherosclerosis.* The narrowing is no different than the accumulation of "gook" in old pipes, which impedes or stops the liquid from flowing through. When such deposits occur, the heart cannot always pump enough blood through the arteries. When one of the coronary vessels is blocked completely, even for a short time, part of the heart may die and scar tissue may form. This injury is called a heart attack. Scarring is bad because any injured area causes the heart to lose its effectiveness as a pump; there is less muscle to contract to force blood out.

Graphic Aids

Graphic aids, as we will see in Chapter 15, are lists, tables, photographs, drawings, diagrams, graphs, charts, flow sheets, and editorial devices. These help the reader to digest, think about, and visualize facts and ideas. Graphics help the reader to organize details and see them in relationships. By strategic use of space, movement, emphasis, color, shades, and highlights, graphic aids lend truth to the old saying that a picture is worth a thousand words in promoting readability.

Factors to Improve Readability

A number of factors will aid readability of written communications:

1. A thorough knowledge of the matter to be communicated
2. A knowledge of the reader
3. Organization of the text to lead the reader to the thesis sentence (see Chapter 12)
4. Use of format devices like display heads, enumerations, uncrowded spacing, and margins
5. Use of illustrations, tables, and other graphic devices to supplement the text
6. Use of topic sentences in paragraphs
7. Use of structural paragraphs, sentences, and words and phrases
8. Use of advance organizers
9. Use of overviews, if appropriate
10. Use of prompting clues
11. Inserted questions, as appropriate
12. Variety in sentence type and length
13. A clear, concise, direct, and objective style of writing
14. Relegation of long derivations, computations, and highly technical matter to the appendix

Some Advice on Style

Style is the way you put your thoughts into words, sentences, paragraphs, and groups of paragraphs to convey information to your reader. Effective technical style consists of these elements:

1. Clarity
2. Precision
3. A logical construction of words into sentences, paragraphs, sections, and chapters
4. Directness in statements of thought.

To achieve these ends, you must do the following:

a. Balance the length and type of your sentences (using more complex than simple or compound sentences).

b. Select, whenever appropriate, the simple word rather than the obscure, the familiar rather than the unfamiliar term. (But if complex and unfamiliar terms are necessary to meaning, logic, and clarity of the thought to be communicated, you should use them—with appropriate definitions or explanations.)

c. Maintain a balance of active verbs with passive verbs. (Explanation depends on facts. Verbs express the relationship among facts. When the action is more important than the actor, you should use the passive construction. Active verbs give more force, directness, and clarity to expression. But your reader is the determining factor. If you know the reader prefers an impersonal style, let your writing reflect it. If you have a choice, use a more personalized approach. By doing so, you will be more concise and to the point because you are expressing yourself in a style that is you, not in the style you think you ought to use.)

d. Finally, make sure that every word, sentence, and paragraph you write is not merely capable of being understood but incapable of being misunderstood.

No one, least of all the technical writer, should be open to the criticism Alice in Wonderland received:

> "Speak English!" said the Eaglet. "I don't know the meaning of half those long words, and what's more, I don't believe you do either."

Chapter in Brief

We have examined the definition of and the characteristic qualities and elements of technical style. In the process, we analyzed the use of the active and passive voice, types of sentences and their capacity for expression, the design of paragraphs, and the role of the topic sentence. We turned next to reader analysis and the factors that improve readability.

Chapter Focal Points

- Technical style
- Clarity and precision
- Objectivity
- Active and passive voice
- Sentence types
- The Paragraph
- The Topic sentence
- Reader analysis
- Readability

Questions for Discussion

1. Locate two technical articles on a subject of interest. Each article should be directed to an audience with a different level of understanding of the subject matter. Prepare to analyze these articles on the basis of the following questions:

   a. What similarities in style and what differences do you find in the two articles?

   b. What audience is each trying to reach?

   c. Is each successful in the factor of readability for the intended audience? Why and how?

   d. Does each article achieve its purpose?

   e. What writing devices do the authors use to achieve or attempt to achieve their purpose?

   f. Are there differences in terms used? Average length of sentence? Paragraphs? Use of personal pronouns? Use of abbreviations? Article organization? Format?

   g. What graphics help maintain readability and interest?

   h. How successful is each article in achieving its purpose? (Do you know at the end what the author has attempted to communicate?)

2. Classify the following titles as to the level of intended readership: general reader, technical, specialist, executive.

   a. Engineering Tomorrow's Dinner

   b. Embryogenesis in Carrot Culture

   c. Coal in the United States: A Status Report

   d. The Architecture of Cognition

   e. Carbon Dioxide and Our Changing Climate

   f. Corporate Computing from the Top Down

    g. Software That Helps Homework

    h. How to Win the Budget Battle

    i. Fourier Transform Infrared Spectrometry

    j. Detecting Deception from Verbal, Visual, and Paralinguistic Cues

3. From any of your textbooks or professional periodicals, locate four paragraphs that have the following:

    a. The topic sentence at the beginning of the paragraph

    b. The topic sentence at the end of the paragraph

    c. The topic sentence in the middle of the paragraph

    d. No topic sentence within the paragraph (the topic sentenced is inferred)

    e. A pointer sentence

    f. Both a topic sentence and a pointer sentence

    Which paragraph construction do you find easiest to follow? Why?

4. Read the four paragraphs aloud in class. Which constructions do your classmates find the clearest to understand?

5. Locate a passage of bureaucratic writing—regulations of various types, leases, laws, legal agreements, insurance policies, bank documents, etc. Analyze the characteristics that interfere with the flow of the message. Identify the stereotype phrases and gobbledygook words the passage contains. In what voice is it written? Any personal pronouns? Any ungrammatical constructions? Does the punctuation help or hinder readability and understanding?

6. Locate a paragraph that you feel uses the passive voice structure effectively. Try to rewrite the passage in the active voice. Bring the two versions to class and have the class judge their effectiveness. Opinions should be substantiated.

7. How would you rewrite the following announcement?

> The Student Chapter of the Society for Technical Communication is an organization of interested students in college or in a university dedicated to development of further knowledge about careers in technical writing enhancement of necessary technical writing and computer skills and further contacts for the beginning careerist. The student chapter is now open to accepting applications for new memberships. Application forms are obtainable from any present member. Dues are modest for students in this prestigious organization.
>
> The meeting will be on October 1, at seven o'clock PM in Room 17 the Student Center. An interesting program is being planned with Montgomery F. Blakestaff of and also president, Blakestaff & Associates main speaker with a very interesting topic. Interested students are invited to attend and questions are permitted.

Assignments

1. The sentences or paragraphs below lack clarity, precision, directness, and balance. Rewrite them in accordance with the principles in this chapter.

a. The purpose of this report is to offer the State of Ohio ways and resources to assist those in charge of developing procedures to increase the effective utilization of computer-based information system.

b. Beginning management accountants may often advance to chief plant accountant, chief cost accounting, budget director, or manager of internal auditing though starting as ledger accountants, junior internal auditors, or trainees for technical accounting positions with salaries just as good as engineers starting out.

c. It is concluded that the use of flexitime scheduling by companies will continue to bring about an increase in productivity by increasing employee morale and job satisfaction, the quality of labor, organization of work, and less absenteeism. Although drawbacks were mentioned about flexitime scheduling the rate of success and positive points clearly outnumber the problems or drawbacks. We will continue to see flexitime scheduling in the future because young workers of today want individual say and input into the company and by implementing flexitime scheduling this will be made possible.

d. On any given day, one can turn on the "idiot tube" (television) and be assaulted by any number of twenty- to sixty-second messages and longer which preach of the unfulfillment in our life because the lack of a specific product or service.

e. Adults, humans who have accrued a certain number of years of existence, must do all that they can to inform children in the area of sexual behavior and protect them from future problems.

f. As you requested here is the report on the problems faced by book and clothing buyers of retail and wholesale firms. These buyers also provided me with a number of solutions to these problems. I have included them also, as you requested. This paper will be limited to the major problems of book and clothing buyers and will not include specific details of individual situations faced by book and clothing buyers such as the management situation at their firms.

g. A clothing buyer must be able to work together with her store-department coordinators. The buyer should show them how to coordinate outfits for display. The perspective customer needs to know how the outfits go together. A clothing buyer spends a great deal of time working on advertisements for the items he has purchased. People need pictures to entice them to buy clothes if they're expensive, so the buyer must work up lay-outs of the different items he has purchased. He or she should also work up layouts for the floor of the store.

h. A 20 milliamp (ma) signal shall be available to control the additive system pumps from the control panel, automatically, or with a manual override.

i. A surgically implanted device devised to normalize the rhythm of a patient's failing heart will significantly reduce the death of high risk cardiac patients is called an implantable automatic defibrillator because it prevents sudden death caused by VA's (ventricular arrhythmias), which are rapid erratic heartbeats that cripple the heart.

j. Each pump's fabrication shall consist of ASTM 316 stainless steel where in direct contact with chemical additives and all piping connections shall be flanged to conform to ANSI Standard B16.5 to joint with interface connections.

2. a. Lawyers are the worst offenders in the use of a very specialized professional language in their writings. The writing of most lawyers is shackled by the gobble-

dygook of legal terminology and by the use of the third person and passive voice. Below are examples of an article by a lawyer writing in an engineering magazine. How would you rewrite the excerpts to make sense for a reader? Keep in mind that the qualities of technical style are clarity, precision, conciseness, and objectivity.

> The argument has been made that the contractor is a third party beneficiary of the promise of the engineer or architect to the owner that he, the architect, will act in a certain way with respect to the contractor, such as, for example, timely approval of plans. . . .
>
> Another interesting possibility presents itself with respect to the plaintiff contractor's problem of avoiding the exculpatory and notice provisions of a construction contract. These are the "fine print" clauses that attempt to shift onto the contractor all responsibility for everything that does or does not happen. Such provisions, either by terms or because a plaintiff contractor fails to comply with them, often effectively deprive a plaintiff contractor of the fruits of what otherwise could have been a successful suit against the owner. It has been suggested that when the actions of the engineer or architect have given rise to the cause of action, and the owner has protected himself as noted, there may exist enforceable legal liability against the engineer or architect independently of the owner under conditions where the architect will not receive the benefit of the protection afforded the owner [6:49–50].

   b. If possible, check your revision with a lawyer to see whether your rewriting is accurate and conforms with the legal nuances in the original. Obtain the lawyer's opinion about the specialized terminology of the legal profession. Is there anything wrong in the original quotation? What is the lawyer's opinion of legal writing when done by lawyers for laymen, as well as by lawyers for lawyers?

   c. Locate a Supreme Court decision and compare its writing style and clarity with that in the quoted material. An alternative reference might be the decision of U.S. District Court Judge John M. Woolsey lifting the ban on the book *Ulysses* by James Joyce [9:ix–xiv].

3. Locate and read a short article in a professional journal of your field that you find easy to read and understand. Analyze the paragraph and sentence structure. How many simple sentences does it contain? How many complex? How many compound? How many complex-compound? Does it have any fragmentary (incomplete) sentences? Are there any ungrammatical sentences? Note how the construction leads you to the point of the thesis of the article. See how the relationship between sections, paragraphs, and sentences is handled. First, go through the article paragraph by paragraph and underline the thesis sentence. Second, draw a box around the transitional paragraph. Then, circle each transitional sentence or phrase. Are there places in the article in which transitions are missing—where the flow of information stops? Or did the logic of the construction move the information unimpeded in that instance? What insights about writing for the reader did you gain from this exercise?

4. Find a paragraph in one of your texts that you find difficult to understand—not because of its technical substance but because of its expression. What readability principles did it violate? Rewrite the paragraph to make it more readable.

5. Write your instructor a biographical letter. The purpose of this assignment is to help open communication channels. Include details not only on your background and ambitions but also let him or her know how you feel about writing. Let your hair down so you can be helped. Let your instructor know the following:

   a. You enjoy writing or you hate it, or you are afraid of writing, or you haven't thought much about it.

   b. You would like to be able to write better, but you recognize you have certain problems. (Identify them.)

   c. You are taking this course because: it's required, or because you recognize you need it, or the computer scheduled you in it.

   d. How do you feel about being in the tech writing course?

   e. Have you given any thought to the reader of your writing? Do you think the reader should influence your writing? Why?

   f. How long do you spend on your writing assignments? Do you plan ahead or do you wait to the last minute? Do you feel pressure about writing? More or less than from assignments in other classes?

   g. About your writing assignments: Do you worry about them? Do you spend more time worrying than you do on the writing?

   h. Do you get writing blocks? What do you do about writing blocks? Are you aware that professional writers get writing blocks and they worry about them—even more than most students?

   i. Do you write at one sitting? Or do you doodle around? Do you try to write sections you are most comfortable with or do you try to follow the logical sequence of the subject?

   j. Have you had other classes in writing? Other than Freshman Composition? Have they helped? How, if you answer, yes? Why not, if you answer, no?

   k. What do you want this class to do for you?

   l. How can your instructor help you?

References

1. Dewey, John. *How We Think.* Lexington, Mass.: D.C. Heath, 1923.

2. *The D(ratted) P(rogress) Report.* E. I. Dupont de Nemours and Company, Explosives Department, Atomic Energy Division, Technical Division, Savannah River Laboratory, July 12, 1954.

3. Flesch, Rudolf. *The Art of Readable Writing.* New York: Harper and Brothers, 1949.

4. Huxley, T. H. *Essays.* New York: Macmillan, 1929.

5. Huxley, T. H. *An Introduction to the Study of Zoology.* New York: D. Appleton, 1938.

6. Jarvis, Robert B. "The Engineer and Architect—As Defendants." *Civil Engineering,* April 1961.

7. Jeans, Sir James. *The New Background of Science,* Cambridge: Cambridge University Press, 1933.

8. Kafka, Franz. "The Metamorphosis." In *World Masterpieces,* vol. II, revised, edited by Bernard M. W. Knox. New York: W.W. Norton, 1965.

9. Joyce, James. *Ulysses*. New York: The Modern Library, 1940.

10. Lehmann, Phyllis. "Food and Drug Interactions." *FDA Consumer*, April 1988.

11. Perrin, Porter G. *Writer's Guide and Index to English*. 3d ed., Glenview, Ill.: Scott, Foresman, 1959.

12. Sprat, Thomas. "The Vanity of Fine Speaking." In *A Science Reader*, edited by Lawrence V. Ryun. New York: Rinehart, 1959.

13. Stein, Gertrude. "Four Saints in Three Acts." In *This Generation*, revised ed., edited by George K. Anderson and Eda Lou Walton. Glenview, Ill.: Scott, Foresman, 1949.

5

Automating the Writing Process

Chapter Objective

Provide computer literacy in word processing and desktop publishing.

Chapter Focus

- History and development of the writer's tools
- The Computer as a word processor
- How word processing works
- Desktop publishing

*I*n the beginning, we are told, was the Word. Without the "Word," indeed, there would be no beginning. *Pre*historic humans groped their painful way to being *historic* humans when they began to share their experience with their fellows. Memory was the first tool for recording information. The written word was the development that gave both durability and precision to communication. The idea of using marks to represent and communicate thoughts and spoken words was our greatest invention. The first writing was pictorial representation; it emerged about 25,000 years ago. From rudimentary pictures, writing developed into more complex ideographic phases, into phonetic systems.

The phonetic alphabet made writing simple by reducing the number of symbols needed and by providing flexibility of expression. As soon as writing developed, discoveries could be handed down, records could be kept, and reports could be made. Barriers to communication between living people or between the past and the present were reduced.

The earliest known writing tool was the bone stylus used by the Sumerians in Mesopotamia about 4000 years ago to engrave symbols on a clay surface. The ancient Egyptians invented the reed pen, ink, and papyrus. Goose quill pens came into use about 1500 years ago and lasted into the nineteenth century. Before the invention of movable type by Gutenberg in the fifteenth century, books were produced by calligrapher-copyists working with quill pens on desktops. The quill pen can be considered the original desktop publisher; it allowed the combination of text, graphics, and color. When Gutenberg printed his Bible, he modeled his 42-line page on the calligraphically produced Bible page.

Traditional writing tools have undergone evolutionary changes. Pencils using lead date to ancient Rome; the graphite pencil came into use about 1850; the fountain pen was manufactured even earlier, in England in 1835; the manual typewriter was invented by a Milwaukee typesetter in 1867; the ballpoint pen first came into being in 1888; and the electric typewriter, the forerunner of the word processor, was introduced by IBM in the 1930s.

Enter Electronics

The electronic digital computer, invented about 1946, is a device that receives information and data (input), stores them, processes them according to instructions (programs), and then transmits the information to a user. Today, the use of micro- and personal computers in processing information is as common to the scene as paper clips. **Word processing** is the term used to describe the automated way textual information is handled by the computer. Essentially, the computer enhances the typewriting process by adding a capability for memory (storage of information), manipulation (editing) of the information/data in storage, and formatted printing of the edited information.

Word processing in its most basic form originated with the automated typewriter. This machine used paper tape, which recorded text as punched holes when the typewriter was operated. The punched tape was then rerun through the typewriter to produce identical copies of the text. In 1964, IBM introduced a more useful machine, which featured magnetic tape cartridges to store text while the user typed words on a page. Eventually IBM replaced the tape with a card for input and storage, an approach similar to what is now called a floppy disk or diskette. The MagCard could hold one page of text at a time before its contents had to be erased and a new page typed. The MagCard contained integrated circuits (chips) rather than transistors, a factor which meant that the machine had a higher storage capacity, was faster and cheaper to use. More important, copy recorded on it could be edited. But a major limitation was that the user could not be sure that edited changes were being made correctly until the page was printed. Storage was also a problem, because lengthy manuscripts required many, many MagCards. Next in the evolutionary process came micro- and personal computers (PCs) whose display screens, with appropriate word processing programming, overcame the editing problem: both text and changes could be seen before the printout stage.

Word Processing

What is word processing? It is not a device into which you feed words and by some magical means they flow out in beautiful prose or iambic pentameter. Word processing is a **software program**—instructions that direct the operations of a computer—that permits the writer to keyboard (type) words into the computer. The words do not go directly on paper, but are stored on a magnetic or optical disk (or other storage device) and are projected on a monitor—a display screen, similar to a television screen. Using appropriate keystroke commands, the writer can correct errors; insert, change, or delete words, sentences, or whole paragraphs; move large blocks of text around; check spelling, grammar, or even style (depending on the type of software program); paginate and insert footnotes; and establish or change the format of the page on which the text will appear.

The writer's text appears and exists only as electronic bytes (characters) on the monitor screen until the writer keys the text material for permanent storage or printing. So, you can see, the word processing system is a useful and helpful technology, allowing the writer full control and exercise of creativity in the composition process.

The Word Processing System

The word processing system usually consists of a general purpose computer, and software. The system has several major components: the central processing unit (CPU), the keyboard, the disk, the display screen, and the printer.

The *CPU* is the heart of the system. It is the hardware portion containing the logic for the editing functions and the means for storing and retrieving text, and it serves to control the system. Some of the logic is in the electronic circuits;

increasingly, however, much of the logic is accomplished by software (instructions to the computer on how to program or process the data). These programs are brought into action by the keyboard.

The *keyboard* is similar to that of the conventional typewriter, but has a number of additional keys that are used to perform editing, formatting, and control functions.

The *disk* is the medium that stores the text after its creation or editing. The most common type is the flexible or "floppy" disk or diskette.

The *display screen* is a cathode-ray tube similar to the TV picture screen. On it are displayed text and graphic matter produced by the user at the keyboard.

The **printer,** a device similar to a typewriter, is used to print the word processing system's output. There are two major categories of printers: dot matrix and letter quality. Least costly and of lesser quality, the dot-matrix printer has from nine to twenty-four steel pins arranged in a row. As the printhead moves across the paper, the inked pins are fired by electromagnets in different patterns of dots. This matrix of inked dots forms characters on the paper (Figure 5.1). It can form simple graphics and charts, as well.

Letter-quality printing can result from three approaches. Two are similar in quality: the daisywheel and print thimble. Each provides fully formed characters of almost print typeface quality and can produce good quality graphics. (Figure 5.2) Of still superior quality is the laser nonimpact printer. It produces images on paper by directing a laser beam into a drum, leaving a negative charge in the form of a character to which a positively charged toner powder will stick. The toner powder is transferred to paper as it rolls by the drum and is bonded to the paper by hot rollers. The laser printer, far more expensive and speedier than the others, produces superior quality text and illustrations.

A **terminal** is a device for both sending and receiving data and information. It is usually a keyboard and screen housed together with associated electronics with which the user can enter information or commands and have these displayed on the screen. The terminal may be connected to a remote computer system or be part of a "stand-alone" word processing system (Figure 5.3). If the terminal is used with a remote computer, the system may be configured to operate as either a shared-logic, a distributed-logic, or a time-shared word processor.

The **shared-logic** word processor is designed to accommodate several users at the same time to share the processing, storage, printing, and other computer capabilities in conjunction with word processing. (Figure 5.4.)

The **distributed-logic** word processor is similar to the shared-logic system, except that parts of the memory and control circuits are taken out of the main electronic package and put into the individual entry point or work station.

The **time-shared** word processor is part of a large computer system, with the processing capability rented out to customers or made available to different members of an organization at various locations. Input is done by means of a terminal connected to the computer and output is routed to a letter-quality printer. This system is also based on the shared-logic approach.

Figure 5.1
A Dot Matrix Character Representation

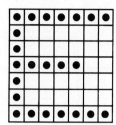 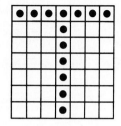

This is an example of dot matrix printing.
Dot matrix printing may also be available
in italics and boldface type.
This is an example of letter quality printing∅
Letter quality printing resembles typewriter
type but may also come in italics and boldface∅

Figure 5.2

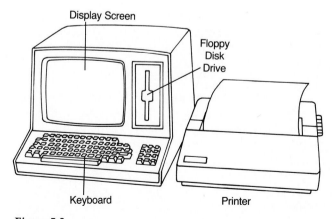

Figure 5.3
*Example of a stand-alone word processing system. Memory, Central Processing, and input-
output controls are contained on printed circuit boards within the unit.*

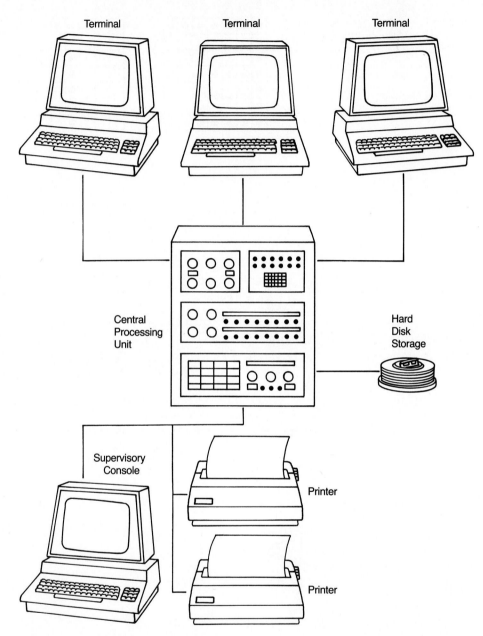

Figure 5.4
Example of a shared-logic word processing system

How the Word Processor Works

Like humans, word processing software packages have diverse talents but share many features and functions. Functions basic to all are text editing and formatting.

The **text editing** function includes the ability to enter (keyboard) words into the computer, which commits them to temporary memory or, if you desire, permanent storage; further, it allows you to make deletions or insertions, and to move words, phrases, or blocks of text from one place to another. When you enter the text, it is displayed on the screen and is stored in the computer as a **document,** or **file.** Your software allows you to recall it on the screen to make whatever revisions you want.

The software's **formatting** function helps you to design the document's appearance on the screen and on the printed copy. It allows you to establish left and right margins, line spacing, tabs (indentations), column alignments, placement of headings, and page length. You can underline words, make them boldface, italicize them, and place headers (headlines) at the top of the page and footers (page numbers) at the bottom of the page. You can also have the pages numbered automatically. Your formatting instructions are communicated by the computer (software) before the printing begins.

There are two approaches to the way word processors manipulate documents: menu-driven and command-driven. **Menu-driven** means that the operator (you, the writer), have a choice of operating options listed on the screen. (This list is called a *menu.*) You select the choice you want either by moving the cursor (a blinking character on the screen that indicates where the next character will appear) to the selection or by typing the number of the selection you want to be operated. In **command-driven** word processors, choices are not presented but the program requires the operator to enter or *command* the program to make the desired operation. The operator needs to know the operation commands or must consult a manual. Many persons prefer menu-driven programs because they seem easier. Some software programs use a combination of both.

Text Editing Features

The word processor can help you write by making your rewriting and revising tasks easier and less cumbersome. It gives you the ability to make changes in your manuscript without retyping what is already acceptable. Such a capability is especially useful to a technical writer. Modifications frequently become necessary because of changes in specifications, or because progress in research and development has altered processes and procedures, requiring insertion of new data or formats. The writer may be a member of a team, and the written data and information from others may need to be merged into the manuscript. Word processing permits the writer to use what will remain the same and enter revisions and editorial modifications without keyboarding the entire document.

Quick and graceful revisions are accomplished with the following features:

The **cursor** is the box or blinking character line that indicates the user's present location on the screen. The cursor is controlled by either keys on the keyboard or by a **mouse,** a small input device that allows the writer to control the cursor without using the keyboard. Some commands can make the cursor jump to the first or last character on display screen or to the start or end of a line. Such short cuts speed the editing process.

The **status line** can appear at the top or bottom of the screen. It provides information about the document (file) you are working on, the page, line, position, or column, amount of available memory, and designation of the disk drive being used.

A **window** is a section of the screen, usually with a border around it. Some word processing programs can display two or more windows at the same time. The purpose of the window is to let you see two portions of the document at the same time or see two portions of two separate documents. Windows can display not only portions of specific documents but also menu options, messages, and warnings.

Editing text is entered in either the **insert mode** or **exchange (replace) mode.** When you enter additional text in the insert mode, all text to the right of the addition shifts to make room for the new material. In the exchange mode, newly inserted characters take the place of or overwrite existing material occupying the same place.

Wordwrap is a very useful feature because the system automatically moves a word to the beginning of the next line if it does not fit within the set margins at the end of the original line.

Some programs have an automatic **hyphenation** capability for properly breaking words at the end of a line. If the position of that word changes during the editing process causing the hyphen to be unnecessary, the hyphenation feature automatically removes it.

Most computer screens display 25 lines of 80 characters each. When a document is too long or its lines are too wide to be viewed as a whole, the **scroll** feature gives the user the capability to view it entirely. This feature causes the text on the screen to move up, down, or across. Scrolling from side to side is sometimes called **panning.**

The **delete** feature allows you to remove a character, phrase, or block of text. The remaining text shifts to fill the space left by the deletion.

Search and replace is a useful editing feature. For example, say you have just finished a lengthy report and then discovered that a recurring key formula is incorrect, or that you have consistently misspelled a person's name. By using search and replace, all occurrences of the string of wrong characters (the erroneous formula or misspelled name) automatically can be correctly replaced.

A **cancel** or **undo** feature allows you to cancel the action of a previous instruction.

Block editing allows large units of text, from a paragraph to several pages, to be moved, copied, deleted, saved, or changed as a unit.

Formatting Features

To **format** a document is to specify the way it will look on screen or when it is printed. Some parameters are specified before you begin inputting the document, some during the keying and editing process, and some when the document is printed. **Default settings** are the parameter specifications the system uses automatically when you do not choose a particular setting. Software formatting features include the following:

Character enhancements let you select character size and typeface (font), boldface type, italics, expanded characters, underlining, or a combination of these and other display features.

Margin settings are the specifications for allocating space on the left, right, top, and bottom of a printed document.

The **justification** feature allows you to align the text flush at the left and/ or right margin.

Centering allows you to place a heading or a partial line of text in the center of a blank line.

The **line spacing** feature allows you to single-, double-, or triple-space your lines.

Page break is the feature that controls the number of lines to a page. After you reach the set number of lines, the line that follows is automatically placed on the next page.

The page number is automatically and sequentially printed on each page of the document with the **automatic page numbering** feature.

A **header** is the text heading that appears at the top of a page. Often it is the title of the document or of a section, or of a chapter.

A **footer** is information that automatically appears at the bottom of a page, often the page number.

Specialized Software Features

The following programs augment the capabilities of word processing systems. Usually each of the listed features requires a separate diskette for its operation.

The **glossary** lets you automate some word processing tasks; for example, commonly used phrases kept in storage can be inserted by command. With the glossary feature, keying a time-consuming phrase can be reduced to one or two keystrokes.

Some programs have a built-in **spelling checker** to locate and correct misspelled words. The checking can be limited to the page involved or can cover the entire document. The list of words the spelling checker uses for comparison is called a **dictionary.** You can add to or delete from, and sometimes create, a dictionary.

An **outliner** can help organize a writer's thoughts and written text. You can enter ideas as they occur; with this software feature, your ideas can be organized and reorganized logically as headings and subheadings.

A **grammar checker** is a program that locates standardized grammar and punctuation errors in the document.

A **thesaurus** is a program tool that provides synonyms for a given word in a document. Sometimes the synonym will have a list of synonyms to help you find the most appropriate word.

Desktop Publishing

Desktop publishing is the process that gives the document's originator the capability to combine the use of a "desktop" computer and appropriate software to write, edit, illustrate, design the page layouts, and to produce the final document's reproduction copy. The term was coined as a marketing slogan by the president of a software company,[1] which pioneered in developing a page layout design for printing documents.

The concept, however, is as old as the practice of the fifteenth-century calligrapher-copyists. It can be traced also to eighteenth-century America when Ben Franklin published *Poor Richard's Almanac*. Franklin did the writing, editing, layout, typesetting, and printing himself. Technical writers in the recent past (and some today) produced typewritten, camera-ready documents before computer technology speeded up the process.

Both desktop and traditional publishing have similar basic steps—generation of text and illustrations, compilation into pages, and printing of the document. Sitting at the monitor screen, the writer uses the desktop publishing system software to write and edit the document. To design pages, the system's **template** software lays out the text and headings, inserts pictures and charts, enlarging or shrinking them as desired or as necessary. The writer selects the type font style for text and headings. If the terminal is linked into a network of computers, the writer-operator can shift pages around electronically to allow insertion of text contributed by others.

For example, let us briefly examine a popular desktop publishing program, Aldus PageMaker. Figure 5.5 illustrates the PageMaker **window**—a boxed area on the computer screen—that displays the five major components that control *PageMaker's* functions: (1) Menu Titles. These are used to select options available from different menus. Each menu contains different items to be used to create documents. (2) Title Bar. This element identifies the title of the document and its disk location. When *PageMaker* is first started, the Title Bar indicates that *PageMaker* is in operation; the word, "Untitled," in the Title Bar indicates that a name for the document as yet has not been specified. The Title Bar also contains the Control Menu Box and the Minimize and Maximize arrows. (3) Tool Box. This component contains several crucial text and graphic tools used for designing and editing a document. The *A* in the Tool Box is used for entering and editing a document; the diagonal line, perpendicular line, rectan-

---

[1]Paul Brainard, president of Aldus PageMaker.

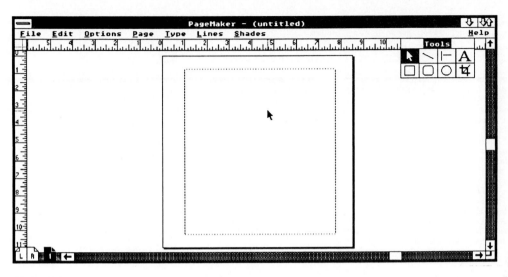

Figure 5.5
An Aldus PageMaker Window

gle, rounded rectangle, and the circle are used for drawing graphic objects; the cropping mark is used for adjusting graphics imported from other sources into the document. (4) Scroll Bars. These allow different parts of the page to be viewed by movement up, down, and across the screen page. (5) Publishing Area. This is the place where both text and graphics are displayed (the large dotted rectangle). The Publishing Area includes the pasteboard (area surrounding the page on the screen), the page image, column indicators, and rulers [4:47]

Figure 5.6 illustrates the process whereby *PageMaker* automatically wraps text around a graphic. The boundary is represented by the dotted lines on all sides of the graphic. This boundary can be adjusted to any style and size within the page area. The boundary is created by pointing the cursor onto the desired boundary area with the mouse (cursor control device). *PageMaker* software then wraps the text about the graphic. [4:27]

As shown, desktop publishing makes electronic images of what the printed document will look like, thus eliminating costly preparation of reproduction copy and trips to the print shop. Once the design is done, the laser printer produces master copy for ultimate offset printing, or the system can print the final copy in black and white at the rate of about two pages per second (Figure 5.7).

As can be seen in Table 5.1 (p. 158), desktop publishing can offer advantages over traditional publishing. Probably the most important are savings in time and money, as shown in Table 5.2 (p. 159).

A young technology, desktop publishing has several limitations. Use of color to date (1991) is much less professional in quality than with traditional printing. Printing speed is slow compared to the offset process. Professional print shops have a greater selection of type styles and can do a better job

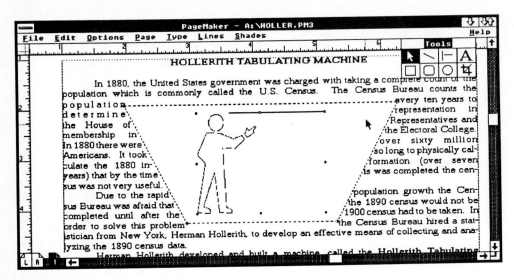

Figure 5.6
Wrapping Text Around a Graphic

manipulating spaces between letters. Laser printers in the desktop system do not produce as sharp an image as do professional print shops. Laser printed pages come out at 300 dots per inch (dpi); professional printing, with approximately 2500 dpi, achieves greater crispness and clarity. Halftones and other finely detailed illustrations are better served by traditional printing.

Startup costs for a desktop publishing system can be high. At the low end of the scale, you can expect to pay about $6,000; the components would include a computer with a minimum of one megabyte storage capacity; a graphics board; a good monochrome display screen; a 20-megabyte hard disk; a laser printer; software to control fonts and to permit the combining of word processing and graphics; and software for creating graphics and illustrations. At the luxury end of the scale, a stand-alone system might cost $100,000 to $200,000. Of course, there are many options in between.

In most desktop publishing systems, the basic software determines and presets the appearance of the printed page. This condition is called WYSIWYG— What You See Is What You Get (a term popularized on one of the early *Saturday Night Live* TV shows). The writer, selecting from available menu options, makes format, illustration, page layout, and typographic decisions. The result of the decisions becomes the WYSIWYG printed document. Usually, the template menus have little or nothing to do with the purpose of the document or its writing. The template nature of desktop publishing can be a major shortcoming. Unless the technical writer has an interest in graphic arts and printing or special training and experience, the document can encounter usability and readability problems. Without such interest and background, a writer's desktop printed

A newsletter style is adapted here for a sales department's monthly bulletin. The rules, bold heads with tag lines, and initial caps dress up and give a newsy image to what could be a pedestrian report.

The highly structured name-plate shows another way of using the open space at the top of the grid. (See the blueprint detail below right for specifications.)

The strong headline treatment requires white space around the various elements and makes separation of headlines in adjoining columns essential.

The initial caps are 60-point Helvetica Condensed Black. Often found in the editorial pages of magazines to highlight points of entry on a page, initial caps are generally underused in business publications. They are especially effective in complex pages such as this sample.

The wide, 3-pica paragraph indent is proportional to the initial cap. Be aware that very narrow letters (such as I) and very wide letters (M and W) will not conform to this proportion. The price of a truly professional document is editing to avoid these letters. Really.

The body text is 10/12 Bookman.

The illustration was created with PageMaker's drawing tools. The box around it uses a 2-point rule and a 10% shade. The headline is 10-point Helvetica Black Oblique, and the descriptive lines are 9/9 Helvetica Light with Helvetica Black numbers. The caption is 10/12 Bookman italic, centered.

The centered folio (10-point Helvetica Light) and the flush right continued line (9-point Helvetica Light Oblique) are aligned at their baselines 2 picas below the bottom margin.

THE GREAT OUTDOORS STORE

Month in Review

July 1989

BRISK START FOR CAMPING AND BACKPACKING EQUIPMENT

Fewer travelers abroad spurs sales in do it yourself activities.

Lorem ipsum dolor sit amet, consectetuer adipiscing elit, sed diam nonummy nibh euismod tincidunt ut laoreet dolore magna aliquam erat volutpat. Ut wisi enim ad minim veniam, quis nostrud exerci tation ullamcorper suscipit lobortis nisl ut aliquip ex ea commodo consequat. Duis autem vel eum iriure dolor in hendrerit in vulputate velit esse molestie consequat, vel illum dolore eu feugiat nulla facilisis at vero eros et accumsan et iusto odio dignissim qui blandit praesent luptatum zzril delenit augue duis dolore te feugait nulla facilisi.

Lorem ipsum dolor sit amet, consectetuer adipiscing elit, sed diam nonummy nibh euismod tincidunt ut laoreet dolore magna aliquam erat volutpat. Ut wisi enim ad minim veniam, quis nostrud exerci tation ullamcorper suscipit lobortis nisl ut aliquip ex ea commodo consequat. Duis autem vel eum iriure dolor in hendrerit in vulputate velit esse molestie consequat, vel illum dolore eu feugiat nulla facilisis at vero eros et accumsan et iusto odio dignissim qui blandit praesent luptatum zzril delenit augue duis dolore te feugait nulla facilisi. Nam liber tempor cum soluta nobis eleifend option congue nihil imperdiet doming id quod mazim placerat facer possim assum.

Lorem ipsum dolor sit amet, consectetuer adipiscing elit, sed diam nonummy nibh euismod tincidunt ut laoreet dolore magna aliquam erat volutpat. Utwisi enim ad minim veniam, quis nostrud wisi enim ad minim veniam, quis nostrud exerci tation ex ea commodo

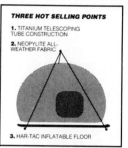

THREE HOT SELLING POINTS

1. TITANIUM TELESCOPING TUBE CONSTRUCTION

2. NEOPYLITE ALL-WEATHER FABRIC

3. HAR-TAC INFLATABLE FLOOR

The newly introduced two-person model X-2000-2 tent

WEATHERMAN PREDICTS ANOTHER HOT SUMMER. EXPECT BOOST IN WATER SPORTS GEAR

Jump in on the action in the new colorful inflatable water toys.

Consequat. Duis autem vel eum iriure dolor in hendrerit in vulputate velit esse molestie consequat, vel illum dolore eu feugiat nulla facilisis at vero eros et accumsan et iusto odio dignissim qui blandit praesent luptatum zzril delenit augue duis dolore te feugait nulla facilisi.

Lorem ipsum dolor sit amet, consectetuer adipiscing elit, sed diam nonummy nibh euismod tincidunt ut laoreet dolore magna aliquam erat volut. Lorem ipsum dolor sit amet, consectetuer adipiscing elit, sed diam nonummy nibh euismod tincidunt ut laoreet dolore magna

1

Continued on page 3

Blueprint detail:

	4 pt rule	4p
12p	Co. name 24 pt Machine	
	.5 pt rule	2p6
	Pub. Name 48 pt Bookman Ital.	5p6
3p	Date 12 pt Bookman	2p
	1p	
5p3	Head 14/13 Helv. Cond. Black — 4 pt rule	
	2p3	
	Tag line 11/13 Helv. L. Oblique	
	1p 2 pt rule	
	Top line of body text	

Figure 5.7

A PageMaker *page created by grid design. A grid represents the specifications creating the underlying structure of page composition.* [7:65]

Table 5.1
Traditional Publishing versus Desktop Publishing (DTP)

Step	Traditional Publishing	Desktop Publishing
Writing/Editing	Author generates the text by typewriter and prepares rough art. Editor makes blue pencil corrections. Author or typist retypes. If changes are extensive, several iterations take place.	Author generates text on Word Processor. Author/Editor key(s) changes or corrections on file diskette.
Redacting	Redactor prepares copy and art for printing.	Author uses DTP software to prepare pages. Text elements automatically take type specifications.
Typesetting	Typesetter rekeys manuscript, following redactor's instructions. Text is typeset into long galleys in proper column width. Author/Editor correct proof; return copy to printer for corrections.	Done above.
Illustrations	Artist takes author's roughs and creates charts, graphs or drawings, using pens, camera, etc.	Author using DTP graphics software creates charts, etc.
Page Design	Designer creates blank layouts on pasteup boards using blue pencil to mark location of text.	DTP software has templates to design pages and scale graphics to proper size.
Pasteup	Pasteup artist cuts galley proofs to page size. Artist leaves proper size "windows" to accommodate drawings and photos. Indicates scale factors to be used.	Done in previous step.
Negatives	Printing camera operator shoots negative image of each page. Blank spaces may appear on negatives of some artwork.	Complex art and pages may be prepared by specialist as in traditional printing.
Stripping	Pre-press specialist assembles and positions artwork on pages.	Complex artwork may be positioned as in traditional printing.
Printing	Negatives of pages are grouped into a "signature" of 4, 8, 16, or 32 pages and a large plate is made. Publication is printed on large sheets which are trimmed, folded, cut, collated and bound into the printed document.	Reproduced copy printed by DTP, but actual document could be printed in quantity by traditional printing.

document can become a printed horror with such defects as unsuitable margins, inappropriate type fonts, too many lines per page, confusing heads, and misplaced illustrations.

Desktop publishing technology has provided a capability to create publications from start to finish at one work station, and has created the opportunity

Table 5.2

Typical Cost Savings Using DTP in the Preparation of a Monthly Technical Bulletin

Task	Traditional	DTP	Savings per Issue	Savings/Yr
Writing @$30/hr	40 hr.	40 hr.	0	0
Typing @$12/hr	16	0	$192.00	$2304.00
Editing @$35/hr.	20	8	420.00	5040.00
Redacting @$25/hr.	8	0	200.00	2400.00
Typesetting @$60/hr.	8	0	480.00	5760.00
Illustrations @$25/hr.	12	8	100.00	1200.00
Pasteup @$25/hr.	12	8	100.00	1200.00
			$1,492.00	$17,904.00

to have one person handle the entire publication process. In the real world, it's unlikely that any one person has the necessary expertise to do a document's content research, its writing, keyboarding, editing, illustrating, redacting, page layout design, and laserographic printing.[2] Nonetheless, desktop publishing has entered our workplace; its technology is shaping both our communication practices and printed products. In its early existence, it is more prevalent in the production of newsletters, brochures, memos of substance, reports, instruction manuals, and printed matter that requires timely updates and revisions. As the technology matures, desktop publishing will underlie most printed communication.

You and Word Processing

There is a firm and growing relationship between computer technology and the technical writing function. A writer's primary task is to generate text for a *specific purpose*. The computer is a tool that lightens the technical writer's tasks. It helps

---

[2]In most DTP systems, at present, the writer makes typographic and format decisions within the menu of that system. The results of those decisions are displayed in their final format on the screen. There are some systems, however, that allow the writer to intervene with an embedded code to manipulate the final printed product for a more sophisticated layout and typographic design than the menu of the system will permit.

the writer to input text more efficiently, it aids in interposing material from different databases; it allows more facile editing and revision by providing such aids as electronic thesauri, spellers, and grammar and style analyzers; it provides an easy storage capability; and it enables proficient printing. Word processing, once the exclusive domain of the typist, has become a communication tool of every professional. Word processing is useful in any activity that requires the creation, editing, and production of large quantities of textual material. Companies and organizations that need to comply with paper-intensive reporting procedures and that need to generate much correspondence, reports, and documents have installed computerized word processing systems. These organizations have found that the technology lowers costs, improves the appearance and quality of their documents, and promotes efficiency.

The writer now has an opportunity to make format, illustration, and typographic decisions, which influence the readability and usefulness of the text for its intended purpose. In these early years, the relationship can become strained, because, first, the technology—still new—is in a state of flux, and, second, technical writers, in addition to learning communication skills, need to be trained to take advantage of the increasingly available enhanced computer technology.

Perhaps I have presented you a happy road to heaven, omitting the bumps and bruises on the way. I recognize that the transition from writing with a pencil to a computerized word processing system can be difficult. Old habits are not easy to break, especially in the formation of thoughts and ideas, in the description of things and processes, and in the development of insights. The composition process, with which this book is concerned, is often painful and frustrating. Intruding a new approach requiring many details of concentrated effort can and often does interfere with the composition process. Even some professional writers compulsively prefer a pad and pencil. They find it simple to erase or scratch out inappropriate words, phrases, thoughts, and ideas. Why work with a contraption that, when a wrong key is struck, erases hours or days of painful effort that can be difficult or impossible to recapture? I surmise that the ancient scribe who was handy with the stylus had similar fears when confronted with the newfangled technology of reed pen and papyrus. I also realize that not all students have access to computers. The technology is here, however, and it is used by all segments of our work society. At the very least, you should know about it and be ready to incorporate it in your professional activity.

What about creativity? Word processors, indeed, have proved in theory and practice that creativity need not be impeded. Word processors free writers from many of the physical and psychological burdens that are inherent in the composition process. As we struggle with the problems in our experience that require written communication, our minds generate ideas. We try to set them down rapidly. Often thoughts come in unrelated order; they need sorting out and restating. The word processing technology readily captures and stores our thoughts. The system allows us to move them about, to change them, and to place them where they more effectively express the idea we need to state. Word processors allow us to tinker creatively, to test ideas, to place them on hold, to recall them easily, and then to find just the right place to fit them in.

Some professional writers who own word processors don't like to "compose" with them. They prefer to handwrite their first drafts and to use the word processor in the revision stages, thus gaining time to free up their "creative" efforts. The word processor makes their editing more efficient and the preparation of the final manuscript a tolerable chore.

So, if you have the opportunity to use the word processing technology, try it! You may like it and it may make you more professional.

Chapter in Brief

Here we have examined the relationship between computer technology and the technical writing function. The chapter showed how that technology can advance and lighten technical writing tasks. For that purpose we emphasized word processing software features and desktop publishing.

Chapter Focal Points

- Technical writing tools
- The computer as a word processor
- Helpful word processing programs and features
- Desktop publishing

Discussion Questions and Assignments

1. Whether or not you have experience with computer technology, go to your college library and browse through the current issues of any or all of the following periodicals:

Byte	*MacWorld*
Computerworld	*On Computing*
Computing	*PCWorld*
Creative Computing	*Personal Computing*
Desktop Publishing	*Popular Computing*
Digital Research News	*Publish!*
Interface Age	*Seybold Report on Word Processing*
Info World	*Word Processing News*
MacUser	*Words*

Read the articles dealing with word processing technology. Discuss the articles in class.

2. Examine the advertisements. Many computer magazines provide reader service cards keyed to inquiry numbers in the ads. Circle the numbers for word processing technology and mail. Within a few weeks you will begin receiving product catalogs and brochures. Establish a file for future use.

3. Note the terminology used in the articles and ads. Develop a glossary that includes abbreviations and acronyms. It should include:

Alarm signal	Justification
Alphanumeric	Laser printer
Architecture	Macro
Bit	Microprocessor
Block Move	Monitor
Byte	Mouse
Central Processing Unit (CPU)	Output
Chip	Pagination
Computer friendly	Peripherals
Cursor	Pitch
Daisywheel	Pixel
Desktop publishing	Program
Disk, diskette	Proportional Spacing
Disk drive	Search
Document storage	Scroll
Electronic mail	Random Access Memory (RAM)
Electronic bulletin board	Read Only Memory (ROM)
Ergonomics	Thimble
Fax	Video Display Unit (VDU)
Format	Word
Function key	Wordwrap
Interface	Work station
Input	

Turn in your glossary to your instructor for comment.

4. Write two letters:

   a. The first to a manufacturer of a dot matrix printer. Indicate you are interested in obtaining a word processor and that you would like information about his product's capabilities, characteristics, and costs. Ask what advantages the dot matrix may have over letter quality printers.

   b. Write to a manufacturer of letter quality printers, including laser printers. Ask for similar information.

   The information you receive will be helpful in assignment 9.

5. Visit a computer store. (College computer centers are usually too busy to allow visitors and to offer explanations.) Computer and word processor sales sources are listed in the yellow pages of your telephone directory under Computers or Data Processing Equipment. They also advertise in the business section of your newspaper. Try to make your visit at the least busy time of day and week so sales personnel have more time available to talk. Look around. Tell the salesperson you are interested in a word processing system, but are at present "window shopping." The salesperson will be happy to show you around and explain the various systems. The salesperson will want to know your specific needs. You will not be lying if you say you are a student who has many letters and papers to write and that you are looking for a system to meet those needs, a system which you can augment later and which will be compatible with future computer and software requirements. Note that various systems combine display and keyboard, the disk drive and the central processing unit; others have

separate components. Note the types of printers and packages. Ask questions about advantages and disadvantages. Ask about the software and how the various functions are accomplished. Examine the manuals for both the hardware and software. The salesperson will demonstrate the systems or can arrange for you to attend a demonstration. Request descriptive brochures and price information. Tell the salesperson you want to study the material and will be in touch.

6. Before you leave, ask the salesperson for the name, address, and phone number of local computer users, clubs, or groups. Personal computer users informally band together—usually they are owners of the same system—to share experiences. User groups offer members a chance to discuss problems, share new programs and the latest information on new developments, as well as a chance to buy peripherals. User groups are a great source of information on merits and drawbacks of various systems. Arrange to attend a meeting of a users' group. You will find the users friendly and full of valuable information. Make notes of what is pertinent to your needs. (See assignment 9.)

7. Study the material you have received and the notes you have taken. You will find you now have a fair grasp of word processing technology. There will be specifics and factors requiring more information. You may need to do further research in computer magazines or question one of the friendly members of the users' group. Armed with your knowledge make a second visit to the computer store or to another store. Ask for another demonstration. Examine the characteristics of components of the different systems and their capabilities. Pay attention to and make notes on:

Screen displays, monitors—their size and the image they reflect;
Keyboards—note capabilities for control, for scroll, for editing, pagination, etc., for placement of keys;
Storage capabilities—types of disks, and search capabilities;
Printer—dot matrix, daisywheel, thimble, laser, size of page;
Graphic capabilities;
Desktop publishing capabilities;
Portability—advantages and disadvantages.

Ask the salesperson about:

Training classes
Prices, services, maintenance costs, software, warranties
Capabilities for correspondence, forms/records, long documents, graphics, storage

Take copious notes and ask for descriptive brochures.

8. Now that you have accumulated all this information, it is time to analyze it to make a "buying" decision. Your best procedure for organizing the factors in your comparison is to chart out or tabularize criteria to be considered. They are:

Architecture (technical specifications),
Word Processing features and functions,
Additional features and functions,
Ease of learning,

Vendor service and support, and

Costs (including maintenance).

Your assignment is to develop the chart. An appropriate format is shown in Table 5.3.

Table 5.3
Word Processing Systems Comparative Chart

	Product A	Product B	Product C
Architecture			
Stand Alone			
Shared Logic (computer interface)			

Terminal
      keyboard
      screen			

Printer
      Dot Matrix
      Letter quality (type)			

Storage Capacity Disk type			
WP Features and Functions			

Display
      Scrolling
        Horizontal
        Vertical			

Paper size			
Print image			
Multicolumn format			
Format control			

Automatic Features
      Word wrap
      Line spacing
      Pagination
      Margin Adjust
      Type variation
      Header/footer
      Revision marking
      Centering
      Justification
      Math
      Move/Delete/Add			

9. With your chart in hand you have the basic data for making a decision. Assume that you are:

an officer in a campus organization, or a member of a community action group, or an employee in a small business.

Table 5.3
(continued)

	Product A	Product B	Product C
Additional Features			
Graphics			
Computer Interface			
Desktop publishing			
Photocomposition Interface			
Ease of Learning			
Training Classes			
Training Package			
Users' Groups			
Built-in Aids			
Vendor Service & Support			
Consultant Support			
Vendor Hotline			
Problem Solving			
Vendor Newsletter			
Availability of Maintenance			
Costs			
Capital Expenses			
Freight Charges			
Installation			
Peripheral Expenses			
Supply			
Training			
System Maintenance			

Write a report recommending either the purchase of a particular word processing system and your justification for the recommendation, or a report delaying or rejecting such a purchase with the justification for that decision. The report may be in the form of a memorandum. It should include:

the background and reason (purpose) for your investigating the possible purchase of a WP system by your organization;

how you went about your investigation;

what information you discovered—types of systems, vendors, services, and costs;

your analysis of the information you accumulated; and finally,

your conclusions and recommendations with their justification.

Chapters 11, 12, 13, and 14 will be helpful to you in the writing.

Desktop Publishing Bibliography

1. Bove, Tony, Cheryl Rhodes, and Wes Thomas. *The Art of Desktop Publishing: Using Personal Computers to Publish It Yourself.* 2d Ed. Toronto: Bantam, 1987.

2. Burns, Diane, S. Vennit, and Rebecca Hansen. *The Electronic Publisher.* 2d Ed. New York: Bantam, 1988.

3. Cavuoto, James, and Jesse Berst. *Inside Xerox Ventura Publisher. A Guide to Professional Desktop Publishing.* Torrance, CA: Micro Publishing and Thousand Oaks, CA: New Riders, 1989.

4. Erickson, Fritz J., and John A. Vonk. *Easy PageMaker. A Guide to Learning PageMaker for the IBM PC.* New York: Macmillan, 1991.

5. Kleper, Michael. *The Illustrated Book of Desktop Publishing and Typesetting.* Blue Ridge Summit, PA: TAB Books, 1987.

6. Seybold, John, and Fritz Dressler. *Publishing from the Desktop.* Toronto: Bantam Computer Books, 1987.

7. Shushan, Ronnie, and Don Wright. *Desktop Publishing by Design.* Redmond, WA: Microsoft Press, 1989.

8. Ulick, Terry. *Personal Publishing With the Macintosh: Featuring PageMaker 1.2.* Indianapolis: Hayden Books, 1987.

9. The Wait Group. *Desktop Publishing Bible.* James Stockford (Editor). Indianapolis: Hayden Books, 1987.

10. Will-Harris, Daniel. *Desktop Publishing with Style: A Complete Guide to Design Techniques & New Technology for the IBM and Compatibles.* South Bend, IN: And Books, 1987.

PART THREE
Technical Writing Fundamentals

6

Basic Expository Techniques in Technical Writing—Definition

Chapter Objectives

Provide algorithms for and proficiency in the expository process of definition.

Chapter Focus

- Formal and informal definitions
- Synonyms and antonyms
- Algorithmic methods for developing definitions
  Formal and informal definitions
  Illustration
  Stipulation
  Operational definitions
  Expanded definitions

*B*ecause a primary aim of technical writing is to establish communication between at least two participants—a writer and a reader—effective writers are always aware of this dualism and direct their writing to a particular reader-audience. As we have seen in Chapter 4, ways of bringing understanding to the reader vary with the particular situation and reader background and interest. The fundamentals and elements of effective communication are the same in the technical field as in other fields. However, the special and basic techniques that characterize technical writing are *expository*. The present and following three chapters are concerned with exposition. The specialized style characteristics of technical writing that promote effective communication were dealt with in Chapter 4.

Exposition is explanation and/or instruction. It is the term for the kind of writing we use to explain facts or ideas. Exposition aims to bring about an understanding of something. The expository techniques of technical writing may be grouped into three major categories: definition, description, and analysis. In explaining facts and ideas of a specific or technical nature, the technical writer uses these methods more often in combination than separately. However, for purposes of illustration and understanding, we will examine these methods individually. You will probably notice that there is an overlapping of methods among the illustrations.

The Nature of Definitions

A **definition** is an explanation of an object or idea that distinguishes it from all other objects or ideas. Definition is basic to knowledge. Infants, for example, learn about their environment by gradually defining the objects in it. A mother points to herself and says, "Mama." After a time, the infant will begin, in imitation and understanding, to point and say, "Mama." Using this process, the parent will teach by definition through pointing to the infant's food, to common articles of usage, and to common experiences. Certain dangerous or delicate articles become "no-no's"; others are "nice-nice." Through this rudimentary instructional process, infants begin to perceive and define their environment.

Similarly, adults attach importance to the names of objects and ideas in their environment. We seem to understand what we are familiar with and what we can name. Anthropologists tell us that our belief in the potency of names was one of our earliest human traditions. As a new activity achieves maturity, it also attains a nomenclature. Systematically, every component is named. In order for technicians, engineers, or scientists to understand a mechanism, they usually must know the names of its various parts.

However, we should not oversimplify. Understanding an instrument, species, or concept is more than just being able to give it a name or list its parts. We must be able to relate it to similar classes of instruments, species, or concepts and distinguish it by those significant characteristics that make it different from other members of its class. Through the process of definition, we become more aware of the exact nature of things that exist or could exist in our world.

The derivation of the word *definition* will help us understand it. The Latin word *definire* is its origin; *de* means from, and *finire* means to set a limit to or to set a boundary about. To define, then, is to delimit the area of meaning of a word, term, or concept. Definitions are verbal maps that indicate or explain what is included within a term and what is excluded. The latter is often communicated by implication.

Methods of Developing Definitions

The Formal Definition

A definition is **formal** when it has a prescribed form consisting of three parts. The first part is the *term*—the word, object, idea, or concept to be defined. The second is the *genus*—the class, group, or category in which the term belongs. The third part is the *differentia*—the characteristics that distinguish the term from other members of the genus. The differentia excludes all other members of the genus except the term being defined.

1. A *rectangle* is a four-sided figure having all its angles right angles and, thus, its opposite sides equal and parallel.

   "Rectangle" is the term; *"four-sided figure" is the* genus; *"having all its angles right angles and, thus, its opposite sides equal and parallel" is the* differentia.

2. An *emetic* is a substance or drug that induces vomiting, either by direct action on the stomach or indirectly by action on the vomiting center in the brain.

   "Emetic" is the term; *"substance" is the* genus; *"that induces vomiting either by direct action on the stomach or indirectly by action on the vomiting center in the brain" is the* differentia.

3. An *electric cell* is a receptacle containing electrodes and an electrolyte, used either for generating electricity by chemical reactions or for decomposing compounds by electrolysis.

   "Electric cell" is the term; *"receptacle" is the* genus; *"containing electrodes and an electrolyte used for generating electricity by chemical reactions or for decomposing compounds by electrolysis" is the* differentia.

4. A *mouse* is an electronic device that looks like a small box and plugs into the computer operating system to allow the user to interact with the information on the screen without using the keyboard.

   "Mouse" is the term; *"electronic device" is the* genus; *"that looks like a small box and plugs into the computer operating system to allow the user to interact with the information on the screen without using the keyboard" is the* differentia.

The formal definition is not an academic exercise. It is closely related to the scientific process of classification. It utilizes a logical method of analysis to place the subject to be defined into a general class (genus) and then differentiates it from all members of its class. The term to be defined should not be repeated in the genus or differentia. For example, to define chess as a game played on a chessboard tells the reader very little about the game. To be satisfactory, the genus must identify the term precisely and completely. If the genus is too general it complicates the identification because the differentia must be expanded. The differentia must be broad enough to include everything the term covers and specific enough to exclude everything that the term does not cover. It fences in the meaning of the term by listing qualities, giving quantities, making comparisons and/or itemizing elements.

Definitions should be stated in simpler or more familiar language than the term itself; otherwise the purpose of the definition is lost. A classic example of this fault is Samuel Johnson's definition of a cough: "A convulsion of the lungs, vellicated by some sharp serosity." Also, definitions should not be phrased in obscure or ambiguous language. Definitions should be impersonal, objective, and should *describe*, not praise or condemn, the matter being defined. Johnson's definition of oats as "a grain eaten by horses and in Scotland by the inhabitants" reflects a personal bias. These stated principles apply, of course, not only to the formal but also to all methods of definition that follow.

The Informal Definition

The strict format of the formal definition is not appropriate or efficient in all situations. The **informal definition** uses the shortest and simplest method for identifying or explaining the matter to be defined. The method may involve substituting a short, more familiar word or phrase for the unfamiliar term or using special expository devices, such as antonyms or illustrations, that give the reader a quick recognition of the term. Informal definitions are used in less formal writing, that for the general public and not for the technical specialist.

Defining by Using Synonyms and Antonyms

If you were asked, "What is a microbe?" you probably would answer, "a germ." *Germ* means the same as *microbe*. The two words are used interchangeably. They are **synonyms**. *Double* means twice: *paleography* means ancient writing; *helix* means spiral. A known word is substituted for an unknown word in definition by *synonym*.

Closely associated with this technique is the use of an **antonym** for aiding definition. *Down* is the opposite of up. *Abstruse* means not obvious; *indigenous* means the opposite of foreign; *empirical* means not theoretical.

Both synonyms and antonyms are simplified forms of definitions that frequently set rough, approximate boundaries of meaning rather than complete and exact limits. They are techniques more appropriate for impromptu and informal situations than for more formal requirements. Children or lay persons do not usually require the complete and specialized details for their understanding. For example, if a child asks, "What is penicillin?" her informational needs would be satisfied with "a substance that kills germs." She might actually be confused by technically more accurate information such as:

> Penicillin is an antibacterial substance produced by microorganisms of the Penicillium chrysogenum group, principally penicillium notatum NRRL832 for deep or submerged fermentation and NRRL 1249, B21 for surface culture. Penicillin is antibacterial toward a large number of gram positive and some gram negative bacteria and is used in the treatment of a variety of infections. [15:1206]

Consequently, definition by synonyms and antonyms is used in the technical article aimed at the general reader.

Defining by Illustration

Offering an illustration is a primary means for defining an unfamiliar thing or concept. We have seen at the beginning of this chapter that adults use pointing as an effective way of explaining things to children: *"This* is an oak leaf." *"This* is a Phillips screwdriver." *"This* is a ten-penny nail." Definition by pointing or illustrating makes the task easier for both adult and child.

Dictionaries, too, resort to definition by illustration. For instance, in defining *emu*, the dictionary says it is "a large, nonflying Australian bird, like the ostrich, but smaller." Included is a drawing of the bird that gives the reader details of the bird's appearance. The dictionary defines the *trapezium* as a "plain figure with four sides, no two of which are parallel." Again, an illustration is included and shown in Figure 6.1.

Figure 6.1
A Trapezium

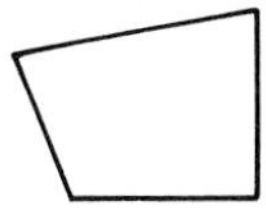

Defining by Stipulation

In Chapter 6 of Lewis Carroll's *Through the Looking Glass*, Humpty Dumpty announces:

> "There's glory for you!"
> "I don't know what you mean by 'glory'," Alice said.
> Humpty Dumpty smiled contemptuously. "Of course you don't—till I tell you. I mean 'there's a nice knockdown argument for you.' "
> "But 'glory' doesn't mean 'a nice knockdown argument,' " Alice objected.
> "When I use a word," Humpty said in rather a scornful tone, "it means just what I choose it to mean—neither more or less."
> "The question is," said Alice, "whether you *can* make words mean so many different things."

"The question is," said Humpty Dumpty, "which is to be master—that's all." [3:214]

Humpty Dumpty is using definition by *stipulation*—that is, he attributes to a word or term a specific meaning he wants it to have. Notice how the writer of the National Science Foundation Report, *Science, Technology and Innovation* attributes special meaning to the term *event* and stipulates particular meanings to types of events:

In the first paragraph, the writer specifies what he intends the term event *to mean.*

Throughout this report the term "event" is used in a special and technical sense. The innovative process comprises myriad occurrences, some of which happen sequentially, and some concurrently at different places. From these occurrences, one can identify some that appear to encapsulate the progress of the innovation. These special occurrences are the "events" in the technical sense just referred to. Their selection reflects the best judgment of the investigators, and is necessarily somewhat arbitrary.

In the second paragraph, he recognizes that there are types of related occurrences in the operation of an event that need explaining. In the third and fourth paragraphs, he differentiates between a significant event and a decisive event.

To clarify further how the study proceeded, other terms associated with the "events" are defined below.

A significant event is an occurrence judged to encapsulate an important activity in the history of an innovation or its further improvement, as reported in publications, presentations, or references to research. Generally these events follow one another in historical sequence, along channels of developing knowledge. Significant events include other classes of events.

A decisive event is an especially important significant event that provides a major and essential impetus to the innovation. It often occurs at the convergence of several streams of activity. In judging an event to be decisive, one should be convinced that, without it, the innovation would not have occurred or would have been seriously delayed.

In the fifth and sixth paragraphs, the writer explains what a nontechnical event is. The definitions of the terms significant, decisive, *and* nontechnical *are needed to explain what the writer stipulates* event *to mean. With this background, the reader can follow with understanding the occurrences of innovation.*

Since science and technology lie at the focus of the investigation, the great majority of significant events are technical in nature. However, a few events that did not involve science and technology were important enough to be included among the significant events. These events are termed nontechnical.

A nontechnical event is a social or political occurrence outside the fields of science and technology. For example, a war or natural disaster would be a nontechnical event, in contrast with a management venture decision within a technical organization, which would be classed as a technical event. [13:2–3]

The Operational Definition

In the scientific and technical fields, the **operational definition** has become a very useful technique. It offers meaning of a term, not by classifying it into a genus and then isolating its distinguishing characteristics from other members of its class, but by describing the activities, procedures, or operations within which the term operates.

Rapoport, in an article called "What is Semantics?" explains this approach:

An operational definition tells *what to do* to experience the thing defined. Asked to define the coefficient of friction, the physicist says something like this: "If a block of some material is dragged horizontally over a surface, the necessary force to drag it will, within limits, be proportional to the weight of the block. Thus the ratio of the dragging force to the weight is a constant quantity. This quantity is the coefficient of friction between the two surfaces." The physicist defines the term by telling how to proceed and what to observe. The operational definition of a particular dish, for example, is a recipe. [11:128–29]

The following example is both a formal and operational definition. The genus, *voiceless labiodental fricative,* requires a qualifying and detailed differentia that determines how the term being defined, [f], operates (is formed):

> The sound [f] is a voiceless, labiodental fricative continuant, formed by placing the lower lip lightly against the upper teeth, closing the velum, and forcing breath out through the spaces between the teeth, or between the upper teeth and the lower lip. We call [f] labiodental because of the contact between the lower lip and the teeth; a fricative, because a characteristic of the sound is the audible friction of the breath being forced past the teeth.
>
> [f] occurs in the beginning, middle, and end of words, and is spelled *f, ff, ph,* or *gh,* as in *fife, offer, photograph* and *cough.* Occasionally in substandard speech it may be replaced by the first consonant of *thin,* most commonly *trough,* but sometimes in other words as well. Ordinarily [f] causes little difficulty to the native speaker of English. In some foreign languages, notably Japanese [f] is formed by the lips alone without the aid of the upper teeth. This weaker [f] may sound slightly queer in English, or the listener may fail entirely to hear it. [14:38]

The Expanded Definition—Amplification

Sometimes a synonym, antonym, or formal definition adequately explains the meaning of a thing or idea. At other times, especially if an idea or a complex object is being explained, the definition must be developed by the use of details, examples, comparisons, and other explanatory devices. The **expanded definition** is frequently an amplification of the formal definition. How it is developed depends on the nature of the concept and the writer's approach to it. Most expanded definitions follow the structure of the paragraph. They will begin with a topic sentence, structured as a formal definition or as a statement of the topic for discussion. Many expanded definitions are not longer than a paragraph, but some may require several hundred or several thousand words. In the example below, a formal definition begins the explanation of *innovation.* Not a simple process, the term's formal definition requires amplification through the use of several expository devices:

> Innovation is a term that describes certain activities by which our society improves its productivity, standard of living, and economic status. Basic to the progress of innovation are the tools, discoveries, and techniques of science and technology. . . .

The first paragraph is a formal definition of innovation.

The second paragraph begins the amplification of the definition by specifying the requirements of an innovation.

The third paragraph differentiates scientific discovery from innovation.

The fourth paragraph provides elaboration on the role time plays, specifically the preconception period and postinnovative periods.

The fifth paragraph traces the development of an innovation from conception to postinnovative periods, by using the heart pacemaker as an example.

When inventions or other new scientific or technological ideas are conceived, they do not immediately enter the stream of commercial or industrial application. In fact, many never get beyond the stage of conception, while others are abandoned during the period of development. But some go through a full course of gestation, and finally emerge as new and useful commercial products, processes, or techniques. Such advances are called innovations.

Innovation should be distinguished from scientific discovery, although relevant discoveries may be incorporated into the innovation. Innovation should also be differentiated from invention, although an invention frequently provides the initial concept leading to the innovation. Nor is innovation merely a marginal improvement to an existing product or process. Rather it is a complex series of activities, beginning at "first conception," when the original idea is conceived; proceeding through a succession of interwoven steps of research, development, engineering, design, market analysis, management decision making, etc.; and ending at "first realization,"[1] when an industrially successful "product," which may actually be a thing, a technique, or a process, is accepted in the marketplace. The term "innovation" also describes the process itself, and when so used, it is synonymous with the phrase "innovative process."

As so defined, innovation extends over a bounded interval of time (the innovative period) from first conception to first realization. Implicitly, therefore, we have defined two other periods of time—the "preconception" period, which precedes the time of first conception, and the "postinnovative" period, which follows the time of first realization. During the preconception period science and technology develop the foundation for the innovation. In the postinnovative period improvements of the innovation are made and marketed, and the technology diffuses into other applications. The postinnovative period is sometimes called the period of "technological diffusion," although the two terms are not synonymous, because diffusion is only one of the activities of this period.

Let's look at an example—one of the innovations studied—the Heart Pacemaker. For this innovation, the preconception period saw advances in electricity, especially electrochemistry, in cardiac physiology, and in surgery and intracardiactherapy techniques. But first conception did not occur until 1928 when Dr. Albert S. Hyman conceived the idea of periodic electrical stimulation of the heart by means of an artificial device, an idea for which he filed a patent application in 1930. The innovative process for the pacemaker proceeded between 1928 and 1960; during this period, batteries were upgraded, the transistor was invented, materials technology enjoyed rapid development, and surgical techniques were advanced. The year 1960 marked the first implantation of a pacemaker in a human patient, and marketing of the device began soon thereafter; 1960 is therefore the date of the first realization for this innovation. Since that date—in the postinnovative period—heart pacemakers have become more sophisticated and work on further improvements continues.

---

[1]Also termed *culmination* in the historical accounts.

Our study concentrated on the innovative period. But to understand the background from which innovation evolves, and to appreciate the impact of innovation on society, we consider also the preconception and postinnovative periods.

The sixth paragraph is a structural paragraph, telling the reader that an innovation has been traced from preconception to postinnovative periods. Such a time interval, the writer states, is part of the innovative process.

The preconception period presents problems, because it is historically open-ended. At what point in history should one start? Electrical stimulation of muscular activity is important to the development of the heart pacemaker. Should one then go back to Galvani's experiments with frog's legs? Since some time horizon has to be chosen, we selected 1900. The continuity of history makes it difficult to close one's eyes completely to pre-1900 events. So some especially significant scientific and technical events that occurred before 1900 are included in the historical record. [13:1–2]

The seventh paragraph stimulates the reader to consider the fact that all progress is based on cumulative experience. Often, the spark of preconception can be traced to much earlier innovations and discoveries. The insight in this paragraph is a good way to complete the writer's definition of innovation.

The Expanded Definition—Example

Closely related to definition by illustration is definition by example. In the scientific and technical fields, examples are necessary to bring understanding to the uninformed. They are especially useful in explaining abstract and conceptual matters. In his book *Electrons, Waves, and Messages,* Pierce uses some everyday experiences to exemplify and quickly define some very abstract processes:

> If you touch a steam radiator, your hand is heated by *conduction;* if you hold your hand over the hot air arising from the hot air register, your hand is heated by *convection;* if you hold your hand in the beam of heat from a reflector-type electric heater, your hand is heated by *irradiation;* it is warmed despite the fact that the air surrounding it is cool. [10:181–82]

Even though conduction, convection, and irradiation are complicated processes, they enter into our everyday experience and can be identified for us by instances within our experience. However, to define another type of abstract concept, in the same book, Pierce turns to another exemplifying illustration to help define communication theory, a very difficult concept:

> In communication theory, information can best be explained as choice or uncertainty. To understand this, let us consider a very simple case of communication. For instance, if you want to send a birthday greeting by telegraph, you may be offered a choice of sending one of a number of rather flowery messages, perhaps one from a list of 16. Thus, in this form of communication, the sender has a certain definite limited choice as to what message he will send. If you receive such a birthday telegram, there is some uncertainty as to what it will say, but not much. If it was chosen from a list of 16 standard messages, it must be one among the 16. The received message must enable the recipient to decide which among the 16 was chosen by the sender. [10:245]

Although this is not the whole story of communication theory, the example gives the reader a generalized view of the essential characteristics.

The Expanded Definition—Analogy

Analogy is a comparison of two things alike in certain respects. Analogies are particularly useful to explain an unfamiliar object, idea, or process by comparing it to other more familiar objects, ideas, or processes. The process of catalysis is imaginatively and sufficiently explained for the general reader by its comparison to a folk-dance leader and a pole-vaulter.

> "All right, folks," calls the folk-dance leader, "don't be bashful! Switch your partners, form your squares, watch my feet, and follow me!"
>
> Round the room whirls the nimble dance instructor, mingling shy singles, regrouping wallflower couples, breaking the ice, putting hand in hand, calling the turns. By the end of the evening, no one is a stranger, and many go home with partners first met on the dance floor. The dynamic leader has catalyzed new bonds of friendship—and more—without getting personally involved.
>
> That is the very essence of a catalyst, a substance that trail-blazes new and faster pathways of molecular interaction, making and breaking chemical bonds more easily than if nature took its course, and itself emerging from the reaction unchanged. A single molecule of catalyst may transform several million molecules per minute of raw materials.
>
> How? A catalyst mobilizes energy at the scene of the action. Compare a high-jumper, who can clear little more than his own height, with a pole-vaulter, who can soar more than twice as high, magnifying his force via a bamboo pole—the catalyst—which remains as it was.
>
> Nature moves in mysterious ways; among the most mysterious is catalysis. From capturing the oxygen we breathe to fermenting the beer and wine we drink to cracking the gasoline we burn in our cars, nearly all the processes of life and of industrial chemistry alike are helped to happen by catalysts. [8:78]

The writer must use analogies with care because the superficial or unknowledgeable reader will not discern likeness in objects, ideas, or processes that are outwardly dissimilar.

The Expanded Definition—Explication

Explication is explanation or interpretation of the terms used in a definition. This is illustrated in the following definition:

> **Sediment** is any loose or fragmented material. Loose sand, shells, leaves, and mud are all examples of sediment. All sediment has a **source** (place of origin), where it may have been produced by the life cycles of plants or animals (e.g., shells, leaves, logs), or by **chemical weathering** of rocks (chemical disintegration and decomposition), or by **physical weathering** (mechanical breakdown) of rocks.

All sediment has a **provenance,** which is the particular place or region from which its components were derived. The particular rocks or other materials that are weathered are called **parent materials.** Sediment is also commonly **eroded** (physically removed) and **transported** (carried to other locations) from its provenance by water, ice, or wind.

The process of transportation also sorts various densities and sizes of sediments from one another, a natural phenomenon known as **sorting.** Poorly sorted sediments are composed of many different sizes and/or densities of grains mixed together. Well-sorted sediments, however, are composed of grains that are of similar size and/or density.

Well-sorted sediments are usually composed of *well-rounded* grains, because the grains have been abraded (worn down) and rounded during transportation. Conversely, poorly sorted sediments are usually *angular* (have sharp corners), because of the lack of abrasion during transportation. *Roundness* is the sharpness of corners on a grain of sediment, viewed in profile (side view). Terms used by sedimentologists to describe roundness are well-rounded, subrounded, subangular, and angular. [1:33–34]

The Expanded Definition—Derivation

Some terms may be explained effectively by discussing the origin of the term. I have used this approach in this text at several points. My discussion of definition as an expository technique began with the etymological origin of the word. Here is another example:

The word *biology* is derived from two Greek words, *bios,* "life" and *logos,* "word" or "discourse," and so "science." In its narrower sense, biology may be defined as the science of life, that is, the science which treats of the theories concerning the nature and origin of life; in a broader sense, biology is the sum of zoology (Greek *zōon,* "animal," + *logos,* "science"), or the science of animals, and botany (Greek *botonē,* "plant"), or the science of plants. It is in this broad sense that the word is used in this volume. [12:1]

Sentence one gives the derivation of the word. Sentence two gives both a restricted and wider meaning, providing details of the wider definition and its Greek origins. Sentence three is a transition between the definition and what will be discussed in the rest of the text.

The Expanded Definition—History

Frequently, knowing the history of a concept helps one understand the concept more fully. Consider the following definition of *eutrophication*:

What do we mean by eutrophication? The word is due to C.A. Weber, who, in 1907, described the nutrient conditions determining the flora of German peat bogs as "nährstoffreichere (eutrophe) dan mittelreiche (mesotrophe) und zuletzt nährstoffearme (oligotrophe)," a sequence that expressed the changes as a bog built up and was raised about the surrounding terrain, so being more and more easily leached of its nutrients.

At first the bog vegetation was what Weber called *eutraphent,* requiring high concentrations of essential elements in the soil solution; at the end of the process an *oligotraphent* flora covered the bog, composed of species tolerating very low nutrient concentrations. In Weber's case the process that took place in going from a less elevated, less leached to a more elevated, more leached bog would not be called oligotrophication, if anyone had occasion to use the word.

In 1919 Einar Naumann, who knew of Weber's work, employed the terms in a discussion of the phytoplankton of Swedish lakes. He originally used them to describe water types, so that springs, streams, lakes, or bogs would contain oligotrophic, mesotrophic, or eutrophic water, according to the concentration of phosphorus, combined nitrogen, and calcium present. Originally no estimates, however, could be given as to what these concentrations were. Naumann throughout his works gives the impression that he liked to draw limnological conclusions, expressible in schematic terms, merely from looking at lakes. Weber's original words *oligotrophic* and *eutrophic* were, in fact, now redefined for limnologists in terms of the appearance of a lake in summer. A lake containing eutrophic water is *"sehr stark getrübt oder sogar vollständig verfärbt"* as the result of a very dense population of algae. The unproductive oligotrophic lake did not support such a population and in consequence remained much less turbid and either blue, if it were unstained by peat, or brown in peaty montane areas. Even today, transparency and color are the simplest indicators of the nutrient condition of a lake, though of course they must be used with great discretion if no other information is available.

Much of the work that was being done by other limnologists in the first third of the twentieth century contributed to the development and ultimate confusion of Naumann's typological scheme.

The great difference between the lakes of the lands bordering the Baltic and those of more mountainous districts notably the Swiss Alps, had been known for some time. Wesenberg-Lund, perhaps the greatest of limnological naturalists, in the general part of his work on the biology of the fresh water plankton of Denmark, had discussed the biological characteristics of the Baltic lakes, of which the phytoplankton consisted largely of diatoms and blue-green algae.

At the same time the English algologists W. and G.S. West were studying the lakes of northern and western Britain, finding a diversified desmid flora in the phytoplankton. Teiling observed comparable assemblages in the mountains of Scandinavia and spoke of a Caledonian type of phytoplankton in contrast to the Baltic type. It was reasonable to regard these two types of plankton as regional expressions of the eutrophy of the waters of the Baltic Lakes and the oligotrophy of those of the mountainous parts of Europe. [5:269]

Thus, tracing the historical development of the concept provides the reader with a fuller understanding of the process of eutrophication.

The Expanded Definition—Analysis

A reader can better grasp a complex subject if it is broken up into component parts. He can then digest one portion before proceeding to the next. This approach is effective in defining various steps in a process, especially if succeeding steps increase in complexity. This definition by analysis is used in the following explanation of the nature of language:

What makes human language unique? According to Erwin-Tripp (1964), three features characterize human language: (1) the combination and recombination of a limited number of elements, (2) the creation of arbitrary meanings for combinations in a social group, and (3) reference to distant objects and events and to intangible concepts. Thus human language is both flexible and expandable [Figure 6.2]. The first characteristic, combination and recombination, makes it possible to create sounds, words, and grammatical rules; so we can invent words to describe, for instance, scientific developments. Because we can refer to distant objects and intangible concepts, we can discuss anything from international law and territorial limitations to trade rights and human relations problems. Human language is so flexible that we can use the same language system to teach the three R's, history, science, literature, and social science.

All languages have two aspects: *structure,* or the basic units, words, sounds, and rules and *meaning,* or the message that the language communicates. There are three basic structural units of language. **Phonemes** are the fundamentals. They include the vowels and consonants. A phoneme is the smallest unit of sound which can distinguish one word from another. A **morpheme** is the smallest unit that can convey meaning. It is usually made up of several phonemes. **Graphemes** are letter-like written symbols—the vertical, horizontal, or oblique strokes—which distinguish the phonemes. While it is agreed that children's language development proceeds from the simple to the complex, there are several different theories of language development. [7:50–51]

To provide an understanding of the uniqueness of human language, the authors call attention to three essential analytical characteristics. This is followed by an amplification (explanation) that tells why each characteristic is essential.

Further analysis follows: Language must provide two other qualities—structure and meaning. Structure is partitioned into phonemes, morphemes, and graphemes. To provide meaning, the second essential quality of language, the authors use a process diagram to show how meaning is conveyed by the various media of human communication.

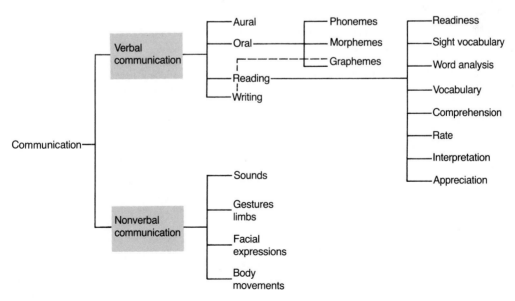

Figure 6.2
The Nature of Human Communication

The Expanded Definition—Comparison and Contrast

What is difficult to understand can be effectively explained by comparing and contrasting it with something less difficult or more familiar. Bronowski develops an effective definition of scientific creativity using this method.

In paragraph 1, Bronowski states his thesis that though three terms are often associated with scientific progress—discovery, invention, and creativity—they are different processes and are not on an equal level. By comparison and contrast he will show why.

The most remarkable discovery made by scientists is science itself. The discovery must be compared in importance with the invention of cave-painting and of writing. Like these earlier human creations, science is an attempt to control our surroundings by entering them and understanding them from inside. And like them, science has surely made a critical step in human development which cannot be reversed. We cannot conceive a future society without science.

In paragraph 2, he states Columbus discovered the West Indies and Alexander Graham Bell invented the telephone. These achievements, he says, cannot be considered as creations because the West Indies and the technology for the telephone were already there.

I have used three words to describe these far-reaching changes: discovery, invention, and creation. There are contexts in which one of these words is more appropriate than the others. Christopher Columbus discovered the West Indies, and Alexander Graham Bell invented the telephone. We do not call their achievements creations because they are not personal enough. The West Indies were there all the time; as for the telephone, we feel that Bell's ingenious thought was somehow not fundamental. The groundwork was there, and if not Bell then someone else would have stumbled on the telephone as casually as on the West Indies.

In paragraph 3, Bronowski calls Shakespeare's Othello a genuine creation because Shakespeare created the play from his own insights and genius.

By contrast, we feel that *Othello* is genuinely a creation. This is not because *Othello* came out of a clear sky, it did not. There were Elizabethan dramatists before Shakespeare, and without them he could not have written as he did. Yet within their tradition, *Othello* remains profoundly personal; and though every element in the play has been a theme of other poets, we know that the amalgam of these elements is Shakespeare's; we feel the presence of his single mind. The Elizabethan drama would have gone on without Shakespeare, but no one else would have written *Othello*.

In paragraphs 4, 5, 6, and 7, Bronowski calls attention to other discoveries and inventions. But can a scientific theory ever reach the fullness of a creative work such as Othello? Most nonscientists contend, no. Science engages only part of the mind, they say; creation engages the whole mind. The personal qualities of the artist are not part of the scientific endeavor. But Bronowski does not accept this contention. There is mediocrity in the humanities as well, and Bronowski provides examples.

There are discoveries in science like Columbus's, of something which was always there: the discovery of sex in plants, for example. There are tidy inventions like Bell's, which combine a set of known principles: the use of a beam of electrons as a microscope, for example. In this article I ask the question: Is there anything more? Does a scientific theory, however deep, ever reach the roundness, the expression of a whole personality that we get from *Othello*?

A fact is discovered, a theory is invented; is any theory ever deep enough for it to be truly called a creation? Most nonscientists would answer: No! Science, they would say, engages only part of the mind—the rational intellect—but creation must engage the whole mind. Science demands none of that ground swell of emotion, none of that rich bottom of personality, which fills out the work of art.

This picture by the nonscientist of how a scientist works is of course mistaken. A gifted man cannot handle bacteria or equations without taking fire from what he does and having his emotions engaged. It may happen that his emotions are immature, but then so are the intellects of many poets. When Ella Wheeler Wilcox died, having published poems from the age of seven, *The Times* of London wrote that she was "the most popular poet of either sex and of any age, read by thousands who never open Shakespeare." A scientist who is emotionally immature is like a poet who is intellectually backward: both produce work which appeals to others like them, but which is second-rate.

I am not discussing the second-rate, and neither am I discussing all that useful but commonplace work which fills most of our lives, whether we are chemists or architects. There are in my laboratory of the British National Coal Board about 200 industrial scientists—pleasant, intelligent, sprightly people who thoroughly earn their pay. It is ridiculous to ask whether they are creators who produce works that could be compared with *Othello*. They are men with the same ambitions as other university graduates, and their work is like the work of a college department of Greek or of English. When the Greek departments produce a Sophocles, or the English departments produce a Shakespeare, then I shall begin to look in my laboratory for a Newton.

Literature ranges from Shakespeare to Ella Wheeler Wilcox, and science ranges from relativity to market research. A comparison must be of the best with the best. We must look for what is created in the deep scientific theories: in Copernicus and Darwin, in Thomas Young's theory of light and in William Rowan Hamilton's equations, in the pioneering concepts of Freud, of Bohr, and of Pavlov. [2:59–60]

In paragraph 8, Bronowski makes the point of his thesis. In science, as in the humanities, there are works of mediocrity of discovery and of invention as well as of creativity.

The scientific creativity of a Copernicus, Darwin, and Freud compares with the creativity of a Shakespeare.

The Expanded Definition—Distinction

The essence of definition is differentiation. An extended definition may be composed entirely of the distinguishing characteristics of various concepts or things, as is the case with the following consideration of what constitutes published information and what constitutes unpublished information:

The traditional system of formal scientific publication constitutes the most highly-organized channel of scientific communication, having evolved over the years into a complex integrated series of operations intended to insure that a given piece of information once deposited in the system, is permanently available to all who need and look for it. The sequence for new information begins with primary publication of the original paper; *secondary* publication as an abstract follows; and appropriate reference to the paper made in standard indexes to the literature, represent what may be called *tertiary* publication.

In paragraph 1, three types of published information are identified.

This orderly sequence is the formal channel for disseminating new knowledge, and is the traditional mainstay for all scientific communication. However, there have always existed other less formal channels through which information is disseminated and exchanged among scientists, practitioners, teachers and students in the biomedical sciences. In the past few decades, these alternative and supplementary channels have multiplied in numbers and are handling increasing volumes of information.

Biomedical information transmitted by any method other than the formal system of scientific publication has been termed "unpublished information." This terminology assumes a restricted and special definition of publication that differs from common usage. Most scientists regard any information recorded in the form of the written word and disseminated more or less widely as having been published.

In paragraphs 2 and 3, alternatives for formally published information are indicated.

In paragraph 4, what is accepted as published information by librarians and scientific journals is contrasted with what is accepted by the discipline of law.

By the lawyer, publication is construed more broadly, and does not necessarily require that information be printed. Librarians and documentalists, in contrast, use the term in a much more restricted sense. To them, information is "published" only when it has appeared in those scientific periodicals covered by the standard bibliographic services, or in books obtainable by the usual library purchasing or borrowing practices. Several of the more conservative scientific societies and journals also honor the strict definition and allow authors to cite as references only information to which these criteria apply.

In paragraph 5, the writer identifies several synonyms of unpublished information—ephemeral, exotic, and informal. Unpublished information, though perhaps produced by some reproduction process, is either too short-lived or inaccessible. Formal or published information by contrast has more enduring worth and is readily available through normal publishing channels. That is the distinction which documentalists make between published and unpublished information.

Printed information not falling in this category is often referred to as "ephemeral" or "exotic" and is considered "unpublished." Since this distinction is in essence, an operational definition of publication, what is included in the category of unpublished information changes as practices of libraries and bibliographic services change. Defining "published" information in this way, i.e., in terms of ready availability and ease of finding emphasizes an important point—if information is to be useful for more than a very short time, it must be incorporated into some system capable of finding and supplying it to a user when it is needed. Using this restrictive definition, however, causes confusion in many minds when information is classified as "unpublished," although it has been printed and distributed widely. For want of better terminology, the adjective "informal" will therefore be used here to distinguish all those media methods for transmitting biomedical information other than the conventional scientific periodicals and monographs, or other bound volumes, issued by scientific societies and commercial publishers. These "informal" channels will be contrasted with the "formal" channels—the traditional system of scientific publication described above. This distinction roughly parallels the documentalist's delineation of "published" from "unpublished" information. [9:7–8]

The Expanded Definition—Elimination

In some instances, definitions may be effectively developed by the process of elimination—demonstrating what something is by the enumeration of what it is not. Huxley does just this in his essay "War as a Biological Phenomenon":

> War is not a general law of life, but an exceedingly rare biological phenomenon. War is not the same thing as conflict or bloodshed. It means something quite definite: an organized physical conflict between groups of one and the same species. Individual disputes between members of the same species are not war, even if they involve bloodshed and death. Two stags fighting for a harem of hinds, or a man murdering another man, or a dozen dogs fighting over a bone, are not engaged in war. Competition between two different species, even if it involves physical conflict is not war. When the brown rat was accidentally brought to Europe and proceeded to oust the black rat from most of its haunts, that was not war between the two species of rat, nor is it war in any but a purely metaphorical sense when we speak of making war on the malaria mosquito or the boll weevil. Still less is it war when one species preys upon another, even when the preying is done by an organized group. A pack of wolves attacking a flock of sheep or deer, or a peregrine killing a duck, is not war. Much of nature, as Tennyson correctly

said, is "red in tooth and claw"; but this only means what it says, that there is a great deal of killing in the animal world, not that war is the rule of life.

. . . So much for war as a biological phenomenon. The facts speak for themselves. War far from being a universal law of nature, or even a common occurrence, is a very rare exception among living creatures, and where it occurs, it is either associated with another phenomenon, almost equally rare, the amassing of property, or with territorial rights. [6:26–27]

Definition by a Combination of Methods

The more complicated a term or concept is, the more involved the explanation must be. No one system of definition may be adequate, and a combination of several techniques may be called for. We have already seen this in some of the previous examples. In the following definition of *waves*, negative details, history, example, listing of characteristics, explication, illustration, reiteration, operational analysis, comparison, and description are interwoven for a very graphic definition:

What are waves? They are not the earth, or water, or air; steel, or catgut, or quartz; yet they travel in these substances. Nineteenth-century physicists felt constrained to fill the vacuum of space with an ether to transmit electromagnetic waves, yet so arbitrary a substance seems more a placebo to quiet the disturbed mind than a valid explanation of a physical phenomenon. When we come to the waves of quantum mechanics, the physicists do not even offer us a single agreed-upon physical interpretation of the waves with which they deal, although they all agree in the way they use them to predict correctly the outcome of experiments.

In paragraph 1, Pierce indicates that complex concepts do not lend themselves to simple definitions.

Rather than asking what waves are, we should perhaps ask, what can one say about waves? Here there is no confusion. One recognizes in waves a certain sort of behavior which can be described mathematically in common terms, however various may be the physical symptoms to which the terms are applied. Once we recognize that in a certain phenomenon we are dealing with waves, we can assert and predict a great deal about the phenomenon even though we do not clearly understand the mechanism by which the waves are generated and transmitted. The wave nature of light was understood, and many of its important consequences were worked out long before the idea of an electromagnetic wave through space was dreamed of. Indeed, when the true explanation of the physical nature of light was proposed, many physicists who recognized clearly that light was some sort of wave, refused to accept it.

In paragraph 2, Pierce starts his definition by recalling how difficult the concept of light—a type of wave—was to comprehend.

We can study the important principles of waves in simple and familiar examples. As we come to understand the behavior of these waves, we can abstract certain ideas which are valid in connection with all waves, wherever we may find them. Such a study is the purpose of this chapter.

Suppose we watch the waves of the sea from a pier. Let us imagine that today the waves are particularly smooth and are very regular in height. We see a certain number of crests pass us each second—let us say a number f. This number f if the *frequency* of the waves. Frequency is reckoned in *cycles per second*, or *cycles* for short. The cycle referred to is simply a complete cycle of change; the departing of a wave crest, the passing of the trough, and

In paragraphs 3 and 4, Pierce takes the approach of using a familiar example to help explain the concept—how ocean waves appear. To help explain their operational occurrence, he turns to mathematics. He compares the ocean cycle of crests and troughs to an electric current.

finally, the arrival of the next crest. As a complete wave, from trough to crest again, passes us, the height of the water goes through a complete change, from high to low to high again.

A cycle is a complete cycle of change, at the end of which we are back to the original state. It is the same in the case of 60-cycle electric power. The 60-cycle electric current alternates in the direction of flow and goes through a complete cycle of change 60 times a second. Broadcast waves reach a receiver about a million crests per second; some television waves a hundred million crests a second, and radar waves leave the radar antenna and are reflected back again at the rate of billions of waves or cycles per second.

In paragraphs 5–9, Pierce develops step-by-step the mathematics of wave cycles, adding other categories— light and radio waves. By means of the previously explained mathematics, he shows waves operate under the same principles.

The waves in the ocean each take several seconds to pass us, so the frequency of the ocean waves is a fraction of a cycle per second. We can if we wish measure, instead of the frequency, the time between the passing of the two crests; this is the *period* of the wave, which we call T. We see that

$$T \text{ is the } reciprocal \text{ of } f, \text{ that is } T = \frac{1}{f}$$

Looking out at the waves, we may estimate or measure the distance between the crests of the waves; this is the wave length, which is always denoted by the Greek letter λ (lambda). Among radio waves from radar to broadcast, the wave length ranges from a little over an inch to around 1,000 feet.

The time between the passage of wave crests is T. In this time the next crest must travel just one wave length λ, to reach the position of the preceding crest. Thus, the wave travels with a velocity, v, which is the distance of travel λ, divided by the lapsed time, T, so that

$$v = \frac{\lambda}{T} = \lambda f$$

Thus, we can express λ in terms of f and f in terms of λ by using the velocity, v:

$$\lambda = \frac{v}{f}$$

$$f = \frac{v}{\lambda}$$

Light waves and radio waves are both electromagnetic waves, and for such electromagnetic waves traveling through space, the velocity, v, is the velocity of light:

$$v = 186,000 \text{ miles per second}$$

$$v = 3 \times 10^8 \text{ meters per second}$$

In paragraphs 10–12, Pierce amplifies further wave cycles, describing their shape

Waves have various shapes. We have been considering smooth rollers which come in one after another, regularly spaced. We can also have a single wave or a short *train*, or waves such as those caused by throwing a single stone into a pond. There is a reason, however, for considering a particular regular, smooth persistent kind of wave called a *sinusoidal* wave.

The waves we consider are waves of what we call *linear* systems. We will see what this means later on. Now, we will merely say that, while for some linear systems a wave of any form travels along preserving that same form, in many other linear systems a wave of arbitrary form will change form as it travels. Consider, however, a wave such that as it passes a given point, the height of the water or the magnitude or *amplitude* of some other significant quantity, varies sinusoidally with time with some frequency f. If this is so at any point in any linear system, the wave will also vary sinusoidally with time, with the same frequency f, at any other point. Strictly, the term *frequency* should be applied to sine waves only.

A sinusoidal variation can be understood in terms of a crank on a shaft which rotates at a constant rate, f turns per second, as shown in [Figure 6.3]. The height, h, of the end of the crank above the level of the center shaft varies sinusoidally with time. If we plot height or amplitude vs. time in seconds, as in [Figure 6.4], we get *sine curve* or *sine wave*.

Figure 6.3

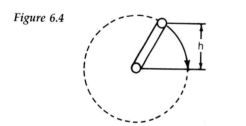

Figure 6.4

This is the way we will talk about the way waves vary with time. As long as the wave travels with constant velocity, this is also the way the height above the mean or zero level, which is called the *amplitude,* varies with distance.

In instances in which a wave other than the sine wave will change in form as it travels, sine waves of different frequencies (and hence of different wave lengths) travel with different velocities. . . . [10:82–85]

(curvature), magnitude (amplitude), and frequency. Using illustrations, he charts the mathematical operation and appearance of the wave cycle. The familiar example of a crank shaft helps to illustrate the rhythmic curvature variations of the height of a series of wave crests.

In paragraph 13, Pierce notes that the same principles apply to waves of different frequencies and velocities.

From the various examples quoted, it can be seen that there are many methods of achieving definition. The effective writer does not choose his or her method haphazardly, but chooses a method appropriate to bringing understanding to a particular reader. In report writing, as in other types of writing, definitions are given in the introductory section so that the reader can understand what is necessary to follow the main body of the writing.

General Principles of Definition

1. Definitions are needed when an unfamiliar term is used; or a familiar term is given a specialized meaning; or a term is given a stipulated meaning.
2. Definitions should include everything the term means and exclude every thing the term does not mean.
3. Definitions should not include the term being defined or any variant form of it in the genus (class or family to which it belongs) or in its distinguishing qualities or traits.
4. Definitions should include the essential qualities of the term defined.
5. Definition of terms in which magnitudes are the essential differentia should give the essential measurable quantities involved.
6. Definitions not measurable by quantities but limited or bounded by other terms that are closely related should include essential similarities and differences of the term defined and the terms bordering it.
7. Definitions should be stated in simpler or more familiar language than the term being defined.
8. The audience for whom the definition is intended should determine the definitional approach to be taken.
9. An expanded definition should employ expository devices such as examples, comparisons, contrasts, details, distinction, analysis, analogy, history, and so forth to promote clarity, meaning, and interest. Each expanded definition should contain a logical or formal sentence definition of the term being amplified.
10. When appropriate, illustrations should be used to promote clarity and help the reader visualize the subject being defined.
11. Definitions should not be phrased in obscure or ambiguous language.
12. Definitions should describe, not praise or condemn, the matter being defined.
13. Definitions may appear anywhere in the text, as a footnote, in a glossary at the end of a text, in an appendix, or in a special section of the introduction to a document. Where it appears depends on its importance to the text and on the knowledge of the intended reader.

Chapter in Brief

In this chapter we began examining the basic expository methods that characterize technical writing. We started the explanatory analysis with the role and techniques of definition. We explained the various ways to develop definitions and provided examples of each.

Chapter Focal Points

- Formal and informal definitions
- Synonyms and antonyms
- Defining with graphic illustrations
- Defining by stipulation
- The ten categories of the expanded definition
- Combining the various techniques to develop a definition

Questions for Discussion

1. Find five formal definitions in any of your other textbooks. Chart the definitions into their three parts.

2. Your instructor will divide the class into small groups. Each group will write single sentence formal definitions for the terms that follow. (Your instructor may want to substitute other terms.) Be sure to include the genus in the definitions. Avoid repeating the genus in the differentia.

   a. gimlet (the hand tool)
   b. gimlet (the cocktail)
   c. file (the hand tool)
   d. file (for data or information)
   e. Oedipus complex
   f. herbicide
   g. solar cell
   h. clone
   i. modem
   j. IQ
   k. host (biology)
   l. joystick (airplane)
   m. joystick (computer game)
   n. scroll (religious document)

   o. scroll (computer)
   p. fault (geologic)
   q. wordwrap
   r. interferon
   s. artificial intelligence
   t. extraterrestrial
   u. chip (computer)
   v. chip (verb)
   w. chip (poker)
   x. chip (a small piece off some thing) [keep it clean]
   y. chip (flaw)
   z. chip (fragment)

Assignment

1. Write expanded definitions of 200–350 words for three of the terms in discussion question 2, using some of the expository devices discussed in the chapter. Your instructor may assign different terms for this exercise.

References

1. American Geologic Institute and the National Association of Geology Teachers. *Laboratory Manual in Physical Geology*, 2d ed. Columbus, Ohio: Merrill, 1990.

2. Bronowski, J. "The Creative Process," *Scientific American,* September 1958.

3. Carroll, Lewis. *The Complete Works of Lewis Carroll.* New York: Random House (n.d.).

4. Chief of the Bureau of Weapons Systems Fundamentals. *Basic Weapons Systems Components.* Washington, D.C.: U.S. Government Printing Office, 1960.

5. Hutchinson, G. Ewelyn. "Eutrophication." *American Scientist,* May–June 1973.

6. Huxley, Julian. "War as a Biological Phenomenon." *On Living in a Revolution.* New York: Harper & Row, 1942.

7. Magoon, Robert A., and Karl C. Garrison. *Educational Psychology: An Integrated View.* 2d ed. Columbus, Ohio: Merrill, 1976.

8. "New Trails to Chemical Productivity." *Frontiers of Science.* Washington, D.C.: National Science Foundation, February 1977.

9. Orr, Richard H. "Cummunication—'Unpublished Information.' " *Industrial Science and Engineering,* October 1960.

10. Pierce, John R. *Electrons, Waves and Messages.* Garden City: Hanover House, 1956.

11. Rapoport, Anatol. "What is Semantics." *American Scientist,* January 1952.

12. Rice, Edward Loranus. *An Introduction to Biology.* Boston: Ginn, 1935.

13. *Science, Technology, and Innovation.* Columbus, Ohio: Battelle Columbus Laboratories, February 1973.

14. Thomas, Charles Kenneth. *An Introduction to the Phonetics of American English.* New York: Ronald Press, 1947.

15. *Van Nostrand's Scientific Encyclopedia.* New York: D. Van Nostrand Company, 1958.

7

Basic Expository Techniques in Technical Writing—Description

Chapter Objective

Provide algorithms for and proficiency in the expository process of describing mechanisms and organisms.

Chapter Focus

- Purpose of the description process
- Algorithms for describing mechanisms and organisms
- How illustrations help the description process
- Style considerations

The primary aim of science is the concise description of what can be known in the universe. The technical writer, therefore, is concerned with exact representations of phenomena. To achieve exactness, the technical writer often uses drawings so that the reader can visualize what is being described. But drawings and photographs cannot give complete representations. Although they help the reader see what something looks like, they cannot answer certain questions:

1. What is it?
2. What is it for?
3. What does it do?
4. How does it do it?
5. What happens after it does it?
6. What is it made of?
7. What are its basic parts?
8. How are they related to make it do what it does?

To answer these questions fully, technical writers must be able to visualize clearly the thing or process being described. They must have a thorough command of the appropriate nomenclature. They must know the components and their interrelationships and proportions. Finally, they must know the "why": the purpose of the thing or organism in question, the end it serves.

Describing a Mechanism or Organism

Complete knowledge of the thing described is fundamental to writing a technical description. Being able to draw the matter to be described indicates ability to visualize it completely and comes only with intimate familiarity and understanding. Nomenclature is important and inherent in technical descriptions. Every part has a name and should be correctly labeled. Part identification is best accomplished through labeling on the drawing. However, mere drawing is not full description because the sketch or diagram cannot indicate many essentials and attributes, nor can it always show relationships or structure.

It should be apparent that you, the technical writer, in order to give complete details, need to have the subject before you. Drawings are two-dimensional; descriptions need to be three-dimensional. You must be concerned not only with proportions but also with shapes, materials, essences, finishes, connections, relationships, purposes, and actions of the various components. Size, proportions, materials, finishes, and compositions of essences are fairly straightforward. Connections and relationships, however, need precise definition. How an element is fastened or connected needs to be examined very carefully. Is one component riveted, bolted, screwed, coupled, molded, wired, etc., to another?

The precise location of components in relation to each other is important. At what angle is one element joined to another? Is it under or above? Inside or outside? Beside, parallel to, diametric to, and so forth? These details are significant to the *what* and *how* of the matter described.

In such analysis, one other element remains: the *why*, or the purpose. What does the thing or organism do? How and why does it do it? What is the significance of its action? The combination of the *what, how,* and *why* becomes the complete technical description.

Algorithm for Organizing the Written Description[1]

After becoming familiar with the mechanism or organism, your task is to organize the data into a written description. Written description usually falls into an arrangement of major sections. Section I is the introduction; it consists of a general description of the mechanism, phenomenon, or organism. Section II contains the main functional divisions of the subject, its main parts, and the principle under which they operate. Section III, the concluding section, shows how the subject operates by taking it through a cycle of operation. The three sections with their components might be considered as an outline for a technical description.

I. Introduction
   A. Definition of the device
   B. Purpose
   C. Generalized description, including perhaps an analogy or comparison with a familiar matter
   D. Division of the device into its principal functional portions
II. Principle or theory of operation, if the device is a complicated or complex one. Detailed description of the portions, major assemblies, sections, and so forth.
   A. Division one—what the part is, its purpose and appearance, including an analogy or comparison or division into subparts
      1. Subpart one
         a. Purpose
         b. Appearance
         c. Detailed description
            (1) Shape, size, relationship to other parts, method of connection, material, finish
            (2) Function
      2. Subparts two, three, and so forth
   B. Division two, and so forth

---

[1]Don't be afraid of the term **algorithm**. It is borrowed from the discipline of mathematics, where it means a rule or procedure to solve a recurrent problem. In this text, *algorithm* means a method or procedure to solve a communication problem.

III. Brief description of the device in a cycle of operation
    A. How each division achieves its purpose
    B. Causes and effects of the device or organism in operation
    C. Summary or other generalized conclusions

This outline considers the essentials included in a technical description. It is not prescriptive. Some elements might need further explanation. Major headings and subheadings are helpful road signs to the reader and serve as transitional devices.

The Introduction

Technical descriptions usually begin with a formal definition of the matter to be described. The definition is followed with a statement of the purpose or use of the matter and then with a very general description of its appearance. This initial description is general; its purpose is to give the reader a visual image of the subject by describing its size, shape, and appearance. An analogy can help readers understand something that might not be familiar to them. The major components or assemblies of the subject are then listed. This listing leads readers to the middle portion of the report.

Major Divisions and Principle of Operation

A technical description of more complex devices often includes the principle or theory of operation. The description of the major components can be arranged in three ways. One arrangement may follow the order in which a viewer's eye might see the matter described. For example, in the case of a claw hammer, the viewer might first see the head, then the peen, the face, the neck, and finally the handle. Each part, then, would be described in the sequence from which it is viewed.

Another way of arranging the description of the parts may be according to the sequence in which the parts are assembled. For example, a car door lock cylinder might be described in the order its various parts are arranged in the car door; the retaining clip, the pawl, the cylinder housing, the lock cylinder, the cylinder cap springs, the cylinder cap, and the cylinder housing scalp.

A third method of arrangement that might be appropriate is the order in which the parts operate, as in the example of a duo-brake on pages 198–200.

Which type of description to use depends on the reader's requirements. If the reader is interested only in a general knowledge and understanding, the first arrangement—according to the visual perspective—would be appropriate. If the reader needs to assemble the object, the second order should be used. If the reader needs to know how to operate the mechanism, obviously the third order is called for.

After fully describing all the parts, your next step is to explain how the complete unit functions by taking it through a cycle of operation. The concluding section is, in effect, a description of a process in condensed form (see Chapter 8). This section should show how each division achieves its purpose. Emphasis is placed on the action of the parts in relation to one another. Special applications, variations, cost, and other pertinent general details are appropriate for the terminal section.

Use of Illustrations

Drawings and photographs can indicate precisely the size and shape of things. They need to be integrated with the text. Illustrations are fully discussed in Chapter 15, but for present purposes, the following suggestions will be helpful:

1. All drawings should be made to scale and fully dimensioned.
2. All illustrations should be properly numbered and each should have its own captioned title.
3. Even where there is only one illustration, it should have a title.
4. Text matter in the illustration should be kept to a minimum.
5. Descriptive matter needed to explain an illustration should be incorporated.
6. Standard abbreviations should be used.
7. Where space permits, words should be spelled out completely.
8. Reference letters or numbers in illustrations make it easier for the reader to correlate the drawing with the text.
9. When the purpose of a technical description is to enable the reader to reproduce or fabricate a device, complete dimensions should be indicated in drawings and critical dimensions given in the text description.

Style in Technical Descriptions

The two major requirements in a technical description are completeness and clarity. The requisites of technical style discussed in Chapter 4 are certainly necessary. Conciseness, precision, and objectivity are the keystones. Description of an existing mechanism or organism is usually written in the present tense. On occasion, a more appropriate tense is the future or the past, but whatever tense is used, it should be consistent and logical throughout the description. A writing pitfall in taking a mechanism or organism through a cycle of operation is lack of parallel grammatical structure (see pages 547–548 in the Reference Index and Guide to Grammar, Punctuation, Style, and Usage). Finally, always keep in mind the reading audience. Are you writing for experts? For well-informed individuals? For the general public? Do the readers need to know only in a general way what the matter described looks like and how it functions, or do they need to be taught how to manufacture or to operate the device? Your purpose is going to determine the organization and the amount of detail necessary. In a generalized approach, for example, the organizational divisions of the text are going to be the same as in a specific approach, but the details are going to be fewer and not as exacting. Drawings will be generalized and functional rather than specific and detailed.

Examples of Technical Descriptions

The following description of the venturi carburetor by a student in a technical writing class is an example of an effective technical description, meeting, on the whole, the requirements set forth in this chapter.[2] It would be enlightening to check this description against the outline of a detailed, typical technical description.

The Simple Venturi Carburetor

The simple venturi carburetor is used primarily as a fuel metering device for small, internal combustion engines such as are found on modern, self-powered lawn mowers. This type of carburetor, which is only remotely similar to a complicated automotive carburetor, consists basically of a venturi, a fuel supply line and needle jet, a butterfly valve, and sufficient mounting lugs so that the unit may be adequately secured to the engine on which it is to be used.

The Main Casting

The main casting, the diagonal pattern area [Figure 7.1a], which is formed of 195-TS-62 aluminum, is 5 inches in length, and has machined surfaces on all of the inner areas. The venturi entrance angle 0 is 5.38 degrees and the throat exit angle α is 7.12 degrees. The mounting flange is ¼-inch thick, and two ⁵⁄₁₆-inch diameter holes are drilled in this area for mounting bolts, as shown in [Figure 7.1b]. Butterfly shaft holes are drilled to a size of ⅛-inch in diameter, and are located in the unit's central, vertical plane, ¾ of an inch inward from the flange mounting surface. The fuel pick-up tube mounting hole is ⅛-inch in diameter and is also located in the central, vertical plane of the carburetor. This hole is drilled through the bottom of the casting at the mid-point of the throat section. The needle jet hole is radially drilled into the pick-up tube mounting lug, as shown in [Figure 7.1b]. This hole, which is ⁷⁄₃₂-inch in diameter, is then threaded with a ¼-inch National fine tap which corresponds to the needle jet's adjustment threads.

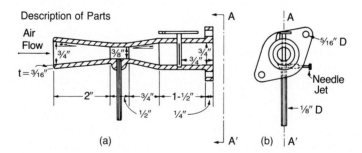

Figure 7.1

---

[2]"The Simple Venturi Carburetor" by Joe Marcus was written to meet an assignment for writing a technical description.

The Pick-up Tube

This unit is a ⅛ + 0.001 inch diameter (outside) brass tube which is two inches long, and is press fit into the casting due to the 0.001 inch interference. The inside diameter is ¹⁄₁₆-inch, and a ¹⁄₃₂-inch diameter hole is drilled radially through one wall of the tube ⁷⁄₁₆ of an inch down from the top surface which eventually is flush with the inside surface of the throat. This hole is the one through which the needle jet point protrudes so that the air-fuel mixture may be altered.

The Needle Jet

The needle jet, as shown in [Figure 7.2], is formed of brass material employing the dimensions shown. The threads indicated are ¼-inch fine to correspond with the threads cut in the main casting needle jet hole.

Figure 7.2

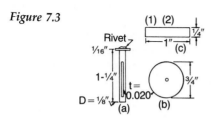

The Butterfly Valve

The butterfly valve consists of a ⅛-inch diameter by 1¼-inch long brass shaft with a diametral slot cut in the location shown. This slot and the circular butterfly plate are 0.020 inch thick and ¾ of an inch in height and diameter respectively. A hole [Figure 7.3a] is drilled after the shaft is installed in the main casting, and it corresponds with the hole in the butterfly plate. A rivet then joins the plate and shaft, employing this hole, which is ¹⁄₁₆-inch in diameter. A small shank is machined in the shaft which provides a retaining shoulder for the throttle linkage, [Figure 7.3c]. The stub extending from this shank is ¹⁄₁₆ of an inch in diameter and ⅛ of an inch tall; it is used as a rivet to secure the throttle linkage to the shaft. This throttle linkage plate is ¹⁄₁₆-inch thick steel plate 1-inch long and ¼-inch wide. Holes *(1)* and *(2)* are ¹⁄₁₆-inch in diameter, and are used for throttle rod attachment and mounting to the shaft respectively.

Figure 7.3

The Mechanism in Use

The nature of the engine with which this fuel metering device is to be used is such that a flow of air will almost continuously be drawn through the carburetor's throat. The amount of flow entering the carburetor is also a function of how much the butterfly valve has been opened. From fluid mechanics it can be shown that as a fluid (gas or liquid) passes through a venturi, its velocity increases: but its pressure decreases to some value below atmospheric. This negative pressure is greatest at the point in the throat where the fuel pick-up tube is located. This differential pressure, P (atmospheric) minus P (throat), will cause the gasoline, into which the tube has been submerged, to be forced up the tube and into the air stream. The needle jet, previously described, is used to accurately control the amount of gas flow through the pick-up tube. As the gasoline enters the air stream, it is vigorously mixed with the air flow and is therefore made compatible with combustion requirements.

The following is a more generalized description, that of a duo-servo brake taken from an automobile service manual. It would be interesting for you to check it carefully with the specifications for a technical description in this chapter. Note that the description is more generalized. Dimensions are offered only where the knowledge is useful to the reader—the serviceman. Note, too, that the descriptions of the brake and its components otherwise are quite detailed. The writing is directed to aid the garage mechanic to service and repair the brake. You may note a grammatical slip in the writing—a dangling participle (first sentence last paragraph).

The Duo-Servo Brake

General Description

The brakes used on both front and the rear of some Chevrolet models are the Duo-Servo, single anchor type, which utilize the momentum of the vehicle to assist in the brake application. The self-energizing or self-actuating force is applied to both brake and shoes in each wheel in both forward and reverse motion.

Each brake [Figure 7.4] has one wheel cylinder located near the top of the brake flange plate, just below the anchor pin. Each wheel has two shoes with a pull back spring installed between each shoe and the anchor pin, to hold the upper ends of the shoes against the anchor pin when the brakes are released. The lower ends of the shoes are connected by a link and helical spring. The link is made up of an adjusting screw, riding a socket at one end, and threaded into a pivot nut at the other. The outer ends of the socket and the pivot nut are notched to fit the webs of the brake shoes, providing freedom of motion between the link and the shoes. The spring is stretched from one shoe web to the other, crossing over the notched head of the adjusting screw. It bears against one of the notches in the head, and thus acts as a lock for the adjusting screw. Bonded brake linings are used and brake drums are 11″ in diameter. The front brakes are 2″ wide, while the rear are 1¾″.

In each brake assembly, the linings for the front and rear shoes differ in length, the secondary facing being 2½″ longer than the primary, because in operation, a greater force is applied to the secondary facing than to the primary.

The brake flange plate has six bearing surfaces, three for each shoe, against which the inner surfaces of the shoes bear to maintain alignment. Slightly below the center of each shoe web is a hole through which a hold-down pin is inserted. A spring, fitted over the outer end of the pin holds the shoe against the braking surfaces. At the top of the brake, where the

1. Backing Plate
2. Anchor Pin
3. Guide Plate
4. Secondary Shoe
5. Pull Back Spring
6. Primary Shoe
7. Pull Back Spring
8. Hold Down Spring
9. Hold Down Pin
10. Adjusting Screw
11. Adjusting Screw Spring

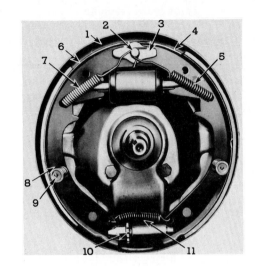

Figure 7.4
Duo-Servo Brake

shoes butt against the anchor pin, a guide plate separates the pull back springs from the shoe webs, and assists in keeping the shoes properly aligned. The brake mechanism is effectively sealed against the entrance of dirt or mud by a joint between the brake flange plate and the drum. The outside edge of the flange plate fits over the edge of the drum which has annular grooves located between two flanges.

The outer flange is of a larger diameter than the inner one, so that dirt and moisture which collect in the groove are thrown off the larger flange by the centrifugal force of the rotating drum, thus keeping foreign matter away from the drum-to-flange plate joint.

Operation

When the brakes are applied, the pistons in the wheel cylinder, acting on the brake shoes, through the connecting links, force the shoes against the drum. Since the shoes float free in the brake, the force of friction between the shoes and the rotating drum, turns the entire assembly in the direction of the wheel rotation. The front or primary shoe moves downward, and the back or secondary shoe is carried upward until its upper end butts against the anchor pin. The friction between the moving drum and the stationary shoes now tends to roll both shoes toward the drum with increased pressure. The secondary shoe pivots on the anchor pin at the top, and the primary shoe tends to turn out the adjusting link at the bottom which is held stationary by the secondary shoe. This self-energizing effect greatly increases the pressure of the shoes against the drum and reduces the physical force required on the brake pedal.

Inasmuch as the brake shoes are freely connected at the bottom by the adjusting link, the self-energizing or friction force which is applied to the primary shoe by the brake drum is transmitted to the secondary shoe through the link. The effectiveness of the secondary shoe is nearly doubled, because the total force applying this shoe becomes the sum of the force which is received from the primary shoe, and the self-energizing effect that is derived from the rotating drum.

When backing the car, the brake action is reversed. The rear shoe becomes the primary shoe, and the front shoe becomes the secondary, butting against the anchor pin during braking and being forced against the drum with great pressure. [1:5–12]

Description of an Organism

The algorithm for describing an organism is similar to that for describing a mechanism. The following excerpt and illustrations [Figure 7.5] from a National Cancer Institute publication are an example of a description of an organism [2:2–3].

The Pancreas

The pancreas is a thin, lumpy gland about six inches long that lies behind the stomach. Its broad right end, called the head, fills the loop formed by the *duodenum* (the first part of the *small intestine*). The midsection of the pancreas is called the body, and the left end is called the tail. (See illustration, Figure 7.5).

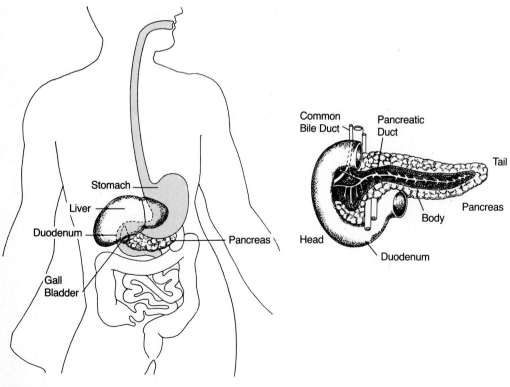

Figure 7.5

The pancreas produces two kinds of essential substances. Into the *bloodstream* it releases *insulin*, which regulates the amount of sugar in the blood. Into the duodenum it releases *pancreatic juice* containing *enzymes* that aid in the digestion of food.

As pancreatic juice is formed, it flows through small ducts (tubes) into the main pancreatic duct that runs the full length of the gland. At the head of the pancreas, this main duct joins with the *common bile duct,* and together they pass through the wall of the duodenum. (See illustration.) The common bile duct carries *bile* (a yellowish fluid that aids in the digestion of fat) from the *liver* and *gallbladder* to the duodenum.

Algorithms for Writing Descriptions of Mechanisms and Organisms

1. Begin with a definition of the matter to be described.
2. State its purpose, what it does or what it is used for.
3. Describe what it looks like. For complex or unfamiliar matters use an analogy.
4. Use illustrations to help the reader visualize the appearance, size, dimensions, and shapes both of the item as a whole and of its components.
5. Explain why (the theory) the device or organism operates or behaves the way it does.
6. Use steps 1–5 above to describe each main functional division of the item, its parts, and the principles under which they operate.
7. Consider your reader's requirements.
8. Finally, state how the mechanism or organism functions by taking it through a cycle of operation.

Chapter in Brief

In this chapter we examined the purpose and process of technical descriptions. We presented algorithms for describing both mechanisms and organisms.

Chapter Focal Points

- Purpose of technical descriptions
- Algorithm for describing a mechanism
- Algorithm for describing an organism
- Role of illustrations in technical descriptions
- Style considerations in technical descriptions

Assignments

1. Choose any of the common objects listed below and write a paragraph description of 100–250 words to enable a reader to accurately visualize the subject; include a drawing.

a. Flattop Philips screw

b. Hair curler

c. Dendrite

d. Scalpel

e. Diskette

f. QWERTY keyboard

g. Cell (biology)

h. Whistling tea kettle

i. Hand calculator

j. Eye

k. Carpenter's hammer

l. Stapler

2. Choose a simple mechanism or device within your field of interest. Prepare an outline similar to that discussed in the chapter. Before preparing the outline, make a drawing (or as many drawings as necessary) to familiarize yourself with the subject. Compare your outline to the drawings. Rewrite the outline as necessary. When you are satisfied that the outline is accurate, write the description. Use major headings for each major division and subdivision heads for each subdivision. Include drawings necessary to supplement the description.

3. Consider what you need to do to revise your description for:

a. A reader who needs to use the device, or

b. A reader who needs to fabricate the device.

Revise your description for either purpose a or b.

References

1. *Chevrolet Passenger Car Shop Manual*. Detroit, Michigan: General Motors Chevrolet Division, 1955.

2. *What You Need To Know About Cancer of the Pancreas*. Washington, D.C.: National Cancer Institute, 1979.

8

Basic Expository Techniques in Technical Writing— Explaining a Process

Chapter Objective

Provide algorithms and proficiency in explaining a process

Chapter Focus

- Algorithms for organizing and writing the explanation of a process
- How illustrations help to explain a process
- Style considerations

*E*xplaining a process is describing, narrating, or instructing how something is done, how an occurrence has happened, or how an effect has taken place. Certain events or activities are connected in a significant sequence and create an observable or measurable effect that, if repeated with the same quantitative and qualitative ingredients and in the same sequence, should produce the same or similar results. Some **processes** are due to natural—physical and biological—factors, some are due to societal—political, sociological, economic—forces, and some are due to psychological factors. There are processes resulting directly from human participation in scientific, technological, and industrial activities, as well as those resulting from commonplace tasks. All of these processes result from the procedures, or sequence of events, by which products are made, occurrences take place, or results are achieved. Technicians, engineers, computer systems analysts, biologists, physicians, farmers, geologists, social scientists, home economists—practically everyone—are concerned daily with processes. On occasion all of us must explain procedures so that others may reproduce the same result, occurrence, or product.

Algorithm for Organizing the Explanation of a Process

The explanation of a process in which the reader must take part requires more specific details than the explanation of an activity or occurrence in which the reader does not take part but needs to understand the occurrence. In other words, the written description of the process depends, as does every type of writing, on the intended reader's requirements. For readers who must perform the process, the technical writer will need to include every detail. If the reader needs only a general knowledge of the principles involved and will not need to perform the process or to supervise its performance, the writer need not go into specific details that may only confuse the reader. Instead, the technical writer should emphasize the broad principles and give a generalized account of the steps and sequences.

Intimate knowledge of the process to be described is fundamental. This involves not only complete understanding of how the process works or an ability to perform it but also a knowledge of the appropriate terminology and a logical organizational plan for the explanation. The organizational pattern for explaining an industrial process might be as follows:

I. Introduction
  A. Definition of the process—general information as to why, where, when, by whom, and in what way the process is performed, carried out, or occurs
  B. Fundamental principle of operation, unless explained above
  C. List of the main steps, events, or sequences in the action
  D. Requirements for the performance of the process or the occurrence of the event

       1. Principal tools, apparatus, and supplies

       2. Special conditions required

  II. Description of the steps or analysis of the action

     A. First main step (or sequence of events)

       1. Definition

       2. Special materials, apparatus, and so forth

       3. Division into substeps

          a. Details of the first substep—action, quality of action, reasons for results of the action

          b. (etc.) Description of other substeps, as needed

     B. (etc.) Description of other main steps or sequences as required

 III. Conclusion (a summary statement about the purpose, operation, evaluation of the whole process)—the process is taken through a complete cycle of operation or through to its final stage

Introduction

The introduction serves to define what is being done or what happens. Inherent within it are such questions as:

What is the process?

Who performs it?

Why is it performed? or

How does it happen?[1]

What are the chief elements or steps in what happens or in what results?

What is the consequence or result of the process?

Not all of the questions need necessarily be answered, nor is the order of the listing significant. Reader level and reader requirements determine the depth of detail used or omitted.

    As in the case of the technical description, the explanation of the process may begin with a definition; then it is followed by a generalized description to give the reader an immediate visual image of the nature of the process.

    The introduction should include any special circumstances, conditions, and personnel requirements. Frequently, it is necessary to explain why the process is being described. Sometimes the reason is contained in the definition. Sometimes the purpose is so obvious that it is unnecessary to explain as, for example, the opening of a jar of pickles or the stapling of a sheaf of papers.

    The introduction might also include the necessary materials, tools, and apparatus required for the process. The tabular form lends convenience and clarity. It is helpful to the reader to get a bird's-eye view of the general procedure

---

[1]For processes occurring without human intervention.

before getting into the details. The introduction, accordingly, will end with a listing of the chief steps in the process.

In an explanation of a simple process, the introduction may consist of no more than a paragraph or two. In a complex process, the principle of operation or theory may occupy several hundred words. Drawings are frequently used to help explain what tools, instruments, or materials are used.

Main Steps

The main body of the explanation may well have a paragraph of details for each major step. Each step is taken in sequential or chronological order, defined, described, and explained. Steps may be arranged as a list, each item preceded by graphic bullets, sequential numbers, or letters, or they may be written or keyboarded in sentence form with or without graphics, numbers, or letters. Whether in an itemized list or in sentence form, steps are listed in parallel grammatical structure and in correct sequence for the process. Materials, conditions, tools, and apparatus necessary in the performance are included in the explanatory unit of each step. Each descriptive unit has a topic sentence indicating the subject of the step. The details clarify the action in their relation to the process as a whole. All essential details of action should be included in the explanation. Whenever a nut and bolt are removed, the explanation must be complete enough to show their replacement. Each essential action should be specified in the terms of who, what, when, why, and how. Each substep within a major step constitutes a process within itself. Substeps should be properly introduced, and, if necessary, subdivided so that the reader will understand and will be able to perform the entire action. In the description of any action, the reader needs to understand and to visualize the activity. Qualitative conditions as well as precise quantitative factors are essential to obtain the required result.

Concluding Section

After the step-by-step explanation of the process is completed, you should summarize the whole process so that the reader can see the activity as a whole and not as a series of separate steps. Requirements of the reader may necessitate additional information, such as an evaluation of the process, a discussion of its importance, an underscoring of important steps, equipment, materials, and precautions to consider, and if appropriate, alternative steps that might be taken if certain difficulties are encountered.

Use of Illustrations

The reader will frequently be helped by drawings that show different stages of the process. Drawings can verify relationships and show shapes, which can be difficult to describe precisely in language. Dimensions are most conveniently indicated in a drawing. In complex processes or those in which several steps or actions are going on simultaneously, a flow diagram will be necessary to show the relationship of the several phases. Drawings must be integrated with the text and referred to specifically by figure number and with callouts (see Chapter 15) to help the reader follow the complete process in text and illustration.

Style Considerations

The style considerations discussed in Chapter 4 are applicable and should be followed in explaining a process. Depending upon the point of view and the formality of the writing situation, processes may be explained in the first, second, or third person, in the active or passive voice. Many processes are described for the express purpose of directing others to replicate the particular activity. In those instances, the active voice and imperative mood are required. However, the style must be consistent. If the explanation of the process is part of a larger report or a handbook of instructions, it must be integrated stylistically with the larger work.

Examples of Processes

Three examples of process descriptions written by experienced professional writers follow. The first example is from a commercial technical manual explaining the process of removing a sample of a steel roof deck from an existing roof for calorimeter fire hazard evaluation. The process, though not complex, involves critical details. The explanation is intended to instruct safety or maintenance personnel who may or may not have professional technical training. Its details are very specific and precisely enumerated. The illustration is integrated into the text and adds clarity to the explanation of a critical step in the process. Note the conformity of this process explanation to the specifications of this chapter.

Steel Roof Deck Sample for Calorimeter Evaluation

It is sometimes necessary to remove a steel roof deck sample from an existing roof for calorimeter fire hazard evaluation.

The following materials, equipment, and tools are necessary for removing one sample.

Materials

2—4½ ft. × 5–ft. × ¾–in. sheets of plywood (half table tennis table size).

4–½ × 7–in.-long (minimum) carriage bolts with nuts and washers (7 in. is a minimum length for 1½ in. deep deck. For a deeper deck, increase the bolt length by the increase in deck depth over 1½ in.).

Equipment

Staging to support roof sample.

Waterproof coverings.

Tools

1 Heavy duty saber saw with 7-in. (min.) metal cutting blade.

1 Crescent wrench (8 or 10 in.).

1 Heavy duty drill with ⁹⁄₁₆ in. bit.

1 Ruler.

Sample Area

When selecting a sample area, make certain that it is between roof supports so that the sample will not be fastened to the supporting steel.

After the sample area has been chosen, remove any stock or equipment from the area immediately below and for a 10-ft. radius of the area. Equipment that cannot be moved should be covered with waterproof covers.

Wrap the sample (including the bolted-on plywood) to protect it from the weather and ship to the following address:

Chief Materials Engineer

Factory Mutual Research Corporation

1151 Boston-Providence Turnpike

Norwood, Massachusetts 02026

Note: Deliver to Bldg. 14

Place one sheet of plywood over the other and drill four ⁹⁄₁₆ in. holes through both sheets 12 in. in from each corner measured diagonally [Figure 8.1].

Leave one piece of plywood in the building; take the second piece to the roof. The tools and materials should be brought to the roof at this time also.

Attaching and Cutting the Sample

1. At the point on the deck from where the sample is to be taken, place the 4½-ft. × 5-ft. sheet of plywood. Make certain the 5-ft. dimension is parallel to the ribs of the steel deck.

2. Using the plywood as a template, drill four ⁹⁄₁₆-in. holes through the roofing and steel deck.

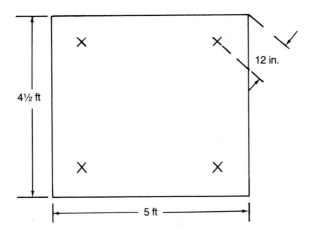

Figure 8.1

3. Place the bolts through the plywood and holes.

4. From the underside, place the second piece of plywood over the bolts and tightly fasten with the washers and nuts.

5. Provide support for the sample from the underside so that when the roof sample is cut, there will be no collapse.

6. Cut around the full perimeter of the plywood with the saber saw.

7. Remove the sample carefully so as not to disturb any of the components.

Roof repair work should be started as soon as the sample has been removed to avoid the possibility of water damage.

This data sheet does not conflict with NFPA standards. [3:1–2]

Next, the ensuing two examples illustrate descriptions of biological processes that occur without human intervention.

Cell Division

Introduction

One of the most fundamental of all cellular processes and one in which the tissue culturist is particularly interested is that of cell division. In fact, the process to be described is nuclear division. The cytoplasm usually follows the nucleus in dividing (but this does not happen invariably).

Three types of nuclear division have also been described. Amitosis is a simple division of a nucleus without formation of chromosomes. It is very uncommon and probably not a normal process. Consequently, it will not be discussed further.

Meiosis and mitosis are both characterized by the appearance of chromosomes which are segregated into two groups, one group going to each of the daughter cells. Meiosis is a special case, occurring in the formation of germ cells. During meiosis the normal (diploid) number of chromosomes is halved so that each daughter cell contains a haploid number. In mitosis, on the other hand, the cell at the time of division has material for a double set of chromosomes. On segregation of the chromosomes each daughter cell is left with the normal (diploid) number. This is the normal type of cell division occurring in somatic cells during growth and is almost the only type of cell division seen in tissue cultures. All subsequent discussion will be concerned only with mitotic division.

The early stages of mitosis are characterized by the appearance of the chromosomes. These structures carry the genes, which are mainly responsible for transferring information about hereditary characteristics to the daughter cells. Each cell has a double set of chromosomes and the number is characteristic of the species. The number of chromosome sets is referred to as the "ploidy" of the cells. Thus, normal cells with two sets are referred to as diploid, cells with three sets are triploid, with four sets tetraploid, and so on. Germ cells with half the diploid amount are referred to, somewhat inconsistently, as haploid.

The stages in cell division are as follows [Figures 8.2 and 8.3]:

Main Steps

1. *The Intermitotic Phase* (Interphase). This is also, mistakenly referred to as the resting phase. In growing cells most of the synthesis of new materials proceeds during this period. When the cell contents have been doubled the cell then enters prophase.

2. *Prophase.* This is, as a rule, the earliest recognizable stage of the cell division. Within the nucleus the chromatin condenses into a continuous skein and begins to break up into chromosomes. In the meantime, the centrosome divides and the two portions move in opposite directions. Between these two poles, the structure of the spindle (rather like visualized magnetic lines of force) appears and chromosomes arrange themselves on it. The nucleoli disappear and the nuclear membrane disintegrates. At the end of this phase, the cell has the typical appearance of metaphase.

3. *Metaphase.* The chromosomes are arranged in the equator of the cell. By time-lapse cinemicrography that can be seen to be highly active at this stage, often apparently rotating en masse. Quite suddenly the cell enters anaphase.

4. *Anaphase.* The chromosomes split longitudinally and the halves (chromatids) move in opposite directions toward the poles. When they are completely separated, they are again regarded as chromosomes. As cytoplasmic division begins, the cell enters telophase.

5. *Telophase.* The chromosomes at each pole come together and begin to fade and disappear. Simultaneously a new nuclear membrane forms around each nuclear zone. A constriction appears in the cytoplasm between the two nuclei. It gradually deepens until the two new cells are separated. The cell surface "bubbles" very actively indeed at this stage and as the activity decreases the new daughter cells spread out over the surface on which they are growing.

Concluding Section   The duration of cell division varies with different cells. With avian fibroblasts, it can be quite rapid and can be watched with the eye over a period of about twenty minutes (at 38°C). The interphase period in rapidly growing cells is of the order of eighteen hours or less. [2:15–17]

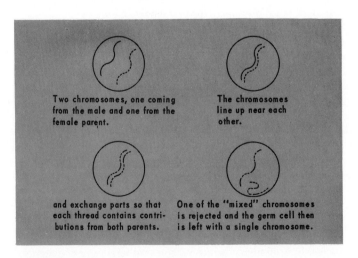

Figure 8.2
The Reduction Process by Which Germ Cells Are Formed

Figure 8.3
Fertilization of the egg by the sperm restores two chromosomes of each kind to each cell (only one chromosome in each germ cell is shown).

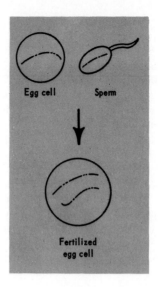

One of the more effective explanatory devices is the *analogy*. The writer makes use of something familiar or known to the reader to help explain the workings of something little known or complex. The process of how the debilitating disease of multiple sclerosis afflicts a patient is graphically and clearly portrayed by the analogical explanation below:

Imagine for a moment a complicated telephone switchboard in a large organization. Hundreds of wires run from the central console to all floors. These wires carry messages from the outside world and from one office to another. Data, reports, instructions—converted into electrical impulses—race along the wires all day, tying the many units of this complicated enterprise together in a communications network. Most of the wires are protected by a sheath of insulating material that keeps the electrical impulses from leaking away or being short-circuited.

Let us suppose that, for some unknown reason, patches of this insulation begin to disintegrate and sections of bare wire are exposed. Very slowly, as the disintegration progresses, strange things start happening in the network. Some offices get a scratch buzz when the receivers are raised. There are complaints of garbled messages. Messages transmitted from the switchboard wind up at the wrong offices. In some sections of the building, the phones go dead. The functioning of the entire system becomes erratic, and the baffled repairmen are unable to explain what has got into the insulation.

This is a highly simplified mechanical analogy for an advanced case of multiple sclerosis—an analogy that conveyes none of the human anguish caused by each attack of the disease.

Multiple sclerosis attacks the insulation around the message-bearing central nervous system of man. Almost all the information he receives from the outside world and from inside his own body, as well as all the conscious

Introduction
The first four paragraphs serve as the introduction containing the requirements under section I of the outline (see pages 204–205). Note that the description of this process does not begin with a definition. To provide an understanding of what multiple sclerosis (MS) is, the writer uses the analogy of a telephone switchboard. Having given the reader a general understanding of MS, the writer proceeds to explain the particulars of how this dread disease operates.

and unconscious control he has over his own movements and behavior, man owes to this system that consists of the brain, the spinal cord, and the various nerves that connect the brain and the spinal cord with the sense organs, glands, and muscles of his body. Although such chemicals as hormones and such purely physical factors as temperature help to integrate the various tissues and systems of the body into a smoothly operating unit, the central nervous system is the most important means by which the body coordinates its activities and responds to environmental changes.

Main steps of the process. Paragraphs 5–10 describe and analyze the process occurring within a person afflicted with MS.

The easiest way to understand how multiple sclerosis affects this communication system is to take a microscopic look at its enormously complex tissues. Such a close-up reveals that the central nervous system is built of cells, as are other parts of the body. The basic units are the nerve cells, or neurons. In addition, there are many neuroglial cells, or glia, which correspond to the connective tissue cells found in other organs. Together with the blood vessels, the glia pervade the substance of the brain and spinal cord, and serve as the "glue" and structural supports that hold the delicate nerve network together. There are many more glia than neurons, but it is the neurons that conduct messages to and from the brain and from one part of the brain to another.

A neuron is a highly specialized unit with a central body containing a nucleus and cytoplasm. What distinguishes the neuron from other cell types are the various threadlike appendages or processes extending from it. Each neuron has relatively short processes called dendrites, which have many fine extensions like the branches of a tree, and a single fiber called the axon, which usually does not branch out except at the very end, although it may be two or three feet long. Both the dendrites and the axon have a property not shared by most other cells in the human body: the capacity to transmit electrical impulses or messages. The dendrites, transmitting nerve impulses toward the cell body, and the axon, conducting the impulses to the dendrites of other cells or to muscles and glands, are in effect the wiring of the body's communication system.

If a section of an axon is properly stained and placed under a microscope, two parts become apparent—a central core, or axis cylinder, surrounded by a sheath of fatty tissue called myelin, which is formed by processes of neuroglial cells that wrap around the nerve's fibers. Unstained, myelin is whitish in appearance. Even to the naked eye, certain areas of the brain and spinal cord appear to be white; this *white matter* contains the nerve fibers with their myelin sheaths. What is popularly known as a *nerve* is really a bundle of axons with their myelin sheaths, all bound together by a packing of connective tissue much as a bunch of wires is wrapped together to form an electric cable [Figure 8.4].

Because the fatty material is indeed a good insulator, scientists presume that the sheath restricts a nerve impulse to a distinct pathway and prevents the leakage of electric potential to surrounding tissue. However, since the chemical composition of the myelin is much more complex than an insulator needs to be, some researchers suggest that the sheath also helps to nourish or maintain the integrity of the axon, which stretches far from the nucleus so vital to the life of the entire cell.

There is also evidence that the sheath speeds up the transmission of the nerve impulse. Myelinated nerve fibers transmit at the rate of 100 to 120

Figure 8.4
Effect of Demyelination

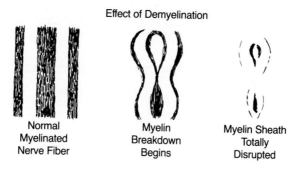

Effect of Demyelination

Normal
Myelinated
Nerve Fiber

Myelin
Breakdown
Begins

Myelin Sheath
Totally
Disrupted

meters per second, or 225 miles an hour—fast enough for the message to travel from head to toe of a six-foot man in 1/50 of a second. Along a non-myelinated nerve, the impulse travels at a speed of only 10 to 20 meters per second.

The living web of the central nervous system is in ceaseless activity, with passages of sensory information from the outside and from all parts of the body along sensory nerves to the spinal cord and brain, and an unending stream of return messages speeding along motor nerves to the muscles and glands. The brain is the central switchboard. The spinal cord, a somewhat flattened cylindrical bundle of fibers running from the base of the skull to the lower back through a bony canal formed by the vertebrae, is the main trunkline for sensory and motor impulses.

If the central nervous system of a person suffering from multiple sclerosis could be spread out for inspection and the impulses made visible as light traveling along the nerves, it would be easy to spot discolored patches where the myelin sheath has disappeared or been destroyed, exposing the nerve fibers. The disease is termed *multiple* because the patches, or plaques as they are called, often are widely scattered and affect many areas of the central nervous system—the spinal cord, the cerebellum, the brain stem, the cerebrum.

The plaques, which vary greatly in size and shape as well as location, appear as clearly defined islands where the soft myelin has degenerated and later will be replaced by hard, semitransparent scar tissue. These are islands of sclerosis, from a Greek word meaning "hardening."

At first only the sheath is affected. The axons still transmit the messages, although there may be malfunctions indicating that a weakening or partial blockage of the nerve impulse has occurred. As the plaques develop and scar tissue forms, the impulses find it increasingly difficult to travel past these hard spots; ultimately, nonconducting scar tissue may completely block the impulses. [1:2–5]

Concluding section. Paragraphs 11–13 and the illustration summarize what takes place in the MS debilitating process.

The explanations of two technical processes (Figure 8.5), written by students, exemplify the principles discussed in this chapter. For a valuable learning exercise, check these explanations against the preceding outline and text material to see which guidelines were followed and which were not.

How to Obtain a Velocity Profile of
Water Flowing in a Pipe

The process of obtaining a velocity profile of water
flowing in a pipe consists of taking velocity measurements by
means of a U-tube manometer at specified distances across the
pipe. The velocity is measured by a pitot tube. This is a tube
which is situated in the pipe so that it is pointing in the
direction of flow. The flowing water is brought to a stop in
front of the tube. This results in an increase in pressure of the
fluid in the tube. This increased pressure is transmitted inside
a rubber tube in a U-tube manometer where the pressure
differential pushes the water level down one leg and up in the
other. The difference in the height of these columns of water in
the U-tube manometer is an indication of the velocity of the
water in the pipe.

The process of measuring the velocity of water flowing in a
pipe requires the use of the following pieces of equipment:

U-tube manometer

Rubber tubing

Pitot tube assembly with adjustment screw, pointer and
        scale

Pipe flanges

Rubber gaskets

Wrenches

Data paper and pencil

French curve

Scale

The steps to be followed are: (1) placing the pitot tube
assembly in the pipe, (2) assembling and checking the equipment,
(3) taking the velocity data, (4) plotting the results.

Figure 8.5a
Sample Explanations of Technical Processes

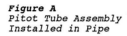

Figure A
Pitot Tube Assembly
Installed in Pipe

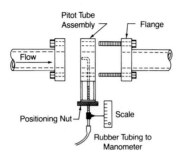

Placing the Pitot Tube Assembly in the Pipe

1. Choose a joint in the pipe and screw the pipe flanges on the pipe at this point. Bolts through the flanges will hold the pipe together with the pitot tube assembly between them.

2. Insert the pitot tube assembly in the pipe making sure that the pitot tube is pointed upstream. See Figure A.

3. Slip the bolts through the holes in the flanges and pitot tube assembly. See Figure A.

4. Place rubber gaskets between the pipe flanges and pitot tube assembly, then screw on the nuts and tighten them with a wrench. Care must be taken here to be sure that the rubber gaskets do not slip and cause a leaky joint.

Assembling and Checking the Equipment

1. Select a U-tube manometer with two foot legs and a scale between them. This size of manometer will permit a large range of flow rates. See Figure B.

2. Fill the manometer legs with water until the top of the meniscus in both legs is at the zero mark of the scale.

Figure B
U-Tube Manometer

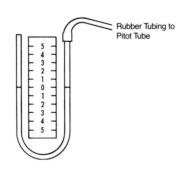

Figure 8.5a
(continued)

3. Attach one end of the rubber tubing to a leg of the manometer, and the other end to the pitot tube assembly.

4. Select a flow rate and allow water to flow in the pipe.

5. Check the equipment for plugged lines and leakage by turning the screw clockwise to advance the pitot tube into the stream flow.

6. Check that the pointer indicates zero on the scale when the pitot tube is against the near side of the tube.

Taking the Readings

1. Observe and record the difference in levels of the liquid in the manometer when the pitot tube pointer indicates zero. There will be a slight difference in levels indicated since the pitot tube has the same diameter and will project into the flow area.

2. Advance the pointer by turning the screw clockwise to 0.05 inches and read, then record the difference in levels in the manometer liquid again. The difference will be greater since the velocity is greater in this part of the pipe.

3. Continue taking readings every 0.05 inches across the diameter of the pipe.

4. Repeat the process of determining the velocity by the difference of liquid levels in the manometer every 0.05 inches on the return trip through the pipe. This procedure will give two values at each point and their average will give a more representative value.

Plotting the Results

1. Determine the average of the pressure difference at corresponding points in the pipe.

2. Set up a Cartesian coordinate system with the difference in levels of the manometer (Δ_η) as abscissa and the distance from zero (Δ_x) as ordinates. See Figure C.

3. Plot the points and connect with a smooth curve. This curve represents the velocity profile of the water in the pipe. See Figure C.

As may be seen in Figure C, water does not flow at a constant speed across the diameter of a pipe. The velocity is zero at the inner edge of the pipe and maximum in the center of the pipe. This can be explained by the fact that there is relative motion between the water and the wall of the pipe. The water must therefore shear and the velocity profile shown indicates the symmetrical distribution of shear stress and velocity.

Figure C

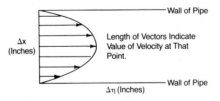

Figure 8.5b
(continued)

```
              How to Determine the Internal Resistance
                      of a D.C. Voltage Source

        A manufacturer does not always provide specific
information on the internal resistance of a D.C. voltage source.
You can determine the unknown resistance by making several simple
voltage measurements using ordinary laboratory equipment.

        For the measurements, you will need the following
laboratory equipment:

        A high impedance D.C. voltmeter
        A calibrated variable resistor
        Four insulated conductors

        The steps followed are these: (1) connecting the
voltmeter, (2) measuring the open-circuit voltage, (3) connecting
the variable resistor, (4) measuring the voltage across the
variable resistor.
```

<u>Connecting the Voltmeter</u>

```
1.    Estimate the magnitude of the internal resistance of the
      D.C. source, and choose a voltmeter with a resistance of at
      least twenty times this estimated value.

2.    Connect one insulated conductor to the voltmeter terminal
      marked (+).

3.    Connect one insulated conductor to the voltmeter terminal
      marked (KV). K is the voltage magnitude corresponding to
      full scale deflection of the meter. The scale chosen should
      be large enough to measure the total source voltage. See
      Figure D.
```

<u>Measuring the Open-Circuit Voltage</u>

```
1.    Connect to one source terminal the free end of the
      conductor attached to the (KV) terminal of the voltmeter.

2.    Touch the other source terminal with the free end of the
      conductor attached to the (+) terminal of the voltmeter,
      and observe the voltmeter needle. If the voltmeter needle
      moves to the left, the terminal is negative.
```

Figure D

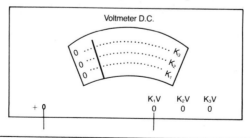

Figure 8.5b
(continued)

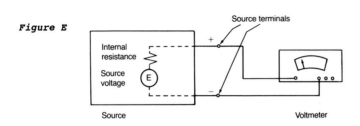

Figure E

3. Having determined the source polarity, connect the (−)
    terminal of the voltmeter to the positive terminal of the
    source and the (KV) terminal of the voltmeter to the
    negative terminal of the source.

4. Read and note the voltage indicated on the appropriate
    voltmeter scale. The voltage thus read is effectively the
    source voltage. See Figure E. Since the voltmeter
    resistance is so much larger than the internal resistance
    of the source, the voltage drop across the internal
    resistance is insignificant compared to the voltage drop
    across the meter. Also the large impedance limits the
    current, thus making the voltage drop across the internal
    resistance of the source even more insignificant.

Connecting the Variable Resistor
1. Do not change the voltmeter connections.

2. By means of the other two conductors, connect the variable
    resistor between the two source terminals. See Figure F.

Measuring the Voltage Across the Variable Resistor
1. Adjust the variable resistor manually until the voltmeter
    indicates a voltage exactly one-half that indicated for the
    open-circuit condition previously noted.

2. Note the resistance of the variable resistor at that
    setting. That value of resistance is the internal
    resistance of the source. The high impedance voltmeter will
    draw only an insignificant current when in parallel with a
    low resistance, so therefore will indicate the actual
    voltage across the low value resistor. Since the sum of the
    voltages around a closed loop is zero, and since one-half
    of the source voltage now appears as a voltage drop across
    the variable resistor, the other half must be a voltage
    drop across the internal resistance of the source. With the
    same current through both resistors and equal voltage drops
    across both, the resistors must be equal in magnitude.

Figure F

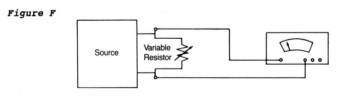

Figure 8.5b
(continued)

Principles for Explaining a Process

The explanation of a process, especially one that involves the fabrication of a product or the conducting of an experiment or activity that leads to a specific result, should include the following information:

1. The definition of the process
2. A description of the time, setting, performers, equipment, materials, and preparations
3. An indication of the principles behind the operation
4. The listing of the major steps in sequential order
5. A step-by-step account of every action and, if appropriate, inclusion of description of apparatus, materials, and special conditions
6. Details under major steps, arranged in sequential order
7. Drawings to aid the explanation of crucial steps of an action
8. A concluding section, which may be a summary or an evaluation of the process

The major elements in the organization of the description of a process are:

1. The introduction (items 1–4),
2. A step-by-step description of actions taken by the performer, or the steps in the development of the process (items 5–7), and
3. The conclusion, which summarizes the process and comments on its importance or usefulness (item 8).

Chapter in Brief

Chapter 8 continued the examination of the basic expository techniques of technical writing, focusing on how to describe processes, both those resulting directly from human participation and those occurring without human intervention. We presented an algorithm for accomplishing this descriptive technique.

Chapter Focal Points

- Algorithm for organizing and writing a description of a process
- Use of illustrations in process descriptions
- Style in process descriptions

Assignments

1. Write a detailed explanation of an experiment or process you have performed. In writing your exposition, use the first person active voice. Consider the following questions:

   a. What was the experiment or process you performed?

   b. Why did you perform the experiment or process?

   c. What tools or instruments did you use?

   d. What materials did you use and where did you get them?

   e. How did you proceed in performing the experiment or process? What steps did you follow?

   f. What was the result of your work?

2. Write a set of directions for a young or nontechnical reader on any of the following:

   a. How to fold a sheet of paper into a ship.

   b. How to start a fire without matches or a lighter.

   c. How to do a trick of magic.

   d. How to miter a board.

   e. How to prepare a specimen for a microscope.

   i. How to clean a computer's flexible disk drive.

3. Write a description of a process within your field of professional interest whose occurrence is the result of a natural, psychological, sociological, political, or human interaction factor. Follow the organizational pattern explained in this chapter. You may choose from any process identified below or any your instructor may assign.

   a. Volcanic eruption

   b. Home equity loan

   c. How a space rocket works

   d. Scoring in professional football

   e. Perception of depth

   f. Homogenization (physical or social)

   g. Role relationship within a group

   h. NOW bank checking account

   i. Photosynthesis

   j. Operant conditioning

   k. Electric storm

   l. City Manager form of government

References

1. Bardossi, Fulvio. *Multiple Sclerosis: Grounds for Hope.* Public Affairs Pamphlet No. 335A. New York: Public Affairs Pamphlets, 1971.

2. Grant, Madelein Parker, *Microbiology and Human Progress.* New York: Rinehart, 1953.

3. "Steel Roof Deck Sample for Calorimeter Evaluation." *Factory Mutual System Loss Prevention Data,* July 1972.

9

Basic Expository Techniques in Technical Writing—Analysis

Chapter Objective

Provide algorithms for and proficiency in the expository process of analysis

Chapter Focus

- Analysis defined
- Classification
- Partition
- Synthesis and interpretation in technical writing

Dictionaries define **analysis** as a systematic and logical process of separating or breaking up a whole into its parts, so as to determine their nature, proportion, function, or relationship. The word is derived from the Greek, *lyein*, meaning "to loosen," and the Greek prefix, *ana-*, meaning "up." To analyze, then, is to loosen or break up a subject into its logical entities.

The process of analysis is fundamental to all scientific and technical activity and to reporting and communicating such activity. It is the process of dividing a problem into its component parts. In chemistry, analysis plays an important role in the separation of compounds and mixtures into their constituent substances to determine the nature or the proportion of the constituents. In mathematics, the process is used to aid in solving problems by means of equations and to examine the relation of variables, as in differential and integral calculus. In medicine, analysis of symptoms plays an important part in the proper diagnosis of a disease. In logic, analysis is used to trace things to their source and to resolve knowledge into its original principles. Analysis aids clarity of thought by breaking down a complex whole into as many carefully distinguished parts as possible and helps determine how the parts are related within the whole. In technical writing, the process of analysis helps writers understand the subject under investigation. It enables them to see component parts and identify the relationship of those parts to each other and to the whole. Analysis helps writers to distinguish and group together related things; to select data essential to the subject and problem; to select out of a mass of data the relevant material and to eliminate the irrelevant. In short, analysis is the process that helps writers understand their material and organize it into a logical order for efficient communication.

The basic operational element in analysis is division. When a heterogeneous assortment is divided into categories or classes, the process is known as **classification**. When a whole is divided into its parts, the process is called **partition**.

Classification and partition are closely related to definition. Definition, as seen in Chapter 6, is actually a form of analysis that resolves a subject into its component parts by (1) identifying the class to which a subject belongs and (2) differentiating or distinguishing the characteristics by which the subject is set apart from other members of its class. To define is partly to classify and partly to partition. Analysis by classification examines one arm of definition—the genus or class. To *classify* is to determine the whole of which the subject is part. Analysis by partition examines the other arm of definition—the differentia or distinguishing elements. To *partition* is to determine the parts of which the subject is a whole. Classification defines a subject by revealing its essence through comparison; partition defines a subject by listing the details or parts of its essence.

Classification

All science starts with classification. Our universe is much too vast and complex to be examined as a whole; so scientists select manageable portions for observation and investigation. Thus, science is frequently defined as the branch of study concerned with the observation and classification of facts.

A group of things that have a defined characteristic in common is called a *class*. Examples are compounds containing carbon, things made of iron, substances which are transparent, compatible computer systems, coordinate numbers, and scientific theories. It should be noted that membership in a class does *not* imply being *exactly* like all members of the class but being alike only with respect to the *specific* quality or characteristic on which the classification is based. One object can be a member of a large number of different classes, one for each of the qualities it possesses. For example, a dog is a member of the Vertebrata, the Mammalia, the Carnivora; it is furbearing, friendly to children, and vicious to intruders, and if the dog is young enough, it can be taught new tricks. Fogs may fit the classification of a colloid, a liquid, a gas, water, air, and, at times, ice crystals. Classifications can be structured in an infinite number of ways. The basis for the classification is always the practical consideration—the use to which it will be put.

Classification has become a fixture in the operation of some of the sciences, especially botany and zoology:

> Taxonomy, the science of classification, attempts to arrange organisms in relation to one another. The theory of organic evolution has greatly enriched this science, so that today the orderliness exhibited by living forms and recorded by the taxonomists has come to imply hereditary relations. In other words, classification of higher plants and animals is based on kinship and reveals lines of descent. . . . The fact that both birds and bats possess wings and fly is a similarity of no taxonomic importance. A study of the embryology and adult anatomy of these two animals reveals two similarities and significant differences. Both birds and bats possess a segmented vertebral column and a dorsal hollow nervous system. These characteristics together with others, place both these animals in the major phylum, Vertebrata. Feathers place birds in the class, Aves, and the possession of mammary glands and hair places bats within the Mammalia, which also includes man. Bats possess claws with forearms modified for flight, which classifies them in the order Chiroptera. Man, with nails, is admitted to the order Primates. . . .

Classification of Three Well-Known Animals

Taxonomic Group	Blue Jay	Little Brown Bat	Man
Phylum	Chordata	Chordata	Chordata
Sub-Phylum	Vertebrata	Vertebrata	Vertebrata

Classification of Three Well-Known Animals

Taxonomic Group	Blue Jay	Little Brown Bat	Man
Class	Aves	Mammalia	Mammalia
Order	Passiformes	Chiroptera	Primates
Family	Corvidae	Vespertiliomidae	Hominidae
Genus	Cyanoeitta	Myotis	Homo
Species	cristata	lucifugus	sapiens
Scientific Name	*Cyanoeitta cristata*	*Myotis lucifugus*	*Homo sapiens*

Man and bat are both vertebrates. They are also mammals. At this point, however, their lines of descent diverge and we must arrange them in separate orders. If we study birds and bats in detail, we find that although they share a vertebrate ancestry they diverge from a common reptilian stock on the main trunk vertebrate line at this juncture. They must therefore be grouped directly into separate classes. In a word, bats and man are much closer relatives than birds and bats, in spite of the fact that birds and bats appear to have more in common with each other than either group seems to have with man. The similarities seen in birds and bats, however, are superficial compared to the closer relationship between man and bats revealed by their embryological development and their adult anatomy. The adult modification of forearm and skin flap used by the adult bat for flight is a species specialization just as is the erect posture seen in man.

The science of classification makes it possible for scientists working with bats to communicate with one another concerning a particular bat. An unknown bat may be identified by referring to the key of nomenclature, and if a new form is found the taxonomist is free to decide whether the new characters warrant establishing a new species or whether these new identifications merely indicate a new strain of a previously described species has been found. The formulation of rules for international nomenclature in taxonomy was an early step toward meeting the growing need for an easy and accurate exchange of essential information among all countries of the world. [4:154–55]

In their book, *An Introduction to Logic and Scientific Method,* Cohen and Nagel have this to say:

There is a general feeling shared by many philosophers, that things belong to "natural" classes, that it is by the nature of things that fishes, for instance, belong to the class of vertebrates, just as vertebrates "naturally" belong to the class of animals. Those who hold this view sometimes regard other classifications as "artificial." . . . All classification, however, may . . . be said to be artificial, in the sense that we select the traits upon the basis of which the classification is performed. . . .

Various classifications . . . may differ greatly in the logical or scientific utility, in the sense that the various traits selected as a basis of classification differ widely in their fruitfulness as principles of organizing our knowledge. Thus the old classification of living things into animals that live on land, birds that live in the air, and fish that live in water gives us very little basis for systematizing all that we know and can find out about these creatures. The habits and the structure of the porpoise or the whale may have more significant features in common with a hippopotamus or the horse than with a mackerel or a pickerel. The fact that the first two animals named have mammary glands and suckle their young, while all species of fish deposit their eggs to be fertilized, makes a difference which is fundamental for the understanding of the whole life cycle. In the same way the fact that some animals have a vertebral column, or, to be more exact, a central nervous cord, is the key which enables us to see the significance of the various structures and enables us to understand the plan of their organization and functioning. Some traits, then, have a higher logical value than others in enabling us to attain systematic knowledge of science. [1:233–34]

Using Classification as an Expository Technique

Classification is a useful technique in exposition. It enables writers to systematize widely diverse facts for presentation in an orderly arrangement so that readers can follow easily and understand the relationship of these facts. Classification may occupy an entire volume, as a textbook on botany or on types of motors, or it may form a chapter, a section of a chapter, or a paragraph in a chapter. Whenever writers must answer the question, Where does this species or this idea belong? they classify. The analytical process of classification helps writers organize their material for logical presentation. It enables them to examine their material in proper proportion: to differentiate the important from the unimportant, to give proper weights and values to items within the material, and to structure the presentation in parallel fashion or in subordinate fashion.

The recognition of similarities and differences among experiences through classification is a basic process in all scientific research. It enables useful insights that make it possible to structure classes in such a way that mere membership in a class reveals probable attributes. In other words, certain characteristics appear to be associated with each other. Thus, a class of objects with the properties of strength, susceptibility to corrosion, and capacity to be magnetized can also be assumed to be good conductors of electricity and will probably have the other properties of iron and steel. The process of classification, as we shall see later, helps the mind to interpret facts and data through interference.

Here is how the process of scientific classification enables the scientists inductively to identify an unknown—in this case, bacteria:

Let us suppose that we have collected a large number of observations concerning the structural and physiological characters of an unknown organism. The data that we have are listed below:

Shape:	Long rods; occur singly and in chains
Motility:	Actively motile
Gram Stain:	Gram-positive
Spore Formation:	Definite spore, central in position
Oxygen Requirement:	Grows best in presence of atmospheric oxygen; facultative anaerobic growth
Agar Colonies:	Spreading, grayish, irregular borders, adherent
Gelatin Stab:	Surface growth; liquefaction
Litmus Milk:	Alkaline; peptonized

Dextrose Phenol Red
    Agar Stab: Acid; no gas
Saccharose Phenol Red
    Agar Stab: Acid; no gas

Grows best at 37°C. Grows also, but less well, at room temperature. Organism was obtained by rubbing sterile, moist swab over new petri dish just unpacked from straw wrappings.

Here, then, are our data. We next consult *Bergey's Manual of Determinative Bacteriology.* . . . To the experienced bacteriologists this organism at once suggests that it is a member of the family Baccillaceae, because it is a rod-shaped, spore-forming bacterium; and of the genus *Bacillus*, because it grows aerobically. By following out further details, we find that it corresponds to the detailed description of *Bacillus subtilis*. The inexperienced person who does not carry this information in his mind as part of his usual equipment can turn to the key and follow the logical procedure in tracing out the identification of an unknown organism. . . . [4:205–6]

Algorithms for Classification

1. Know your subject.
2. Define the term (subject) to be classified. The purpose of classification is to bring clarification and order to a number of elements that are diverse and perhaps confused in the reader's mind. Grouping things together may be meaningless to a reader unless he understands what you are talking about in the first place. This requires careful identification and definition of the term. It may also mean sufficiently limiting the term so that it is useful to the purpose of the reader.
3. Select a useful and logical basis for the classification.
4. Keep the same basis for all of the items within a grouping. Changing the basis invites confusion. For example, a division of the cat family into lions, tigers, leopards, lynxes, pumas, jaguars, and cats friendly to children does not follow a consistent principle of grouping.
5. Keep groupings, though related, distinct. Classes must not overlap. Classification of a technical report as examination, investigation, recommendation, research, or formal indicates overlapping because a single report could easily fit all the listed categories at the same time. Reduction of phenomena into a logical and orderly system frequently can offer difficulties. Species do not always seem mutually exclusive. There are times when the subject seems to be neither fish nor fowl. The duckbill platypus is such an anomaly. This zoological curiosity has a beak, webfeet, and lays eggs. These factors would serve to classify it as a bird. However, it also has the reptilian characteristic of poisonous burrs, which are similar to the fangs of a snake. Because the furry duckbill platypus suckles its young, it has been classified as a mammal. This feature is significant in the evolutionary processes of animal forms, and from that approach, its classification as a mammal is justified by zoologists.

6. Be sure to include all members in the classification. In other words, the classification should be complete. If incomplete, it does not serve as the "umbrella" to cover every member that logically belongs under it. For example, a classification of semiconductors would be incomplete if it did not contain thermisters.

7. Have at least two classes on each level of division. Since classification entails the breaking down of a grouping into its member parts on a logical basis, there must be at least two member parts in any such breakdown.

8. Arrange classifications in a logical order, convenient for the reader to follow. Two formats lend themselves to such a presentation—tabular and outline. Tables facilitate classification of data. Outlines enable readers to see relationships and help reveal how subspecies are related to the parent genus.

Partition

Partition is a form of analysis. It is the division of a whole into parts. It is exactly the process used by an automobile mechanic in taking a motor apart, by a chemist in breaking down a substance into its components, and by a biophysicist in breaking down the cell into its minute elements.

For example, a typical microcomputer system has the following components:

Basic computer (containing the microprocessor and primary storage)
Keyboard
Disk drives
Monitor (display screen for soft copy output)
Printer (for hard copy output)
Serial or parallel interface card (to permit printer, modem, and other peripheral connections)
Connecting cables (to connect hardware components)
Software (for operation and application)
Modem (for telephone telecommunication)

The vertebrate eye may be partitioned into the conjuctiva, aqueous humor, pupil, lens, iris, cornea, vitreous body, layer of rods, optic nerve, retina, pigment layer, choroids, and sclera.

A formal report has a cover, a title page, an abstract or an executive summary, a table of contents, an introduction, a body, a terminal section, a bibliography, and, frequently, appendixes.

Partition, although employing a logical division, is not the same as classification. Within the process of partition, a complete entity is broken up into its components. Within classification, a general group is logically divided into the various species that make up the class. For example, mollusks, arthropods, and vertebrates all have eyes or visual organs. The eyes are quite different types of sensory organs, but they do perform the seeing function for the animal of which they are a part. However, when an eye is partitioned, any of its parts— whether the pupil, lens, iris, retina, vitreous body, or optic nerve—is not, of itself, the eye.

Partition may be applied to abstract concepts, as well as to physical entities. In communication theory, for example, the abstract concept of the communication unit is divided into the source, transmitter, receiver, and destination. The practice of psychiatry might be divided into the prevention, diagnosis, treatment, and care of mental illness and mental defects. The study of color divides into shade, hue, and intensity.

This analytical process is basic to all fields of activities. The logician, Crumley, has observed:

> Partition is employed by the builder in laying out his work; it is indispensable to the playwright in fashioning his plot; it is an aid to the lawyer in drawing up his brief, to the orator in marshaling his argument, to the painter in balancing his composition, and to the musician in apportioning his theme. [2:85–86]

In summary, classification is a means of analyzing or explaining a plural subject. It examines relationships by determining similarities within the various species of its subject. Partition, on the other hand, analyzes and explains a single subject by examining its various components. The analytical approach used in partitioning is similar to that used in classification.

Both classification and partition are obviously related to formal logic. Classification uses reasoning from the particular to the general (induction); partition uses reasoning from the general to the particular (deduction). The identity of a class can be established by examining and grouping particulars together on the basis of a common element. The entity of particulars can be established by showing that the particulars belong to a class whose identity has been established. Often deduction (partition) is possible only after the identity of a class has been established by a previous induction (classification). In practice, this interrelationship is exemplified by a physician who diagnoses a disease by classifying the symptoms, then detects other diseases through the deduction of the diagnosis.

Algorithms for Partitioning

1. Be consistent in your approach or viewpoint in dividing your subject. Your viewpoint should be made clear to your readers. A logical approach to the partition might (as in the description of a mechanism) be from the viewpoint of the observer's eye, from the sequence by which the various parts are arranged, or from the way the parts function.

2. Separate and define clearly each part of the division. Parts should be mutually exclusive.

3. List all parts or explain any incompleteness. Within a generalized description—one representative of a class—partition may be limited to major, functional parts. In such a situation, the partition must be preceded by a qualifying statement such as "the chief functional parts are. . . ."

4. Check your stated partition for unity. The mechanism, organism, thing, or concept must be properly introduced and defined. The point of view from which the partitioning is approached must be explained. The analysis follows with a logical listing, description, and explanation of parts and their relationship. The concluding element of analysis by partition will show how the various parts relate to each other and to the whole.

Partition as an expository technique may comprise an entire monograph or it may be included as a section or a chapter in a book or a paragraph or two within a larger piece of writing. Maintenance manuals are characterized by their numerous uses of the partition explanatory technique.

Use of Synthesis and Interpretation in Technical Writing

Closely related to analysis are the two processes of synthesis and interpretation. **Synthesis** is the antonym of analysis; it is the putting together of parts or elements so as to form a whole. While opposed to analysis, the process of synthesis is frequently complementary to it. This is borne out graphically in the science of chemistry, which utilizes both processes of taking apart and putting together. Analysis and synthesis help in arriving at interpretation. Analysis breaks data down and arranges them into the logic demanded by the situation. The scientist or the technical writer may rearrange or recombine the various elements of the data in order to arrive at the meaning or explanation of the data through inductive and deductive inferences. **Interpretation** affords explanations, meaning, and conclusions regarding problems associated with the data. Analysis and synthesis provide a logical process for achieving appropriate insights, conclusions, and generalizations.

In his book, *How We Think*, Dewey shows how interpretation evolves through the operation of analytic and synthetic thought:

> Through judging [judgment], confused data are cleared up, and seemingly incoherent and disconnected facts are brought together. This clearing up is *analysis*. The bringing together, or unifying is *synthesis*. . . .
>
> As analysis is conceived to be a sort of picking to pieces, so synthesis is thought to be a sort of physical piecing together. . . . In fact, synthesis takes place wherever we grasp the bearing of facts on a conclusion or of a principle on facts. As analysis is *emphasis*, so synthesis is *placing*; the one causes the emphasized fact or property to stand out as significant: the other puts what is selected in its *context*, its connection with what is signified. It

unites it with some other meaning to give both increased significance. When quicksilver was linked to iron, tin, etc., as a *metal*, all these objects obtained new intellectual value. Every judgment is analytic in so far as it involves discernment, discrimination, marking off the trivial from the important, the irrelevant from what points to a conclusion; and it is synthetic in so far as it leaves the mind with an inclusive situation within which selected facts are placed. . . .

The analysis that results in giving an idea the solidity and definiteness of a concept is simply emphasis upon that which gives a clew for dealing with some uncertainty. If a child identifies a dog seen at a distance by the way in which the animal wags its tail, then that particular trait, which may never have been *consciously* singled out before, becomes distinct—it is analyzed out of its vague submergence in the animal as a whole. The only difference between such a case and the analysis effected by a scientific inquirer in chemistry or botany is that the latter is alert for clews that will serve the purpose of sure identification in the *widest possible area of cases*; he wants to find the signs by which he can identify an object as one of a definite kind or class even should it present itself under very unusual circumstances and in an obscure and disguised form. . . .

Synthesis is the operation that gives extension and generality to an idea, an analysis makes the meaning distinct. Synthesis is correlative to analysis. As soon as any quality is definitely discriminated and given a special meaning of its own, the mind at once looks around for other cases to which that meaning may be applied. As it is applied, cases that were previously separated in meaning become assimilated, identified, in their significance. They now belong to the same kind of thing. Even a young child, as soon as he masters the meaning of a word, tries to find occasion to use it; if he gets the idea of a cylinder, he sees cylinders in stove pipes, logs, etc. In principle this is not different from Newton's procedure in the story about the origin in his mind of the concept of gravitation. Having the idea suggested by the falling of an apple, he at once extended it in imagination to the moon as something also tending to fall in relation to the sun, to the movement of the ocean in the tides, etc. In consequence of this application of an idea that was discriminated, made definite in some one case, to other events, a large number of phenomena that previously were believed to be disconnected from one another were integrated into a consistent system. In other words, there was a comprehensive synthesis.

It would be a great mistake, however, as just indicated, to confine the idea of synthesis to important cases like Newton's generalization. On the contrary, when any one carries over any meaning from one object to another object that had previously seemed to be of a different kind, synthesis occurs. It is synthesis when a lad associates the gurgling that takes place when water is poured into what he had thought was an empty bottle with the existence and pressure of air; when he learns to interpret the siphoning of water and the sailing of a boat in connection with the same fact. It is synthesis when things themselves as different as clouds, meadow, brook, and rocks are so brought together as to be composed into a picture. It is synthesis when iron, tin, and mercury are conceived to be of the same kind in spite of individual differences. [3:126; 129–30; 157–59]

Examples of Classification

Fog

Condensation and consequent formation of water droplets (or ice crystals) in the air at the earth's surface will produce a fog. Fogs are classified in many ways. One of the simplest is the use of formation cause or processes as a basis for differentiation among the various types.

1. Advection fogs are fogs that owe their existence to the flow of air from one type of surface to another. Surface temperature contrast between two adjacent regions is necessary in causing the formation of advection fogs.

   a. The usual type of advection fog is formed when relatively warm and moist air drifts over much colder land or water surfaces. Examples of this type are found over land when moist air drifts over snow-covered areas, or over water when moist warm air drifts over currents of very cold water. The latter happens with the southerly or easterly winds blowing from the gulf stream over the Labrador Current.

   b. Coastal and lake advection fog forms when warm and moist air flows offshore onto cold water (summer) or when warm, moist air flows onshore over cold or snow covered land (winter).

   c. Sea smoke, arctic fog, or steam fog form in very cold air when it flows over warm water.

2. Radiation fog is the type that develops in nocturnally cold air in contact with a cool surface. Radiation fog forms over land and not over water because water surfaces do not appreciably change their temperature during hours of darkness.

3. Upslope fog is caused by dynamic cooling and air flowing uphill. Upslope fog will form only in air that is convectively stable, never in air that is unstable because instability permits the formation of cumulus clouds and vertical currents.

4. Precipitation fog forms in layers of air which are cooler than the precipitation which is falling through them. The greater the temperature difference between a relatively warm rain (or snow) and the colder air layer, the more rapidly will the fog develop. Fogs associated with fronts are largely precipitation fogs. Visibility in fogs varies from a few feet up to a mile. Often in fog blankets, all ranges of visibility are present. [7:682–83]

Types of Fault Movement in Earthquakes

Most but not all earthquakes are associated with observed faults. A fault is a fracture zone along which the two sides are displaced relative to each other. Most faults are readily recognizable by trained geologists. Not all faults are active, but there is considered to be no satisfactory method of predicting the probable future activity of a given fault in a precise sense.

Displacement along a fault may be vertical, horizontal, or a combination of both. Movement may occur very suddenly along a stressed fault, producing an earthquake, or it may be very slow, or what is called "creep," unaccompanied by seismographic evidence. Permanent displacement of ground during an earthquake might be several inches, or it might be tens of feet.

Sentence 1 defines fog operationally. Sentence 3 indicates the basis for the classifications.

Under Classification 1 are grouped the advection types of fogs. Note that the classification begins with a definition in order to explain the basis for all types within it.

Classification 2 is also defined.

Classification 3 is defined operationally.

Classification 4 also is defined operationally. The final paragraph offers a characteristic general to all groupings of fogs.

This example on fault movement follows the principles of classification and offers an interesting and frequently used technique in technical writing—illustration. Sentence 1 sets the background for the subject to be classified. Sentence 2 defines the subject. The text and caption provide background information but the illustration explains graphically and effectively the similarities and differences.

Types of fault movement [are shown in Figure 9.1]. a) Names of some of the components of faults. b) Normal fault, in which the hanging wall has moved down relative to the foot wall. c) Reverse fault, sometimes called thrust fault, in which the hanging wall has moved up relative to the foot wall. d) Lateral fault, sometimes called strike-slip fault, in which the rocks on either side of the fault have moved sideways past each other. It is called left lateral if the rocks on the other side of the fault have moved to the left, as observed while facing the fault, and right lateral if the rocks on the other side of the fault have moved to the right, as observed while facing the fault. e) Left lateral normal fault, sometimes called a left oblique normal fault. Movement of this type of fault is a combination of normal faulting and left lateral faulting. f) Left lateral reverse fault, sometimes called a left oblique reverse fault. Movement of this type is a combination of left lateral faulting and reverse faulting. Two types of faults not shown are similar to those shown in e and f. They are a right lateral normal fault and a right lateral reverse fault (a right oblique normal fault and a right oblique reverse fault, respectively). [6:1–2]

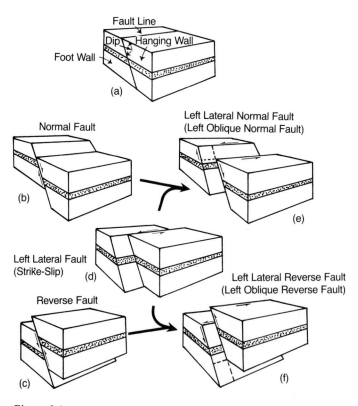

Figure 9.1

Examples of Partition

The descriptions of mechanisms and organisms given in Chapter 7 also illustrate partition. The zones of the earth are partitioned in the following exposition:

Crust of Earth

As it presents itself to direct experience, the earth can be physically described as a ball of rock (the lithosphere), partly covered by water (the hydrosphere) and wrapped in an envelope of air (the atmosphere). To these three physical zones it is convenient to add a biological zone (the biosphere).

The *atmosphere* is the layer of gases and vapor which envelopes the earth. It is essentially a mixture of nitrogen and oxygen with smaller quantities of water vapor, carbon dioxide and inert gases such as argon. Geologically, it is important as the medium of climate and weather, of wind, cloud, rain and snow.

The *hydrosphere* includes all the natural waters of the outer earth. Oceans, seas, lakes and rivers cover about three-quarters of the surface. But this is not all. Underground, for hundreds and even thousands of feet in some places, the pore spaces and fissures of the rocks are also filled with water. This ground water, as it is called, is tapped in springs and wells, and is sometimes encountered in disastrous quantities in mines. Thus there is a somewhat irregular but nearly continuous mantle of water around the earth, saturating the rocks, and over the enormous depressions of the ocean floors completely submerging them. If it were uniformly distributed over the earth's surface, it would form an ocean about nine thousand feet deep.

The *biosphere*, the sphere of life, is probably a less familiar conception. But think of the great forests and prairies with their countless swarms of animals and insects. Think of the tangle of seaweed, of the widespread banks of mollusks, or reefs of coral and shoals of fishes. Add to these the inconceivable numbers of bacteria and other microscopic plants and animals. Myriads of these minute organisms are present in every cubic inch of air and water and soil. Taken altogether, the diverse forms of life constitute an intricate and everchanging network, clothing the surface with a tapestry that is nearly continuous. Even high snows and desert sands fail to interrupt it completely, and lava fields fresh from the craters of volcanoes are quickly invaded by the presence of life outside. Such is the sphere of life, both geologically and geographically it is of no less importance than the physical zones.

The *lithosphere* is the outer solid shell or crust of the earth. It is made of rocks in great variety, and on the lands it is commonly covered by a blanket of soil or other loose deposits, such as desert sands. The depth to which the lithosphere extends downward is a matter of definition: it depends on our conception of the crust and what lies beneath. It is usual to regard the crust as a heterogeneous shell, possibly about twenty to thirty miles thick, in which the rocks at any given level are not everywhere the same. Beneath the crust, in what may be called the *substratum*, or *mantle*, the material at any given level appears to be practically uniform, at least in those physical properties that can be tested. [5:29–30]

This is an interesting example in that it is the basis of a definition of the planet earth, employing the techniques of explication and partition. The terms within the definition in the first paragraph are explicated by the explanations organized in partitioned components of the earth.

Summary of Analysis

1. Analysis requires thinking. It is the systematic and logical process of separating a whole entity into its component parts.
2. The basic operation in analysis is division. When an assemblage of things, ideas, people, or processes is divided into categories, the operation is known as *classification*.
3. When a whole is divided into its parts, the process is called *partition*.
4. Classification defines a subject by revealing its essence through comparison.
5. Partition defines a subject by listing and explicating the details or parts of its essence.
6. Tables, charts, and illustrations are effective aids in showing comparisons, similarities, differences, and component parts in the classification and partitioning processes.
7. Analysis helps writers to organize their subject matter into a rational order for more effective communication.
8. In technical writing, analysis helps both writers and readers to understand a subject; the process helps both to see and understand the ingredient parts and the relationship of those parts to each other and the whole.

Chapter in Brief

In this chapter, the last in our series examining the basic expository techniques in technical writing, we concentrated on analysis, a process that helps writers to understand the substance of their material and to organize it logically for more effective communication. The basic operational element in analysis is division. When a heterogeneous group of things is divided into classes, the process is known as classification. When a whole is divided into its parts, the process is called partition. We also discussed synthesis and interpretation of data, processes for achieving insights, and conclusions.

Chapter Focal Points

- Analysis
- Classification
- Partition
- Synthesis and interpretation

Questions for Discussion

1. What is analysis? From your personal experience, select an illustration of a form of analysis you have undertaken in the laboratory or in a classroom or work situation and one from a simple analysis you have carried out in solving a personal problem.

Present orally or in writing, as your instructor desires, the two problems with their respective backgrounds and point out the ways in which they are similar and dissimilar. What have the two problems in common? Synthesize the results.

2. Locate and discuss examples of classification and partition in two of your texts covering different subject areas.

3. Show the relationship of definition to classification and partition.

4. Analyze the use of space in the building housing your college or your departmental field of study. You will want to consider such elements as adequacy of classrooms, laboratory or other resources, offices, and student study, comfort, and recreation areas. Consider also the flow of traffic in hallways, parking facilities for cars and bicycles, and other factors.

Assignments

1. Write a classificatory analysis of one of the following topics:

   a. Grasses
   b. Pets
   c. Psychosomatic illnesses
   d. Human sense perception
   e. Cells
   f. Drugs
   g. Insurance
   h. Games people play
   i. Forms of government
   j. Computer systems

2. Prepare an expository partition of one of the following topics:

   a. The building you live in
   b. Ten-speed bicycle
   c. Beef Wellington
   d. Management information system
   e. Rattlesnake
   f. Sun dial
   g. Heat pump
   h. Your university/college
   i. Rose bush
   j. Economic recession

References

1. Cohen, Morris, R., and Nagel, Ernest. *An Introduction to Logic and Scientific Method.* New York: Harcourt, Brace, 1934.

2. Crumley, Thomas. *Logic: Deductive and Inductive.* New York: Macmillan, 1947.

3. Dewey, John. *How We Think.* New York: D.C. Heath, 1923.

4. Grant, Madelein Parker. *Microbiology and Human Progress.* New York: Rinehart, 1953.

5. Holmes, Arthur. "Interpretations of Nature." In *The Crust of the Earth,* edited by Samuel Rappaport and Helen Wright. New York: Mentor Books, 1955.

6. "Types of Fault Movement." *Factory Mutual System Loss Prevention Data,* November 1973.

7. *Van Nostrand's Scientific Encyclopedia.* New York: D. Van Nostrand, 1958.

PART FOUR
Technical Report Writing

10

The Role of the Scientific Method in Report Writing

Chapter Objective

Demonstrate that the scientific method, which underlies all research, is also the foundation for the technical report.

Chapter Focus

- Origins of science and learning
- Greek contribution—logic
- The scientific method
- Role of the problem in scientific research
  Types of Problems
- Role of the hypothesis
- Relationship of research to report writing

*I*t is traditional for scientists to report the results of their work. Research is not complete until is has been recorded, reported, and disseminated. The report is the end product in much scientific activity. Since reporting is part of the scientific method, we need to examine the background, meaning, and characteristics of the scientific method.

Science has been defined in several ways. One appropriate definition would be operational. The purpose of scientific research is to discover facts and ideas not previously known. According to many scientists, science always has two aspects. On the one hand, it reveals knowledge and understanding of something previously unknown, mysterious, or misunderstood. On the other hand, it increases our control over nature and the forces at work in nature. This definition implies that science is verifiable and communicable knowledge. The word *communicate* is significant. As Whorf points out, scientific knowledge is, in large part, symbolic knowledge expressed in symbols. Man's ability to communicate enables him to understand and describe the world [4:220].

The Beginning of Science and Learning

Science, as we know it today, dates to the introduction of the experimental method during the Renaissance. There were, however, earlier beginnings.

People learn new facts and ideas through chance, through trial and error, and through experience. Our prehistoric ancestors probably learned about cooking by chance. For example, a primitive man may have come across a stag burned in a forest fire that had been started by a bolt of lightning. He may have bitten into the roasted carcass and found that it tasted better than raw meat. This experience made him want to have all his food acted upon by fire. Consequently, humans may have learned about cooking by chance, but they had to learn the art through trial and error. They soon found that if they were not careful, their meat might either burn into ashes or remain raw.

Primitive people acquired much knowledge by deliberately trying things out. After watching animals eat herbaceous plants, they did the same. Through trial and error, they discovered that some plants were pleasant tasting and others bitter and harmful. They communicated their discoveries to others, and the discoveries became common knowledge. This was how apprehension and comprehension of the world began. Such information was passed on by whatever means of communication available and became a permanent body of useful knowledge.

Information generalized from experience was added to the information derived from chance and from trial and error. Heavy, cloudy skies warned of rain; it was best to seek shelter. Certain shellfish made one sick; it was best not to eat them. Primitive people advanced a notch toward civilization by arriving at useful conclusions based on experience. Generalizing from experience involves the faculty of logic, or reasoning things out, not only from one's own experience but also from the experience of others. Thus, our ancestors began to use the important process of modern scientific inquiry. This process enabled them to acquire more knowledge by examining the experiences of tribe members. It helped them meet potentially harmful situations, advance their well-being, and build up a considerable body of knowledge as a common heritage.

However, let us not think that our primitive ancestors were happily on their way to discovering vacuum cleaners, compact cars, and vitamin tablets. There were many things about themselves and their environment that they could not understand. What they could not satisfactorily explain through experience, they attributed to magic or to supernatural powers. When a tribe was attacked by fierce beasts or by other tribes, when it was struck by plague, or flooded out by a thunderstorm, they wanted to know why such a thing had happened. For answers they turned to priests, witch doctors, or tribal wise men. A priest might call the disaster a punishment for sins, or a failure to honor the tribal gods. The priests or tribal wise men often offered good advice, and people came to trust these chosen authorities. Thus arose a tradition of trust in or reliance on authority, based on the belief that great thinkers were able to discover truth for all times and that humans could best learn about themselves and their world by studying the words of the sages.

This subservience to authority was most influential during the medieval period, when the teachings of the ancients—Plato, Aristotle, and the early church leaders—received more credence than did firsthand observation and analysis of facts.

Greek Contribution: Logic

Aristotle and other Greek philosophers and scientists controlled much of Western thought for almost 1200 years. The Greeks were responsible for developing the mental tools with which we have been able to solve many mysteries of the universe. By developing logic and mathematics, the Greeks were greatly responsible for ending reliance on explanations by magic. Greek philosophers were the first to arrive at the concept of the universe as an ordered cosmos in which everything happens according to the law of cause and effect. They believed in the natural order of things and set out to learn the characteristics of this natural order.

Aristotle believed that one could learn about the unknown by examining the known through a deductive process. The basis of that process is the *syllogism,* a device for testing the truth of any given conclusion or idea. The syllogism is best illustrated by Aristotle's classic example:

> Major premise: 1. All men are mortal.
> Minor premise: 2. Socrates is a man.
> Conclusion: 3. Therefore, Socrates is mortal.

If the first two assertions, or **premises,** can be shown to be true, it must follow that any reasonable person will agree that the third statement (conclusion) must be true. By these principles of classical logic, a conclusion properly deduced from reliable premises is necessarily reliable. This assumes that the premises are true and that every element in the argument has been carefully examined to eliminate all unrelated or false materials. Hence, if it is certain that all acids turn blue litmus paper red, then acetic acid, CH_3COOH, will turn blue litmus paper red. This is certainly true. Given a valid major premise and a related valid minor premise, we can deduce a valid conclusion. The process can be an intellectually stimulating and valuable exercise.

Deductive reasoning, although useful, can also be hazardous because the user can become more involved in the mental processes and in skillful argumentation than in the facts of the case. Medieval scholars paid more attention to Aristotle's methods of deduction than to his caution to observe nature directly. As a result, they arrived at their conclusions and generalizations by reasoning or logic alone. Through syllogism, they deduced conclusions from statements of approved authorities without checking their accuracy and reliability. By skillfully selecting premises, medieval scholars were able to prove almost anything—even how many angels could dance on the head of a pin. The syllogistic approach is limited because only if the premises are true and properly related can we know that the conclusion is true. The truth of premises and their proper relationships can be determined only by observation and experience. Take, for example, this syllogism:

> All creatures that fly with wings are birds.
> The bat flies with its wings.
> Therefore the bat is a bird.

Not all creatures that fly with wings are birds. Among such creatures are insects, some fish, and a few mammals—other than the bat, the flying lemur.

Unless each of the premises is carefully tested and its relationship shown, the conclusion may be wrong. Of course, the basic weakness of the syllogism is that it can guarantee the truth of the conclusion only when it begins with true and properly related statements of fact, but it can provide no assurance of the truth of its premises.

In the sixteenth century, Bacon pointed out the limitations of deductive reasoning. He advocated the need to base general conclusions on facts discovered by direct observation. Bacon urged scholars to ignore authority, to observe nature, to experiment, to draw one's own inferences, and to classify facts in order to reach minor generalizations, before proceeding from minor generalizations to greater ones. In other words, Bacon advocated **inductive reasoning,** or generalizing from examples. He preferred to go from the particular to the general, rather than from the general to the particular, as is the case in deductive reasoning. Induction is based on the assumption that future instances will be the same as those from past experience, such as:

> Water freezes at 32° Fahrenheit;
> Metal expands when heated;
> The sun will rise tomorrow;
> Honesty is the best policy.

This is not to say that research problems can be solved by induction alone. Induction is incomplete until the accumulated data have been analyzed, classified, and related to a guiding principle or problem. Darwin found this out when he followed Bacon's advice. He collected fact after fact in his biological research, hoping the facts themselves would lead to an important generalization. Then he reports:

> In October, 1838, that is fifteen months after I had begun my systematic inquiry, I happened to read for amusement, Malthus "On Population," and being well prepared to appreciate the struggle for existence which everywhere goes on. From my long-continued observation of the habits of animals and plants, it at once struck me that, under these circumstances, favorable variations would tend to be preserved, and unfavorable ones to be destroyed. The result of this would be formation of new species. Here, then, I had at last found a theory by which to work. . . . [2:68]

Darwin, after gathering considerable data and ruminating on his reading of Malthus, deduced what the data might mean. He formulated a hypothesis from the facts he already knew, then investigated further to see whether it would be supported or refuted by additional evidence. Darwin, thus, stumbled on a possible solution to the problem of how evolution takes place and began to test it. By deducing from the evidence, he was able to put his facts into a workable theory. Therefore, he used both inductive and deductive reasoning to arrive at his final conclusions. Darwin's method is a good example of how modern research is frequently carried on.

Scientific Method and Its Approach

What are the distinguishing characteristics of the scientific method? Science begins with observations of selected parts of nature. Although scientists imagine ways the world might be structured, they know that only careful scrutiny will show them if the ways they imagine correspond with reality. They reject authority as an ultimate basis for truth. Where primitive people ascribed anything unusual to special intervention by the gods, today we look for "natural" causes. Though compelled to consider facts and statements issued by other workers, scientists reserve for themselves the decision as to whether these other workers are reputable, whether their methods are sound, and whether the alleged facts are credible.

Before Galileo's time, it was commonly assumed that Aristotle was right in saying that heavy objects fall to the ground faster than lighter objects. This assumption seems logical and reasonable to anyone who thinks about it without bothering to test it by experiment. Galileo, however, refused to accept either logic or authority as the basis for his conclusions. He experimented, in 1589, by dropping cannon balls of various weights and sizes from the leaning tower of Pisa. Perhaps even to his own surprise, he found that the cannon balls fell, except for minor differences caused by air resistance, at the same rate of speed.

The Significance of the Problem to the Investigation

From earliest history, one of our greatest urges has been to *know*. The deep impulse to learn and to share knowledge with others has been one of the most potent factors in our rise from primitive caves to travels in space.

What are the lights in the sky? What makes grass grow? Why is grass green? How long is a river? How deep is the sea? How far away is the moon? These questions were probably asked by our earliest ancestors. The basics of "new" science are found in primitive people's ability to identify and articulate the unknowns of life and in their crude attempts to find solutions to such questions. Primitive savages' recognition of an obstacle, difficulty, or problem and the urge to find answers started them on their way to civilization.

As a primitive people conquered obstacles, difficulties, and problems, they told others of the tribe about their experiences and successes. This communication helped the tribe. The more "research minded" the primitive clan, the greater its prospects for survival. Therefore, civilization grew as humans observed the phenomena of their world, recognized and isolated their problems, investigated them, and arrived at answers. When approaching more difficult obstacles, they had to encourage others to cooperate with them as a team. One person might not be able to get a large log across a raging stream, but two, three, or four might be able to do it. The clan could then cross the stream safely.

Recognizing the role of the **problem** in scientific research is a dominant feature of our modern world. The following episodes illustrate the role of the problem in scientific advancement.

In the fifteenth-century seaport of Genoa lived a young boy who loved to visit the docks and watch the ships come and go. What fascinated him most were the ships leaving the docks and sailing out into the horizon. The boy would watch as the ships grew smaller and smaller, until they disappeared entirely. It was as if they dropped into the ocean.

"What causes the ships to disappear?" he asked himself. "It can't be that they reach the end of the earth and fall straight down, because the ships that disappear return after a year or two." When the boy grew older and became a sailor himself, he was intrigued by another phenomenon—the disappearance of land from the sailor's view as the distance between port and ship increased. After talking to a fellow sailor who had been blown off course and claimed to have reached and returned from a land in the west, Christopher Columbus was strengthened in his hypothesis that the reason ships disappear to the viewer on land is that the earth is not flat but round. The "beyond" in the west was not the edge of the world but the Orient. Columbus sought to prove this hypothesis. If the earth were round, he ought to be able to reach India by following the curvature of the earth. As we know, this hypothesis was accurate—although the North American continent stood in his way.

Paul Ehrlich, the father of chemotherapy, hypothesized that since certain dyes had a special affinity for certain tissues in the body, it might be possible to find substances that would go one step further and seek out germs as well. His faith in the hypothesis made him persist in his investigation despite continued frustration, repeated failures, and friends' attempts to dissuade him from the apparently hopeless task. Eventually, after more than 600 attempts, he came upon his "magic bullet," salvarsan, which could track down the syphilis germ and destroy it without harming surrounding healthy tissues and cells.

In our time, after it was discovered that poliomyelitis was caused by a virus, a doctor in Pittsburgh asked himself, "If polio is caused by a virus, why can't we use immunization techniques against polio and come up with a protective innoculation?" Dr. Jonas Salk and others have fortunately proved this possible.

Recognition of a problem, then, is the starting point of inquiry. In *Logic: The Theory of Inquiry*, Dewey states that "a problem well put is half solved" [3]. In *An Introduction to Logic and Scientific Method*, Cohen and Nagel point out the following:

> It is an utterly superficial view that the truth is to be found by "studying the facts." It is superficial because no inquiry can even get underway until and unless *some difficulty is felt* in a practical or theoretical situation. It is the difficulty, or problem, which guides our research for *some order among facts* in terms of which the difficulty is to be removed. [1:199]

The problem helps researchers decide what direction they must take in their investigation. It is their criterion for what is relevant and what is not; it guides them in selecting the data appropriate to the end result—the answer to the problem.

Types of Problems

Problems that scientists face fall into three major categories: problems of fact, problems of value, and problems of technique.

Problems of Fact

As the term implies, these problems seek answers to what the facts are. Is the earth flat or round? At what temperature will water freeze at an altitude of 10,000 feet? What causes cancer? How much salinity is in Farmer Brown's well? How effective is interferon against viral infections? Problems of fact involve questions of what happens, when it happens, how it happens, and why it happens.

Problems of Value

These prolems deal with what is more valuable, what is preferable, or what ought to be the case (rather than what is the case). Problems of value are involved in setting up standards or criteria. Examples of problems of value that affect scientific research are the determination of criteria or standards of safety, health efficiency, tolerance, economy, etc., within a particular situation. Many problems of value are related to problems of fact.

Problems of Technique

Such problems concern the methods for accomplishing a desired result. How, for example, can a space station be launched in the mesosphere? Problems of technique are usually the concern of applied science or engineering. These problems combine elements of fact and value. Solving a significant problem frequently entails all three categories. For example, in conquering polio, there was first a problem of fact—what causes polio? After discovering that a virus was the cause, the problem of value arose—what known techniques of immunology against viruses are effective against polio? After Salk perfected his vaccine, there arose a technical problem—how to manufacture the vaccine in adequate amounts to supply the need.

Apprehending the Problem

Every field of study has problems requiring investigation. New areas of knowledge open up every day. New discoveries uncover the need for further research. Often, much time and effort go into identifying the problem for research. This is especially true when the researcher is inexperienced or lacks knowledge of the field. Frequently, the researcher is not in a position to make a free choice; a client or supervisor may decide what problem is to be studied. The researcher may be a staff member of an industrial concern, of a government laboratory, or of a research unit in a university. In such a case, the field will probably be designated for the researcher, and the preliminary work of selection already carried through by others. At academic institutions, students are frequently required to pursue selected fields for research. The field of choice may be further limited for the student by the requirement that he or she major or specialize in a particular department. These restrictions can be necessary and desirable, since after graduation the young researcher will certainly be confronted with similar restrictions of choice by the industrial or government employer. So freedom to select a problem for research is frequently limited by the needs of the researcher's position or affiliation. Personal interest is often compromised by the demands of circumstances. However, there are research workers who are aggressive, alert, imaginative, and quick to see fields that are fertile for investigation.

Guides for Selecting a Problem to Investigate

The first step, and an important one, in selecting a problem to research is acquiring a thorough understanding of the known facts and accepted ideas of the field of interest. Researchers who are familiar with their fields and know what research has already been completed will also know something of the many problems that remain. Their acquaintance with the literature often leads to long lists of areas or problems requiring additional research. More experienced researchers, from their own work, can also observe which phenomena are not satisfactorily accounted for by existing knowledge and, therefore, need additional investigation.

The researcher or student who decides to investigate a problem because it interests her or him has a natural incentive to find the answer. An academic advisor or a supervisor may suggest something that opens the researcher's eyes to new possibilities.

Not to be overlooked are the actual and pressing problems arising out of researchers' everyday experiences, problems related to their everyday work in business and industry. When serious problems arise, research often reveals solutions. Thus, actual experience frequently suggests topics for investigation.

Here are some guides to help you define and formulate your problem:

1. Can the problem be stated in question form? Stating your problem as a question is an excellent way to make your problem clear and concise. Stated in question form, your problem obviously requires a specific answer. Finding the answer then becomes the objective of the study.

2. Can the problem be delimited? By deciding in advance the boundaries and considerations of your problem, you will save yourself much useless work.

3. Are resources of information available and the state of the art practical? The previous two points become immaterial if the present state of knowledge is inadequate or the means for research is unavailable. For example, certain research material in the field of aerodynamics may not be available to a researcher because of its classified nature. Government security restrictions may prevent a researcher from obtaining all necessary resources of information for the investigation.

4. Is the problem important or significant? Your problem need not be earth-shaking in its importance, but the question should be worthwhile for the expenditure of time, energy, and funds involved. No one in our day can afford to pursue the problem, as did medieval scholars, of how many angels can dance on the head of a pin!

The Hypothesis—A Tentative Solution to the Problem

The observation that leads a scientist to recognize a problem also suggests the tentative answer. This answer is called a *hypothesis,* or a working guess. Hypothesis is derived from a Greek verb, *hypotithenai,* meaning "to place under." The hypothesis is an explanation placed under the known facts of a problem to account for and explain them. Scientists test their explanation or hypothesis by experiment. If the hypothesis does not meet the test, it is revised in accordance with the new facts revealed by the testing. If the hypothesis is proved by sufficient testing, it becomes one of the accepted generalizations of science.[1] Let us see how this works.

---

[1] The terms *hypothesis, theory, law, generalization,* and *conclusion* are closely related in meaning. They relate to the solution that the investigation has disclosed for the problem studied. The term *hypothesis* is most frequently considered as a provisional working assumption. As a rule, theory is much broader than hypothesis. A series of hypotheses may be investigated, all of which lead to a theory. *Theory* is an umbrella term suggesting a basic condition common to a number of hypotheses. A particular theory about psychology of learning evolves from a series of hypotheses relating to a variety of ways of learning, all of which must be validated. Because knowledge arrived at through the scientific method is subject to revision in the light of new data, a theory is frequently composed of a number of working assumptions. *Law* is a term applied to a theory, generalization, or conclusion purportedly proved as conclusively as the state of knowledge or art permits.

A hypothesis, as stated previously, is an explanation placed under the known facts of the problem to account for and explain them. Recognizing and stating the problem clearly move the researcher along the path where the answer may lie. Identifying clues and recognizing road signs of the problem point toward an answer. The hypothesis is tested by following the directions and picking up the clues. Traveling the road of the hypothesis, the researcher finds out facts about the problem. Some facts do and some facts do not corroborate the hypothesis. Constant examination along the way enables revision to correspond with the facts of the researcher's investigation.

The Researcher as Detective

From watching TV mysteries, we are all familiar with the process just described. A man is found murdered in a hotel room. There has been nothing unusual to call attention to the murder—no one heard any noise or saw any suspicious characters. The problem is easily recognized: Who killed this man? The detective on the scene examines the facts of the problem. The man has been shot through the heart. From the size of the wound, the murder weapon appears to have been a small-caliber revolver. There are no powder burns on the man's clothes, so he must have been shot from across the room. There is no sign of struggle, so the murdered man apparently trusted or did not suspect the individual who shot him. There are cigarette butts in the ash tray—some of a brand found in the pocket of the murdered man's coat and some of another brand. The cigarette butt of the other brand has lipstick on it. Here is the first clue. At some time there must have been a woman in the room, presumably with the man who was murdered. She certainly could tell much about it if she were in the room at the time of the shooting. The detective moves into the kitchenette from the room where the body was found. Two empty cocktail glasses are on the serving table. One contains lipstick on the rim—the same color as the lipstick on the cigarette butt. This fact reinforces the beginnings of a hypothesis: that a woman who smokes a certain brand of cigarettes and who uses a certain brand and color of lipstick should be a suspect. From the question, Who killed this man? certain evidence has been gathered to point toward a possible solution to the problem—a woman is involved! If the detective finds the woman, he may find further data that could lead to the solution.

This problem illustrates how hypotheses are derived from recognition of the problem and of data that can point to the answer. The detective's tentative conclusion that a woman is involved in the murder was reached from the available clues—the data. Similarly, educated guesses lead a researcher down the path of his investigation toward an answer. Answers are not always simple. There are many complexities to throw the researcher off. For example, in the situation just described, the detective might find that the woman in the case has also been murdered and stuck in a closet. All previous clues have been false starts, and he must start over. All of us have seen crime solvers face complications. Frequently, the most obvious answer is the one furthest from the truth. However, there is also a principle of scientific thinking known as Occam's razor, which states that the most simple but adequate explanation is preferable, at least provisionally, to a more complex explanation. Researchers follow all clues until they find the answer to the problem.

The problem and its hypothesis are, then, the starting points in report writing.

Summary of the Role of Hypothesis

1. The investigator, after gathering data or evidence and analyzing it, uses inductive reasoning to see what the data, evidence, and clues add up to. The provisional answer or hypothesis is the first or trial hypothesis.

2. With this trial hypothesis in the background, the researcher next uses deductive reasoning to decide what kind of data, evidence, or fact she will need to test the trial hypothesis. In other words, the researcher determines what should follow logically from the provisional conclusions being tested.

3. From this analysis, the researcher proceeds to apply her hypothesis. She gathers all possible data and tests them to see whether the actual accumulated evidence agrees with her hypothesis.

4. If the evidence fails to support the hypothesis, the investigator either rejects it or modifies it to conform with the evidence she has. If she rejects it, she will analyze the facts and search further until arriving at a second hypothesis. She will test the second hypothesis by comparing it to the new total evidence amassed.

5. In a search which calls for finding facts or information alone, there may be little use for a hypothesis. For example, if the search is historical in nature or if the search is a request for information—e.g., What industries in the United States use robotics? What is the number of robotics in use in the United States?—then a hypothesis is rarely necessary. However, most legitimate research involves interpretation of facts. For example, the knowledge of the number of robots in use in the United States may raise the problem, Why isn't the robot our company manufactures in greater use? After the facts have been discovered, they must be analyzed to find out what they mean, what conclusions should be drawn, and what should be done about those facts. The conclusions and recommendations are usually the chief object of such research.

Summary of the Scientific Method

The scientific method has these features:

1. It concerns itself with problems to be solved. Therefore, the scientific method is purposeful, with specific goals directing the research activity.

2. When the problem is clearly identified and stated, a working hypothesis or hypotheses about the explanation of the phenomenon or solution to the problem is formulated.

3. Establishing a hypothesis or theory is followed by observation and/or experiment to test the hypothesis.

4. Finally, the data testing the hypothesis are recorded, analyzed, and interpreted. The conclusions are reported, published, and disseminated.

Relationship of Research to Report Writing

Report writing is the reconstruction of a purposeful investigation of a problem in written form. Bloomfield has noted that scientific research begins with a set of sentences pointing the way to certain observations and experiments the results of which do not become fully scientific until they have been turned back into language yielding again a set of sentences that becomes the basis for further exploration into the unknown [4:220].

Chapter in Brief

In this chapter we examined the background, principles, and fundamental algorithms of scientific research. The topics covered were the origins of science and learning, the contributions of the ancient Greeks, the problem and its significance in scientific research, the role of the hypothesis, and the relation of research to technical writing.

Chapter Focal Points

• The beginning and development of science and learning
• Deductive and inductive reasoning
• The scientific method
• The role of the problem
• The role of the hypothesis
• Reporting, publishing, and disseminating research results.

Questions for Discussion

1. What is the relationship between the scientific method and ordinary logical thinking?
2. Identify and explain examples of scientific curiosity in children. Is the curiosity of a puppy or kitten scientific? Explain.
3. Explain the process of induction. How does it differ from deduction?
4. Explain the differences among hypothesis, theory, law, generalization, conclusion, results.
5. What is the role of the hypothesis in problem solving?
6. In what ways are the scientific method and technical writing related?

Assignments

1. Over the previous twenty-five years, the sale of raw popcorn has increased more than 1000 percent. The National Association of Popcorn Poppers initiated research to find the reason. They found that the sale of raw popcorn to commercial poppers such as candy stores and movie theaters, dropped 350 percent during the period in question. They also found that sales of raw popcorn in retail outlets, such as drugstores and supermarkets, increased 1200 percent in the same period, accounting for nearly all of the overall increase. How do you account for the increased sale of raw popcorn? How would you go about testing your theory?

2. An elderly aunt is considering buying a new car. She has pared her choices down to two similar automobiles. She has asked you to help her make the selection on the basis of economy and safety. Using the principles of the scientific method described in this chapter, write her a letter offering your evaluative recommendation for the car she should purchase. You may wish to prepare a table comparing the factors involved, including the reasoning behind your evaluation of the factors. (Note the comparison table, Table 15.4 Chapter 15).

3. Read a detective story or watch a television show or movie in which a crime is solved by a detective. Trace the detective's use of the scientific method to solve the crime.

4. Choose a broad field or area of study for investigation. Narrow or localize a topic within that field. Pinpoint a problem within a delimited topic in the form of a question in accordance with the principles discussed in this chapter. Submit the interrogative statement of your problem in memo form to your instructor for approval. Explain why you have chosen the problem. Your reasons should include your interest in the topic, any pertinent or related experience with it, and your capability to perform the investigation.

References

1. Cohen, Morris R., and Nagel, Ernest. *An Introduction to Logic and Scientific Method.* New York: Harcourt, Brace, 1934.

2. Darwin, Frances, ed. *The Life and Letters of Charles Darwin,* vol. 1. New York: D. Appleton, 1899.

3. Dewey, John. *Logic: The Theory of Inquiry.* New York: Henry Holt, 1938.

4. Whorf, Benjamin Lee. *Language, Thought and Reality, Selected Writings of Benjamin Whorf.* Edited by John B. Carroll. New York: John Wiley and the M.I.T. Press, 1956.

Suggested Further Readings

1. Barzun, Jacques, and Henry F. Graff. *The Modern Researcher,* 4th ed. New York: Harcourt Brace Jovanovich, 1985.

2. Toulimin, Steven, et al. *An Introduction to Reasoning.* New York: Macmillan, 1984.

11
Report Writing—Reconstruction of an Investigation

Chapter Objective

Provide the prewriting background for the preparation of a technical report.

Chapter Focus

- Definition of a report
- Facts and opinions
- Types of reports
- Prewriting strategies
- How to search the literature
- Computerized on-line services
- Hypertext
- Research—Observation
- Research—Experimentation
- Research—Interviews and discussions
- Research—Questionnaires
- Systematizing, analyzing, and interpreting data

Decision Making—The Role Reports Play in Science, Technology, and Industry

Our society depends on cooperation. The range of knowledge within the last few generations, and most of all within our own life span, has increased faster than humans are able to assimilate and understand it. No one person, even within the narrowest field, can grasp all there is to know about that field. Today's knowledge and experience represent a coordinated effort. Efficient communication, therefore, has become an important research tool and an important process in the standard operating procedure of every human activity. Today's scientific genius does not work in an ivory tower laboratory. What the scientist does depends on the efforts of the other fine workers. In turn, his or her efforts influence the efforts of others. Scientists of various countries are researching energetically for a cure for cancer, but every competent scientist knows that individual efforts alone are not enough. The researcher must acquire and use the knowledge, experience, and help of others by knowing what they are doing and the significance of what they are doing. The role of reports and technical papers is to relay this information.

Here is where you and I come in. Some of us may be engaged in reporting to the boss how many nuts and bolts are left in the storage bin. But all of us are concerned with organizing the facts of an experience we are engaged in because the facts are important for someone to know. A decision or action may wait upon these facts. The decision may be as commonplace as ordering more nuts and bolts or as dramatic as finding which carcinogen causes which type of cancer.

What Is a Report?

One of my colleagues, a teacher of literature, tells an appropriate story. He was lecturing on Scott's poem, *The Lady of the Lake*, which begins "The stag at eve had drunk his fill. . . ." Halfway through the poem, he paused in his enthusiasm to notice that one student looked puzzled. He decided to take more time in his explanation, but no matter how carefully he pointed out the poet's technique, skill, and meaning in each successive line, the puzzled look on the student's face remained.

He finally stopped and asked the student, "Is there anything in the last few lines that you don't understand?"

"It's not the last few lines," said the perplexed student. "It's way back at the beginning. What's a stag?"

What's a report?

The origin of the word tells us much about its meaning and function. *Report* comes from the Latin *reportare*, meaning "to bring back." A working definition of a report might be: *organized, factual, and objective information brought by a person who has experienced or accumulated it to a person or persons who need it, want it, or are entitled to it.* Reports contain opinions, but the opinions are considered judgments based on factual evidence uncovered and interpreted in the investigation.

Facts—The Basic Ingredients

The basic ingredients of a report are facts. A **fact** is a verifiable observation. Webster says: a fact is "that which has actual existence; an event." Facts are found through direct observation, survey techniques, experiment, inspection, and through a combination of these processes as they occur in the research situation. A fact is different from an opinion. An **opinion** is a belief, a judgment, or an inference—a generalization that is based on some factual knowledge.[1] An opinion is not entirely verifiable at the time of the statement. Sometimes opinions are little more than feelings or sentiments with a bit of rational basis. *Opinion* and *judgment* are synonyms and are frequently used interchangeably. Judgments imply opinions based on evaluations, as in the case of a legal decision. Neither opinions nor judgments are facts and must be distinguished from facts in technical writing and reports.

Fact: Jack Nicklaus has won more golf tournaments than any other golfer.

Opinion: Jack Nicklaus will win the Senior's Open this year.

Fact: A circle is a closed plane curve such that all its points are equidistant from a point within the center. A square is a parallelogram having four equal sides and four right angles.

Opinion: A circle is more pleasing to the eye than a square.

Fact: The water in the metal drum is 30 percent saturated with filterable solids.

Opinion: That water is unfit to drink.

Since facts are verifiable observations, a direct investigation is inherent in their disclosure.

Another point that should be emphasized is that facts, the chief ingredient of reports, are expressed in unambiguous language. The purpose of every technical communication is to convey information and ideas accurately and efficiently. That objective, therefore, demands that the communication be as clear as possible, as brief as possible, and as easy to understand as possible.

Forms Reports Take

Information and facts dealing with a simple problem or situation may be reported simply. Thus, if your supervisor tells you, "Joe, find out how much it will cost to replace the compressor on the number 2 pump," all that is necessary for the most efficient reporting of the information called for, after you have

---

[1]An *inference* is a conclusion derived deductively or inductively from given data. However, an inference is considered a partial or indecisive conclusion. The term *conclusion* is reserved for the final logical result in a process of reasoning. A conclusion is full and decisive.

found the answer, is to pick up the telephone or pencil a note on a memo pad and say, "Boss, a new compressor costs $256.28."

But if your supervisor's supervisor wants to know the cost, you may have to "dress up" your penciled note into the form of a typed memo in several copies, with additional information to provide the background for the statement that a new compressor costs $256.28. If customers or clients get involved, the note may have to be enlarged to include comparative costs of several makes of compressors. And if the same pump has given you trouble for the last two years, you may have to dig up more facts and information—perhaps to find the reasons for the continued difficulty or whether another type of pump or a different system is called for. Then the matter begins to become more complex, and the complications become sections of information and facts to be reported. Recommendations, what to do about it, become the most important part of your report. Also, when a customer is involved, instead of using a memo form, you report in a letter. If a situation gets fairly complicated, requiring extensive investigation and reporting, the report becomes more formal and many of its elements receive formal structuring, format, and placement. You may end up having a letter of transmittal, an abstract, a table of contents, a list of illustrations, an executive summary, the report proper divided into formal sections, an appendix, bibliography, and perhaps an index, with the whole thing bound in a hard cover. In other words, the report writer designs her report for the particular use and particular reader it will have. But we are getting a little ahead of ourselves.

Because of the many and complex circumstances that call for reports, it is difficult to classify reports rigidly. In some organizations little formality is attached to report writing. Each writer puts down what she has to say, often haphazardly. In other organizations, particularly large ones, numerous and often elaborate forms are devised and given names, e.g., record report, examination report, operations report, performance report, test report, progress report, failure report, recommendations report, status report, accident report, sales report, etc. Given five minutes, you probably could think of at least thirty different types of reports you have come across.

Actually, the facts and analyses marshaled to meet a certain reporting situation take many forms. There is no universal "right" form to clothe all reports. "Form," a learned colleague once said, "is the package in which you wrap your facts and analysis. Choose (or design) a package that is suitable for your material, your purpose, and your reader" [1:6].

Major Types of Reports

There are many bases for classifying reports: subject matter, function, frequency of issuance, type and formality of format, length, and so forth. A traditional classification divides reports into two descriptive categories:

1. Informational
2. Analytical

The Informational Report

The *informational report*, as the term implies, presents information without criticism, evaluation, or recommendations. It gives a detailed account of activities or conditions, making no attempt to give solutions to problems but confining itself to past and present information. Many informational reports, nevertheless, contain inferences, which suggest the conclusions the writer would like the reader to reach, e.g., There are fourteen nuts and bolts in the storage bin. *Inference:* We should order another train load. In this category belong the routine daily, weekly, or monthly reports on sales, inventory, production, or progress. Often such reports are mere tabulations and follow a definite pattern or have a preprinted form requiring fill-ins. The best-known example of the informational report is the annual report of a corporation to its stockholders.

The Analytical Report

The *analytical report* goes beyond the informational report since it presents an analysis and interpretation of the facts in addition to the facts themselves.

The conclusions and recommendations are the most important and interesting parts of the report. The analytical report serves as a basis for the solution of an immediate problem or as a guide to future happenings. It is a valuable and frequently used instrument in all types of activity. Emphasis in this text is on the analytical report because the techniques applying to it apply equally well to the informational types of report.

Steps in Report Writing

Report writing is the reconstruction in written form of a purposeful investigation of a problem. While most technical writers do not engage in research, many are assigned to a research team or project to serve as interpreters or communicators of the investigation. The technical writer, therefore, must understand research methods and procedures and be intimately conversant with the research he or she is reporting. Frequently, the writer assigned to reconstruct the investigation by means of a report is in a position to help digest and find meaning in the mass of data the investigation reveals.

Within the writing process itself, the writer follows steps analogous to those followed by the experimental scientist. Just as a scientist observes and hypothesizes, so, by analogy, the technical writer examines the mass of data, reads it initially for meaning, tries to determine the readership, and tries to make sense of the data by way of establishing important ideas. In a second or succeeding phase, the scientist proceeds to experiment. Again by analogy, the technical writer organizes the material and sets up an initial order of continuity of data; he or she may find upon closer scrutiny that such an organization is not effective, and proceeds to rearrange and reshuffle the data to make them more logical and understandable. In this way, the writer is *experimenting* with the data.

This chapter begins a four-chapter sequence on report writing. The approach taken in this discussion assumes that your instructor assigns you to write a formal report, preferably on an investigation of a research problem you are conducting or one with which you are associated. In this context, and with your instructor's approval, you have selected a problem for research. You will identify the question (problem) to be investigated in accordance with the guidelines for selecting a research problem discussed in Chapter 10.

For convenience, let us review the guidelines:

1. The problem can be isolated and stated in question form. (Finding the answer becomes the objective of your research.)
2. The problem can be delimited to meet the circumstance of your research situation.
3. The resources for the data and information required to find the answer are available and practical.
4. The problem is purposeful enough to make your effort worthwhile.

Selecting a Problem for Investigation

As some of you read and then reread the guidelines, you may be troubled. "How do I find a problem to write about?" you may ask yourself. I understand your quandary. The guidelines are helpful, but only to a point; they are meant only to make your selection of a problem more managable. You still may need help to decide on a problematic subject area. It's time to do some brainstorming.

Brainstorming is exploring what is already in your mind. The technique is easy: You list information about a topic in any order as it comes to your mind, writing it down quickly, *uncritically* in list form. You record what you think at the moment without heckling your brain about the worth of the thought. You go on this way until your brain is "stormed out." The result is a series of jottings that may help you select a subject area, which you may then localize further by using the guidelines.

Perhaps, after consideration, you are not too enthused with the first topic you brainstormed. Simply brainstorm again, this time listing subjects you are interested in (or could be interested in, with some background research). The subject should meet the following criteria:

1. The subject is interesting to you.
2. The subject is related to your major field of study.
3. The subject is one you already know enough about to research intelligently.
4. The subject falls within the guidelines.

If you come up with more than one research topic, ask your instructor to help you make your final selection. Now you can be on your way.

The purposeful reconstruction of an investigation involves four major steps, which follow very closely the experience of the investigation:

1. Prewriting—strategies, analysis, and planning
2. Gathering the data—investigating the problem or situation
3. Organizing the data
4. Writing the report

The report writer, whether the original investigator or serving as the interpreter or communicator of the original investigation, must follow all the procedures the researcher followed in arriving at a solution to the problem.

Prewriting—Strategies, Analysis, and Planning

You, the report writer, like the researcher, must ask yourself a number of questions: What is the purpose of the investigation? Who needs the answer to the problem? How will the answer be used? From the writer's standpoint, you must know the answers to such questions because the effectiveness of your communication depends on the answers. You may have to rephrase the questions to: What information is wanted? Who will read the report? For what purpose will it be used? What problem or problems am I expected to solve? These four questions resolve the larger question, What is the purpose of my proposed report?

As the writer, you must be careful to distinguish between the purpose of the research and the problem studied in the research. The purpose of a research project is generally understood as the reason why the investigation has been undertaken; whereas, the problem is what the researcher specifically hopes to solve. In considering the purpose of a research project, you should regard it as the explanation of the possible uses to which the results of a study may be put. Purpose concerns the probable value of the study. It offers the explanation of why the research was undertaken. In short, the problem concerns the *what* of the study, and the purpose concerns the *why*. When you, the writer, can formulate the purpose in your mind—actually write it down clearly and distinctly—you have started on the right road to answering your problem.

Once the purpose has been determined, the next step is to define scope. It, too, is not to be confused with purpose. Purpose defines goals to be reached. Scope determines the boundaries of the ground to be covered. Scope answers the questions relating to what shall be put in and what shall be left out. (Scope is sometimes indicated by a client, if the investigation has been contracted for, but it is usually left to the judgment of the researcher.)

Having problem, scope, and purpose in mind, you are now ready to start on the next important step in your prewriting, analysis, and planning stage: blocking out a plan of procedure. Always keep in mind that the writing of a report is a reconstruction activity. You must go back to the equivalent investigative point where the problem was defined in a clear-cut interrogative statement. With the problem defined, the purpose clarified, and the scope clearly set, you are ready to begin the work of solving the problem. The next step consists of breaking the problem or situation into its component parts. Again, ask yourself questions. What are the elements that make up the problem? Which elements are fundamental? Which are secondary? (Figure 11.1.) You, the writer, might set down on paper an outline or a procedural plan to follow. This outline might have the following scheme:

1. Statement of the problem in question form
2. Purpose or objective of the investigation
   a. Primary purpose
   b. Secondary purpose
3. Major elements
   a. First major element
      1. Component
      2. Component
      3. Component
   b. Second major element
      1. Component
      2. Component
      3. Component
   c. Third major element
      1. Component
      2. Component
      3. Component
   d. Additional major elements, etc.
4. Correlation of all elements into a final combination that constitutes the problem

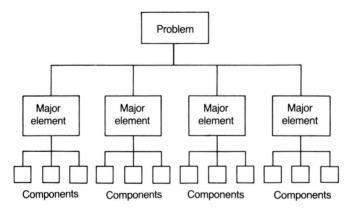

Figure 11.1
Block Diagram Showing Analysis (breakdown of problem into its major elements and constituent components)

The purpose of this type of analysis is to (1) help clarify the problem and what composes it; (2) check and define the scope; and (3) help formulate the task of the research. The analysis should lead to the provisional hypothesis that sends the research on its proper way.

If the problem is a new one, you may have to feel your way at first. As facts accumulate and you reason about the various elements of the problem, you begin to see a logic to them and can arrange them in accordance with the logic of the problem.

In the prewriting analysis stage, no part—purpose, scope, plan of procedure—is final. In the course of the investigation, something unexpected may call for readjusting your approach in any or all of the previous elements. The prewriting analysis has revealed the direction for the writer to follow. You, the writer, gather your information accordingly, or according to any necessary revision in recognition of unexpected turns in the investigation.

Investigative Procedures

There can be no real report without facts. Facts are garnered through research. Research falls into two categories, primary and secondary. *Primary* research is based on an original, first-hand investigation of a problem or situation. *Secondary* research is based on information published by primary researchers. This does not mean that secondary research is necessarily stale, warmed-over information. New, incisive insights on a problem frequently are garnered from published information. Most working research is based on both primary and secondary sources. Ideas, conclusions, evaluations, and recommendations must derive from facts.

Before the writer can begin to organize and construct her report, she must first gather and assemble the facts—her data. Information may be gathered by several methods:

1. Searching all available recorded—electronic or printed—information related to the problem or situation
2. Observing and critically examining the actual situation, condition, or factors of the problem
3. Experimenting (observation carried out under special conditions that are controlled and varied for specific reasons)
4. Interviewing and discussing with experts or persons qualified to give needed data
5. Using questionnaires when interviews are impractical

Searching the Literature

Searching the literature is the method used to learn new facts and principles through the study of documents, records, and the literature of the field. This type of research is used extensively in history, literature, linguistics, and in the humanities. Because it is almost the exclusive research technique used by historians, it is sometimes called the historical method of research. This research method is valuable in all fields whenever knowledge and insights into events of the past are required.

Human knowledge is a structure, which grows by the addition of new material to the store that previously has been gained. An investigator has little chance of making a worthwhile new contribution if he or she is completely ignorant of what is already known about a problem. Before beginning an investigation, the investigator must find out what has been written thus far about the problem. Six hours fruitfully spent in a library or in searching an automated database may save six months in a laboratory.

There are two goals in literature searching:[2]

1. To find out if the information that is the object of the proposed research is already available.
2. To acquire a broad general background in the area of research.

The major secondary research tool available to college students is the library. Efficient literature searching depends on a researcher's knowledge of library reference materials and available automated data and information systems. Library resources include:

Catalog of holdings
The stacks (books on the shelves)
Periodical holdings
Reference books and materials
Literature guides, indexes, and abstract journals
Government documents
Popular press
Audiovisual materials
Computerized databases and information services
Computer software, such as hypertext

Let us now examine some of these resources.

Catalog of Holdings

The catalog is the key to a library's collection of published materials, audiovisuals, and computerized databases. You probably are familiar with the card

---

[2]This statement excludes the historical method of research, which employs literature searching as its investigative technique. The historical researcher puts together in a logical way evidence derived from documents, records, letters, papers, literature, and appropriate electronically stored data, and from that evidence forms conclusions that either establish facts hitherto unknown or offer new insights and generalizations about past or present events, human motives, characteristics, and thoughts.

catalog. It cross indexes the library's holdings by author, title, and subject. Many libraries now produce their catalog listings by computer. It is called a *COM* catalog system, an acronym for *c*omputer *o*utput *m*icroform. **Microform** is the librarian's term for text material greatly reduced on microfilm or microfiche. Microfilm is stored on a reel; **microfiche** is a transparent sheet that may hold images of about 200 or more pages. To read microfilm or microfiche requires special equipment. COM catalogs, like card catalogs, may be searched by author, title, or subject. You read a COM entry the same way you read a card catalog entry. Some libraries have on-line catalogs that can be accessed from a distance by modem, a device that connects computers over telephone lines. (On-line refers to the linkage capability of one computer for communicating with another.)

Some libraries make computer terminals with printout facilities available to students. Entering a subject heading into the system produces a hard-copy list of holdings on that subject. The library may charge for this type of service. When holdings on a subject are simply displayed on a screen, you must copy them yourself, though, of course, this allows you to eliminate entries you deem irrelevant.

Literature Guides and Sources of Published Materials

As an aid to finding bibliographic and other reference materials in a field, guides to the literature of a subject area have been prepared for students and research workers. Sheehy's *Guide to Reference Books* is an excellent first source for finding listing of literature guides, reference works, handbooks, encyclopedias, yearbooks, etc. on any given subject field. Another good source is Walford's *Guide to Reference Material.* For technical subject areas, *Scientific and Technical Information Sources* by Chen covers the full range of reference sources in biological, physical, and engineering sciences. These three sources can help you identify the literature guides for your subject area of interest. Some examples to be found include:

Burman, C. R., *How to Find Out in Chemistry*, London: Pergamon Press, 1965.

Fowler, Maureen J., *Guides to Scientific Periodicals, Annotated Bibliography*, London: The Library Association, 1966.

Jenkins, Frances B., *Science References Sources*, Cambridge, Massachusetts: MIT Press, 1969.

Parke, N. G. III, *Guide to the Literature of Mathematics and Physics*, New York: McGraw-Hill, 1958.

Pearl, Richard M., *Guide to Geologic Literature*, New York: McGraw-Hill, 1951.

Smith, Roger C. and Reid, Malcolm, *Guide to the Literature of the Life Sciences*, Minneapolis, Minnesota: Burgess, 1972.

White, Carl M., *Sources of Information in the Social Sciences*, Chicago: American Library Association, 1973.

U.S. Government Printing Office, *Monthly Catalog of United States Government Publications*.

Bibliographies

Publication details of the literature for a subject area have been assembled and published in bibliographies. Examples include:

Bowker's Medical Books in Print, New York: R. R. Bowker Co., published annually.

DeGeorge, Richard T., *A Guide to Philosophical Bibliography and Research*, New York: Apple-Century-Crofts, 1971.

Irwin, Leonard B., *A Guide to Historical Reading*, Brooklawn, N.J.: McKinley, 1970.

Mellon, M. G., *Chemical Publications*, 4th ed., New York: McGraw-Hill, 1965.

MLA International Bibliography of Books and Articles on Modern Languages and Literature, New York: Modern Language Association, published annually.

Technical Writing: A Bibliography, Washington, D.C.: Society for Technical Communication, 1983.

U.S. Department of Agriculture, *Bibliography of Agriculture*, Washington, D.C., published annually.

Handbooks

Handbooks are compact reference manuals containing the state-of-the-art data and information on particular subjects. Examples include:

Handbook of Abnormal Psychology, Hans J. Eysench, editor. San Diego, California: R. R. Knapp, 1973.

CRC Handbook of Chemistry and Physics, Robert C. Weast, editor. Boca Raton, Florida: CRC Press, Inc., published annually.

Handbook of Physical Constants, rev. ed., Sidney B. Clark. New York: Geological Society of America, 1968.

Poisonous Plants of the United States, John M. Kingsbury, Englewood Cliffs, N.J.: Prentice-Hall, 1964.

Water and Water Pollution Handbook, Leonard L. Ciaccio, New York: Marcel Dekker, 1973.

Encyclopedias

An encyclopedia is a volume or a set of volumes giving information on one special field or all branches of knowledge. Editors and contributors are specialists within specific subject areas. Examples are:

Encyclopedia of Computer Sciences and Technology, New York: Marcel Dekker, 1983.

Encyclopedia of Material Science and Engineering, 8 vols., Cambridge, MA: MIT Press, 1986.

Encyclopedia of Wines and Spirits, Alexis Lichine, New York: Alfred A. Knopf, 1967.

International Encyclopedia of the Social Sciences, David L. Sills, editor, New York: Macmillan, 1968.

Kirk-Othmer Encyclopedia of Chemical Technology, New York: Wiley-Interscience, 1984.

McGraw-Hill Encyclopedia of Science and Technology, New York: McGraw-Hill, 1982.

Miller, William C. and West, Geoffrey P., *Encyclopedia of Animal Care*, Baltimore, Maryland: Williams and Wilkins, 1972.

Books

Reference books and textbooks provide basic material. Monographs are books, articles, or papers written about a narrow, specialized technical area with an indepth treatment of a subject. Searching the subject card index of a good library and then browsing the library shelves can be helpful in locating pertinent books. There are a number of useful publications listing scientific and technical books:

Scientific, Medical and Technical Books Published in the United States of America, National Research Council, Washington, D.C.

The Cumulative Book Index, List of Books in the English Language. This list is issued monthly and then cumulated in semiannual volumes published by the H. W. Wilson Company, New York.

The United States Catalog. This is a comprehensive list of books printed in English, arranged by title, author, and subject.

Books in Print is a listing of books currently available in print, arranged by both subject areas and authors, published by R. R. Bowker Co., New York.

Abstracting and Indexing Journals

Journals that abstract and index publications within a field are the main reliance of investigators seeking papers on scientific topics. These journals provide researchers with an important way of keeping abreast of scientific progress and furnish the investigator with an overview of published material on a given subject by a particular author under specific titles. Indexes usually provide no more than bibliographic references, but abstracts provide a digest or synopsis of the listed literature. Some better-known examples are:

Applied Science and Technology Index covers monthly and cumulatively industrial and trade subjects and provides an annotated description of articles appearing in trade publications.

Biological Abstracts provides worldwide publications coverage of research in the life sciences.

Business Periodical Index is a monthly subject index to all fields of business.

Chemical Abstracts covers worldwide publications; lists by author, subject, and chemical formula. *CA* also includes patents in the chemical field.

Computer Periodical Index is a quarterly index to basic articles in the field.

Congressional Record Abstracts records and updates weekly congressional activities regarding bills and resolutions, committee and subcommittee reports, public laws, executive communications, speeches and inserted material in the *Congressional Record*.

Economic Abstracts International provides coverage of the world's literature on markets, industries, country-specific data, and research in the fields of economic science and management.

The Engineering Index is a monthly listing of published engineering subjects with brief annotations.

Excerpta Medica provides abstracts and citations of biomedical journals published worldwide.

The New York Times Index is a monthly index to news stories appearing in that newspaper.

Psychological Abstracts covers the world's literature in psychology and related disciplines in the behavioral sciences.

Readers Guide to Periodicals Index is an index of general subjects, listing articles from a great number of sources. It has a broad subject coverage of articles appearing in popular publications.

Sociological Abstracts covers the world's literature in sociology and related disciplines in the social and behavioral sciences.

Science Abstracts appears monthly in two sections. Section A, *Physical Abstracts*, covers mathematics, astronomy, astrophysics, geodesy, physics, physical chemistry, crystallography, geophysics, biology, techniques, and materials. Section B, *Electrical Engineering Abstracts*, covers generation and supply of electricity, machines, applications, measurements, telecommunications, radar, television, and related subjects.

World Textile Abstracts covers the world literature on the science and technology of textiles and related materials.

Yearbooks

Yearbooks are annual publications reporting or summarizing events, achievements and statistics for a specific year. Some, like yearly almanacs, are general; others are specific to a subject. Examples include:

Communication Yearbook
Facts on File, cumulated yearly
New International Yearbook
World Almanac
The Yearbook of Agriculture

Yearbook of Education
Yearbook of Forest Products
Yearbook of Labor Statistics

Computerized Databases and Information Services

So much information is currently generated about everything and anything that we speak of an "information explosion." This explosion has resulted in such a mass of published materials that anyone trying to find information by using such traditional methods as card catalogs, literature guides, or subject area indexes often is frustrated by this time-consuming and seemingly endless task. Fortunately, most libraries, including those in colleges and universities, have become computerized. Many college libraries are network members of the On-Line Computer Library Center (OCLC) located in Columbus, Ohio. By using the member library's computer terminal, the student has recourse to the holdings of the entire OCLC network, and the student's own library can order any of these holdings through the Interlibrary Loan System.

Using an on-line computer database requires some training. Your college librarian may either conduct the computer search for you or train you to use the system. The librarian establishes contact with the database via modem by entering the library's identity code, and then enters the keywords describing the type of information you want. The computer responds on screen or with a printout of the titles of the documents filed under your key terms.

There are several thousand commercial computer information services and databases in the United States, as well as bases operated by professional societies, industry associations, government agencies, and corporations. They cover every field, discipline, subdiscipline, and specialized interest groups and subgroups. Below are some of the more prominent systems, listed by field, title, and responsible agency or operator.

Agriculture: AGRICOLA, National Library of Agriculture
Biological Sciences: BIOSIS PREVIEWS, Biosciences Information Services
Business: ABI/INFORM, Data Courier, Inc.
Chemistry: CA SEARCH, Chemical Abstract Service, American Chemical Society
Education: ERIC (Educational Resources Information Center), National Institute of Education
Engineering: COMPENDEX (Computerized Engineering Index), Engineering Information, Inc.
Environment: ENVIROLINE, Environment Information Center, Inc.
Geosciences: GEO-REF, American Geologic Institute
Government research and development reports: NTIS (National Technical Information Service), Department of Commerce

Law: LEXIS, Mead Data Central

Mathematics: MATHFILE, American Mathematical Society

Mechanical Engineering: ISMEC, Data Courier, Inc.

Medicine: MEDLINE, National Library of Medicine

Metallurgy: METADEX, American Society of Metals

Physics: SPIN, Physics Information Notices, American Institute of Physics

Pollution and Environment: POLLUTION ABSTRACTS, Cambridge Scientific Abstracts

Psychology: PSYCINFO, American Psychological Society

Science Abstracts: INSPEC, Institution of Electrical Engineers

Science and Technology: SCISEARCH, Institute of Scientific Information

Among the more prominent commercial information/database systems covering a wide spectrum of subjects are CompuServe Information Services, Columbus, Ohio; The Information Bank of the *New York Times*; the Bibliographic Retrieval Services (BRS); DIALOG, a subsidiary of Lockheed Corporation; and Knowledge Index, an after-hours service offered by DIALOG at a lower cost. Of DIALOG's 320 databases, Knowledge Index offers the 80 most often used. Using simpler operation commands, it is designed for persons with little or no information retrieval skills. BRS also has a reduced service called After Dark, with about 40 databases available on weekends and evenings.

The commercial databases provide instructions in their printed catalogs, as well as on-line indexes to help you select the information appropriate for your needs. Once the database is selected, you enter the search terms of interest. *Search terms* are keywords or phrases to be searched for in the database that identify pertinent documents and information.

Most databases use Boolean algebra to eliminate specific terms or topics and to combine search terms. The major Boolean algebra operators are AND, OR, NOT, IF, THEN, EXCEPT. They designate logical relationship among search terms. For example: to search for information on spaceship disasters, you might use the terms *spaceship* and *disaster*. A search of *spaceship* OR *disaster* would find all references that contain the word, *spaceship*, all that contain the word, *disaster* and all that contain both words. The search would find a mountain of irrelevant information using the OR operator. A search of *spaceship* NOT *disaster* would locate all references containing the word, *spaceship*. In this run, you would also find an overwhelming amount of general information on *spaceships*. By searching *spaceship* AND *disaster*, only the references containing both search terms will be identified. The use of the operator, AND, in this instance provides the best search choice for locating the literature for your topic.

Among DIALOG's special offerings is the complete database of the National Technical Information Service (NTIS). This database contains more than 1,400,000 citations and is updated twice a month. NTIS is the federal government's central technical and scientific service, and provides access to the results of U.S. and foreign government sponsored research and development as well as other types of engineering and scientific activities. It announces annually more than 150,000 summaries of completed and ongoing federal and foreign-sponsored research, and provides complete technical reports for most of the results it announces. DIALOG also offers CD-ROMs (**compact disk**, read-only memory)[3] of the NTIS Bibliographic Data Base, as well as NTIS-created CD-ROMs for AGRICOLA (the bibliographic research publication records of the U.S. Department of Agriculture) and the NIOSHTIC (database of the National Institute for Occupational Safety and Health).

A typical database search from DIALOG is reproduced as Figure 11.2. Note the use of Boolean algebra operators. You can see how simple, efficient, and inexpensive bibliographic searching by computer can be.

There are advantages and disadvantages in using computerized information services and databases. Among the advantages are:

1. Material related to your specific key term is accessed quickly and precisely from a great variety of databases.
2. Many indexing characteristics are available, such as author, time frame, language, type of publication, and journal title.
3. Most human error (mistakes in copying, for example) is eliminated from data retrieved.
4. Current data are available. (Some databases are updated daily, some weekly, some monthly. Published sources take much longer to enter a system.)
5. Printouts are convenient and save time.

Disadvantages to be considered are:

1. Costs can be beyond student range, especially if selected key terms are general or vague. (Every second a database is used costs money.)
2. Chosen subject lacks a standardized indexing term. (It may not be recognized by the system.)[4]
3. Some subjects are difficult to reduce to precise component terms for retrieval (such as those dealing with relationships or ethical problems).
4. Some items within a database may be ten to twenty years old and the information may have become obsolete.

---

[3]Compact disks are being used increasingly to provide information bases on a growing number of subjects. A single optical compact disk, smaller in size than a 45-rpm record, can store the entire *Encyclopaedia Brittanica*. Using a computer, you can access any topic of an encyclopedia within seconds. Many commercial databases are providing specialized subject information on CD-ROMs to customers. Some libraries subscribe to disk-based information and indexing services.

[4]An exception to this problem is the field of chemistry, which has universally adopted the Chemical Society's *Chemical Abstract Service* terminology. The *Chemical Abstract Service* assigns a unique identification number to every chemical substance. Other databases use these identification numbers when dealing with chemical substances.

A Typical Dialog Search

Shown below is a suggestion of the kind of conversation you might have with DIALOG during a typical search. On the facing page is a replica of the actual printout that would be generated at your computer terminal during the search.

From start to finish the search took less than three minutes. During that time more than 15,000 documents were examined and the seven pertinent ones were identified.

PURPOSE OF THE SEARCH: You want to find the sources of recent articles on *the effect of stress on executives.*

WHICH DATABASE TO SEARCH? DIALINDEX, the online subject index, shows you that File 15 ABI/INFORM contains information about articles on business and management.

Here, in effect, is what takes place at your computer terminal.

	What you say to Dialog	How Dialog responds
1	I'd like to search your Business/Economics File 15, please.	What would you like me to find for you?
2	Do you have any articles that include the word *stress* or the word *tension*?	Yes, I have 997 that refer to *stress* and 259 that include a reference to *tension* for a total of 1178 documents that mention either or both terms.
3	How many articles do you have that mention *executives* or *managers* or *administrators*?	I have the following references: *executives*—5,349; *managers*—10,253; *administrators*—648, for a total of 14,962.
4	How many of those articles or documents contain the terms *stress* or *tension* AND ALSO the terms *executives* or *managers* or *administrators*?	Only 311.
5	I'm interested only in *recent* articles. How many of those 311 were published during 1980?	Seven.
6	I'd like the following information about the first of those seven documents: record number, title, journal title, date, pages, and an abstract on the article if available.	Title of the article is "Learning to Handle Stress—A Matter of Time and Training." It's by Dennis R. Briscoe and appeared in Supervisory Management magazine, volume 25, number 2 on pages 35 to 38 of the February 1980 issue. An abstract of the article follows: (For complete text of the abstract please refer to the sample printout on the facing page.)
7	For the remaining six articles please give me only the basic information, no abstracts.	(For detailed response, see printout at right.)
8	Thank you, I'm finished. Please log me out and give me a record of this search and its cost.	This search was made on April 25, 1980 and completed at 3:15:44 P.M. The user's identification number is 3268. Cost for computer time was $3.30. Time required to conduct the search was 0.044 hours. The search was made in File 15. Six descriptive terms were used to make the search. Communications cost (TELENET) was $.22 and the total estimated cost was $3.52.

If you had wished to do so you could have requested that the references be printed offline and mailed to you, typically more cost effective if many references are desired.

Most databases contain abstracts or summaries of the original document such as that shown in our sample search. Often these abstracts provide enough information to answer your question. Should you decide that you want to order the full text of the article abstracted, you can do this easily while still connected to the DIALOG computer through Dial-Order, DIALOG's online ordering system. You simply type ORDER and just the record number from the upper left of each reference.

This search is an example of how simple yet powerful a DIALOG search can be. As you grow in familiarity with DIALOG you'll find yourself taking advantage of the many additional search capabilities that can improve the speed, increase precision, or lower costs.

Figure 11.2
How a Literature Search is Conducted by an On-line Database (DIALOG)

1 *File15: ABI/INFORM 71-80/MARCH*
(Copr. Data Courier Inc.)
    Set   Items   Description.

2 *? SELECT STRESS OR TENSION*
             997   STRESS
             259   TENSION
    1   1178   STRESS OR TENSION

3 *? SELECT EXECUTIVES OR MANAGERS OR ADMINISTRATORS*
        5349   EXECUTIVES
     10253   MANAGERS
       648   ADMINISTRATORS
    2   14962   EXECUTIVES OR MANAGERS OR ADMINISTRATORS

4 *? COMBINE 1 AND 2*
    3   311   1 AND 2

5 *? SELECT S3 AND PY = 1980*
        2347   PY = 1980
    4   7   3 AND PY = 1980

6 *? TYPE 4/7/1*   4/7/1   80005291
Learning to Handle *Stress*-a Matter of Time and Training
Briscoe, Dennis R.
Supervisory Mgmt   v25n2   35-38   Feb 1980

    Management jobs are becoming increasingly stressful. Considering the nature of managerial work today, people can survive only by learning to avoid the situations they find stressful or by adjusting to stress factors. Stress may be defined as the way that people react to the constant changes occurring in and around them. Some stress is used positively to develop abilities and skills, while other types and amounts of stress are dysfunctional. There are 3 important concepts to follow in training *managers* to cope with the stress of their jobs and all 3 must be used to optimize training results and minimize any dysfunctional reactions to stressors: 1. specificity of training, which requires that training be designed to increase the person's ability to manage, 2. magnitude of training, which forces adaptation or learning, and 3. gradual and continual intensification of training, with new levels of learning and adaptation building on prior levels.

7 *? TYPE 4/3/2-7*   4/3/2   80005280
Managing Yourself
Kleiner, Brian H.
Management World   v9n2   17-18,36   Feb 1980

4/3/3   80004717
Managerial/Organizational Stress: Identification of Factors and Symptoms
Appelbaum, Steven H.
Health Care Mgmt Review   v5n1   7-16   Winter 1980

4/3/7   80003034
Getting Out of a Sales Slump
Anonymous
Small Business Report   v5n1   15-16   Jan 1980

8 *? LOGOFF*
25apr80 15:15:44 User3268
$3.30   0.044 Hrs File15 6 Descriptors
$ .22   Telenet
$3.52   Estimated Total Cost

Figure 11.2 (continued)

5. Historical information may be lacking. (Most databases go back only as far as the 1960s.)
6. You pay for irrelevant citations.
7. Browsing—randomly leafing through information—is difficult if not impossible, and is extremely expensive.
8. Thorough searches require the expertise of a librarian.

Costs for using a commercial information/data service are based on connect time (the time the terminal being used is connected to the database computer). Costs range from $15 to $200 an hour. To that, phone charges at the rate of about $10 an hour—the cost of calling coast-to-coast on a computer network, such as TELENET—should be added.

Some universities provide these services to students without charge or at a very modest cost. A computer search provides a more comprehensive listing in a shorter period of time. Material uncovered by the information bases proves to be not only more extensive, but also more pertinent if the research strategy is carefully formulated. Computer searches, however, can also produce an overwhelming amount of irrelevant information that produces more confusion than usefulness.

Suggestions for Literature Searching

In the prewriting, planning stage, you have broken your problem down into its component parts. Next, to find the state of knowledge of the problem and to secure a good general background for pursuing your investigation, you might begin by reading the most general treatments of the problem first, as, for example, in an encyclopedia. This can be followed by a more detailed but still quite broad discussion in a handbook. Next it would be desirable to search the library catalog for books on the subject. If there is a recent monograph on your problem, the library search may end at this point because such specialized books often contain bibliographies sufficient for most purposes.

If you cannot find a book that is complete or entirely up-to-date, you might look for a survey or review article in the professional periodical or trade publication that specializes in the area of your problem. Then appropriate abstract publications should be searched by working backwards in time until the necessary coverage has been obtained or until the year is reached that has been adequately dealt with in a book. Technical and professional papers will usually contain references to earlier works, and in this way, the researcher can be carried backward to pick up references formerly missed.

If you have a computerized information/database available to you, you might begin by querying it for the literature covering the keywords related to your topic or problem. The procedure for using DIALOG Information Retrieval Service described in Figure 11.2 is appropriate. To perform an effective computer search, you need to analyze your topic for terms and concepts that the computer can flag or lock into by searching the vast literature within its store. Be specific in identifying the key terms of your topic.

For example, let us say your research is investigating the role that nutrition plays in the IQ level of children. This problem contains three major concepts:

1	2	3
Nutrition	IQ	Children

Please note that the terms "role" and "level" are not included in your consideration. Why? At this point, they are too general and ambiguous. Their use would confuse the retrieval process by flagging a multitudinous quantity of extraneous literature unrelated to the problem. The exact effect they play in your problem is what is to be determined. The three concepts—nutrition, IQ, and children—form a relationship that lends itself to computer searching, as shown in Figure 11.3.

More effective retrieval can be achieved by expanding each concept with additional or synonymous terms that are relevant to the concept. Each of the concepts may be further expanded as follows:

1	2	3
Nutrition	IQ	Children
Diet	Intelligence	Boys
Food	Genius	Girls
Health	Heredity	Youngster
Nourishment	Mental Age	Juvenile, etc.
Vitamins, etc.	Retardation	
	Imbecile	
	Moron, etc.	

Each group of terms can be used as a set of building blocks to effect the search. What you do is to instruct the computer, using the Boolean operator AND, to print out the citations wherein the three building blocks or concept terms intersect, that is, those citations having the three sets of terms in common. Depending upon what is available in your computerized information/database, you can ask for printouts of the pertinent bibliographic references, for abstracts, or for the full document.

Hypertext

Hypertext is a development in software technology designed to provide on-line nonlinear information. It is a computer system for linking or cross-referencing related text and graphic units into a new document. Some programs, such as **HyperCard** from Apple Computer, have the capability to include animation and

Figure 11.3

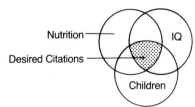

sound segments in the document. A hypertext-produced document in the present state of technology is not designed to exist as a printed text, although you can print out text material you have browsed through and linked together electronically.

The term **hypertext** was coined in the 1960s by a Stanford Research Institute researcher, Ted Nelson, to refer to a proposed computer system to link modules of text. The idea originated in 1945, before there were digital computers. Vanevar Bush, Franklin D. Roosevelt's science advisor, in an *Atlantic Monthly* article envisioned a machine for browsing and annotating scientific literature. Dr. Bush called the machine **memex** (memory extender), and described it in this way:

> The owner of the memex, let us say, is interested in the origin and properties of the bow and arrow. Specifically he is studying why the short Turkish bow was apparently superior to the English long bow in the skirmishes of the Crusades. He has dozens of possibly pertinent books and articles in his memex. First he runs through an encyclopedia, finds an interesting but sketchy article, leaves it projected. Next, in a history, he finds another pertinent item, and ties the two together. Thus he goes building a trail of many items. Occasionally he inserts a comment of his own, either linking it into the main trail or joining it by a side rail to a particular item. When it becomes evident that the elastic properties of available materials had a great deal to do with the bow, he branches off on a side trail which takes him through textbooks of elasticity and tables of physical constants. He inserts a page of longhand analysis of his own. Thus he builds a trail of his own interest through the materials available to him. [2]

Hypertext has three major elements:

Files (also called nodes or fields)
Links (also called buttons) and
User interface.

Files are the data, which can vary in size from a very small document to a multivolume library. **Links** provide the cross-referencing capability for connecting subjects in the files or database. Links can be both the referent in one file and the bridge between it and related information in another file. Links operate as a button, symbol, marker, or designated word. It is the crucial element in obtaining immediate access to appropriate information. Erroneously programmed, it can mislead the user by retrieving wrong, inappropriate information or an overwhelming useless amount. The **user interface** is the structure of navigational commands, formats, and support items that determine the usefulness of the accessed information.

The potential of hypertext software technology to accomplish the purpose of Vanevar Bush's memex has stimulated a rash of off-the-shelf software, such as *Document Examiner, Guide, Houdini, HyperCard, Hyperdoc, HyperTIES, Intermedia, KMS, Navitext,* and *NoteCards*, with new programs being introduced all the time. There is no doubt that you who are students will be involved with hypertext as a document technical writing tool in your professional life.

Preparing a Bibliography

One of the most troublesome problems the researcher has is the recording of information acquired from reading and setting it up in accessible form. This involves keeping accurate and complete bibliography cards and note cards. Bibliography cards provide a complete record of the sources of information used in library searching. The first step in making a literature search is to check with a librarian for help in compiling a working bibliography—a list of possible sources of information. Most libraries have librarians trained in providing help in literature searching. Be sure to take full advantage of such available aid. While searching through indexes, abstracts, card catalogs, and computerized information services, you should keep accurate and complete records of the sources of information. If you have recourse to the services of a computerized database, you can be ahead of the game by having your bibliography printed automatically for you. In listing reference sources, the following elements are recorded:

1. Library call number, usually in the upper-left-hand corner (if the literature is obtained from a library).
2. The author's name, last name first, or the editor's name followed by *ed*. If both the author and the editor are given, list the editor after the book title. The edition number, translator, and number of volumes are listed after the title, if applicable.
3. The title of the book underlined. The title of an article is in quotation marks, and the title of a magazine is underlined if the entry is an article.
4. Publication data:
    a. For a book or a report, record the place of publication, the publisher, and the year of publication.
    b. For a magazine article—the name of the magazine underlined, the date, the volume, and the pages on which the article appears
    c. For a newspaper article—the name of the newspaper underlined, the date, the page number
5. Subject label placed at the top of the card to allow filing of several bibliography cards under one topic.

The abstracts literature provides a quick means for evaluating the potential usefulness of a source item. If available to you, examine the abstract first.

Taking Notes

Most professional research workers take notes on 3″ × 5″ or 5″ × 8″ cards because cards are flexible and easy to handle, sort, keep in order, and later to organize. Three essential parts of the contents of a note card are these:

1. The material, the facts, and the opinions to be recorded (include diagrams, charts, and tables if these are the data to be noted)
2. The exact source, title, and page from which they are taken
3. The label for the card showing what it treats

It is usually a waste of effort to try to take notes in a numbered outline form.

It is best to read the article or chapter through rapidly at first to see what it contains for your purposes and then go over it again and make necessary notes. Distinguish between the author's facts and opinions. Specific points, such as dates and names of persons, should be recorded with particular care. Notes may be recorded either in paraphrase, summary, or direct quotation from the source. The exact wording of the author should be recorded on the note card when the original wording is striking, graphic, or appropriate; when a statement is controversial or may be questioned; or when it is desirable to illustrate the style of the author.

In recording quoted matters, take care to copy accurately the words, capitalization, punctuation, and even any errors that may appear in the original. Use quotation marks to indicate exactly where the quoted passage begins and ends. Ellipsis may be indicated by the use of three dots (. . .) at the beginning or within the sentence, or four dots at the end of a sentence (. . . .). Before you return the reference source, be sure that your bibliography and note cards include every item of information you will need for use in your paper. Also note the library or source where the book or publication is obtainable.

In your library research, you should be concerned with the meaning, accuracy, and general trustworthiness of the material you are reading. In evaluating the reading matter, you should be asking yourself questions related to the author's competence and integrity. How good an observer is the writer? Did he or she have ample opportunity to observe and master the matters he or she writes about? Does the author have a reputation for authoritativeness in the field? How well does the author know the facts of the case? Was he or she personally interested in presenting a particular point of view? Is he or she prejudiced? Is the author trying to deceive? Are his or her observations made firsthand or did he or she receive information from others? Are there discrepancies in the writing that throw question on the reliability of the literature? Are the data sufficient to support the points the author is making? In short, are the data relevant, material, and competent?

Plagiarism

A cautionary word about plagiarism is necessary because student writers—and often professional writers—depend on printed material for much of the information they use. According to copyright law, plagiarism is illegal; it is theft. A writer who steals the substance or the actual expression of ideas from others is a plagiarist. Plagiarism is more obvious when the exact words of another writer are used as one's own.

If you use another person's words or ideas, according to law, you need to give that person credit for it. If you quote word for word 250 or more successive words from a published source, you must obtain formal permission to use that material.

Careless note-taking can often be the cause of plagiarism. Check your notes for phrases and sentences copied word for word. Groups of as few as five successive words from a source should be placed between quotation marks on your note cards. After each bit of information you have recorded on a note card, give the page reference and source from which it comes. Even if you paraphrase the original words, you are still making use of another writer's ideas. I am not suggesting that you should not paraphrase from the reference sources you use to write your report. Far from it: changing another's words to meet the needs and context of your situation is frequently more appropriate and effective. However, the use of another's materials needs to be acknowledged. How you do this is explained under Documentation, Chapter 14.

As soon as the findings of other investigators are exhausted, it is time for you to do your own primary work. Procedure depends, of course, on the technical elements of your problem.

Observation—Examining the Actual Situation, Condition, or Factors of the Problem

A common and traditional means for conducting research is through direct observation—a cornerstone of the scientific method. Observation is careful examination of the actual situation, condition, or factors of the problem. If your problem requires you to use observation in its investigation, it is important to decide whether you will use selective sampling or whether you will cover the entire field.[5] Observation implies selection because our powers are limited. While we might want to make random observations, we do select the conditions of time and place. Frequently, we find it necessary to examine and observe only a small portion of a problem. Inherent in observation is the necessity for recording what we examine and perceive. Human memory is not to be trusted. Observation must be followed by description. Observation should be recorded in precise, exact, and objective language.

Another essential point is that scientific observation tends to be quantitative. Numbers are used as part of the description where possible: how many, at what rate, with what value? The use of numerical measures permits a more precise description and makes possible the application of mathematics. Of course, not all matter and phenomena that are observed are numerical. Qualitative statements have an important role. However, the observer must try to be objective and free of bias. Even though it is perhaps impossible for any observer to be free completely from preconceptions and prejudices, it is important to arrange the conditions of the observation so the observer's bias will not result in distortion. Therefore, your observation should be given to others for checking. If possible, allow another observer or several observers to make independent observations of the same phenomena.

---

[5]For academic purposes you will recall that we are assuming that you, the student technical writer, are conducting the research of the problem to be reported. This assumption is necessary in order that students know the process in thought and activity that researchers experience.

That this is necessary is illustrated by an experiment frequently repeated by psychologists: a man suddenly rushes into a room chased by another man with a revolver. After a scuffle in the middle of the room, a shot is fired and both men rush out again, with barely thirty seconds elapsing for the incident. The psychologists conducting the experiment will ask the group to write down an account of what they saw. These descriptions usually show that no two persons in a group will agree on the principal facts of the incident. A noteworthy feature is that over half of the accounts include details that never occurred. These experiments illustrate that not only do observers frequently miss seemingly obvious things, but also, what is even more significant, they often invent false observations. So in the research situation, the careful observer makes use of instruments wherever possible to aid the observation and promote greater accuracy and objectivity. Photography, the photoelectric cell, and the thermocouple are frequently substituted for direct vision. Sound vibrations are converted into electrical oscillations and are analyzed and measured by instruments having a range far beyond that of the human ear. Thermometers, thermocouples, and pyrometers replace the sense of touch in estimating warmth and coldness. Where a phenomenon occurs rarely, or only for a short time, as in solar eclipses or earthquakes, optical and recording instruments are substituted for direct visual observation. Although many natural phenomena can be observed with the naked eye, many more become accessible with microscopes, binoculars, telescopes, and other optical and measuring instruments. Simple instruments such as meters and stethoscopes aid in direct human observation.

How to Conduct Observations

Here are some helpful suggestions for conducting observations under actual or controlled conditions:

1. Have a clear conception of the phenomena to be observed
2. Secure a notebook or note cards upon which the data are to be recorded
3. Set up entries or headings on the note cards or notebooks to indicate the form and units in which the results of the observation are to be recorded
4. Define the scope of the observations
5. Use care in selecting the basis of sampling, should sampling procedure be used

If your investigation requires a site or field visit:

1. Obtain permission beforehand from proper authorities, or from the owner if the site is private property
2. Arrange access for an appropriate length of time under the proper circumstances for your observations
3. Take along a camera, sketch pad, and tape measure, and arrange before the visit for permission to photograph and to make sketches

4. Take careful notes as you observe and double check any measurements you make

5. At the end of the visit, thank the owner/host in person or by letter

Experimenting

Experimentation is observation carried out under special conditions that are controlled and varied for specific reasons. In an experiment, an event is made to occur under known conditions. Here, as many extraneous influences as possible are eliminated, and close observation is made possible so that relationships between phenomena can be revealed. The sequence of the experimental process begins with the observation of the problem or difficulty. The experimenter formulates a hypothesis to explain the difficulty, then tests the hypothesis by the experimental techniques and draws a conclusion as to its validity.

Galileo is considered the father of the experimental method. According to the physics of his time, which was still the physics of Aristotle, force was defined as that which, acting upon any object, causes it to move or to produce velocity or motion. If, for example, one pushed or applied force to a chair, the chair moved. When one stopped moving it, unless the force of gravity were operating or some other forces were applied to keep it moving, the object or the chair stopped moving.

Even by Galileo's day, the Aristotelian definition of force had become unsatisfactory because it could not explain certain commonplace phenomena. For example, it did not apply to the case of a projectile fired from a cannon. When a cannon fired its shot, all the force was applied at the moment of the explosion and then ceased. Nevertheless, the cannonball continued to move in a great parabola over a considerable distance. Here was an incident that contradicted the Aristotelian explanation of what force was and did.

Galileo set out to find a more plausible explanation. In experimenting with falling objects, he realized that the motion of a body being drawn to the earth by the attraction or force of gravity was not affected by the weight of the object. So he reasoned that force could not be measured as merely proportional to the weight of the body on which the force acted. He thereupon came up with two hypotheses. One was that force might be proportional to the distance through which the object being acted upon travels; he was able to disprove this on mathematical grounds. The other hypothesis was that force would be proportional to the length of time during which it caused an object to move. In other words, the greater the force, the longer the time during which it operates to produce motion.

Up to this point, Galileo worked on this problem using mathematics and logic. He undertook to test his second hypothesis by experimental means. He constructed an inclined board along which he rolled a metal ball. With this equipment, he was able to measure fairly accurately the distances covered by the rolling ball during various units of time; that is, he calibrated the inclined board to measure the time necessary for the ball to travel half the distance, two-thirds of the distance, three-fourths of the distance, and so on, so that he could determine quantitatively the effect of force exerted by gravity. His experiments led him to the conclusion that force was to be defined as not merely that which produces motion or velocity but that which changes velocity as well.

According to Galileo's experimental finding, an object does not necessarily cease to move when the force causing its motion is removed. It ceases only to change its velocity. This now explains the action of the cannonball shot from a cannon. The cannonball's velocity remains constant as it rushes from the cannon's mouth (except for the effect of air resistance) until the force of gravity draws it earthward. Galileo's hypothesis and its experimental proof changed the whole concept of the physical theory of his day. That experiment has been called the foundation of modern mechanics.

Experimentation is not a technique used only by the scientist-researcher developing new instrumentations or discovering new laws in physics or in other natural sciences. Experimentation may also be applied effectively to business and ordinary pursuits. For example, the owner of a hardware store may use an experimental procedure to test the effectiveness of his show window in displaying a new power lawnmower. He might stand outside his store and count the number of persons who look at the show window and note how long they stop. He is merely observing, if he does just that. If he designs three different displays for his lawnmower, each installed for the same period of time on three successive days and, furthermore, if he records not only the number of people who stop to look at the display but also notes the number who enter his store on each occasion to inquire about the product, he is performing an experiment under varied and controlled conditions.

Planning Experiments

Before planning the actual experiment, the investigator should have an understanding of the nature of the problem and any relevant theory associated with it. The **theory** is the explanation of the problem. Of course, the theory or explanation is to be proved, but it serves as a guide to formulating a hypothesis for testing the answer to the problem. The experimenter should analyze the problem and put it into words that express it in its simplest form. Frequently, it is possible to break problems into parts that are more easily answered separately than together.

An experiment should not be conducted without a clear-cut idea in advance of just what is to be tested. The researcher should ask: Why am I doing this particular thing? Will it tell me what I need to know about my problem? The first thing in planning an experiment is to decide the kind of event to be studied and the nature of the variables that previous information suggests might be the controlling ones of the matter being tested. These variables may be divided into those that can be controlled and those that cannot. The ideal experiment is one in which the relevant variables are held constant except the one under study. The effects are then observed.

If observations are to be useful, the matter immediately under consideration must be separated from any that may confuse the issue. In scientific experimentation a device known as *control* is used to avoid such error. Controls are similar test specimens which, as nearly as possible, are subjected to the same treatment as the objects of the experiment, except for the change in the variable under study. Control groups correspond to the experimental groups at every point except the point in question. Controls are frequently used in medical research, as in the test of the polio vaccine where two groups of subjects were used. One group, the experimental one, was given the vaccine, and the other, the control groups, was given a placebo—a pharmaceutical preparation containing no medication but given for its psychological effect.

The use of controls is not always sufficient to insure correct results. A story is told of a test of a seasickness remedy in which samples of the drug were given to a sea captain to test on a voyage. The idea of control was carefully explained to him. When the ship returned, the captain was highly enthusiastic about the results of the experiment. Practically every one of the controls was ill and not one of the subjects had any trouble. "It is really wonderful stuff!" exclaimed the captain. The medical researcher was skeptical enough to ask how the sea captain chose the controls and how he chose the subjects. "Oh, I gave the seasickness pills to my seamen and used the passengers as the controls!"

Rules of Experimental Research

John Stuart Mill, the great English philosopher, set up five rules for evaluating data in experimental research, or to use the philosophical term, "rules for search for causes of phenomena under inquiry." They are useful guides in the design and evaluation of experiments and serve as general principles to aid interpretation of data.

Method of Agreement

The first of these rules is known as method of agreement—recognizing what is similar. It states that if the circumstances leading to a given result have even one factor in common, that factor may be the cause. This is especially so if it is the only factor in common. This principle is universally used but in and of itself is seldom considered as constituting valid proof of cause because it is difficult to be sure that a given factor is really the only one common.

The need for such caution is illustrated by the story of a young scientist who imbibed liberal quantities of Scotch and soda at a party. The next morning he felt rather miserable. So that night he tried rye and soda, again in very liberal amounts. The following day he was again visited by a distressing hangover. The third night he switched to bourbon and soda, but the morning after was no more pleasant. Being a scientist, he analyzed the evidence. Making good use of the method of agreement, he concluded that thereafter he would omit soda from his drinks since it was the common ingredient in his three distressing instances.

Method of Difference

The second rule is known as method of difference—recognizing what is different. It states that if two sets of circumstances differ in only one factor and the one containing the factor leads to the event and the other does not, this factor can be considered the cause of the event. For example, if two groups of experimental guinea pigs are fed identical diets under identical conditions except that one group also receives a certain vitamin, the fact that the guinea pigs in the group receiving the vitamin outgrow the other group is evidence supporting the hypothesis that the vitamin is beneficial to the growth of the guinea pigs. It is evidence but does not conclusively prove the hypothesis. The result may have been due to other factors and circumstances. For example, the second group of guinea pigs might have been sick or been of a different strain and heredity.

Joint Method of Agreement and Difference

The third rule is that of joint method of agreement and difference—recognizing what is similar and dissimilar. This rule states that if in two or more instances in which a phenomenon occurs, one circumstance is common to all of those instances, while in another two or more instances in which a phenomenon does not occur that same circumstance is absent, then it might be concluded that the circumstance or factor present in all positive cases and absent in all negative cases is the factor responsible for the phenomenon.

Principle of Concomitance

The fourth rule is the principle of concomitance—"guilt by association." *Concomitance* is a "hard" word meaning coexisting or occurring with something else. The rule states that when two things consistently change or vary together, either the variations in one are caused by the variations in the other or both are being affected by some common cause. As an illustration of this principle, Mill cited the influence of the moon's attraction upon the earth's tides. By comparing the variations in the tides with the variations in the moon's position relative to the earth, it can be observed that all the changes in the position of the moon are followed by corresponding variations in the times and places of the high and low tides throughout the world, with the high tides always occurring on the side of the earth nearest the moon. These observations have led to the conclusion that the influence of the moon causes the movement of the tides.

Method of Residue

The fifth rule is known as the method of residue—the process of elimination. This rule recognizes that some problems cannot be solved by the techniques called for in the other four rules. The method of residue arrives at causes through the process of elimination. When a specific factor causing certain parts of a given phenomenon is known, this principle suggests that the remaining parts of the phenomenon must be caused by the remaining factor or factors (by the residue).

Keeping Records

Data should be entered in the laboratory notebook at the time of observation of the experiment. Among the most unforgivable practices are dishonesty or carelessness in recording the full procedures, materials, or special elements and factors of the experiment, in neglecting to keep a record of everything done, or failure to take into account every small part of the components and every action of the apparatus. Without records, successes cannot be repeated and failures will not have taught any lessons.

The experimenter sometimes has difficulty finding or knowing what to record. Each record of an experiment should include the purpose of the experiment, the equipment used and how it was set up, procedures, data, results, and conclusions. Sketches, drawings, and diagrams are frequently helpful. It is important to record what is actually seen, including things not fully understood at the time. Poor or unpromising experiments should be fully recorded, even those felt to be failures. They represent an investment of effort that should not be thrown away because often something can be salvaged, even if it is only a knowledge of what not to do the next time. The data always should be entered in their raw form.

Complete records are beneficial because the mere act of putting down on paper the what, where, when, why, and how will generate ideas of how the work can be improved. Where patent questions might be involved, it is desirable to witness and even notarize notebook pages at intervals. The witness should be someone who understands the material but is not a co-investigator.

Interviewing and Discussing with Experts or People Qualified to Give Required Data

While interactive discussion techniques are fourth on the list, you, the writer, may turn to them first. Having ascertained the problem, you may first need to determine what you already know about it. After clarifying your own thoughts, you may find it profitable to turn to other people who are working in the field or on the problem. Frequently, valuable information and shortcuts may be discovered by discussion with experienced and knowledgeable people. If the problem is one which your organization has previously done work in, you would do well to consult the files of your organization or company. You might talk to colleagues and superiors who have had experience in this area. However, you must prepare yourself before beginning any interviews or discussions, especially if you go outside your own organization to obtain information.

How to Interview

The interview may be conducted by letter and by telephone, as well as in person. Letter and telephone interviews are less satisfactory. Direct contact with an individual and a face-to-face relationship often provide a stimulating situation for both interviewer and interviewee. Personal reaction and interaction help not only in rapport but also in obtaining nuances and additional information by the reactions, which are more fully observed in a face-to-face relationship.

Adequate preparation for the interview is a "must." Careful planning saves not only time but also energy of both parties concerned. The interview is used to obtain facts or subjective data such as individual opinions, attitudes, and preferences. Interviews are used to check on questionnaires that may have been used to obtain data, or when a problem being investigated is complex, or when the information needed to solve it cannot be secured easily in any other way. People will often give information orally but will not put it in writing.

Here are points to consider to promote the success of your interview:

1. Make a definite appointment by telephone or letter, explaining carefully the purpose for the interview, the type of information required, and—this is very important—the significance or importance of the contribution the interviewee will make.

2. Select the right person who is in a position to know and has the authority to give the information you need. Find out as much as possible about the person to be interviewed before the interview.

3. Tape the interview if possible. Taping your interview has the advantage of capturing exactly what your expert or respondent says; you are then able more conveniently to check the details and accuracy of the information you obtained. Taping cuts down on note-taking efforts. Be aware that some persons object to being taped, some freeze up, and some will not provide full or candid answers, especially on sensitive or controversial subjects. Always obtain permission for taping prior to the interview.

4. Prepare for the interview. Know the subject matter of the interview so that the interviewee does not feel she is wasting valuable time explaining basics that you should have known beforehand.

5. Arrive on time. Have questions prepared and lead the interview. Do not expect the person interviewed to do your job for you by volunteering all information you may need or pointing out other matters you do not know about or have forgotten.

6. Do only as much talking as is necessary to keep the interviewee talking. Do not contradict or argue with the interviewee, even if you know she is wrong. Sometimes erroneous information can be significant data. Be courteous. If there is a point requiring further explanation, ask a question that might show up the other side to the problem.

7. If the interviewee gets off the subject, be ready with a question that will get her back on the track.

8. When the session is not taped, and the person interviewed says something especially significant or expresses herself particularly well on some point, or if she offers statistics, figures, or mathematical formulas, or other matters that might require careful checking, ask permission to write such statements down.

9. Do not prolong your interview. Should the originally allotted time have passed, remind the interviewee of this fact. It is up to her to ask you to stay longer in order to finish the conversation.

10. Immediately after the interview, record the answers. At a later time it is a matter of courtesy to send the interviewee the record of the interview to give her the opportunity to correct any errors you may have made or permit her to change her mind about anything she has said. If you are planning to quote her, be sure that the interviewee is aware of this and gives you permission to do so.

The interview has the advantage that the interviewer can partially control the situation and can interpret questions, clear up misunderstandings, and get firsthand impressions, which might throw light on data. The interview is a valuable instrument in those situations where the only data available are opinions.

Using Questionnaires

As a research tool, questionnaires require judicious handling. However, there are certain situations where questionnaires must be used and can be used effectively. A much wider geographic distribution may be obtained more economically from a questonnaire. Certain groups within the population are frequently more easily approached through a questionnaire; for example, executives, high-income groups, and professional people. Questionnaires often allow for opinions from the entire family. A questionnaire can be filled out at the respondent's leisure and, therefore, more thought can be given to the response. Finally, by careful sampling methods, a representative sample of a study universe may be made. See Figure 11.4 for a sample questionnaire.

Like the interview, the questionnaire must have a reason for being used. It should be designed to require as little time as possible for completion. A cover letter should be included with a questionnaire, that will motivate the reader to answer the questions. The letter should tell the use to be made of the answers. If possible, the writer should offer a copy of the summary of the results of the investigation and assure the reader that the task of answering will not require too much of his time.

The questions should be phrased in the form of a list, not in lengthy sentences or paragraphs. The wording must be structured with care to insure clarity and completeness. Where possible, questions must be grouped into logical arrangements. The first few questions should be easy to answer and should be interesting in order to secure cooperation. One question should stimulate interest in the next. Questions should be arranged in an order that aids the respondent's memory. The most effective questionnaires are frequently those that are designed to provide a yes or no, a one word, a check-off, filling in a

Dear Mr. Nagel:

May we have just a few minutes of your time and the benefit of your experience in answering a few questions pertinent to the development of a Suggestion System Program?

Your assistance will be especially useful in helping us to plan effectively for a suggestion program. We would greatly appreciate your answering the brief survey which is attached and returning it to us in the attached postage-paid envelope. If you would like the tabulated results of this survey, please sign your name so that we may direct the results to you personally.

Won't you help us benefit from your experience? If we can ever serve you in a similar way, please call on us.

Sincerely yours,

Please Answer All Applicable Questions

1. Age of Company: Less than 5 years _____, 5 to 10 yrs. _____, more than 10 years _____.
2. Approximate number of employees: Under 50 _____, 50–99 _____, 100–250 _____, 251–499 _____, 500 or more _____.
3. Do you have or have you had a Suggestion System in your Company? Yes _____, No _____.
4. Is it in operation now? Yes _____, No _____.
5. If answer in 4 is no, please indicate reasons by checking:
   a. Too costly
   b. Lack of interest of employees
   c. Management indifference
   d. Supervisory resistance
   e. Reward system inadequate or unsatisfactory
   f. Other
6. If you do not have a suggestion System, do you have alternate means of eliciting employees' ideas? Please indicate_____
   _____
7. How long have you had your Suggestion System in operation? _____.
8. Does your program include participation by Engineers _____, Foremen _____, Middle Management (Department Heads) _____, Upper Management (Division Heads) _____?
9. Types of Awards Offered: Cash _____, Certificates _____, Other _____?
10. What is your maximum award _____; minimum award _____; average size of award _____?
11. Who administers your Suggestion System?
    a. Personnel Office
    b. Suggestion System Director
    c. Committee
    d. Other _____
12. Do you consider that your Suggestion System accomplishes its objectives? Yes _____, No _____.
COMMENTS:_____
_____
_____
_____

Figure 11.4
Sample Questionnaire

small circle, or pushing out a blank for the answers. They are best because they are easier for a respondent to reply to and they lend themselves to computer tabulation and analysis. Each question should contain one idea only. If the questionnaire becomes too long or involved, the reader may get discouraged and neither finish nor mail it. Questionnaires should include a self-addressed stamped envelope.

Much of the success of a mail questionnaire depends on its physical appearance, the arrangements of questions, the ease with which answers may be filled in, and the amount of space left for comments. A questionnaire of one page will bring more returns than one of two pages, and a post card is frequently the most effective questionnaire instrument. The sending of questionnaires should be accurately timed so that they reach their destination at a time when favorable replies may be expected. As far as possible, select periods when data or information may be fresh in the mind or when interest in a subject is ripe. Avoid vacation periods, periods when reports are made, and the close or the beginning of the year.

Systematizing, Analyzing, and Interpreting the Data

The data you have obtained from your investigative techniques must be systematized, analyzed, and interpreted in order to answer your problem. The data you have accumulated must be evaluated to ascertain the degree of appropriateness, accuracy, and completeness. If gaps appear in the mass of data acquired, further steps may be needed to obtain the missing information.

How to Systematize the Data

After you have decided that the raw data are complete and satisfactory, the next logical step is clear, methodological arrangement. This means that your raw or basic data must be so organized that they will lead you to the answer to your problem. This type of arrangement is done best by listing the elements of your problem on note cards and then arranging the raw data alongside the listed components of the problem. This systematization is chiefly a matter of linking the data or bases of evidence to the specific elements to which they relate. Make use of the block diagram in Figure 11.1.

How to Analyze and Interpret the Data

The previous step consisted of evaluating, sifting, and arranging the data in an orderly relation to the elements in the research problem. You may now find it desirable or necessary to recast your data in a more refined way in order to derive inferences or conclusions. This is done by tabulation and retabulation of the raw data. Averages in various forms—modes, medians, quartiles, deciles—

may be used to obtain quantitative criteria of mass data.[6] Frequently, only by averages will the complete meaning of mass data be comprehended, and they may then reveal qualitative differences. Data may have to be refined in a statistical form to eliminate the influences of such factors as cyclical or seasonal fluctuations. Intricate relationships may be revealed by tabulating, graphing, and diagramming. This refinement often results in greater insight and suggestions of inferences that may answer the components of the problem. By use of inductive and deductive reasoning, conclusions or answers to the problem can be arrived at.

Interpretation, as was seen in Chapter 9, is the process of arriving at inferences, explanations, or conclusions by examining data and applying to them the process of logical analysis and thought. Interpretation is accomplished when phenomena are made intelligible by relating them to the requirements of the matter or problem being investigated. Individual facts and/or opinions are placed within a deductive or inductive system for inferential generalization. In short, interpretation, by use of observation, logical analysis, synthesis, experience, knowledge, and insight, arrives at explanations and answers to the questions the investigation must resolve.

Applying and organizing the interpreted data constitute the next step, which will be discussed in Chapter 12.

Chapter in Brief

In this chapter we began by defining what a report is, then identified the four steps in report preparation. The chapter focused on a report's prewriting processes, and examined the necessary strategies, planning, and analysis required to reconstruct the investigation being reported. In addition, we explicated the methods for conducting research, and discussed the contribution of computer technology to literature searching.

Chapter Focal Points

- Facts
- Type of Reports
- Prewriting strategies and analysis

---

[6]*Mode* is the value or number that occurs most frequently in a given series; *median* is the middle number in a series containing an odd number of items (e.g., 7 in the series, 1, 3, 5, 7, 15, 21, 34); *quartile* is the designating of a point so chosen that three-fourths of the items of a frequency distribution are on one side of it and one-fourth on the other, *decile* is any of the values of an attribute that separate the entire frequency distribution into ten groups of equal frequency.

- Investigative procedures

  How to search the literature, including computerized information resources

  Observation

  Experimentation

  Interviews and questionnaires

- Systematizing, analyzing, and interpreting data

Questions for Discussion

1. What is a report? What role do reports play in present-day industry and scientific endeavor?

2. The use of the scientific method in problem solving receives much emphasis throughout the pages of this text. Its purpose is to force the technical writer to think somewhat as the researcher thinks. Justify this emphasis.

3. What sources of information are employed in the historical method of research? Its aspects of library searching are useful in what ways to other methods of research?

4. What is the role of the problem in the technical paper or report?

5. Locate and examine a report from business, from industry, from a private, government, or educational research laboratory, and from a university experiment station. What are their differences and similarities? What is their purpose? intended use? readership? Can you classify these four reports? How was the research carried out in each? What methods were employed? Note their format, organization, style of writing, and use of illustrations and appendixes, if any.

6. Discuss the four major steps in report writing. What inference is to be drawn from the fact that the writing process is the last step?

Assignments

1. Select a topic for an analytical, formal report in accordance with the precepts stated in this textbook. Preferably, your topic should be in your major field of study; however, you may select one of practical value affecting your college, community, or yourself. Your instructor must approve the topic you select. To help your thinking, I have listed below some suggestions for types of problems to consider:

   1. Should the Harmony Preschool invest in two computers for its classroom use or buy additional jungle gyms and sandboxes?

   2. How to design a unique anti-icing device for home concrete walks and driveways.

   3. When and how should your township increase its water supply?

   4. Should the Mathematics department add five additional graduate assistants to meet the need for remedial assistance to freshmen students or invest instead in two computer aided instruction machines?

   5. Is high-pressure hot-water heating more economical than steam for heating a given complex of buildings in the university?

6. What is the effect of antihistomine on the antibody titer induced by influenza virus in mice?

7. What environmentally safe measures can be used for campus lawns, trees, and shrubs without resorting to pesticides?

8. What is the feasibility of instituting a child care center on campus for parent students and university employees?

9. Are there effective alternatives to dieting for lowering cholesterol?

10. In what university department can the videotext technology be an advantageous pedagogical tool?

11. What industrial training areas best lend themselves to the hypertext technology?

12. Should student fees be reduced for students not desiring to attend football and basketball games, play performances, musical recitals, and similar events?

2. After you have selected a topic, write your instructor a memo asking his or her approval for your research question. Specify both the audience for whom your report is being prepared and the use for which it is intended. Justify your topic selection by stating the reason for your interest and its importance to the intended reader. Identify what, if anything, you already know about the subject and what you need to learn. Indicate how you will go about obtaining and researching the information.

3. Develop a working bibliography on your topic by searching your library catalog, literature guides, and such reference sources as abstracting and indexing journals, subject bibliographies, handbooks, yearbooks, and, if possible, computer databases such as OCLC. Don't hesitate to seek help from the librarian. Use index cards for your bibliographic entries. Take notes on at least three reference sources. Turn them in to your instructor for comment.

4. You have chosen to write a report on the subject, Panic During Emergencies. In your research you have come across an article by Thomas E. Drabek, "Shall We Leave? A Study on Family Reactions When Disaster Strikes." The article includes some key principles, which you want to include in your conclusions. As you know, plagiarism is using words, ideas, and thoughts you have learned from another source as your own. Dr. Drabek's principles are quoted below. How would you make use of his material without resorting to plagiarism? While material that is considered as common knowledge need not be documented, direct quotations exhibiting a particular style or choice of words that express ideas concisely and effectively require acknowledgment. Even paraphrasing of such material requires documentation.

Warning Responses: Eight Behavioral Principles. [3]

1. Disbelief, not mass panic, is the typical initial public response to disaster warning.

2. A siren—like any other noise—does not constitute a public warning; at best it may alert some, but many will ignore it.

3. Warning messages must include both threat information and directions for adaptive actions.

4. The greater the specificity of the information given, the more likely people are to believe it.

5. Community warning systems must accomplish seven key functions: (1) detection; (2) measurement; (3) collation; (4) interpretation; (5) decision to warn; (6) message content; and (7) dissemination.

6. There are patterned variations in response—women and children are more likely to believe; elderly and ethnic minorities, like males, take more convincing.

7. Warnings received from individuals perceived to be authorities—a uniformed police officer—are more likely to be believed than from other sources, e.g., a relative or a media representative.

8. Typically, groups—be they family or work—receive and process warning messages, not single individuals in total isolation from others.

For this assignment, include introductory material to your presentation of Dr. Drabek's principles.

5. You are working for your college Public Affairs Office. You receive an assignment to analyze some statistics based on a questionnaire sent to the students who graduated ten years ago. Of 63 responses from women graduates, twenty provided data on their present yearly earnings. These reveal:

Betty J,	$17,500.	Donna T,	$23,000.
Vera P,	$20,000.	Catherine P,	$40,000.
Elizabeth K,	$28,000.	Sherry H,	$23,000.
Mary T,	$12,000.	Nancy S,	$50,000.
Nora A,	$36,000.	Karen K,	$40,000.
Beatrice W,	$70,000.	Helen B,	$12,000.
Laura B,	$9,800.	Ginger R,	$100,000.
Madeline L,	$1,850,000.	Jean E,	$17,500.
Terry M,	$15,000.	Virginia T,	$10,000.
Holly G,	$17,000.	Lenore L,	$17,500.

When you totaled the yearly salaries, it came to $2,408,800. Then you divided that amount by 20 to arrive at a mean (average) of $120,440 earned by women ten years after their graduation from your college! This startling statistic could make a wonderful publicity story for your college. But how truthful is that statistic? To get a complete picture, what other factors need to be considered about the data in these returns? How should you handle the statistics to make a more factual and objective presentation? Write a 200–250 word article maintaining the integrity of your data without distorting their validity.

6. a. Prepare a set of questions for an interview with an "expert" on an aspect of the problem you are investigating and reporting.

   b. Design a questionnaire for your use in obtaining information on your problem or an aspect of it.

7. If you are conducting an experiment or are doing field work and observation to obtain a solution to your problem, keep a laboratory notebook or journal to record your data.

8. After you have identified the problem you are to investigate for your report in your preliminary analysis phase, build a block diagram like the one pictured in Figure 11.1. Resolve or factor out the major elements of the problem and their components. Place these elements and components within their proper boxes. This type of block diagram analysis will be invaluable in the prosecution of your research and its reporting in the next stage.

9. Set up a work schedule listing the tasks and time table for accomplishing the identification of your research problem, its investigation, organization, and analysis of data, and the writing of a report on the research.

References

1. Anderson, Chester R.; Saunders, Alta Gwinn; and Weeks, Frances W. *Business Reports*. 3d ed. New York: McGraw-Hill, 1957.

2. Bush, Vanevar. "As We May Think." *Atlantic Monthly*, July, 1945.

3. Drabek, Thomas E. "Shall We Leave? A Study on Family Reactions When Disaster Strikes." *Emergency Management Review*, Fall, 1983.

12

Report Writing—Organizing the Report Data

Chapter Objective

Provide algorithms for organizing the report's data and structure.

Chapter Focus

- The Thesis sentence's role in structure
- Logical methods for organizing a report
- The Psychological method for organizing a report
- Outlining

When the data for the report are in, we face the problem of organizing them into useful, functional form, interpreting and analyzing them to see what they mean. If your prewriting planning and analysis have been carefully done, the work at this stage is almost half accomplished. You ask yourself questions again:

What facts have I found that are significant to my problem?

What facts are most significant?

What is their significance?

How do these facts answer my problem?

What is the answer to my problem?

The answers to these questions become the basis for the organization of the main body of the report.

How to Develop the Report's Thesis Sentence

Having begun with a problem, the investigator has accumulated data for answering it. Raw data of themselves do not make a report. Raw data properly examined, analyzed, and interpreted through the methods of logical reasoning become the material out of which the report is designed. The first step in structuring a report is to develop a clear and concise statement of the answer. The answer may be positive or negative. The statement, explicitly and clearly stated, is the payload you are bringing to your reader. This payload often is called the **core idea** or **thesis sentence** (Figure 12.1). The thesis sentence is the answer to

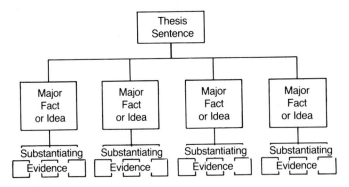

Figure 12.1
Each block of major fact or idea relates to and answers the block of the major element of the problem diagramed in Figure 11.1 (Chapter 11). Substituted for the block of the Statement of the Problem is that of the Thesis Sentence. Replacing the block of the major elements of the problem are the blocks of the major facts or ideas. The blocks of substantiating data or evidence replace the blocks of components of the earlier diagram.

the problem or situation that you have investigated and are reporting. It is what you want your reader to know. The thesis sentence, or core idea, is a single, comprehensive statement in complete sentence form that synthesizes the complete report and presents the problem's answer to the reader succinctly and clearly. A properly formulated thesis sentence is concrete evidence of how clearly and simply you, the writer, see your subject. It is the backbone for unifying all of the elements of the report.

Adequate research requires the clarified statement of where the research has gone, what it has meant, and where it has led. Since clear thinking is the basis of clear writing, the writer must state the answer to the problem clearly and succinctly. So the first step in designing your report is to state clearly and concisely the core idea, or thesis sentence.

For the sake of illustration, we might examine a few hypothetical problems and see how a thesis sentence might be structured from the data:

1. *Problem:*   What causes the diurnal migrations of zooplankton?

    Thesis Sentence: Migration of zooplankton is caused by an interplay of factors such as light, temperature, chemicals, and quest for food.

2. *Problem:*   Could a light, single-engine, pontoon-type aircraft be used effectively for fire suppression in the lake states?

    Thesis Sentence: The Beaver, a light, single-engine aircraft, could be used effectively for fire suppression in the lake states.

3. *Problem:*   Are existing environmental conditions in Decatur County, Kansas, conducive to the propagation of bobwhites in sufficient numbers to support an annual hunting season?

    Thesis Sentence: The environmental conditions in Decatur County, Kansas, are unfavorable to increased population of bobwhites.

4. *Problem:*   Can implanted sex hormones in dairy heifer calves be used to promote maturity, better breeding efficiency, and greater udder development?

    Thesis Sentence: Sex hormone implantation in dairy heifer calves is useless in promoting maturity and is detrimental to breeding efficiency and udder development.

5. *Problem:*   Which of three wiring processes presently used in X Manufacturing Company is most efficient and best suited to the conditions of the factory?

    Thesis Sentence: The printed circuit process is the most efficient and best suited for the special factory conditions of X Manufacturing Company.

The thesis sentence serves to keep you, the report writer, from exceeding your limits. It serves to help you follow consistently your objective, and in certain situations it suggests points for you to include and the arranged order of their sequence. The thesis sentence is the test of your mastery of your subject.

Inability to formulate the thesis sentence is an indication that either you have not completed your investigation or have not been able to interpret and analyze logically the data accumulated in your investigation.

In purely informational types of reports, it might be difficult to synthesize the material of the investigation into the thesis sentence; here a summary statement is called for. But in all analytical types of reports, the thesis sentence is not only practical but necessary.

Although the thesis sentence is necessary in the construction of the report, it may not be found verbatim anywhere in the report itself. It might be compared to a working sketch, which must be drawn (written) in order to point to the directions where the model or composition is going. The thesis sentence obviously cannot be written until all the research has been completed and all of the raw data analyzed in relation to the problem. How else is the answer derived? The answer to your problem becomes the core idea of your report. It is the dominant idea you want your reader to reach after he completes reading your report.

A Suggested Algorithm for Organizing the Report Structure

Write the thesis sentence on a 3″ × 5″ card, then set it on a table before you. A reconstruction process follows: Use other 3″ × 5″ cards and set down the major facts that have led you to the thesis sentence. Write each major fact or idea on a separate card. Place the card with your thesis sentence at the left of the table. Next, make a mental equal sign and arrange the major facts or ideas of the data of your investigation to equal your thesis sentence. What you have arranged before you is your thesis sentence = one major idea + another major idea + another major idea + . . . as many major ideas as are necessary to lead you to the answer to your problem.

As stated above, the synthesizing process has been turned around. In actuality, the formula for the thesis sentence was derived from this sequence:

$$\text{major idea}_1 + \text{major idea}_2 \ldots + \text{major idea}_n \xrightarrow{\text{(yields)}} \text{thesis sentence}$$

In arriving at this formula, you may find it advisable to check on yourself by having each major fact on a separate card, followed by cards containing substantiating proof or data below it. Spread the cards out on a table and rearrange the cards into the best sequence for an outline, beginning with the most logical or valid fact or datum relevant to the answer. Follow sequentially in the order of the logic. The logic for the structuring will be discussed later in the chapter. The major facts become headings in your report. To identify their place in the outline, number each of these and number their substantiating matter.

Patterns for Organizing the Report's Data

I have just noted that the major facts and ideas and their substantiating data are to be organized in a logical sequence. Let us now see how the various kinds. of content matter might lend themselves to effective arrangement for bringing the reader to the answer you, the writer, have reached. In organizing the material of the investigation, you must keep in mind the purpose of the report and the purpose for which the answer to the problem will be used by the reader. A report's organization must be functional. You arrange your data so that they meet the reader's specific needs.

A number of approaches in the organization of the report are possible. The one you choose depends on such factors as these:

1. Purpose
2. Needs of the reader
3. Nature of the material

Temporarily, we are going to skip the elements of introduction and conclusion in the report and examine the part containing the main data, the *body* of the report. There are two major methods for organizing the data contained in the body of the report—the logical and the psychological.

Logical Method of Organization

The logical method builds its case step-by-step within the logical scheme of its algorithm or pattern, leading to conclusions and possibly recommendations. Since problems and their investigations vary in nature, the logic to the organization of the data is going to be different in each report. Accordingly, in the logical method, there are a number of patterns to the organization of the main body of data.

1. Chronological Pattern

The chronological organization algorithm, sometimes called the narrative pattern, is the simplest and often the most obvious way of arranging the data of an investigation. It follows a time sequence. It is the pattern most frequently used in informational types of reports, particularly those of a historical nature. Chronological order may also be used in reports using other organizational patterns where the presentation of a sequence of doing a thing is important to the exposition. In a technical experiment, for example, conditions or results within a sequence must be presented as they occur; otherwise, the reader will be unable to duplicate the work. In an investigation of a historical nature, the chronology of events is of vital importance to the understanding of events; so the presentation of data must follow the proper time sequence. This type of organizational approach does not lend itself well to either complex situations or those situations where nontemporal relationships and special emphasis are important. The chronological approach serves in simple, uninvolved situations where the time order is important and special emphasis on particular matters is not required.

2. Geographical or Spatial

Data may be presented or arranged according to geographical or spatial relationships. For example, a report to the president of a corporation on how to allocate space to the various departments of the corporation in a new building might logically begin with an assignment of the various wings moving east or west, as the case may be, or moving from the basement up to the various floor levels of the building. Or, in his State of the Union address at the start of a new Congress, the president of the United States might review the economic state of the country geographically, beginning with the northeastern states, working south and west across the country.

3. Functional

If you were reporting on the design of a new mechanism or instrument, a logical approach would be to examine each component or assembly by the function it performs in the working of the whole.

4. Order of Importance

It is sometimes appropriate to begin with the data most significant to the problem or situation, then move on to the next most significant, and so forth. This approach would be highly suitable in a situation where the researcher must find the best plan or procedure for accomplishing a certain end. Related to this pattern is the approach of putting first those phases of a project in which the greatest success was achieved and following sequentially to the least successful phase.

5. Elimination of Possible Solutions

Within this pattern, you would examine all possible solutions to a problem, beginning with the least likely and working toward the best possible solution. (This approach is opposite to that of number 4 above). This approach borrows from the narrative technique of building toward a climax.

6. General to Particular

The sixth structural pattern (deductive method) starts with discussion of general principles and then deduces the specific applications from these. In reporting on the efficacy of inaugurating a company training program, for example, you might discuss first the general benefits to be derived from a training program and then arrive at a specific benefit from a specific training, as for example, report writing. You would then show how the class in report writing would add to the general benefits of a training program.

7. Particular to General

This pattern (inductive method) is the reverse of number 6. You might start with a specific training program—again, say, report writing—and then induce general benefits derived from such a specific program.

8. Simple to Complex, or Known to Unknown

This pattern of organization is appropriate to situations where it is advisable to begin with the simple case or known situations and then move to more and more complex or unfamiliar grounds.

9. Pro and Con

In the report that investigates whether to do or not to do something, the material on which an answer is based might appropriately be grouped into data for and data against, or advantages and disadvantages.

10. Cause and Effect

This approach is particularly useful in the exposition of problems that deal with questions like these: What is it? What caused it? or What are the effects of . . . ? The writer may begin with a fact or a set of facts (cause or sets of causes) and proceed to the results or effects arising therefrom. Conversely, the writer may examine and report the causes from which effects arise.

These ten patterns are not mutually exclusive and are frequently used in combinations.

The Psychological Pattern of Organization

With the psychological pattern of organization, the most important data of the report are arranged in the most strategic place—the beginning. The psychological method follows no order of time or sequence in which the data are collected. Usually the conclusions are placed first and the discussion second. This report pattern, sometimes known as the *double report,* originated in business and industry, where busy executives prefer to conserve time and get at once to essential matters. When reading a report, they want to know quickly what the problem is, what the answers to the problem are, and how the answers were derived. The psychological pattern permits the conclusion to be seen first so the major issues can be apprehended immediately. The study of supporting data is usually left to subordinates. The organizational sequence of the psychological pattern is as follows:

A short introductory section states the problem. This is followed immediately by a section of conclusions and recommendations. Because executives reading the report may be interested in knowing how the conclusions were reached, the section on procedure follows immediately. Should they want to check points more carefully, they will continue to read the results based on the procedures. Then, if they want all the details, they will turn to the remaining section, the discussion.

In the psychological organizational pattern, the main body of the report does not contain "arithmetic" or other raw data. This material is put into the appendix. Busy executives want only the significant factors that have resulted in the generalizations making up the conclusions and recommendations. Technical subordinates will check the computations and the raw data in the appendix if the executives want full confirmation of the recommendations on which to act.

Just which of the two major methods of organization to use is determined by the writer after considering all aspects of the purpose and nature of the report, and of the needs of the readers.

The Outline—A Guide to the Writer and Reader

The synthesizing approach we have been discussing in this chapter can conveniently take form first as an outline. The **outline** is a schematic road map of the report and shows the order of topics and their relationships. The outline makes the writing of your paper easier and more effective and enables you to write with confidence and to focus on one stage at a time. You can see how the whole will take shape, and you will not be distracted while writing by the question of whether to put a particular piece of information in here or reserve it for later. The outline, then, permits you, the investigator, to test the adequacy of your data. Subsequently, it offers the reader guideposts or road signs in the form of heads and subheads. The outline is the last step in the organization of the data for the writing of the report. Several types of outlines are used in practice. The simplest is the topical outline. Subjects or topics are noted in brief phrases or single words and numbered sequentially. Another type, the sentence outline, is more specific and provides a complete sentence about each topic.

The Topical Outline—The "Laundry List"

The topical outline is a useful device in thinking through the organization of the paper. To begin, you may want to jot down key ideas and data related to the material of your paper. For example, let us take the case of a geology student whose research problem is tracing the origins of the intrusive body of rocks in Big Thompson Canyon, Colorado. Initially, the student would write down the series of major facts and their components as follows:

1. Introduction
2. Historical background
3. The problem

4. Location and description of intrusive
5. Location and description of Silver Plume Granite outcrop
6. Specimen selection and field trimming
7. Microscopic analysis of specimens
8. The intrusive specimen
9. The Silver Plume specimen
10. Comparison of field and laboratory data
11. Conclusion
12. Recommendations

As you can see, topics are listed in a sequence without any indication of specificity, importance, or relationships, or that some are topics subordinate to major subject heads. In early planning, this type of outline is useful; it sets down on paper matters to be considered for more detailed structuring.

The Detailed Topical Outline

The second phase of outlining would focus on greater specificity, subordination, and relationships. The analysis entailed would reveal a breakdown of greater detail. This phase of the outline might be structured as follows:

1. Introduction
   a. Significance of problem
   b. Scope
   c. Historical background
   d. Definitions
2. Analyzing the Problem
3. The Intrusive
   a. Location of intrusive
   b. Description of intrusive
4. The Silver Plume Granite Outcrop
   a. Location of the outcrop
   b. Description of the outcrop
5. Specimen Selection
6. Field Trimming
7. Microscopic Analysis of Specimens
   a. The intrusive specimen
   b. The Silver Plume specimen
8. Field and Laboratory Data
   a. Similarity of occurrence in field
   b. Comparison of laboratory data

9. Conclusion

10. Recommendations

The Sentence Outline

The second, the detailed topical outline, is an improvement over the first. It is more specific and it identifies relationships, allows some emphasis, and shows subordination of lesser items to more important ones. Nevertheless, it does little more than identify these matters. The writer still faces a painful task of thinking through the ideas and the status of the evidence to be presented. At some point in the compositional process, the writer will have to reason through all points and details of the report. Many writers accomplish this by a third phase of outlining—the *sentence outline.* This form picks up the schematizing where the second phase left off. It places each topic within a complete sentence. Completing the thought of each topic enables the writer to test the context of the topic's environment. Such deeper thought not only reveals flaws in the data as heretofore seen by the writer but also permits deeper probings, examinations, analysis, and interpretation of major points and minor details. The result is clarification of aspects of the problem and derivation of insights not previously attained. A concurrent and important benefit is an improved, orderly outline in complete detail, with properly designed relationships, emphasis, and subordination. Now examine the sentence outline for "A Preliminary Report on the Possible Origin of the Intrusive Bodies in the Big Thompson Canyon" (Figure 12.2).[1]

Ideas and information in a report should be presented in a steady progression so that the reader feels she is getting somewhere as she reads. The reader must have important data clearly pointed out to her. The topical outline does not provide any easy mechanics for this. The outline arranged to show relationships and subordination is an improvement. The sentence outline, because it has put the writer through the chore of clarifying her thoughts about the main facts and their substantiating details, promotes a more logical composition and a more readable and comprehensible report.

The sentence outline shown in Figure 12.2 reveals the conventional structuring of this form. It uses Roman numerals to designate the main divisions and capital letters to indicate major subdivisions under each division. Further subdivision makes use of Arabic numerals, and still further division makes use of lower-case letters. Each entry in the sentence outline is composed in a complete grammatical sentence. This is demanded not only for the Roman numeral headings but also for divisions and subdivisions within them. Since a topic is not divided unless there are at least two parts, logic demands that there be at least two subheads under any division. For any Roman numeral *I*, there should

---

[1]Check this outline with the student report based on it, Figure 13.10.

```
                        Sentence Outline for
                    a Preliminary Report on the Possible
                    Origin of the Intrusive Bodies in the
                          Big Thompson Canyon

Thesis Sentence: The intrusive body in the Big Thompson Canyon,
Colorado, is a recently exposed portion of a larger mass known as
Silver Plume Granite which was intruded into the base of the
Rocky Mountains in Pre-Cambrian times.

  I.   This investigation undertakes to trace the geologic origin
       of the intrusive bodies in Big Thompson Canyon, Colorado.

       A. The purpose of this investigation is to serve as a field
          analysis problem for beginning geology students in
          properly analyzing and identifying geologic features.

       B. The meaning of certain geological terms, though
          frequently used by students, is not always fully
          understood and should be defined.

          1. An intrusive body was once a molten lava, which cut
             across or was injected into overlying sedimentary or
             metamorphic rocks.

          2. An acidic rock, such as granite, is derived from a
             molten source that was high in silica content.

          3. A batholith is an igneous body more than forty
             square miles in area.

          4. A stock is an igneous body less than forty square
             miles in area.

          5. A dike is an igneous body that has cut across layers
             of sedimentary or metamorphic rock.

       C. Field work was conducted by the author and other
          students at the intrusive in Big Thompson Canyon and at
          an outcrop of Silver Plume Granite west of Horsetooth
          Reservoir.

          1. The intrusive in Big Thompson Canyon is located at
             the base of Palisade Mountain on State Highway 34.

          2. The location of the Silver Plume outcrop is midway
             between the west edge of Horsetooth Reservoir and
             the town of Masonville.


                                1
```

Figure 12.2
Example of a Sentence Outline

D. Early preliminary observation and interpretation by geology students brought forth two varying opinions:

    1. That the intrusive was a granite stock;

    2. That the intrusive was a granite dike.

E. Extensive fieldwork and a library search have shown validity to both opinions.

    1. Analyzed samples taken from the intrusive show that orthoclase feldspar, plagioclase feldspar, quartz, and biotite mica are present.

        a. Comparison of these constituents from a known outcrop of Silver Plume Granite shows that they are one and the same.

        b. Variations in color of the outcrops are due to magmatic differentiation.

    2. The Silver Plume formation is composed of small batholiths and stocks.

        a. As the intrusive in question is relatively small in size, it can be assumed that it is an offshoot from a larger, underground body.

        b. Similar small intrusive bodies at other locations bear this out.

F. An analytical table of comparative mineral constituents and a map of the region validating this conclusion will accompany the report.

II. Field work and laboratory experiments combined with material gathered from a library search answered the question about the intrusive body.

A. Structurally, the intrusive and the outcrop of Silver Plume Granite are dissimilar, but of the same geologic period.

    1. The intrusive is of a gray homogeneous granite with very little fracturing.

    2. The Silver Plume Granite outcrop is flesh-colored with large, individual crystals.

    3. Both of the bodies investigated were intruded into a Pre-Cambrian, metamorphic, country rock.

B. Hand specimens were taken from the site of the intrusive and the known Silver Plume outcrop.

    1. A geology pick or prospector's hammer was used to break off good-sized rocks.

2

Figure 12.2 (continued)

2. The rocks were held in the hand and chipped to approximately 1″ × 3″ × 5″.

3. The rocks were chipped in such a manner as to eliminate hammer bruises on the faces and also to expose only unweathered rock material.

C. The hand specimens, as observed under a microscope, were found to differ slightly in physical properties.

1. The specimen from the intrusive was found to be a granite.

   a. The specimen from the intrusive contained orthoclase feldspar, 60 percent; plagioclase feldspar, 5 percent; quartz, 20 percent; and biotite mica, 15 percent.

   b. The color of the intrusive specimen was gray.

   c. The component grains of the specimen were one to five millimeters in diameter.

2. The specimen from the known outcrop of Silver Plume Granite was flesh-colored with grain sizes from 1 to 30 millimeters in diameter.

D. Comparison of field observations and laboratory data with information from T. S. Lovering's professional paper showed that the intrusive is a portion of Silver Plume Granite.

III. This preliminary research of the problem has led to a theory that may be fully validated following a more extensive investigation.

A. The intrusive in Big Thompson Canyon is a portion of Silver Plume Granite that was intruded into the roots of the Rocky Mountains during Pre-Cambrian times.

B. It is recommended that a more complete investigation be undertaken to determine definitively the relationship between the intrusive bodies in Big Thompson Canyon and that of the Silver Plume Granite.

(From a student report by Gerald L. Owens)

3

Figure 12.2 (continued)

be a Roman numeral *II*, for every capital *A*, there should be a *B* in sequence. If only one point follows a main heading, make it part of the main heading:

I.
  A.
    1.
      a.
      b.
      c.
        1)
        2)
    2.
  B.
II.

The sentence outline enables you, the writer, to control the content and structure of your report in order to secure unity and coherence. Though more difficult to construct than the topical outline, the sentence outline is recommended because this method forces you to think your points out more fully and to express them as whole units. Each sentence in the outline can become the topic sentence of a paragraph in the final writing.

The Decimal Outline

The decimal system of outlining uses decimals to show the rank of heads and subheads.

1. First main heading
  1.1 First subdivision of main heading
  1.2 Second subdivision of main heading
2. Second main heading
  2.1 First subdivision of main heading
  2.2 Second subdivision of main heading
    2.2.1 First subdivision of 2.2
    2.2.2 Second subdivision of 2.2
      2.2.2.1 First subdivision of 2.2.2
      2.2.2.2 Second subdivision of 2.2.2

The decimal system enables easier revisions and additions. It can be used with a sentence outline or with a topical outline.

The Format and Organization of the Outline

The outline may be single-spaced or double-spaced, or single-spaced within groups of headings and double-spaced between such groups. The writer should be consistent, however. A very short outline may be arbitrarily double-spaced in order to fill the page. All symbols, letters, and numbers should appear in a straight, vertical line throughout the whole. All capital letters should be indented the same distance, and all Arabic numerals the same distance, as should all lower-case letters. Three to five spaces for each level is conventional. Consistency in the amount of spacing will add neatness of layout and appearance. Excessive indention will look as unattractive as too little or none.

Computer Outliners

Some computer software programs have a feature called an outliner, which will format spacing and place Roman and Arabic numerals in outline format. If you are developing your outline on a computer screen, the outliner feature gives you the convenience of entering your ideas as they occur and then enables you to organize and reorganize them as headings and subheadings until your outline reads the way you want it to. Further, the outliner software allows you to break up headings so that you can see only major topics or allows you to check a list of subtopics under a major head without the distraction of other heads or items diverting your view. At the proper point, the software enables you to convert the topical phrases into complete sentences with convenience.

Should your formal report be lengthy and require an index, you can mark key terms in your sentence outline for later retrieval in alphabetical order for index construction.

Introduction, Body, and Terminal Section

The final draft of the outline can become the basis of the table of contents of your report. During the actual writing of the paper, the outline is used as the plan for composition. Structurally, the outline, like the report, has a beginning, middle, and end, or to use the classical structural terms—introduction, body, and conclusion. The introduction and conclusion are fixed points at the beginning and end of the outline. The experienced writer spends much time on the outline because it saves problems in the later writing of the paper.

Frequently, the body—which will begin with Roman numeral *II*, or if the decimal system is used, with 2—may have two or three major divisions within it. The terminal section usually may be encompassed in one Roman numeral, although it is sometimes advantageous to separate the recommendations from the conclusions by creating another major heading preceded by a Roman numeral. This is appropriate if the recommendations are significant. Too many major heads could be symptomatic of inadequate analysis of the data.

Checking Outline Requirements

After you complete the outline, check it for sequence of major divisions. Are they in the right order? Look over your note material to be sure nothing important has been omitted. Check for duplication of headings. All main headings should add up to the thesis sentence. Are all main headings really main headings or should some of them be subheads? Check your subheads for proper subordination and sequence. Do all subheads within a major heading add up to the major heading? Should some of the subheads be promoted to main heads? Finally, check the wording of all divisions for clarity, correctness, and parallel construction. Usually, there are no more than four or five major heads. If a paper is to be long, the increased length should result from the use of more subheads, rather than from an increase in the number of major headings. The outline should have no fewer than three and no more than five or six major heads, which are, in the conventional outline numbering system, preceded by Roman numerals.

The sentence outline forces you, the writer, to test the adequacy of your data. Not all data included are of equal significance. Evaluating facts correctly means giving them only as much weight as they have value in accomplishing the objectives of the report.

In summary, this type of planning gives the report coherence, continuity, and unity. By focusing on a central idea—the thesis sentence—you succeed in presenting ideas and information in a steady progression. As a result, the reader feels she is getting somewhere as she reads; everything in the report is pertinent to the problem. The outline has enabled you to achieve an overview of the report and to see relationships that are decisive in contributing to the objective. These factors can only help the reader.

Chapter in Brief

In this chapter we discussed how to organize research data and analyses into a report. We stated that composing a thesis sentence, derived from the answer to the chosen problem, is critical to this organization process. The various algorithms or patterns to organize the report material were presented, and we explained how to develop an outline, which serves as the skeleton of the report.

Chapter Focal Points

- Thesis sentence
- Algorithms for organizing the report
- Logical methods
  Chronological pattern
  Geographical or spatial
  Functional
  Order of importance
  Elimination of possible solutions
  General to particular
  Particular to general
  Simple to complex
  Pro and con
  Cause to effect
- Psychological pattern
- Outlining

Questions for Discussion

1. Explain how the thesis sentence is developed and discuss its role in focussing the report's material.

2. Interview two or more prolific researcher/writers on and off your campus. Ask them how they organize their research data for report or professional paper writing. Find out what system of outlining each researcher uses. Find out their reactions to the sentence outline. If a researcher does not use it, would he or she recommend it to a beginning research writer? Do these researchers use the device of a thesis sentence or core idea? Do they use computers in their research? Does the researcher use a computer for writing the report or professional paper? If so, does the researcher record his or her data and results on computer disk? Does the researcher convert any of the computer data and information into the report or paper? How? Do any of the researchers use hypertext techniques for their writing?

3. Analyze four of the reports you located for question 5 in Chapter 11 (page 289) as to type of organizational patterns used by the report writers.

4. Your text claims that the process of outlining promotes coherence, continuity, and unity in your report. Defend or dispute this claim.

Assignments

1. Develop the thesis sentence for the report topic you selected in Chapter 11.

2. After you have derived the thesis sentence, construct a block diagram by using the major facts or ideas and their substantiating evidence within the blocks of the diagram. Use the structure recommended in this chapter.

3. Develop a sentence outline for your report by going through the three phases discussed in this chapter. The work you do in problem 2, above, should simplify your task.

4. Evaluate, criticize, and rewrite the student sentence outline on sagebrush control (Figure 12.3). Lack of knowledge of the subject should not prevent you from correcting the faults in the outline.

Sagebrush Control

Thesis sentence: The most effective method of sagebrush eradication in western Colorado is either grubbing, railing, harrowing, blading, burning, plowing, or spraying.

I. The geographical area studied, species of sagebrush studied, and purpose of this research need to be explained.
    A. The area included in western Colorado is largely covered by big sagebrush (Artemisia tridentata).
        1. The area of western Colorado includes thousands of acres.
        2. This area of Colorado is typical sagebrush land in physical characteristics.
        3. Big sagebrush (Artemisia tridentata) is the dominant species in western Colorado.
    B. The purpose and object of this research are economic in nature.
        1. Valuable land is presently occupied by big sagebrush.
        2. With only moderate costs sagebrush can be eradicated by various methods.
    C. Past and present research in the field is inconclusive for western Colorado.
        1. Past research has been general.
        2. Present research is still inconclusive as applied to small areas.
    D. A restricted approach is taken toward the problem.
        1. The most effective method of eradication only is sought.
        2. Library research, interviews, and field work are the sources of information.
            a. Extensive library research is the basic source.
            b. Interviews with authorities in the field of range proves valuable.
            c. Verification of data through field investigation is necessary.
II. There are several different methods of sagebrush eradication.
    A. Grubbing is one method of sagebrush eradication.
        1. Advantages of grubbing are few.
        2. Disadvantages of grubbing are many.
    B. Railing is a method of sagebrush eradication.
        1. Advantages of railing are many.
        2. Disadvantages of railing are few.
    C. Harrowing is a method of sagebrush eradication.
        1. Advantages of harrowing are moderate in number.
        2. Disadvantages of harrowing are few.
    D. Blading is a method of sagebrush eradication.
        1. Advantages of blading are few.
        2. Disadvantages of blading are many.
    E. Burning is a method of sagebrush eradication.
        1. Advantages of burning are many.
        2. Disadvantages of burning are few.
    F. Plowing is a method of sagebrush eradication.
        1. Advantages of plowing are many.
        2. Disadvantages of plowing are few.
    G. Spraying is another method of sagebrush eradication.
        1. There are many advantages of spraying.
        2. The disadvantages of spraying are few.
III. The most effective method of sagebrush eradication in western Colorado is either grubbing, railing, harrowing, blading, burning, plowing, or spraying.

Figure 12.3

13

Report Writing—Writing the Elements of the Report

Chapter Objective

Provide the means for structuring the elements of the formal report.

Chapter Focus

- Constructing the organizational elements of the report
  The Front matter
  The Text of the report
  The Back matter
  Requirements for final reports

So far, we have been examining the main discussion or body of the formal report. The formal report has other elements, usually arranged in a prescribed form. If you will recall, Chapter 11 stated that there is no universal "right" form governing the arrangement of a report's elements. Elements should be arranged in an order that promotes greatest convenience of use to, and understanding by, the reader. Reader requirements vary as reporting situations and problems vary. Many research and industrial organizations and government agencies have developed style manuals, guides, and specifications to insure standards and promote effectiveness of reports. These guides prescribe forms and arrangements of report elements in keeping with the organizations's requirements.

Specific details of form and content may vary considerably, but generally, elements of a report find arrangement in either the logical pattern of organization or the psychological pattern (the double report), as described in Chapter 12. The present chapter follows the logical method of organizing the report's elements because that approach is universal. Once you, the student, have mastered the ability to structure and arrange the data of the investigation and the elements of the report within the framework of the logical pattern, you will be able to meet the requirements of any type of report pattern and form.

The elements of the formal report are these:

A. Front Matter
   1. Letter of Transmittal
   2. Cover
   3. Title Page
   4. Abstract
   5. Executive Summary
   6. Table of Contents
   7. List of Tables and Illustrations
B. Report Text
   1. Introduction
   2. Body
   3. Terminal Section
C. Back Matter
   1. Bibliography
   2. Appendix
   3. Glossary
   4. Index
   5. Distribution List

Let us examine each of these elements.

The Front Matter

The **front matter** in a formal report includes the introductory elements that introduce, help explain, summarize, and assist the reader locate the report's various sections.

The Letter of Transmittal

The major purpose of the letter or memo of transmittal is to present formally the report to the reader as a matter of record. The letter of transmittal indicates exactly how and when the report was requested, the subject matter of the report, and how the report is being transmitted—as an enclosure of the letter or under separate cover. Its length varies. Frequently, it is necessary to say only, "Here is the report on . . . (The problem is indicated), which you asked me to investigate in your letter of December 21, 19___." If the report is sent to someone within one's own organization, the form used is a memo from the author to the recipient.

Letters of transmittal have been used occasionally to refer to specific parts of the report and to call attention to important points, conclusions, and recommendations. Some letters of transmittal may even go into certain elements that may appear in the introduction, such as statements of purpose, scope, and limitation. The letter of transmittal may mention specific problems encountered in making the investigation. For example, a delay due to a strike or shortage of certain materials may have influenced the performance time, but not the data and their results; this situation may be mentioned in the letter of transmittal. Some letters may even offer acknowledgments.

The placement of a letter of transmittal varies in practice. It is sometimes bound within a report, or it may be placed within an envelope and attached to the outside of the package. In the latter case, the address on the letter envelope may serve as the address for the report packet itself. When the report is sent to the intended reader within one's own organization by hand, or through organization communication means, the memo of transmittal is frequently clipped to the outside cover of the report or placed within the cover for protection. Examine the sample letters in Figures 13.1 and 13.2.

The Cover and Title Page

The formal report usually is bound within a cover. Identifying information is listed on the cover page; this consists of the title of the report and its author. Occasionally, the recipient's name is also included in the cover information.

The purpose of the title page is to state briefly but completely the subject of the report, the name of the organization or person for whom the report is written, the name of the person and/or the firm submitting the report (frequently the address of the report's maker is included), and the date of the report. The title page may also have a number assigned to the report by the

805 Remington Street
Fort Collins, Colorado 80521
December 5, 19__

Professor A. J. Williams
Dept. of Physics
Colorado Polytechnic University
Fort Collins, Colorado 80521

Dear Professor Williams:

In compliance with your request for a full report on the Power Calibration Curve for the CSU AGN-201 Reactor, this report is submitted.

This report defines the problem, develops the method, the theory for determining the power calibration curve, and discusses the results and conclusions of the experimental data.

The power calibration curve was found to be linear with the maximum power level of 101.5 milliwatts at 5×10^9 amps on channel 3 of the reactor control console. This compares favorably with 100 milliwatts determined by the Aerojet General Nucleonics Corporation.

It is recommended that in future experiments the data and data calculations be rounded off to less significant figures than those used in this report. This can be done with sufficient accuracy still maintained.

From the indications of the data in this report, an interesting experiment for determining the effect of foil thickness upon activity may possibly be incorporated in the Ph-124 Experiments in Nuclear Physics course.

Yours truly,

Nathan M. Duran

Encl.

Figure 13.1
Sample Letter of Transmittal

900 South College
Fort Collins, Colorado 80521
January 28, 19__

Dr. Daniel Weisman
Room B-210, Engineering Building
Colorado Polytechnic University
Fort Collins, Colorado 80521

Dear Dr. Weisman:

This report on an automatic parking lot attendant is hereby submitted in accordance with your assignment of last September 28.

The report includes an analysis of the problems involved in such a system, the methods used in solving these problems, and conclusions and recommendations of the author.

Sincerely,

James B. Donnell

Encl.

Figure 13.2
Sample Letter of Transmittal

authorizing organization, as well as the number of the project concerning which the report was prepared. The title page gives the reader first contact with the report. The various elements, therefore, should be arranged neatly and attractively on the page (Figure 13.3).

The Abstract

The abstract has a major role to play in the formal report. It has two important uses.

As Index and Announcement

Reports are frequently placed in the files of the organization's information system or in the information data base of other information services and are published in abstract journals covering the latest literature of a field. To facilitate retrieval of the information of a report in a file, the document is first indexed; that is, the key topics discussed are identified. Many organizations require authors to provide **keywords**, which index or list the most important topics in their reports or papers. The keywords, usually limited from five to twelve in number, are arranged alphabetically and placed below the abstract.

Geologic Report

on

The Physiographical Development

of the

Colorado Piedmont Area

Submitted to Dr. Herman M. Weisman,

Professor of Technical Journalism

Colorado Polytechnic University

Fort Collins, CO 80521

by

Maurice De Valliere,

Geology Student

November 30, 19__

Figure 13.3
Sample Title Page

As a Synopsis

The abstract is a brief factual version of the complete report. It provides the gist of essentials of the investigation. Included are an explanation of the nature of the problem; an account of the course pursued in studying it; the findings or results; and conclusions and recommendations of the investigation, which are summarized in the order of their importance.

Ideally, the abstract is no more than one typewritten page of about 100 to 250 words. It is placed first on a separate page following the title page to save a busy reader from wading through masses of technical data in search for answers to her questions (Figures 13.4 and 13.5).

The Executive Summary

The executive summary, a device based on the psychological pattern of organization (see Chapter 12), has become popular. The use of this element recognizes that busy administrators and decision makers will not wade through entire reports. Abstracts provide a bare inkling of the contents—not enough to base a decision. A synthesis or summary of the important issues is required. Further, the report may be too massive or technically formidable. The *executive summary* recomposes the report into an abbreviated, reordered form that highlights the key factors of the investigation. Prerequisite information is provided; included are purpose of the study; the nature and significance of the problem; the scope; an account of the investigation, its methods and materials; results obtained; conclusions about the results; and recommendations for future course of action.

The executive summary adds greatly to the usefulness of the report. It keys the reader to important issues; it enables the reader to comprehend and focus on the factors of the problem and the means for its solution. The highlights of the report as presented in the executive summary are more easily digested and remembered by the decision maker than is the report itself.

An example of an abstract and an executive summary of the same report are shown in Figure 13.6.

The Table of Contents

The table of contents lists the several divisions and subdivisions of the report with their related page references in the order in which they appear. It is set on the page to make clear the relation of the main and subordinate units of the report. These units should be phrased exactly as are their corresponding headings in the text. A list of tables and/or illustrations follows the table of contents (Figure 13.7).

Writing the Report Text—The Introduction and Its Elements

The elements discussed so far have been preliminary. The report itself begins with the introduction. The purpose of the introduction is to answer the immediate questions that come to the reader's mind: Why should I read the report?

Abstract

Geologic Report on Physiographical Development
of the Colorado Piedmont Area

The Colorado Piedmont is an area that lies topographically lower
than the Colorado Front Range mountains to the west and the Great Plains
Section of the High Plains Province on the east.

During the fall of 19__, the writer, under the direction of professor
D. V. Harrison of the Colorado Polytechnic University Geology Department,
made a detailed investigation of the Colorado Piedmont area for the
purpose of writing a technical report, regarding the origin and possibilities
for future expansion of the area. Library research was also employed.

The conclusions of this investigation are: (1) the Colorado Piedmont
is an erosional feature that has developed upon the Tertiary sedimentary
beds since the end of the late Pliocene uplift, (2) the major erosional process
was stream activity that results in extensive stream capture, and (3) future
expansion of the area is likely to occur along the northwestern border, near
the Wyoming state line, and along the eastern border, toward the Kansas
state line.

Key words: Colorado Front Range; Colorado Piedmont; erosion; High
              Plans Province; physiographical development; Tertiary
              sedimentary beds.

(From a student report by Maurice DeValliere)

Figure 13.4
Sample Abstract

Abstract

An Examination of Some Methods Used by Hunters
for Care of Deer Meat

This paper examines some of the methods used by hunters for caring
for deer meat. The objective was to determine just how a deer carcass was
field-cared for. The data for the study were obtained by direct observations of
deer carcasses and supplemented with questions asked hunters. An
examination of 371 carcasses, over a two-week period, showed the methods of
care given to deer carcasses to vary greatly. Search of the literature showed no
similar study. The extremes of care varied from animals processed to those
that were not field dressed. An average condition of care was determined.
Many of the hunters live very close to the area sampled and do not take
proper care of their deer, as they assume they will arrive home soon enough
to care for it. Time, always a limiting factor, hurries a hunter in all of his
procedures, including proper care of deer meat. Suggestions for proper care
and a step-by-step procedure for field dressing a deer are presented. Ten
recommendations to aid hunters in caring for deer are suggested. Nine bar
graphs, located in the Appendix, diagram the data. A map of the area
sampled and a form used for recording the data are also included within the
Appendix. The most important recommendation for hunters is: Game should
be treated like any good meat.

Key words: Deer, deer meat; field-care, carcasses; game; hunter.

(From a student report by Jack D. Cameron)

Figure 13.5
Sample Abstract

Abstract

The purpose of this report is to develop a guide for planning and selecting the site for the National Academy for Fire Prevention and Control. The investigation determined that the legislation, the Academy's mission, the nature of the training to be offered by the Academy, and the instructional techniques to be used indicate the need for a national campus. The Academy may obtain its physical facilities by building a new structure or by adapting an existing Federal, state, or privately owned facility to the Academy's purpose. Cost constraints favor the latter approach. Site selection criteria are developed and four categories of criteria factors are arranged in a matrix for possible use by the Site Selection Board.

Key words: National Academy for Fire Prevention and Control; Site Selection Board; Site Selection Criteria; fire education and training.

Executive Summary

The National Fire Prevention and Control Administration (NFPCA) awarded a grant to the Academy for Educational Development, Inc. (AED), to develop site criteria and to outline procedures to be followed in selecting the site for the National Fire Academy. The result of AED's effort was to be a report that would serve as a planning document and as a guide to the Site Selection Board.

In preparing this report, AED was guided by the legislation, <u>America Burning</u>, position papers on the National Fire Academy, and other written documents. To determine additional views about the Academy, AED interviewed selected fire service personnel, educational leaders and authorities in the field of fire prevention and control, and others.

In the course of its investigation, AED found that the issues of program and site selection were strongly interrelated; and that the Academy's new site should be identified and designed to meet the Academy's program and operational needs. With this in mind, AED examined the Academy's program and concluded that it should consist of five major components:

1.   Education and training that would focus on management; provide short-term courses lasting from several weeks to two or three months; and reach as many as 400 persons at any one time and 12,000 over the course of a year at the national campus.

2.   Curriculum development for education and training at the national campus and throughout the nation.

3.   Information collection and dissemination, which includes the establishment of an educational research and reference center, including a first-rate library.

4.   Technical and financial assistance to fire service personnel attending Academy courses and courses offered at other institutions; and to fire education and training institutions and organizations, including colleges and universities.

5.   Relationships with higher educational institutions, which would be concerned primarily with the Academy's accreditation of individual programs and with arranging for individual institutions to award credit for the Academy's programs.

1

Figure 13.6
Sample Abstract and Executive Summary

AED further concluded that the legislation, the Academy's mission, the nature of the education to be offered by the Academy, and the instructional techniques to be used (such as simulation) indicated the need for a national campus facility. The Academy may obtain its physical facilities by adapting an existing federal, state, or privately owned facility to Academy purposes; or by building a new structure on land acquired for that purpose. Cost constraints seem to favor the former approach.

After developing space requirements in considerable detail, AED estimated that nearly 240,000 gross square feet of space would be required and that the size of the site should be approximately 100 acres distributed among built-up space, open space (landscaped and natural), a constructed outdoor recreation area, a reserved area for outdoor demonstration, and space for on-grade parking.

The suggested facility plan provides for office space, an auditorium, classrooms equipped with audio-visual materials, seminar rooms, a library, an audio-visual distribution center, lecture and demonstration rooms, dining and residential facilities, indoor simulation facilities, health conditioning facilities, a close-circuit system for instruction, and adequate storage space.

In estimating the Academy's space needs, AED considered a planning option dependent on site location, the need for growth in the Academy's services, and the likelihood that the NFPCA and certain major programs—such as the National Fire Data Center, the Fire Safety Research Office, and the Office of Public Education—could be located in the Academy's national facility. The day-to-day interaction between the Academy, other NFPCA activities, and related activities of other federal agencies will require setting aside suitable space.

In reviewing and evaluating sites and making recommendations to the Secretary of Commerce, the Site Selection Board should, in AED's view, be guided by factors that fall in the following four groups:

A. <u>Critical factors that provide indirect support to the Academy in the accomplishment of its mission.</u> Group A factors indicate the Academy's relationship to professional and governmental activities, and proximity to a diversity of fire research and service activities (sixteen such factors are identified).

B. <u>Physical and geographic factors.</u> Included are access to public transportation arrangements, community support, and a moderate climate (six such factors are identified).

C. <u>Factors related to the actual site of the Academy.</u> Group C includes various land and environmental considerations (seven such factors are identified).

D. <u>Mission accomplishments and support activities.</u> These factors describe what the Academy is expected to accomplish over the years (twelve such factors are identified).

AED has developed a site and facility evaluation matrix that might be used by the Site Selection Board as a tool to determine how well each proposed site meets designated site factors. AED has also included, for the consideration of the Site Selection Board, a detailed checklist of commonly used factors to be used in reviewing specific sites.

The reports concludes with a suggested list of eight steps to be followed by the Site Selection Board in making site recommendations.

2

Figure 13.6 (continued)

Table of Contents

List of Illustrations

Figure 13.7
Sample Table of Contents

Who asked for the report? When? Where did the information come from? Why was this report written? What significance has this report for me?

So the introduction gives the reader a first contact with the subject of the report. The introduction states the object of the report. It may be a restatement of the terms of the original commission either verbatim or as understood by the writer. It should give the background information necessary to understand the discussion that follows. The nature and amount of this information depend, of course, on the intended reader and the reader's knowledge of the subject. In order to follow the report, the reader must know this information:

1. *The purpose of the investigation*
2. *The nature of the problem*
   Here will be found the Who, How, What, Why, and When of the investigation. Some writers distinguish between *purpose* and *objective*, although the terms are often used interchangeably. If there are both immediate and ultimate goals, these must be identified. Either or both of these factors should be explained to the reader. The purpose and the objective concern the *why* of the study, and the problem concerns the *what* of the study. The nature of the problem and the purpose are often grouped together, but they are not the same, and this confusion should be carefully avoided.
3. *Scope, or the degree of comprehensiveness*
   Scope determines boundaries—what considerations are included and what are excluded. The delimitations of the investigation are carefully stated.
4. *Significance of the problem*
   The reader of a report consciously or subconsciously asks herself, "Why should I read this report?" The writer must answer this question in the introduction. He must tell the reader why the report is of significance to her. Needless to say, if the reader does not feel that the report holds any significance for her, she will not read it.
5. *Historical background* (review of the literature)
   The previous state of knowledge and the history of prior investigations on this subject are included. A review of the literature may be desirable to do the following:
   a. Give the reader confidence in the investigator's awareness of the state of the art of knowledge of the problem
   b. Enable the reader to get the necessary background quickly
      Only enough of the past should be included to make the present understandable. Some reports continue previous investigations; the report must then provide the historical background needed to orient the reader properly. In beginning a report, the reader may have the following questions in mind: What has given rise to the present situation? What occasions have given rise to similar situations? What significance has been attached to them in the past? Are there any conflicting views on this problem? A review of present conditions may also be necessary.

6. *Definitions are given of new or unusual terms or those having a specialized or stipulative meaning*

   Terms used only once are better defined in the context; those appearing constantly throughout the report should be defined in the introduction. If many such terms are to be used, you may include a glossary in the appendix.

7. *A listing of the personnel engaged in the investigation, together with a brief sketch of their background and duties*

   This will often give the reader more confidence in the data and conclusions obtained. The reader wants to know who the people are who obtained the data and have assurance that they possess the requisite qualifications for working on the problem.

8. *The plan of treatment or organization of the report*

   This element gives the reader a "quick map" of the major points, the order in which they will be discussed, and perhaps the reason for the arrangement. It serves also as a bridge between the introductory section and the body and discussion sections.

9. *Methods and materials for the investigation*

   This item is included in the introduction only in reporting those investigations where procedures and materials are simple and inconsequential to the data obtained, as is illustrated in the sample introduction from a student report on how hunters field-care their deer meat (Figure 13.8). In this instance, the method of investigation is a simple interview and questionnaire, supplemented by observation. The materials were forms for recording the data, a clipboard, and a pencil. However, in those investigations in which the data obtained are directly related to laboratory or experimental methods, or are dependent on elaborate procedures and materials, this element is placed in the body section.

   These eight or nine points are general to most formal reports, although not absolutely necessary to all reports. There is no special significance in the order of their listing. Introductions may begin with the significance of the problem or with the historical background. In certain situations where the problem may be new and complex, definitions of terms may begin the introductory section. These eight or nine elements (or those that are used) are frequently organized as subheadings within the introduction. The purpose of the introduction is to orient the reader. Placed within it is the information the reader needs to follow the main body of the report easily and confidently.

Writing the Report Text—The Body of the Report and Its Elements

Although the term *body* as a heading is seldom used, it is the major section of the report, which presents the data of the investigation. In its stead, words descriptive of the material contained in the section are used. The body section or sections include the theory behind the approach taken, the apparatus or methods used in compiling the data with necessary discussion, the results ob-

I. Introduction

Objective and Scope

This paper examines some of the methods used by hunters for the care of deer meat. The objective was to determine just how a deer was field-cared for. The data for the study were obtained by direct observations of deer carcasses and supplemented with questions asked hunters at Ted's Place in the fall of 1986. Ted's Place is located at the junction of Highways 14 and 287 in North Central Colorado. A Colorado Game and Fish Department's permanent big game check station is located here. Since all hunters, successful or not, are required to stop, it is an ideal location for a study of this kind.

An examination of 371 carcasses, over a two-week period, showed the methods of care given deer to vary greatly. The procedure used in determination of the methods proved to be satisfactory; however, many limitations were encountered. The inability of one man to check all of the hunters and their deer during a rush period at the check station was the greatest difficulty encountered.

Also the great majority of the hunters shot their deer in the northeastern mountains of Colorado, game management units eight, nine, and nineteen. Therefore, this limited sample may be biased as only one location within the state was observed. However, the data will hold true for this specific northeastern area.

Time, as in the case of much research, was limited. However, 16 percent of the total deer checked through Ted's Place were observed. Since only 10 percent of a population is usually required for a statistical analysis, the data obtained were considered sufficient.

Review of the Literature

A thorough search of the literature uncovered many techniques and methods concerning the care of deer meat. However, this search showed no similar work on an actual examination of carcasses to determine the care given deer meat by hunters and the condition in which it was brought through a check station.

Much of the literature emphasized the great loss that occurs annually by improper care of game meat, but no specific study was found concerning examination of the field-dressed carcass.

At this date the author is unaware of any similar work in this specific field.

Methods and Materials

A direct observation of the deer carcasses and oral questions asked the hunter, make up the data presented. By the use of a prearranged form, as shown in the Appendix, the information desired could be readily recorded. All information, observations, and questions were recorded on the form. Columns on the form were arranged to facilitate recording maximum information in a minimum of time.

As a hunter's car approached the station the form was readied. While the Colorado Game and Fish employee was checking the license, most of the direct observations could be made and recorded. When the employee was finished, the hunter was asked the following questions:

1

Figure 13.8
Sample Introduction

1. Did you save the heart, liver or kidneys?

2. What method was used to bring your deer from the field to your vehicle?

    Only when the carcass was obscured by camping equipment and tarps were additional questions needed. These would include questions concerning the following:

1. Sex

2. Appendage removal

3. Splitting of the pelvic and/or brisket bone

4. Tarsal and metatarsal gland removal

    The majority of the carcasses were located so the flesh could be felt to determine coolness. Although this is a relative factor, 18 percent of the carcasses were not cooled as the flesh was still warm to the touch. Many of these instances were very recent kills.

    Cooperation from the hunters was very satisfactory. Less than ten hunters were curious about the questioning and asked for an explanation. The other 360 hunters evidently took the questions for granted as they assumed the author to be a Colorado Game and Fish Department employee.

    The only materials used in this study were forms for recording the data, a clipboard, and a pencil.

(From a student report by Jack D. Cameron. This paper was prepared as a class assignment in technical writing. All data presented are original.)

2

Figure 13.8 (continued)

tained from the procedure, and an analysis of the results. In a scientific or technical investigation, the theory determining the procedure is explained and the method of procedure is recounted in chronological order; apparatus, materials, setups, and so forth are described. The reader will place greater confidence in a report's conclusions if she knows how they were determined. Duties of the personnel are presented, and evidence is organized so that the reader can follow the thinking in orderly fashion. These principles are illustrated in the "Report on the Possible Origin of the Intrusive Bodies in Big Thompson Canyon," Figures 13.10 and 13.11, at the end of this chapter.

The body is usually the largest section of the report. The various organizational approaches for the body, which were determined when the outline was developed (see Chapter 12), were discussed earlier. The pattern of organization chosen depends, of course, on the nature of the problem. The reader is guided through the body of the report, which may be structured into more than one section, by the major and secondary headings as set down in the outline. These headings will indicate the relationship of the main and subordinate units of the organizational plan. They should be in agreement with the table of contents. Paragraphs, which represent units of the organizational plan and serve to bridge, introduce, and conclude topics, help to provide a coherent text leading the reader through the discussion of the investigation.

Examine the body sections of the report, "Possible Origin of the Intrusive Bodies in Big Thompson Canyon." The writer reports on a field investigation. The chronological pattern of organization is the appropriate one for this situation. The report writer begins with a field description of the geologic formation being investigated. He then moves to examine and describe a nearby known outcrop. Specimens are selected from both the known and unknown formations. These are compared and analyzed. Analysis reveals some similarities. The structure of the body of this report follows the chronological sequence of the events of the investigation. The terminal section follows with the report writer's preliminary conclusions on the origin of the unknown outcrop and with the recommendation that further studies be made.

Writing the Report Text—The Terminal Section

The terminal section may be either a summary of major points of the body, if the report is purely informational, or conclusions based on analysis of the data following logically from the evidence given in the body. The conclusions should, whenever appropriate, be listed numerically in the order of their presumed importance to the reader. Recommendations follow next. Conclusions and recommendations may be placed in one section or in separate sections. **Conclusions** deal with evaluations as to the past and present of a situation, whereas **recommendations** offer suggestions as to future courses of action. Recommendations are often the most important part of the report, and their adoption or rejection depends on how they are presented. Like the conclusions, they should be positive statements. They should suggest specific things to be done or a course of action to be followed and should, if appropriate, include an estimate of the cost involved. Some types of reports do not require recommendations, but usually the section on conclusions and recommendations is the most important part of the report. It is what the client is paying for. Read Figure 13.9 for an example of a conclusion and recommendations.

The Back Matter

The back matter is a descriptive term for the several supporting and ancillary data that supplement the report's information.

The Bibliography

The bibliography is a list of references identifying literature that has contributed to your report. Its placement follows the report proper. It includes references to both published and unpublished material, such as notebooks, reports, correspondence, and drawings. The items listed need not necessarily have been referred to in the text of the report. The conventional listing of the bibliography is alphabetical. The items should be numbered sequentially (see Chapter 14, pages 382–384 for examples).

The Glossary

The glossary is an alphabetical list of terms with their definitions, which are highly technical or which are used in a specialized way. A glossary is unnecessary if your report does not use highly technical terms or if your readers are familiar with the terms you use.

The Appendix

The chief purpose of the appendix is to gather, in one place, all data that cannot be worked into the body without interrupting the flow of the report. If the body is self-contained, the detailed data in the appendix provide points of reference when questions arise. An appendix may not be necessary in a very short report. However, the appendix is indispensable to the type of report that uses considerable statistical information. Sometimes, all 8½" × 11", or larger, charts, tables, diagrams, photographs, and other illustrations are placed in the appendix of a typed or photocopied report. The following items are usually saved for the appendix:

1. Charts
2. Tables
3. Computations and data sheets
4. Diagrams and drawings
5. Exhibits, graphs, maps, photographs, letters, and questionnaires
6. Records of interviews and other similar matters serving as data that are not found in the literature or are not revealed through the methods and procedures of the investigation.

Conclusions and Recommendations

Through the design and construction of the prototype model, it has been shown that an automatic parking lot attendant system is feasible. The prototype has been tested and operates as described in this report. The same circuits that operate the readout can be used to supply control information to the coin-operated gate. The finished model is a compact unit weighing approximately 30 pounds and measuring 18 x 11½ inches around the base. It is 10 inches high. For the proposed system to become operational, several difficulties present in the prototype must be eliminated. The prototype has no provision for altering the fee rate structure which is built into it. This could be accomplished by redesign of the printed circuit boards using removable jumper leads, whereby various combinations of digital readouts could be arranged. There are instances of price inequities in the present model such as in the example on page 11 of this report. In the prototype, a carry-over between rate change periods sometimes results in such a discontinuity of the accumulated fee. In the example given, the accumulated fee at 6:00 a.m. was 0.75. At the changeover to the day rates, this was picked up at 0.80 due to the fact that no 0.75 combination is available on the day rate photocell disk. Ideally, the card would have been picked up at 0.75 and continued from that point. These small inequities may or may not be considered serious. If the inequities are considered undesirable, it may be necessary to attack the method of decoding the card from a different viewpoint.

The handmade rotary stepping mechanism which moves the photocell disk is not completely positive in its action. There are commercial rotary stepping solenoids available on the market which should eliminate this problem, and it is recommended that such commercial components be used in the production system.

The author is of the opinion that the elimination of these difficulties may be accomplished in the production engineering phase and the system will be acceptable.

(From a student report by James B. Donnelly)

Figure 13.9
Sample Conclusion and Recommendations

The Index

An index is necessary only in voluminous reports where the alphabetical listing with page references of all topics, names, objects, etc., is useful for ready reference. Normally, the table of contents performs this function adequately.

The Distribution List

The distribution list not only provides the names of persons and offices to receive the report but also serves as a control device to ascertain that the confidence of the sponsor or client is not violated. Distribution lists may be tacked to the inside of the front cover or placed at the very end of the report.

Requirements for Final Reports

Final reports follow the substance and format of formal reports. Specific requirements include the following elements:

1. Cover
2. Title Page (includes the title of the report; grant, contract, or project number; the performing organization, author, organization for whom the report was prepared; and the date)
3. Executive summary
4. Table of contents (including list of tables and illustrations)
5. Text of the report
   A. Introduction
      1. Purpose of the study
      2. Nature of the problem
      3. Scope
      4. Significance of the problem
      5. Historical background (Review of the literature)
      6. Definitions
      7. Personnel making the study
      8. Organization of the report
   B. Account of the study
      1. Methods and Materials
      2. Theory (if appropriate)
      3. Problems
      4. Results obtained
      5. Discussion of reports
      6. Analysis and significance of results
   C. Conclusions
   D. Recommendations (include ways to utilize and diffuse results)

6. Bibliography
7. Appendix materials
8. Abstract (if needed for depositing report with the National Technical Information Service (NTIS)

422 West Laurel
Fort Collins, Colorado
November 30, 19___

Professor Weisman
Department of English
Colorado Polytechnic University
Fort Collins, Colorado 80521

Dear Professor Weisman:

In compliance with your term assignment to complete a technical paper, I hereby submit my report on the Possible Origin of the Intrusive Bodies in Big Thompson Canyon.

This report includes field work, description and location of the subject, laboratory research, and library research.

From research, laboratory, and field work, I found that the intrusive is an offshoot from a mass called Silver Plume Granite.

It is recommended that for additional study, thin sections of the intrusive should be used, thereby more accurately determining the percentage composition.

Sincerely yours,

Gerald L. Owens

Enclosure
GLO/lo

Figure 13.10
Example of a Student Report
(see the sentence outline for this report in Chapter 12, Figure 12.2).

Preliminary Report

on the

Possible Origin of the Intrusive Bodies

in

Big Thompson Canyon

Submitted to
Professor H. M. Weisman
of
Colorado Polytechnic University
Instructor
Technical Writing

By
Gerald L. Owens
Fort Collins, Colorado
November 30, 19__

Figure 13.10 (continued)

Abstract

The intrusive body which lies at the base of Palisade Mountain on Highway 34 west of Loveland, Colorado, is an igneous mass that has been intruded into a metamorphic country rock. The country rock, which is essentially mica schist, covers a wide area with no other intrusives in evidence. How does this one small body of granite come to be in this location, and is it related in any way with known bodies of granite in the front range?

From the investigation, as to mineral constituents and structural correlation as compared with a known outcrop of Silver Plume Granite, it was found to be an offshoot from a main mass of Silver Plume Granite. The granite was intruded into pre-cambrian country rock during pre-cambrian times. Locally, the granite changes color due to varying percentages of mineral constituents. Individual crystal sizes depend on the rate of cooling in conjunction with the amount of mineralizers present.

For further study, it is recommended that thin-section studies be made of the intrusive in order to determine more accurately the percentage composition of the body.

Figure 13.10 (continued)

Table of Contents

Figure 13.10 (continued)

INTRODUCTION

West of the city of Loveland, Colorado, in the front range of the Rockies, lies the Canyon of the Big Thompson River. The river, while the mountains were being thrust upward, cut a narrow, deep canyon through the ancient metamorphic rock. At the base of Palisade Mountain, where construction work on Highway 34 exposed the flanks of the mountain, lies an area in which igneous material has been intruded.

An occurrence of this nature is not uncommon in a mountainous region. However, the fact that there are no other igneous bodies within miles led to the question of origin. Where did this lava originate? What were the factors of its intrusion into the metamorphic country rock? Could this body be correlated in any way with known igneous bodies in the front range? If there is a connection, would comparison of constituents of the two prove that they are of the same age?

SIGNIFICANCE OF PROBLEM

The problem and its solution is of most interest to geology students, for it will show them how important correlation and scientific investigation are to the study of geology. As in every unusual geologic feature, there is a subtle challenge to every geologist that says, "Come, find me out." The intrusive has, many times in the past, issued this challenge; but, due to other obligations, few students have tried to work it out. To benefit those who have the desire to know, but who do not have the time to spare, this report is submitted in hopes that it will give them insight into the processes of such an investigation.

SCOPE

Limited available laboratory time has necessitated investigation of only two major areas. One area deals with field observations in respect to correlation of the known and unknown bodies; and the other, in laboratory analysis of the hand specimens.

HISTORICAL BACKGROUND

There have not been any professional papers written on this particular intrusive body, but there are professional papers about the Silver Plume Granite formation. According to my advisor, Mr. Campbell of the Geology Department, there have been a few papers written about the intrusive by students in the past.

DEFINITIONS

The definition of terms in relation to description is a necessity for the non-technical as well as technical reader.

Magma—A magma is a naturally occurring molten rock mass in or on the earth which is composed of silicates, oxides, sulfides, and volatile constituents such as boron, fluorine, and water (1:177).

Igneous Rocks—Igneous rocks are those rocks which have formed in or on the earth's crust by solidification of molten lava (5:121).

Intrusive Body—An intrusive body is an igneous rock mass which has solidified within the earth's crust (1:31).

Acidic—Acidic refers to rocks which are very high in silica content (2:321).

1

Figure 13.10 *(continued)*

Metamorphic Rocks—Metamorphic rocks are rocks that were originally of igneous or
sedimentary origin. Tremendous pressure and heat generated during mountain building
transformed them into new minerals with a foliated outward appearance (3:35).
Country Rock—A country rock is defined as any rock that is penetrated by an intrusive,
igneous body (5:92).
Stock—A stock is an intrusive body less than 40 square miles in area (3:77).
Batholith—A batholith is an intrusive igneous body more than 40 square miles in area (3:78).
Dike—A dike is a rock of liquid origin that intruded along faults and slips of a formation and
cut across bedding planes (3:78).

ORGANIZATION

The manner in which material in this paper is presented will bear some explanation. In
order for the reader to more fully understand the processes and geologic theories, it was
necessary to include, at some points in the body of the report, certain evaluations of the
material presented.

ANALYZING THE PROBLEM

The intrusive body in Big Thompson Canyon offered no tangible clues as to its origin.
That it was a granite rock intruded into a metamorphic country rock was apparent, but how
would this help? The only method to follow was to find out whether or not there were any
known intrusive bodies in the area; and if there were, could they be correlated in any way?

From notes taken on a geological class field trip, it was found that there was a Silver
Plume Granite outcrop near Masonville, Colorado. Although this was approximately 12 airline
miles distant from the intrusive, it was the only one in the area that was positively known.

From this start it was decided that samples should be taken from both bodies and
analyzed. Also, any material in the library that pertained to igneous rock in the front range
should be studied.

With the thought that correlation between the two bodies might answer the question,
work was begun.

LOCATION AND DESCRIPTION OF INTRUSIVE

In order for this report to be of value to those who wish to conduct subsequent studies
or check the validity of the existence of the intrusive, direction will be given as accurately as
possible.

2

Figure 13.10 (continued)

LOCATION OF THE INTRUSIVE

Starting from the south city limits of Fort Collins, Colorado, proceed south to Loveland, Colorado. At the junction of U.S. Highway 287 with State Highway 34, turn west on Highway 34. A general store is located near the entrance of Big Thompson Canyon called the "Dam Store." Check the mileage on the car's odometer at this point and proceed approximately eight miles west up the Canyon (see Appendix 1). On the north side of the highway stands a small white sign which reads "Palisade Mountain Elevation 8,258 Ft." (See Figure 1). At this point the small bodies which make up the intrusive extend parallel to the highway for a distance of 150 yards. The main intrusive is 15 feet to the northeast of the sign in Figure 1 and is readily identified by its clean, gray color.

Figure 1

DESCRIPTION OF THE INTRUSIVE

The larger of the four small intrusive bodies will be described as it is representative of the other three.

The intrusive extends vertically from the level of the highway approximately 25 feet to its uppermost contact with the metamorphic country rock. Horizontally, from contact to contact, the distance is 40 feet (see Figure 1, Appendix II).

Externally, the intrusive is a light-gray, fine-grained granite which lies in perfect contact with the dark-gray, foliated, micaceous schist (see Figure 2, Appendix II). Upon closer examination, small black books of biotite mica are seen to be incorporated into the mass. Cigar-shaped portions of the country rock, ranging from two to six inches long and one block three feet by six feet long, are evident in the mass, thus indicating that the intrusive cooled before complete melting had occurred. On a smaller intrusive, to the west, a pegmatic zone surrounds the intrusive at its contact with the country rock. The zonal growth of the large

3

Figure 13.10 (continued)

crystal grades inward for two feet where it again assumes the smaller size of crystals of the mass proper. This zone of pegmatite granite was the result of mineralizers, or volatile constituents, which were trapped in this area. The fluidity of the cooling mass was higher than normal for this particular intrusive; and, subsequently, crystal growth was quite rapid as proved by the large size of the individual crystals.

LOCATION AND DESCRIPTION OF THE SILVER PLUME GRANITE OUTCROP

For the reasons mentioned in "Location and Description of Intrusive," the following directions for finding the Silver Plume Outcrop will be given as accurately as possible.

LOCATION OF THE OUTCROP

Starting from the south city limits of Fort Collins, Colorado, proceed south on U.S. Highway 287 for three miles and then turn west on County Road Number 186, which goes to Masonville, Colorado. Upon reaching the location of the South Horsetooth Dam, check the odometer reading on the car and proceed for approximately four miles. At this point the road will be rising to the top of a narrow canyon. At the crest of the hill on the north side of the road lies a reddish-brown granite mass which has been partially exposed by construction work. This is the outcrop of Silver Plume Granite (see Appendix I).

DESCRIPTION OF THE OUTCROP

The outcrop is approximately 15 feet high by 50 feet long from east to west. Erosion has worn down the metamorphic rock into which the granite was intruded, but now a soil mantle covers it on the topside to a depth of several inches. Considerable chemical and mechanical weathering has taken place as the mass is fractured and crumbly in places. Also, a reddish film of iron has been deposited by circulating ground water.

Upon closer examination of a freshly chipped surface, crystal sizes appear to be fairly uniform with an occasional crystal approaching 15 millimeters in length by 10 millimeters in width. Color is pinkish-gray due to the flesh-colored feldspar and the black biotite crystals.

SPECIMEN SELECTION AND FIELD TRIMMING

SPECIMEN SELECTION

Specimens were taken from the intrusive and the known outcrop of Silver Plume Granite with an eye towards specimens that would be representative of the whole mass, as well as specimens that indicate unusual facets of its formation. Specimens were taken from the centers of the two bodies and also along the contacts with the country rock.

FIELD TRIMMING

It is sometimes necessary to use a three- to four-pound striking hammer to obtain rocks from a solid mass, or even to break up large chunks of rocks. However, there is generally enough rock material lying on the ground from which a specimen can be chosen. If possible, rocks should be chosen that approximate the

4

Figure 13.10 (continued)

3"x5"x1" size that hand specimens should be. In any event, a rock should be chosen that has at least one right angle and/or two flat sides.

A geology pick and a pair of heavy work gloves were used to chip the specimens down to size (see Figure 2). While holding the rock in one gloved hand, the geology pick was used to strike a glancing blow, making sure that the chips would fly away from the operator. This operation may be understood more readily with a diagram (see Figure 3). As indicated by the drawing, an imaginary line should be visualized down the center of the rock. Small fragments indicated by the dotted lines should be chipped off. The blows should be struck down and away so as not to crack the specimen in half. When one side has been trimmed off flat, the specimen should be turned over and chipped on the other side. Considerable patience is needed as these rocks are brittle and fracture very easily.

Figure 2   Figure 3

MICROSCOPIC ANALYSIS OF SPECIMENS

THE INTRUSIVE SPECIMEN

The specimen from the intrusive was observed under fifty power magnification. The predominant mineral was orthoclase feldspar with its white color and irregular grains showing many of the faces with nearly perfect right-angle cleavage. The next mineral, in order of abundance, was quartz. This mineral is the clear, colorless variety which has no crystal form as it was the last mineral to crystalize out in the cooling intrusive. It filled in all pore spaces left by the other crystals. Biotite mica, with its black, splendent luster and flexible folia, was regularly spaced in the matrix. The relatively small size of the crystals made them unnoticeable beyond a few feet. Infrequent, but quite large, crystals of plagioclase feldspar were observed. These were identified by their gray-white color and twinning striations on the cleavage surfaces. As evidenced by the high percentages of quartz, this rock can be termed acidic (see Table 1, Appendix III).

THE SILVER PLUME SPECIMEN

The specimen from the Silver Plume outcrop was also observed under fifty power magnification. The orthoclase feldspar was the most abundant mineral found. It varied from white to flesh-colored with euhedral, crystal faces. The quartz, next in abundance, which filled in the remaining spaces, was rose-colored to colorless. The color could have been imparted to the quartz by a minute quantity of cobalt. The biotite mica present was in small, black masses which were quite close together. A very small quantity of white plagioclase feldspar, showing excellent twinning striations, was present. This specimen represents a magma which was high in silica content. Therefore, this rock is also termed acidic (see Table 2, Appendix III).

5

Figure 13.10 (continued)

COMPARISON OF FIELD AND LABORATORY DATA

SIMILARITY OF OCCURRENCE IN FIELD

Comparison of the types of country rock penetrated by the respective bodies showed that they are of the same age and the same rock type. The country rock in both localities is a pre-cambrian, metamorphic rock called "mica schist." Both of the bodies are granite, varying only in percentages of minerals present, as well as in color. This follows quite well with T. S. Lovering's professional paper entitled Geology and Ore Deposits of the Front Range, U.S. Geol. Surv. Professional Paper, 223, page 28. The following is the section written on the Silver Plume Granite:

> In the front range are a large number of small stocks and batholiths generally intruded at the same time.... Most of the stocks and batholiths are pinkish-gray, medium-grained, slightly porphyritic biotite granite, composed chiefly of pink and gray feldspars, smoky quartz, and biotite mica; but muscovite is present in some facies.... The percentage of biotite varies from place to place....

The intrusive bodies are very small when compared to the size of a stock (3:77), or a batholith (3:78). They do, however, compare favorably with the definition for a dike (3:78). For the most park, dikes are rather small in a cross-sectional area but are known to extend up to 100 miles in length (3:78). For both of the bodies, which the author will call dikes, the rock that the molten lava passed through would have some bearing on the composition. This could be one of the reasons why there is a difference in color and percentage of composition. From this, then, it is logical to assume that both bodies are offshoots from a larger mass. They probably followed fractures or faults caused by crystal movements during the time of the pre-cambrian, metamorphic country rock.

COMPARISON OF LABORATORY DATA

The specimen of the intrusive, in comparison with that of the Silver Plume Granite, shows that they are very close in chemical composition. The intrusive specimen had 60% orthoclase in comparison with the Silver Plume's 40%. Quartz in the intrusive averaged 20%, while in the Silver Plume it was 30%. For biotite mica, the percentage was 15% in the intrusive and 25% in the Silver Plume. Both of the bodies had 5% plagioclase. For a better visual comparison, see Appendix III.

CONCLUSION

The intrusive bodies in Big Thompson Canyon are of the same geologic age as that of the Silver Plume Granite. They both were intruded into a pre-cambrian, metamorphic country rock during pre-cambrian times. The country rock, into which the granite was intruded, is a mica schist, which is quite common in the front range.

6

Figure 13.10 (continued)

Chemically, the intrusive and the Silver Plume Granite are the same. Although there is some variation in color and percentages of the constituents, this is within the somewhat broad range given for the Silver Plume Granite (4:28). Locally, the percentage can change due to an increase or decrease in amount of constituents. Color variations are due to minute inclusions of a metallic ion or molecule derived from the walls of the conduit or feeder pipe.

RECOMMENDATIONS

It is recommended that a larger number of Silver Plume Granite specimens be analyzed. These should come from different outcrops. Also, to more accurately determine the constituents and their percentages, thin-section studies of the specimens should be made.

APPENDIX I

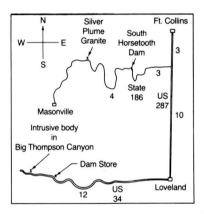

Figure 1
Illustrations Showing Work Sites (Not drawn to scale)

7

Figure 13.10 (continued)

APPENDIX II

Figure 1
Photo showing relative size
of intrusive in Big Thompson Canyon

Figure 2
Photo showing contract between
country rock and granite intrusive

8

Figure 13.10 (continued)

APPENDIX III

Tables of Comparative Constituents

Table 1

Specimen From Intrusive

Mineral	Percentages	Color
Orthoclase Feldspar	60%	White
Quartz	20%	Clear, glassy
Biotite mica	15%	Black
Plagioclase Feldspar	5%	Gray-white

Table 2

Specimen From Silver Plume

Mineral	Percentages	Color
Orthoclase Feldspar	40%	Flesh-colored to white
Quartz	30%	Rose to clear, glassy
Biotite mica	25%	Black, vitreous
Plagioclase Feldspar	5%	White

BIBLIOGRAPHY

1. Bowen, N. L. (1922). "The Reaction Principle in Petrogenesis." <u>Journal of Geology</u>, Vol. 30, pages 177–98.
2. Grout, F. F. (1941). "Formation of Igneous Looking Rocks by Metasomatism. A Critical Review and Suggested Research." <u>Bul. Geol. Soc. Am.</u>, Vol. 52, pages 1525–26.
3. Longwell, Chester R., and Flint, Richard Foster. <u>Introduction to Physical Geology</u>, New York: John Wiley and Sons, 1955, 432 pages.
4. Lovering, T. S. <u>Geology and Ore Deposits of the Front Range</u>, U.S. Geol. Surv., Professional Paper, 223, page 28.
5. Pirrson, Louis V., and Knolph, Adolph. <u>Rocks and Rock Minerals</u>, New York: John Wiley and Sons, 1957, 365 pages.

9

Figure 13.10 (continued)

STUDY TO ESTABLISH THE EXISTING AUTOMATIC FIRE SUPPRESSION TECHNOLOGY FOR USE IN RESIDENTIAL OCCUPANCIES

A SUMMARY REPORT

Prepared for:

U.S. DEPARTMENT OF COMMERCE
National Fire Prevention and Control Administration
National Fire Safety and Research Office
NFPCA Contract #6-35587

Prepared by:

Peter Yurkonis
Rolf Jensen & Associates, Inc.
Deerfield, Illinois 60015

Figure 13.11
Example of a Professional Technical Report

U.S. DEPARTMENT OF COMMERCE

NATIONAL FIRE PREVENTION AND CONTROL ADMINISTRATION

National Fire Safety and Research Office

STUDY TO ESTABLISH **THE EXISTING AUTOMATIC FIRE SUPPRESSION TECHNOLOGY FOR USE IN RESIDENTIAL OCCUPANCIES**:

A SUMMARY REPORT

Figure 13.11 (continued)

Table of Contents

Figure 13.11 (continued)

EXECUTIVE SUMMARY

In 1977 Rolf Jensen & Associates, Inc., along with Schirmer Engineering Corporation, completed a study of the existing automatic residential suppression system technology under contract to the National Fire Prevention and Control Administration (NFPCA).

The ability of an adequate suppression system to cope with extinguishing a residential fire was obvious. The real problems were to identify what was an adequate system and how we could get <u>adequate</u> systems <u>into</u> the Nation's homes.

Thus, NFPCA funded this study to establish the extent to which currently available automatic suppression systems may reduce the national residential fire loss. The purpose was to identify user needs, how suppression systems can be improved to be more cost-effective, how business organizations can influence use, and what is needed to encourage greater use.

Residential fires (in single and multi-family dwellings) were characterized from available fire record and fire research information. All known types of suppression systems were evaluated for potential in residential fire extinguishment. Three specific systems were selected and evaluated in depth to determine the factors that would influence their acceptance by users (occupants or owners of residences) and to determine how their use would be influenced by insurance companies, regulatory agencies, developers, designers, builders, installers, and manufacturers. Existing and future design technology was studied, and factors of installation cost and related benefits were established. Operations research methods were used to evaluate data as a basis for measuring the overall impact of suppression systems on the national residential fire loss and to establish the relative importance of all input factors on this loss.

The analysis of users' needs showed that voluntary installation will occur only if the cost of the system can be recovered by other savings resulting in a net zero cost and if the public is educated to be more aware of the residential fire threat and the value of suppression systems.

Installation costs may be lowered if system design technology improves and if results by current research show residential fires will be controlled by less total water at lower flow rates. To achieve more cost-effective design, lower-cost piping methods and regulations mandating use of meters, backflow preventers and prohibiting integrated use with potable water systems must be eliminated.

Installation costs for existing buildings must be reduced to a tolerable level, probably requiring developing of flexible piping or self-contained systems to reduce or eliminate the cost of related painting and patching to conceal pipe.

Installation benefits may be improved by zoning benefits (less land or frontage), lower municipal service costs (water supply, fire department), insurance credits, building code trade-offs (multi-family only), increase in home valuation, easier access to financing, lower interest rates and property or income tax incentives. The potential impact is a reduction of the national residential fire loss of at least 50 percent in loss of life, injury and property damage if sprinkler systems can be installed in all residences by 1990.

1

Figure 13.11 (continued)

THE PROBLEM

Residential fire in the United States results in the annual death of about 68 percent of the 7500 Americans who die in building fires and a property loss exceeding one billion dollars.[1] Each year, approximately 4 million families will experience a minor fire in their residence while approximately 800,000 will experience a serious fire. The probability is high that most people in the United States will face an accidental fire in their home. Fire will intimately touch the lives of almost every person in the United States.

Much has been done in an attempt to solve the residential fire problem. During 1965 the National Fire Protection Association (NFPA) created a Home Fire Alarm Standard which encourages installation of residential smoke and fire detectors. Several model building codes and some states and municipalities now require that such detectors be installed in homes and apartments. The fire safety community is presently searching for better test methods and data on which to base fire detector installation and design.

In 1965 Chandler [2] stated several reasons why residential fire detection systems would not be widely used in the home. He cited conditions such as:

- lack of an appropriate installation standard;

- lack of equipment designed specifically for residential use;

- lack of a distribution and service network;

- excessive cost; and

- general apathy on the part of the public.

Concurrent with the work described herein, Johns Hopkins University's Applied Physics Laboratory analyzed the impact of fire detection, suppression, and remote alarm systems as if they had been installed in the locations of all fatal fires within the State of Maryland during a one-year period.[3] The results of this study indicate that a residential suppression system would probably have saved 92 percent of the fatalities and 97 percent of the injuries. Fire detector and remote alarm systems would have saved 89 percent and 90 percent of the fatalities respectively and 92 percent of the injuries. The impact on property losses is more defined. A suppression system would save 88 percent of the total property losses against 72 percent for a detector system and 79 percent for a remote alarm system. While the

---

[1] While these statistics are based on 1976 NFPCA studies, the full report uses 1974 loss data; however, the results and conclusions are the same.

[2] Chandler, Lee T., "Fire Detection Systems for the Home: The Electrical Industry's View," January 1965, Boston, MA.

[3] "Fire Fatalities Case Studies—Determine the Impact of Smoke Detectors, Automatic Suppression Systems, and Automatic Direct Alarm Systems on Fire Injuries and Property Loss," Johns Hopkins University/Applied Physics Laboratory, 1977. This work is not included in our complete report.

2

Figure 13.11 (continued)

distinct advantage of an automatic suppression system is evident, residential fire detectors are being marketed on a nationwide basis as each of the above problems has been overcome.

Automatic suppression systems have many advantages over detection systems alone. They are less occupant dependent—residents do not have to initiate an action for fire suppression to begin. The fire will be less toxic because there will be fewer combustibles burning and consequently fewer products of combustion given off. Adequate suppression systems will prevent flashover, which will make fires safer for the fire services to extinguish. Yet, Chandler's arguments on residential use of <u>detection</u> systems currently apply to <u>suppression</u> systems; their use is negligible.

One of NFPCA's objectives is to develop automatic extinguishing systems that will be low-cost, acceptable to users, and provide protection to both existing dwellings and new construction. This "Study to Establish the Existing Automatic Fire Suppression Technology for Use in Residential Occupancies" is an effort to evaluate the feasibility for a residential automatic suppression system. This paper is a summarized version of the complete report written by Peter R. Yurkonis of Rolf Jensen & Associates, Inc. The complete report is available from the NFPCA, NFSRO, P.O. Box 19518, Washington, D.C. 20036.

THE STUDY PROJECT

This study was planned to determine whether any automatic suppression systems can significantly reduce the national residential fire loss in single and multi-family dwellings. If so, a further objective was to characterize such systems and establish user needs to accomplish adequate public acceptance.

SELECTION OF CANDIDATE SYSTEMS

Fire record data and fire research data were studied to establish a typical residential fire scenario as a basis for selecting candidate suppression systems likely to be effective in fire extinguishment and which would lack undesirable side effects (high costs, toxicity, injury to residents, property damage, etc.).

The following candidate systems using the following agents were evaluated:

- Carbon dioxide
- Halon 1301
- High expansion foam
- Multi-purpose dry chemical
- Water

On the basis of analysis, agent/systems other than water were excluded from practical use by a combination of factors including:

- Cost
- Agent toxicity
- Agent-induced damage
- Practicality/agent leakage
- Aesthetics

3

Figure 13.11 (continued)

· Maintenance
· False activation

For example:

· Halon, carbon dioxide, high expansion foam, and dry chemical systems' costs are more than $2.60 per square foot (vs. sprinklers at $1.15 or less).
· Carbon dioxide systems produce a lethal atmosphere at an effective extinguishing concentration
· Dry chemical and high expansion foam systems cause considerable damage and reduced visibility when they operate.
· Total flooding Halon and carbon dioxide systems require a reasonably tight enclosure to be effective. Leakage from open windows, doors, or attic vents could be excessive.
· All candidate systems require some special efforts to make them aesthetically acceptable to the homeowner.
· Candidate systems other than sprinklers are too susceptible to false actuation and require excessive maintenance.

A suppression system using water as the agent and similar to available automatic sprinkler systems based on the standard NFPA 13 or 13-D was selected as the most likely candidate suppression system for use in residences. See Figure 1.

FIGURE 1

SELECTED STUDY SYSTEM

BASICALLY .... NFPA 13—D

· 0.10 GPM/SQ.FT. (ONE SPRINKLER)
· 10 MIN. WATER SUPPLY
· 25 GPM. FLOW
· 250 GAL. STORED WATER
· STANDARD SPRINKLER
· 256 COVERAGE
· OMIT ATTIC, CLOSET, HALL, BATH
· LOCAL ALARM

4

Figure 13.11 (continued)

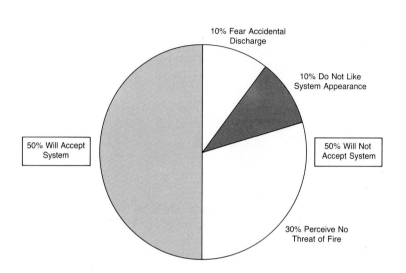

Figure 2

USER NEEDS

The next step was to evaluate acceptability of the candidate systems by a series of telephone interviews which were conducted in:

· Chicago
· Atlanta
· Philadelphia
· Dallas/Ft. Worth
· Los Angeles

The sampling was evenly divided between the five geographic regions. Interviews were conducted with the owners of single family residences (400 interviews), tenants in multiple family residences (400 interviews), and either owner or tenants living in rural single family residences (200). The questions asked were varied depending upon the group and responses of the persons interviewed. This allowed a certain flexibility to explore different concepts, which could not have been done had a formal list of questions been used.

Users interviewed showed a general lack of awareness or concern for the residential fire problem.

5

Figure 13.11 (continued)

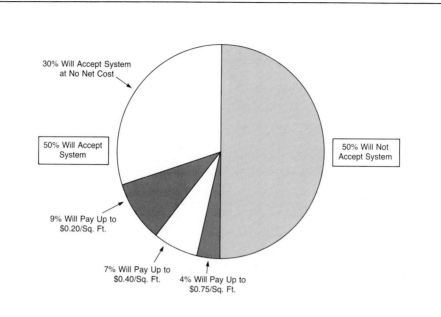

30% Will Accept System at No Net Cost

50% Will Accept System

50% Will Not Accept System

9% Will Pay Up to $0.20/Sq. Ft.

7% Will Pay Up to $0.40/Sq. Ft.

4% Will Pay Up to $0.75/Sq. Ft.

Figure 3

Approximately 50 percent indicated that they would accept a residential suppression system. Fifty percent would not accept a system; 10 percent rejected systems because of appearance, 10 percent because of fear of accidental discharge, and 30 percent because of the lack of a perceived fire threat. See Figure 2.

Cost was a major factor in systems acceptance; 40 percent said they would pay if the cost was $.20 per square foot or less; 60 percent would not pay for a system but would accept it at no cost. See Figure 3.

Seventy (70) percent favored a requirement for systems in new residences if their property taxes would decrease.

Routine maintenance of an automatic suppression system will be acceptable to 85 percent of those who accept a system.

The false system actuation tolerance expressed by those interviewed was approximately 90 percent.

6

Figure 13.11 (continued)

FACTORS INFLUENCING USE

Personal interviews were conducted with representatives of private and government organizations to determine pertinent reactions influencing the use of automatic residential fire suppression systems. No attempt was made to achieve a statistically valid sample; the plan being primarily to establish the views and opinions of those contacted. The organizations contacted were:

- Fire insurance industry
- Financial institutions
- Real estate concerns
- Residential builders/developers
- System designers and installers
- Trade unions
- Architects/engineers
- Manufacturers of automatic suppression systems equipment
- Water utilities
- Health departments
- Fire departments
- Building departments

SYSTEM INSTALLATION COST

From these interviews, factors which influence the cost, benefits, or practicality of installation were identified. See Figure 4, 5, and 6. Some of these are:

System installation cost can be reduced by

- developing methods to reduce the labor needed to install a system.
- developing system components designed for use in residential occupancies.
- elimination of requirements for meters and backflow preventer.
- simplifying plan approval process.
- encouraging self-installation by owners.

Installation benefits for system installation may occur from

- lower land, frontage, and municipal service costs.
- building code trade-offs (multi-family).

LONG RANGE BENEFITS

- potential homeowner's insurance credit of 5 to 10 percent (possibly 25 percent if loss experience is favorable).
- increased value of home.
- reduced property taxes.
- increased Federal income taxes (less property tax and interest to deduct).

7

Figure 13.11 (continued)

FIGURE 4

SYSTEM COST STUDIES

SINGLE-FAMILY

2000 SQ. FT.—ONE STORY
2400 SQ. FT.—TWO STORY
2000 SQ. FT.—SPLIT LEVEL
[ALL 3-BEDROOM WITH FULL OR PART BASEMENT]
MULTI-FAMILY

[600 SQ. FT. APARTMENT]

FIGURE 5

PROBABLE COST
OF
STUDY SYSTEMS

USE	CONTRACTOR INSTALLED	OWNER INSTALLED
SINGLE-FAMILY		
NEW—URBAN	$0.65	NA
NEW—RURAL	$0.85	NA
EXISTING—URBAN	$1.10	$0.35
EXISTING—RURAL	$1.10	$0.55
MULTI-FAMILY		
NEW	$0.45	
EXISTING	$0.55	

FIGURE 6

ACTUAL COST COMPARISON
CASE #1: 2000 SQ. FT. NEW SUBDIVISION

SYSTEM COST—$1190

POTENTIAL BENEFITS—INITIAL

· LOT SIZE		$500
· STREET FRONTAGE		$600
	TOTAL	$1100

POTENTIAL BENEFITS—ANNUAL

· HOMEOWNERS INSURANCE PREMIUM CREDIT		$38
· REDUCED MORTGAGE INTEREST		$11
· REDUCED PROPERTY TAXES		$43
· INCREASED FEDERAL INCOME TAX		($11)
	TOTAL	$81

8

Figure 13.11 (continued)

Identified practicality of installation factors are

· manufacturers' indicated concern for product liability.

· union jurisdictional problems.

· access to insurance and financing in certain areas may be increased.

· residential builders and developers may offer systems as sales incentives.

IMPACT EVALUATION

The developed data provided a base for analysis. The residential loss prevention potential of each candidate system was evaluated by means of a logic tree method. See Figure 7 and 8. The portion of each tree identified by a heavy black line identifies variable factors which influence the use or value of residential fire suppression systems. Other factors were held constant in this study.

The logic tree approach permits a quantitative evaluation of the effect of an automatic residential suppression system on the national residential fire loss. By constructing graphical probability statements at input levels, one can then compute resulting probabilities at intermediate and output levels of the logic tree. It is rec-

9

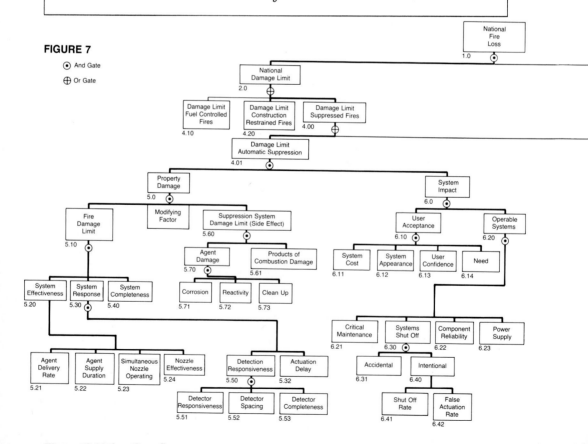

Figure 13.11 (continued)

ognized that the accuracy of this approach is limited by the accuracy of available input data. Nevertheless, the approach enables a reasonable approximation of:

· The relative ability of each type of automatic suppression system to prevent residential fire-related losses. Systems can be arranged according to their loss prevention effectiveness.

· The reduction in the national residential fire loss caused by automatic residential suppression system (or any other fire control or prevention system).

· The relative importance of system design factors and use influencing factors identified in this study and listed on the tree.

The result of the analytical study and the computation methods are described in detail in the original report. The reader who desires to review or critique this analysis is directed to that report. The important findings are shown on Figure 9, 10, and 11.

Figure 9 shows the reduction in the national fire loss resulting from use of a Level I or NFPA 13-D sprinkler system. The full report also contains analyses of two higher quality levels of systems achievable with current and future technology.

10

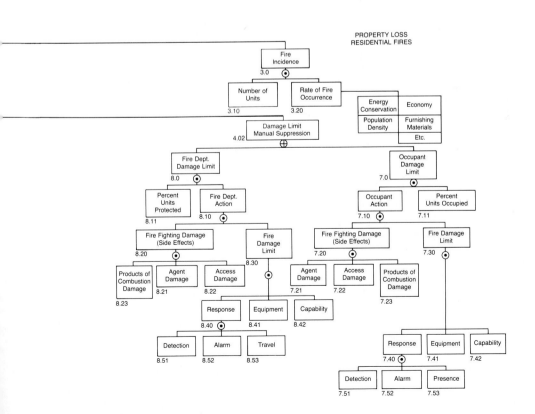

Figure 13.11 (continued)

The degree of reduction in the national fire loss will depend on the number of residences protected with such systems. If <u>all</u> residences are so equipped, the reduction will exceed 50 percent. If systems are required, reduction will exceed 50 percent. If systems are required in all new residential construction between now and 1990, the national residential fire loss in 1990 can be reduced 20 to 35 percent depending upon the cost of the system installed. With voluntary installation of systems (available at an installation cost of $.20 per square foot), the 1990 national residential fire loss can be reduced by approximately 10 percent.

Figure 10 shows the projected property loss that may be expected in 1990 if there is no change in the degree to which sprinkler systems are installed in residences. It also shows the reduction in this property loss for various degrees of installation. It must be emphasized that the numbers thus generated are dependent upon the data base and assumptions described in the complete report.

Figure 11 shows the estimated 1990 experience for fire related injuries and fatalities on the same basis as described previously. Again it must be emphasized that the results are based on the input data and assumptions described in the complete report. The cost projections (related to injuries) assume that a non-hospitalization injury will cost $25, a moderate injury with brief hospitalization $2000, a serious injury, including hospitalization and rehabilitation, $100,000 and a death $1,000,000. Since no recognized validated study objectively establishing

11

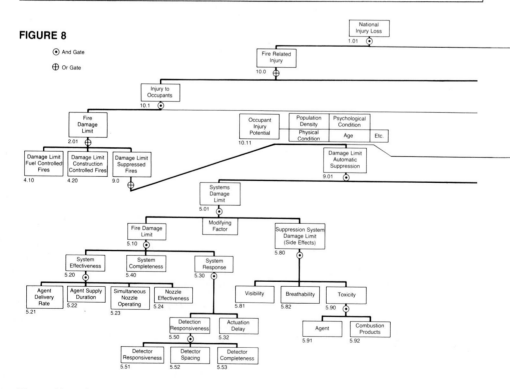

FIGURE 8

Figure 13.11 (continued)

these costs was found, they were developed as a part of the study on the basis of the described assumptions. They should be considered in that light.

FUTURE TECHNOLOGY

Phase 3 consisted of determining how an automatic residential sprinkler system might be improved to make it more effective in system performance, ability or cost. Numerous possibilities were identified including flexible piping systems, faster-response sprinkler detection mechanisms, modifications in equipment design, etc. Primarily these are responsive to the installation cost and practical problems identified in the user needs and influencing organization surveys. Ongoing work funded by NFPCA will logically contribute in this area and should be studied by the reader of this report.

CONCLUSIONS

The following are excerpted from the complete report which contains the supporting rationale and data base.

Automatic residential sprinkler systems, which are available today (Level I system) can significantly reduce the national residential fire loss. If <u>all</u> residences are equipped with such systems, it is estimated that the reduction can be more than

12

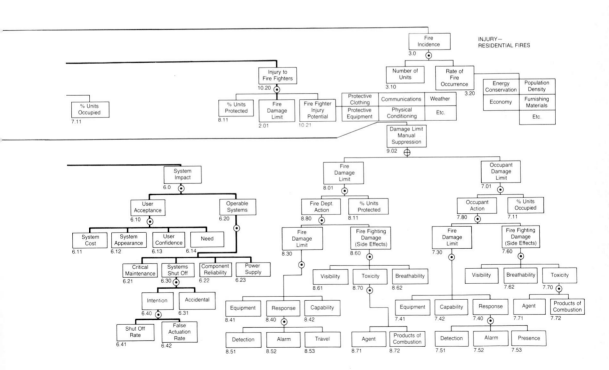

Figure 13.11 (continued)

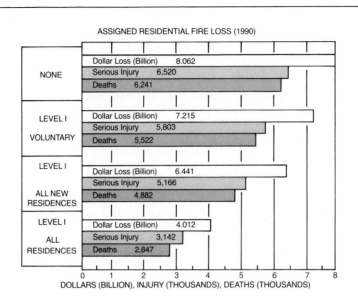

ASSIGNED RESIDENTIAL FIRE LOSS (1990)

NONE	Dollar Loss (Billion) 8.062
	Serious Injury 6,520
	Deaths 6,241
LEVEL I	Dollar Loss (Billion) 7.215
VOLUNTARY	Serious Injury 5,803
	Deaths 5,522
LEVEL I	Dollar Loss (Billion) 6.441
ALL NEW RESIDENCES	Serious Injury 5,166
	Deaths 4,882
LEVEL I	Dollar Loss (Billion) 4.012
ALL RESIDENCES	Serious Injury 3,142
	Deaths 2,847

DOLLARS (BILLION), INJURY (THOUSANDS), DEATHS (THOUSANDS)

Figure 9

FIGURE 10

ESTIMATED 1990 RESIDENTIAL
PROPERTY FIRE LOSS

ASSIGNED
PROPERTY LOSS

NO SYSTEM	SYSTEM
	$1,130,270,000
LEVEL 1—VOLUNTARY	$1,076,415,000
LEVEL 1—ALL NEW RESIDENCES	$1,009,480,000
LEVEL 1—ALL RESIDENCES	$825,707,500

50 percent. Figure 9 is a summary of the calculated losses of all systems evaluated in depth by the previously described methods.

Sprinkler systems, as currently marketed, will not achieve this goal unless they can be installed at a significantly lower cost. In existing residences this means patching or painting, walls, floors, or ceiling must be avoided.

13

Figure 13.11 (continued)

FIGURE 11

ESTIMATED 1990 RESIDENTIAL
INJURY FIRE LOSS

SYSTEM	FATALITIES	SERIOUS INJURIES	COST
NO SYSTEM	6241	6520	$6,932,348,775
LEVEL 1—VOLUNTARY	5522	5803	$6,138,824,450
LEVEL 1—ALL NEW RESIDENCES	4882	5166	$5,432,347,100
LEVEL 1—ALL RESIDENCES	2847	3142	$3,186,901,325

The least expensive sprinkler systems available today are not economically feasible in new single family residences because the related benefits for reduced construction costs, fire losses, insurance costs, and other benefits do not justify the investment in the view of the potential users.

In multiple family residences and new single residential developments (i.e., large project), sprinkler systems may be cost-effective where building code land use or other alternatives permit construction cost savings comparable to the sprinkler system.

The major factors which may improve system acceptance are, in order of priority:

- Development of a "Package System" for retrofit which can be installed without opening the finished walls.
- Development of installation incentives to reduce net cost, such as tax or insurance savings.
- Development of flexible piping to reduce the labor needed to install a system.
- Development of system components designed for use in residential occupancies.
- Removal of restrictions to installations—such as requirements for metering, backflow prevention, and excessive permit fees.
- Development of more efficient (less water, better extinguishment) and responsive (faster operating) sprinklers.
- Improvement of the appearance of sprinklers.
- Showing the public that there is a residential fire problem <u>and</u> that a suppression system will control residential fire without excessive water damage.

A self-contained system is needed, one that can be installed as easily as a picture can be hung on a wall. Until such a system is available, automatic fire suppression

14

Figure 13.11 (continued)

systems will not be provided in existing residential units except by legislative action.

RECOMMENDATIONS

While the complete report contains many recommendations, only the most important are given here, arranged in priority order. They represent opinions of the contractor regarding areas in which specific programs could be established to improve the acceptance and lower the cost of automatic residential sprinkler systems.

SYSTEM EFFECTIVENESS

Continue and expand ongoing studies to develop low-cost residential sprinkler systems having greater extinguishing effectiveness, faster detection capability, and easier installation characteristics. Include studies to establish minimum application rates, water supply (duration) needs, and efficiency of water distribution under actual fire conditions.

NFPA STANDARDS

Encourage the NFPA to incorporate revisions to NFPA Standards 13, 13-D, and 14 facilitate the installation of more effective residential suppression systems.

SYSTEM QUALITY ASSURANCE

The sale of systems needs to be regulated to assist the public in determining what constitutes an adequate residential fire suppression system. To assure the purchase of adequate systems, control of system installation is desirable. Failure to provide adequate control could result in sub-marginal systems which could put the entire residential suppression system program in jeopardy. Control should consist of two parts:

- Third party evaluation is needed of system components critical to the operation of the system.
- Total system evaluation is needed to ensure the adequacy of system design.

PUBLIC AWARENESS

An education program should be initiated and funded by NFPCA in order to instill in the public an awareness of the fire problem and confidence in an automatic suppression system as a solution to the problem.

REGULATORY AGENCIES/BUILDERS-DESIGNERS

A specific education program should be formulated by NFPCA and pursued with code enforcing officials, fire departments, builders, and designers to enhance their understanding of automatic residential suppression systems.

SYSTEM INCENTIVES

A national policy statement is needed to encourage development of:

15

Figure 13.11 (continued)

- Building code incentives that encourage the installation of automatic suppression systems in multi-family housing.
- Land use trade-offs when systems are used.
- Tax incentives (on Federal, state, and local levels) to encourage system use.

BACKFLOW PREVENTION/METERS

Discussion should be initiated with health authorities and water departments to determine if the current regulations for metering and backflow prevention are needed. The development of products which accomplish metering or unwanted backflow of water without the disadvantages of severe pressure loss or excessive cost should be encouraged.

STANDARD RESIDENTIAL FIRE TEST PROCEDURE

A project to develop and promulgate a standardized residential fire test procedure is needed. It is recommended that this be a funded ad hoc committee with the secretariat located at the National Bureau of Standards and that it include interested groups such as users, researchers, appliers, testers, manufacturers, and sellers.

PACKAGED SYSTEM

Support a research program to develop a cost-effective packaged or self-contained system that can be installed in an existing building without the necessity of breaking into walls, floors, or partitions (retrofit use).

16

Figure 13.11 (continued)

Chapter in Brief

In this chapter we learned to write each element of the formal technical report.

Chapter Focal Points

- Letter of transmittal
- Cover and title page
- Abstract
- Executive summary
- Table of contents
- Report text—Introduction, body, terminal section
- Bibliography
- Appendix
- Glossary
- Index
- Distribution list
- Requirements for final reports

Questions for Discussion

1. What is included in the front matter and what is the purpose of those elements?
2. What is an abstract? What is a synopsis? What is a summary? Answering these questions might require library searching. Why does a report require an abstract? Should the abstract of a report be composed primarily to follow the requirements of publication in an abstracts journal, or should your abstract be composed to meet the requirements of your intended reader?
3. Conclusions and recommendations are frequently grouped together within the terminal section of the report. Are they the same? What is their relationship? How are they derived?

Assignments

1. Write the first draft of the report on your chosen research problem, based on the outline(s) you have previously prepared in problem 3, Chapter 12. Your instructor may wish to break this assignment into stages:
   a. First write the introduction to your report. Turn it in to your instructor for evaluation and comment.
   b. Write the body section(s) of your report. Your instructor may again offer comments on this stage of your report.
   c. Write the terminal section. Your instructor may again offer comments.
2. Prepare a letter of transmittal for your report; include reference to the commission of the investigation (by your instructor) and other pertinent details. Transmit your report to your instructor.
3. Write an abstract and executive summary for your report.

Classroom Project

Under supervision of your instructor, exchange with your fellow students the first drafts of your reports, or any of their elements, for constructive comments and suggestions. Present these comments and suggestions orally in class. Your instructor may also ask you to write these evaluations and turn them in.

14

How to Write and Edit the Report

Chapter Objective

Instill students with confidence in their capability to write the formal report by taking them through the steps of the writing and editing processes, including the use of word processing technology.

Chapter Focus

- The first draft
- The revision and editing process—What to look for and do
- Revising and editing with a word processor
- Format and mechanics
- Headings
- Documentation
- Bibliography
- Proofreading
- Government specifications

We can now turn to the encoding–composition process. The actual writing begins after you have completed your research, accumulated your data, and organized your work into an outline. Before beginning to write, you might profit from reviewing your outline. You will find that your concept of the report has been growing and ripening in the recesses of your mind, subconsciously, for the most part. A final review will increase your awareness and freshen your viewpoint. It will also test your outline. Your review may stimulate the check of a note here or datum there and perhaps the revision of a point here or there. You are likely to experience a certain stimulation and excitement as you prepare to do the actual writing. As you review, thoughts about various aspects of the report will come rapidly into your mind. You may want to jot these thoughts down. After you have reviewed the outline, data, and notes, you will be ready for the actual process of the writing. Keep your outline before you and begin the writing. (Professional writers are the first to admit that writing is never easy, but writing can become easier with proper preparation and experience.)

Writing the First Draft

Whether you use a pencil, typewriter, or word processor, your first draft should be written[1] in one sitting, if possible; so allow yourself at least several hours for this phase. Your report will have more life and will represent the sense of your material more closely if you write rapidly than it will if you pause to perfect each sentence before going on to the next. Save problems of spelling and grammar for the revision. To allow for revision, leave plenty of space in your rough draft between lines, between paragraphs, and in margins. If you follow a good outline that breaks your report up into logical stages, you will not be burdened by the strain of trying to keep a great deal of material in mind at one time. The outline permits you to concentrate on key aspects of your report. Some people find it is easier to write the body first and the introduction and the conclusion last. Others have a vivid interest in a certain aspect of the research and find they can "lick" an entire report by writing those parts first which come easiest to them. Thus, writing in stages is like hacking off small bits of a whole. Because of the outline, the writer can do this without losing continuity. It is probably better to make the first draft full and complete, even though you may feel it is wordy, since it is always easier to scratch out material and explanations than it is to expand. Your rough draft is a production of ideas rather than a critical evaluation of those ideas. The logical flow of ideas will be interrupted if you try to evaluate your material critically as you write; punctuation, sentence form, and grammar also slow down the flow of writing. You should not worry about such mechanical details in the first draft. You will have ample time at a later

---

[1]In computer terminology, the process of entering data is called either writing or inputting.

stage to verify, check, correct, and revise. It is important that at first you write and write fully on all the aspects of the investigation that you have previously outlined.

The Revision and Editing Process

Professional writers know that papers are rewritten, not written. After the rough draft has been completed, you ought to plan for a "cooling off" period. Plan for at least three readings of your draft for revision and editing.

The First Reading—The Revision/Editing Process

If possible, let at least two or three days elapse before you review your first draft. An interval of time allows you to approach what you have written, not from the closeness of the first heat of the writing, but from an objective distance. In this reading, take the viewpoint of your intended reader to see whether your report's objectives are being met and whether the problem is being answered. You should read your rough draft through from beginning to end without pausing for revisions. This first reading is necessary for an overview; it should not be used for checking details. You may note points by making marks on the margins of the draft for attention at a later time. At this reading, you should ask yourself questions such as these:

1. Does the thesis of the report come through clearly?
2. Have I covered all points essential to the objectives or purposes of the report?
3. Do they come through clearly?
4. Do the various parts of the report fit together smoothly?
5. Is the information adequate and arranged effectively for content and organization?
6. Does extraneous material confuse the issues?
7. Do my conclusions flow logically from my data?
8. Do my recommendations flow logically from my conclusions?

If your conclusion does not have the point you want your reader to get, or if the conclusion is not based on adequate and clear data, you will have to revise your report until its text provides the answers to the preceding questions. This brings us to the "second reading" phase.

The Second Reading—The Revision/Editing Process

The "second reading" is a figurative expression. This phase concentrates on the process of bringing clarity and cohesiveness to your report. You will be concerned with the *content matter*, its *organization*, and its *written expression*. Now is the appropriate time to pay attention to the items you have marked on the margins of your manuscript during the first reading.

The Content Matter

Here you are concerned with the eight questions you raised in your first reading phase. You sharpen your focus with these questions:

1. What additional information does the reader need to reach my conclusions and to accept my recommendations?
2. Is the information accurate, valid, or appropriate for my conclusions?
3. What information did I include that is not pertinent and that may confuse the reader?

In the heat of writing a first draft, it is not unusual for points or even topics—despite a good outline—to be left out. Or, the reading reveals that an explanation is inadequate. Then you may need to provide:

a. additional examples to clarify or reinforce a point
b. additional instructions, diagrams, or tables to visualize a point
c. additional definitions
d. additional details on the who, what, when, where, why, or how of the situation

You are responsible for the accuracy and validity of your information. Check your:

numerical data

facts

graphics

quotations

Relate your conclusions to your data and information. Are your judgments based on solid facts and/or expert opinion? Are there contradictions in the evidence you have presented? If some of your data are based on interviews or references, have you provided the required documentation?

Often in the heat of the writing we empty ourselves on the page, setting forth everything we know. Aspects of the situation have been interesting for us, but details are not always pertinent or appropriate. If we include everything, if we do not sort out the extraneous, the unimportant details clog the reader's mind and interfere with the important matter that leads to your conclusions and recommendations.

The thesis sentence is your test. You ask yourself: Does this item of information focus on or advance my thesis sentence? If it does, include it; if it does not, be ruthless—no matter how much you like the look of your sentences—cross them out. Be concerned to recognize not only what is superfluous but also what is redundant and repetitive.

The Third Reading—Organization, Language, and Style

To check your report's organization, you need your outline before you. You have made revisions for content in your second reading phase. Now you are concerned with the arrangement of the details of your text material, with the emphasis not only on words, sentences, and paragraphs, but also on sections and how they are connected or related to each other. The revisions made in the second reading phase affect arrangement of text within sections or subsections of your report. There should be little, if any, radical reorganization of the outline material. The cohesiveness of the report was implemented in the outlining stage of the writing. Now, you are mainly concerned with pointing up the text to lead the reader through your investigative experience from its initiation through its various steps to the logic of your conclusion—the thesis sentence. Here are questions you should ask yourself about the overall organization:

1. Is my conclusion—thesis—logical, reasonable, substantiated?
2. Are my recommendations reasonable, practical?
3. Is the problem I am investigating clearly stated for the reader in the Introduction?
4. Are the main sections of the report prepared for in the Introduction? Are they inherent in the investigation of the problem or situation?
5. Are the main sections cohesive?
6. Does each section have a heading?

Here are questions you should ask yourself about the organization of each section:

1. Is the point of each section clearly indicated? (Each section, as your outline directs, has a thesis sentence.)
2. Are the subsections evident or announced?
3. Do I have a heading for each subsection?
4. Do I have connections between the subsections (transitional words, phrases, sentences)?
5. Do I have a summary at the end of the section in which the main point of the section is reaffirmed?

Language and Style

Your final concern in your "third reading" phase is language and style. You should review the material in Chapter 4 on style, sentences, and paragraphs. Now check for the following:

1. *Paragraphs.* Do the paragraphs hang together? Do all paragraphs have a topic sentence? Do the paragraphs have unity? Should any paragraphs be combined? Do any paragraphs need further development?

2. *Sentence structure.* Are all sentences complete grammatically? Do any sentences have to be reread for meaning? If so, break up or rewrite. Are sentences punctuated correctly?

3. *Style.* Is the style consistent and appropariate? Is the writing objective, concise, and clear?

4. *Word choice.* Can any deadwood be removed? Are the words as exact and meaningful as possible? Are there wrong, inexact, or vague words? Are there clichés, jargon, shoptalk? Are there any nonstandard abbreviations? Are there any inconsistencies in names, titles, symbols? Are words spelled correctly?

The writer's best tools in these matters are a dictionary and a good handbook of grammar. Consult the reference guide and index in the back of this text for more specifics on grammar, punctuation, style, and usage.

Revising and Editing with a Word Processor

If you have entered your report on a word processor, your revision and editing chores are simplified; nonetheless, the same factors and principles of grammar, logic, and composition apply. Certain time-consuming tasks are made easier because of the versatility of word processing programs. For example:

Delete and *Insert* commands. By positioning the cursor under the exact place on the screen at which a revision is to be made and keying the appropriate command, you can delete a character, a word, a part of a line, an entire line, or a paragraph. With an insert command, you can similarly add material.

Macro. A **macro** is a sequence of key strokes that you record and save in a file (computer memory) so that you can use the sequence again whenever you wish, just by typing the "name" that you assigned to the sequence. The macro is simply a short cut for entering frequently used data, commands, or a combination of the two. Macros range from very short, simple entries to elaborate chains.

Block moves and *block editing*. Block moves are useful commands which allow you to mark off a block of text ranging in length from one character to an entire document. You can mark a block as a separate unit to be saved in permanent memory, sent to the printer (of your computer), copied, deleted, moved to another location in the document, or converted from lowercase to uppercase, or given other enhancements, for example, italics or boldface.

Find and Replace. This command allows you to search for and correct specific changes on command; for example, change all abbreviations of "NSF" to read "National Science Foundation."

Merge. Merge means putting together data from two or more files in memory. For example, you may have entered data into memory in the form of charts or tables. By appropriate command, you can merge the charts and tables into the proper place in your report's text.

Formatting and Reformatting. Rewriting with a typewriter because of changes due to revisions means retyping the entire report manuscript. Word processing saves you from this tedious job. It enables you to make revisions and format changes without having to retype the entire manuscript. Rewriting often involves not only changes in text but also in style and format. Style considerations include:

Capitalization
Italics
Boldface
Centering
Use of quotation marks

Format considerations include:

Margins: left, right, top, and bottom
Space between lines and between paragraphs
Space around illustrations and tables
Paragraph space indentation
Number of lines to a page
Widows and orphans on a page. (**Widows** are the last line of a paragraph at the top of a new page; **orphans** are the first line of a paragraph isolated at the bottom of a page.)
Placement of page numbers
Placement of footnotes and headers. (A **header** is a section heading that is repeated on consecutive pages.)

Word processing technology readily takes care of these formatting requirements, assists with other structural considerations, such as placement of graphics, and provides several convenient writing aids:

Grammar and Style Checkers. These programs help point out such problems as:

Faulty subject-verb agreements
Faulty pronoun-object agreement
Wrong number in predicate nominative
Incorrect formation of the possessive
Careless punctuation and proofreading, such as missing periods and unpaired quotation marks
Clichés and hackneyed expressions
Confusion of words like its and it's and whose and who's
Prolix constructions, like "all of the" instead of "all the"
Barbarisms, like "irregardless"

Some grammar/style checkers keep track of statistics on average word and sentence length and readability formula computations. (See pages 133–137 for comments on readability formulas.) Actually, grammar checkers compare what you have written to a collection of rules to determine if your writing disagrees with any of them, but style templates may not always be appropriate. For example, if you begin a sentence with "And," it will be flagged as improper, even though you may have purposely begun the sentence with the conjunction *And* for emphasis to suggest irony.

Thesaurus. This is a dictionary program that suggests synonyms. When a writer is stumped for a word, the computerized thesaurus can help. By placing the cursor on the word to be replaced and making the appropriate keystrokes for the thesaurus software, a list of synonyms is presented on screen. Sometimes the synonyms have lists of synonyms.

Ideally, computerized writing aids would not only provide grammar and punctuation corrections, word spellings, and style improvements, but would also determine: whether a writer conveys the meaning intended; whether a writer provides a reader with a coherent logic in and flow of text; and whether a writer provides understandable descriptions and appropriate examples. Unfortunately, as of yet, computers have not reached such analytical sophistication. But they can provide the inexperienced writer with helpful aids to catch slips in expression and ungrammatical constructions.

Printing the Report

After the final revision process, writers using a word processor may have choices for the report's printing. The diskette may be sent to the print shop if it has the capability to use the electronic manuscript as direct input. If the computer you are using has a printer attached, you can print the report directly. If you or your organization has a laser printer, you have recourse to the desktop publishing route. With appropriate graphic-oriented page composition software, you can convert the report manuscript into a handsomely printed formal report.

Format and Mechanics for the Typed Report

Most students depend on their typewriters for their written work. What follows are several standard format and mechanics considerations common to most formal reports. In general, these matters apply equally to typed and word processor-produced reports.

Use white unruled 8-½" × 11" bond paper of 20 lb., 25% cotton rag stock. Be sure to have sufficient carbon or duplicated copies for filing and reference. The report should be typewritten or printed with a black-ink ribbon, on one side of the page only. Leave ample margins on all sides of the page. Margins on the left should be no less than 1¼", preferably 1½" to facilitate binding.

Margins on the right should be no less than ¾". The margin should be no less than 1" at the top. The bottom margin, including footnote space, if there are footnotes, should be no less than 1", preferably 1¼".

Manuscripts for publication are always double-spaced. Reports may be either double-spaced or single-spaced, depending on the conventions of your organization or specifications of your client. Any material that contains equations, superscripts, or subscripts is easier to read if it is double-spaced. Reports written as classroom assignments should be double-spaced to facilitate marking and correcting. If the typescript is single-spaced, use double spaces between paragraphs. Two spaces separate paragraphs of double-spaced copy. Paragraphs, whether single-spaced or double-spaced, are indented five spaces.

The appearance of your report page will be improved if you follow these rules:

1. Do not start the first sentence of a paragraph on the last line of a page.
2. Do not place a heading at the bottom of the page with less than two lines of text to follow.
3. Avoid placing the last line of a paragraph at the beginning of a page.

Text pages of the report are numbered with arabic numerals. The placing of numbers on the page should be consistent. They may be centered at the top or placed in the upper right-hand corner. Numbering at the bottom of the page, either in the center or at the lower left, should be avoided because these numbers may be confused with footnote material. Prefatory or preliminary pages of the report (Title Page, Abstract, Table of Contents, List of Illustrations, Foreword) are numbered with small Roman numerals. Should blank pages be used for the sake of appearance, they are counted but not numbered. The Title Page is counted as the prefatory page (i) but is not numbered. Final assignment of numbers to pages might be delayed until all the pages are typed, unless the exact numbers of illustrations and their placement are known and planned beforehand. However, tentative numbers might be written in the upper right-hand corners very lightly in pencil.

How to Handle Equations

Mathematical equations are generally centered on a page. Lengthy equations should be typed completely on one line rather than broken into two lines:

Poor:

. . . . the cut-off frequency of a rectangular guide is $f_c = \dfrac{c}{\lambda c} =$

$$\frac{c\sqrt{\left(\dfrac{m}{a}\right)^2 + \left(\dfrac{u}{b}\right)^2}}{2}$$

Preferred:

. . . . the cut-off frequency of a rectangular guide is

$$f_c = \frac{c}{\lambda c} = \frac{c\sqrt{\left(\frac{m}{a}\right)^2 + \left(\frac{u}{b}\right)^2}}{2}$$

Each new line of an equation should be positioned so that the equal signs are aligned with the equal signs of the preceding line:

$$\frac{\Delta Z_0}{Z_0} = \frac{1}{4\pi^2} \frac{w^2}{D^2 d^2}$$

$$K = \frac{\Delta Z_0}{2Z_0}$$

Many mathematical symbols must be penned by hand. The typist should leave the necessary space for hand lettering by the author. All complete equations are conventionally numbered consecutively within each chapter for easy reference purposes; the numbers appear within parentheses flush with the right margin of the text.

Example:

$$VSWR = \left(\frac{P_{s\ max}}{P_{s\ min}}\right)^{1/2} \tag{1}$$

$$P = c_1 s_{in}^{2} \left(\frac{2\pi x}{g}\right) \tag{2}$$

A short formula within the text is set off by commas, as for example:

It can readily be seen that when $\lambda = 2a$, $\alpha = 90°$, or the waves are not propagated down the guide. . . .

Equations—no matter how short—containing fractions, square root signs, sub- or super-numerals, or letters require extra space above and below the text line and, therefore, should not be included in a text line but centered and placed on a line by themselves, as illustrated in preceding examples.

In summary:

1. Line up equal signs in a series of equations.
2. Keep all division lines (fraction bars) on the same level with equal signs.

3. Divide equations only after plus or minus signs but before equal signs in second line.
4. In dividing equations, line up the second line with the equal sign or the first plus or minus sign in the first line.
5. Parentheses, braces, brackets, and integral signs should be the same heights as the expressions they enclose.

How to Handle Tables

You will help achieve clarity in tables if you observe the following rules:

1. Center all numbers within the column.
2. Align numbers by the decimal points or commas, if any. Otherwise, align numbers by the right-hand digits.
3. Use headings whenever possible in order to avoid repetition in the body of the table.
4. To indicate subdivisions within the tables, use single, horizontal rules below the heading at the bottom. When there are more than two columns, use a single, vertical rule. Omit vertical rules when there is sufficient white space between columns.
5. Number tables consecutively with Roman numerals.
6. Center table numbers at the top and type the title in all caps.
7. Type the first word of a column head in initial caps, but type the other words in the column head in lower-case.
8. Insert short tables within the text. Place longer tables on separate pages. Where there are a number of large or oversize tables following sequentially within a paragraph or page of text, place such tables in the appendix to avoid confusion.
9. Read Chapter 15, for a fuller treatment of tables.

How to Handle Headings

Headings are mapping devices; they help make your report more readable. The features of the report are marked by sectional and subsectional titles or heads. Your headings correspond to your major divisions and subdivisions in your outline and in your table of contents. As mapping devices, headings serve to show relationships and subordinations; they should clearly indicate the logic of relationship and subordination throughout the report. A single system is recommended here, although in actual practice a number of conventions of showing this relationship and subordination are used (Figure 14.1).

FIRST-ORDER HEADING

SECOND-ORDER HEADING

<u>Third-Order Heading</u>
    <u>Fourth-Order Heading</u>. Text of paragraph follows on the same
line as the fourth-order heading.

Figure 14.1
How to Handle Headings

First-order Headings

First-order headings are written in all caps and centered two inches from the
top of the page. Such a head constitutes a major text division and corresponds
to the Roman and Arabic numerals of your outline. First-order headings begin
on a new sheet of paper, and the text follows four typewriter spaces below.

Second-order Headings

Second-order headings are placed flush with the left margin, two spaces below
the last line of the preceding paragraph and are typed in all caps. They correspond to the capital letters in your outline and indicate major subdivisions of a
section. The text follows two spaces below.

Third-order Headings

Third-order headings are typed with initial caps of each word and are placed
flush with the left margin and are underlined.

Fourth-order Headings

These headings are indented five spaces and are also typed with initial caps.
They are the same as the third-order headings except they are part of the paragraph of the text. The text follows on the same line.
    Some reports follow through on the numbering system of the outline to
conform with the various heads and their subheads of various rank. Certain
organizations and many government agencies call for numbering of heads and
subheads. The decimal system of numbering is frequently used.

Documenting Your Report's References

Why Documentation Is Necessary

Documentation is required in reports, professional papers, and other serious types of writing for three reasons:

1. To establish the validity of evidence. All important statements of fact not generally accepted as true, as well as other significant data, are supported by the presentation of evidence for validity if the exposition within the text itself does not offer the proof of the data or the facts. Direct reference to the source is provided so that the writer's statements may be verified by the reader if she so chooses, or if the reader wishes to extend her inquiry into the borrowed matter beyond the scope of this particular writing.
2. To acknowledge indebtedness. Each important statement of fact, data, or information, each conclusion or inference borrowed by the writer from someone else should be acknowledged. Also, a citation is desirable when a conclusion or idea is paraphrased or its substance is borrowed and presented.
3. To provide the reader with information she might need or want about the subject matter a writer has borrowed or obtained from another source.

When any report, professional paper, book, or other type of serious writing contains information obtained from other publications, books, articles, and reports, such sources should be indicated under a list of references or a bibliography. Previously, I indicated that your report should have a bibliography if it uses information obtained from other sources. The bibliography follows the last page of the report and precedes the appendixes. The bibliography, as you will recall, is a list of the sources arranged alphabetically and numbered sequentially.

A Simple Documentation System

Scholarship has a long tradition of providing documentation for borrowed sources. The cited source material is documented in a footnote at the bottom of the page of the text upon which the borrowed material appears. However, this system of documentation is not as efficient as it might be. In this book, I have used what I consider a very efficient and very simple system of documentation. This system, instead of using a footnote at the bottom of the page, integrates the documentation reference within the text line following immediately the matter or source to be documented. The documentation reference begins with a bracket, then lists the sequential number of the bibliographic reference source being used, followed by a colon and the page numbers of that bibliographic reference, and then is closed by the bracket; for example [7:113–19].

Let me illustrate again how this works. In Chapter 3, I borrowed material liberally from scholars in semantics, communications, linguistics, and anthropology. On pages 224–225, I quote from W. A. Sinclair's book, *An Introduction to Philosophy*. In the system I am recommending, instead of a superscript following the words "is very close" I have a bracket, then the number 7, which is the sequential appearance of Sinclair within the list of references at the end of the chapter, followed by a colon and 113–19 at the end of the phrase, then close bracket: [7:113–19]. This tells the reader that this material was borrowed or quoted from the bibliographic source appearing in sequence in the references and that the matter quoted appears on pages 113 to 119 of that source.

Now let us take another example. Again in Chapter 3, beginning on page 106, you will discover a quotation taken from Wilbur M. Urban's book, *Language and Reality*. Following that, I paraphrase additional material. After the quoted material, there is a bracket, followed by 8, indicating the sequential listing in the references for Urban's book, and following the 8, there is a colon with 107–15 following it with a bracket and then a period. After the word *latent* there follows [8:115–16]. Now for the sake of further example and illustration, suppose that the paraphrase was of material that Urban had in several other chapters. You would thus have [8:115–16; 231–35; 412–18]. This reference documentation indicates that the paragraph is a paraphrase of ideas expressed by Urban on the pages listed.

Now suppose you are going to paraphrase some ideas you have obtained from several sources—how would you do this in this system? Again, this is a very simple matter. For the sake of example, let us suppose that I am paraphrasing material from several sources within one paragraph. I will offer documentation at the end of the paragraph for the information I have presented. Following the information I have borrowed within that paragraph, I would begin with a bracket, then with the most important source; let us say it is source number 3 in my bibliography or references. I would have [3:8–12; 23:420–33; 13:56–57; 8:16–17; 43–50]. This documentation will tell the reader that the material of the paragraph preceding was borrowed from the sources indicated and from the pages listed of those sources.

This system actually is simple, efficient, and provides the reader with all the needed information.

Using Footnotes

Although the above system has eliminated the need for footnotes to document borrowed reference material, footnotes may still be used and required in the following instances:

1. To amplify the discussion beyond the point permissible in the text
2. To provide cross references to various parts of the report

For example, on page 7 of Chapter 1, I have a footnote of the first category. This footnote amplifies the information presented in the text about sense perception. If this footnote of about twenty-five words were placed in the text, it would interrupt the continuity of the text matter for the reader. Although interesting, related, and amplifying, this information is not essential to the understanding of the text. Therefore, it is placed in the footnote.[2]

In the second type of footnote, where cross references to various parts of the report are given, the writer is permitted to refer to material appearing in other parts of the report, such as the appendix or matters appearing in earlier or later portions of the report. This type of footnote aids the flow of the text, helps in clarity, and sends the reader, if she so chooses, to related information she may want to examine at that particular point.

I do want to make the point that documentation systems vary widely from college to college, discipline to discipline, journal to journal, and organization to organization. Journals and publishers within the same discipline may differ in the style they use to list references and where they place footnotes. Generally, in scientific fields, references are placed at the end of the paper or publication under headings such as "List of References" or "Literature Cited." The term, "Bibliography," is usually reserved for a listing of literature pertinent to the topic reported but not necessarily mentioned in the paper. Thus, in this text I use the term "References" at the end of chapters to identify sources cited within those chapters. A bibliography appears at the end of the book to identify sources providing supplementary information appropriate to technical writing, the subject of this text.

In the fields of chemistry and physics, it is customary to omit the title of the article. Names of journals are given in abbreviated form, for example.

M. Heaven, T. A. Miller, V. E. Bondybey, *J. Chem. Phys.* 80, 51 (1984)

Other journals, especially in the biological sciences, require not only inclusive pagination but also full titles of articles. The style manual of the Council of Biology Editors provides the following as a proper example of a citation for biological journals (1):

> Steele, R. D. Role of 3-ethylthiopropionate in ethionine metabolism and toxicity in rats. J. Nutr. 112:118–125; 1982.

Note its stylistic characteristics:

1. Only the first word in the title is capitalized.
2. The volume and year but not the month are identified.
3. Full pagination is given.
4. *Journal of Nutrition* is abbreviated.

---

[2]Footnotes, as the term implies, are notes at the foot of the page, which the reader can read if she wants to or after she reads the first line, can see whether she wants to continue or not. The flow of the main text is not impeded by this device.

5. The title of the article does not have quotation marks around it nor is the title of the journal in italics.

6. A semicolon is used to separate the year of publication from the rest of the citation.

You can see that the number of documentation systems used in the various fields make the process more complicated and confusing than it ought to be. The system recommended in this text is used most often in the social sciences. Simple and efficient, it provides the reader all the information he or she needs. Nevertheless, it might be well for you to become familiar with how the field of your discipline handles references and footnotes. Toward that purpose, look forward to an assignment at the end of this chapter.

Bibliography

The bibliography, as I previously stated, is a list of references used by the writer in his report. When the list of references is large, it is sometimes convenient to classify items according to types of references, listing in separate categories all books, periodical literature, publications of learned societies and organizations, government publications, encyclopedia articles, and manuscripts and unpublished materials. In most instances, the alphabetically and sequentially numbered bibliography is the most convenient.

Each bibliographic reference should list the complete elements that will help to identify the source. These bibliographical elements are author, surname first; title; place of publication; publisher; date; page numbers. In the case of a report or government publication that is classified, the security classification will be indicated. Listed below are examples of various bibliographic references and the conventional way of indicating them in a bibliography.

1. *Anglo-American Cataloging Rules*, prepared by the American Library Association, the Library of Congress, the Library Association, and the Canadian Library Association, 1967, 400pp.

2. Cain, Sandra E., and Jack M. Evans. *Sciencing, An Involvement Approach to Elementary Science Methods*, 3d ed. Columbus, OH: Merrill, 1990, pp. 84–91.

3. *Calloway Workshop, 1990.* Kansas City, MO: Calloway Productions, 1991.

4. Clayman, Charles B., Ed. *The American Medical Association Medical Encyclopedia*, vol. 1. Pleasantville, NY: The Readers' Digest Association, Inc., published with permission of Random House, 1989, pp. 158–59.

5. Council of Biology Editors. *Style Manual*, 5th ed. Washington, DC: American Institute of Biological Sciences, 1983.

6. Directorate of Small and Disadvantaged Business Utilization, Office of the Secretary of Defense. *Guide to the Preparation of Offers for Selling to the Military.* Washington, DC: U.S. Government Printing Office, n.d.

7. Duran, Lise W., and Larry R. Pease. "Tracing the Evolution of H-2D Region Genes Using Sequences Associated with a Repetitive Element." *The Journal of Immunology*, July 1, 1988, pp. 295–301.

8. Duran, Lise W., and E. S. Metcalf. "Analysis of the murine *Salmonella tythimurium*-specific B-cell repertoire." Postboard presentation at the Annual Federation of American Societies for Experimental Biology, Chicago, April 11, 1983.

9. Flower, Linda and John R. Hayes. "Plans that Guide the Composing Process." in *Writing: The Nature, Development, and Teaching of Written Communication*, Carl H. Frederickson and Joseph F. Dominic, (Eds.). Hillsdale, NJ: Lawrence Erlbaum, 1981, pp. 55–57.

10. Gilsdorf, J. W. "Writing to Persuade." *IEEE Transactions on Professional Communication*, PC-30.2, June, 1987, pp. 68–73.

11. Green, Bonnie L., "Overview and Research Recommendations." *Role Stressors and Supports for Emergency Workers, Proceedings from a 1984 Workshop by the Center for Mental Health Studies of Emergencies and the Federal Emergency Management Agency*, DHHS Publication No. (ADM)85-1408. Rockville, MD: National Institute of Mental Health, 1985, pp. 1–20.

12. *Health Show*. Transcript, ABC Television Program, August 26, 1990.

13. Human Relations Commission. *Human Rights Laws*. Rockville, MD: Montgomery County Government, Maryland, n.d., 58pp.

14. Luhn, H. P., "Selective Dissemination of New Information With the Aid of Electronic Processing." *H. P. Luhn Pioneer in Information Science, Selected Works*. Clair K. Schultz, (Ed.). Yorktown Heights, NY: Sparten Books, 1968, pp. 246–54. (Originally published by IBM Corporation, Advance Systems Development Division, Yorktown Height, NY, November 30, 1959.)

15. *The New York Times*. pp. 1, 26, December 21, 1991.

16. Norstrom, David M., Gerald A. Francis, and Rolland D. King. *Lifts and Wheelchairs, Securement for Buses and Paratransit Vehicles, A Companion Document to the Advisory Panel Accessible Transportation Guidelines Specifications*. (Prepared for the Architectural and Transportation Barriers Compliance Board.) Columbus, OH: Battelle Memorial Institute, n.d.

17. Oren, T. "The Architecture of Static Hypertexts." *Hypertext '87 Papers*. (Proceedings of a conference held at the University of North Carolina, Chapel Hill.) Chapel Hill, NC: University of North Carolina, 1988, pp. 291–306.

18. Stone, Richard. "An Artificial Eye May be Within Sight, Work Progresses on a Prosthesis for Reading." *The Washington Post*, August 20, 1990, p. A3.

19. *TDD (Telecommunication Devices for the Deaf)*. Final Report prepared for the U.S. Architectural and Transportation Barriers Compliance Board, Woodstock, VA: Applied Concepts Corporation, August 3, 1984, 113pp.

20. Tracey, J. R. *The Sequential Topical (STOP) Storyboarding Method of Organizing Reports and Proposals, A Group Book-building Technique*. Fullerton, CA: Hughes Aircraft Company, November, 1968.

21. Weisman, Alan. *La Frontera, The United States Border with Mexico.* (Photographs by Jay Dusard.) San Diego, CA: Harcourt Brace Jovanovich, 1986, pp. 123–44.

22. Weisman, Harlan F. "Acute Infarct Expansion and Chronic Enlargement." Paper presented at the Post Graduate Seminar on Ventricular Remodeling after Myocardial Infarction, sponsored by the Council on Clinical Cardiology at the American Heart Association Annual Scientific Sessions, November 12–16, 1989.

23. Weisman, Harlan F. et al. "Soluble Human Complement Receptor Type 1: In Vivo Inhibitor of Complement Suppressing Post-Ischemic Myocardial Inflammation and Necroses." *Science*, July 13, 1990, pp. 146–51.

24. Williams, Abbi J. Interview on Real Estate Pre-settlement Procedures, held at Beckett, Cromwell & Myers, Settlements Ltd., September 3, 1990.

25. Wolfe, W. M. (Moderator). "How Can Effectiveness of Analysis Centers Be Measured." Addendum to the *Proceedings of the Ad Hoc Forum for Information Analysis Center Managers, Directors, and Professional Analysts*, held at the Battelle Memorial Institute, Columbus, Ohio, November 9–11, 1965. (Unpublished manuscript, n.d., 53pp.)

Proofreading

After your report has received its final typing, you must proofread the copy. You are making a quality control check of the typing. In this proofreading, you may find not only typographical errors but also others that you somehow missed earlier. Here is what you need to do in the proofing or the final draft:

1. Place the draft from which the final copy was typed next to the final copy. Use the index finger of each hand to make a word by word comparison between the two.

2. Examine each page to be sure that all corrections have been made and that the page is clean and neat in appearance.

3. Check that headings and text have consistent and proper spacing.

4. Read the text carefully to ensure that no words or lines have been omitted or that extraneous words or text lines or material have somehow been added.

5. Note any corrections lightly in pencil above the line. Use proofreading symbols if you are acquainted with them (Figure 14.2).

6. Check for proper pagination, numerical sequences (for section, figure, and footnote numbering), capitalization, consistency of style, and spelling.

Editors use printers' conventional proofreaders' marks to edit a manuscript. Figure 14.2 shows the range of proofreading marks. Editors use a selected number of these marks to indicate changes and instructions. All use of proofreaders' marks requires a double marking—one in the body of the text at the point of correction, and a corresponding mark in the margin, which calls attention to the fact that there is a correction to be made at this point. A correction marked only in the text and not in the margin will be missed by the printer.

Reports Prepared to Government Specifications

Contractors doing research for the government are invariably required to furnish reports covering such investigations. Procedures governing the preparation of the reports are determined by specifications. While the Department of Defense and other branches of government have their own specifications guiding the preparation of reports, the report itself conforms with the principles covered in this text. The purpose of specifications is to establish procedures and standards by which the government may insure the quality of the preparation and be assured that all the information required is covered competently by the report. The contract will indicate in detail the type, frequency, and number of copies of the report to be furnished. The specification defines the basic procedures, the content, editorial standards, preparation of completed (reproduction) copy, preparation of illustrative materials, security standards, and distribution. Reports to the government may vary in size from a one-page form, monthly progress report to final reports of many hundreds of pages. Government specifications will indicate all the elements the report should contain. Such definitions insure uniformity and quality for the many thousands of projects being carried out for the government. Although the specifications establish standards, patterns, and criteria, their requirements actually are not formidable to the writer. The specifications determine the what, how, and when of a contract. They will define what goes into a progress report, a quarterly report, or a final report; how it is prepared; and how and in what quantities it will be forwarded or distributed.

Chapter in Brief

We examined the activities involved in the processes of writing, revising, editing, and printing the formal report. Included in our examination were matters of content, organization, language, and style. How computer technology helps in composition, revision, editing, and printing was also discussed. Format and mechanics considerations were then detailed.

Chapter Focal Points

- The Writing process
- The Revision and editing process
- Revising and editing with a word processor
- Software aids in revision and editing
- Format and mechanics
- Proofreading

✕	Change bad letter	⌄²	Superior figure
⋃	Push down space	⋀₂	Inferior figure
⑤	Turn over	⌐	Move to left
ℓ	Take out	⌐	Move to right
⋀	Left out—insert	*out, s.c.*	Out, see copy (be sure MS is returned if this is used)
#	Insert space	⌷	Em quad space
✓	Equalize spacing	⏤ₘ	One-em dash
⌣	Less space	𝓗	Paragraph
◯	Close up	*no 𝓗*	No paragraph—run in
⌐	Raise (show in some manner how much to raise or where to raise)	*w.f.*	Wrong font
⌞⌟	Lower	••••••	Let it stand
≡	Straighten lines	*stet*	Let it stand
⊙	Period	*tr.*	Transpose
⋀	Comma	*cap.*	Capital letters
:/	Colon	*sm. cap.*	Small caps
;/	Semicolon	*l.c.*	Lower case
⋎	Apostrophe	*ital.*	Italics
⋎⋎	Quotation	*rom.*	Roman
=	Hyphen	*b.f.*	Bold face

Many of the proof readers marks may also be utilized in marking manuscripts.

Capitals	Comparative Data	COMPARATIVE DATA
Caps and Small Caps	COMPARATIVE DATA	Comparative Data
Small Caps	Comparative Data	comparative data
Italic	Current Notes	*Current Notes*
Italic Caps	Current Notes	*CURRENT NOTES*
Bold Face	News and Notes	**News and Notes**
Bold Face Italic	News and Notes	***News and Notes***
Bold Face Italic Caps	News and Notes	***NEWS AND NOTES***
To change Caps to Caps and lower case	CONCLUSIONS	
To change Caps to Caps and small caps	CONCLUSIONS	
Flush left	⌐	
Flush right	⌐	
Indented matter	⌐ ⌐	
Center	⟨COMMENT⟩	
Run in	COMPARATIVE TECHNICAL DATA	

◯   *Close up space.* Use when your typewriter skips a letter or when a letter is deleted in a word.

Open a space. Insert a space between words run together.

Figure 14.2
Proofreaders' Marks and How to Use Them

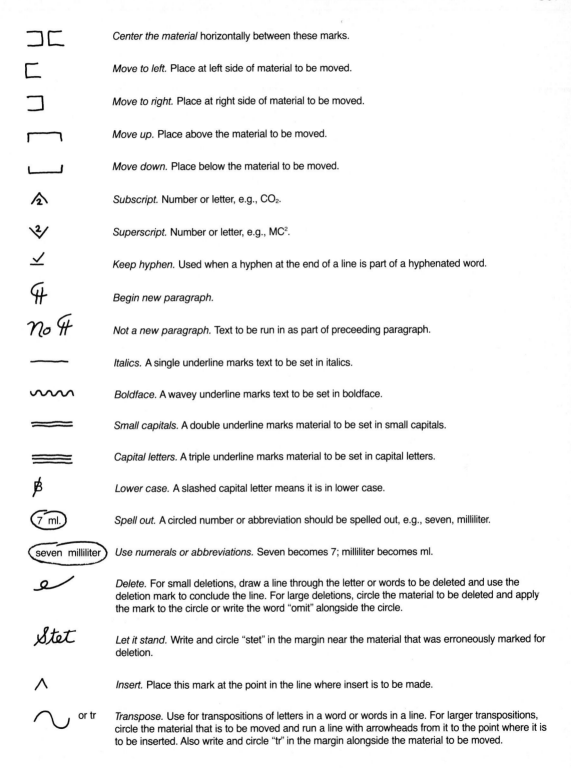

⊐⊏   *Center the material* horizontally between these marks.

⊏   *Move to left.* Place at left side of material to be moved.

⊐   *Move to right.* Place at right side of material to be moved.

⌐¬   *Move up.* Place above the material to be moved.

⌐¬   *Move down.* Place below the material to be moved.

⋀   *Subscript.* Number or letter, e.g., CO_2.

⋁   *Superscript.* Number or letter, e.g., MC^2.

⌣   *Keep hyphen.* Used when a hyphen at the end of a line is part of a hyphenated word.

¶   *Begin new paragraph.*

no ¶   *Not a new paragraph.* Text to be run in as part of preceeding paragraph.

———   *Italics.* A single underline marks text to be set in italics.

〰〰   *Boldface.* A wavey underline marks text to be set in boldface.

═══   *Small capitals.* A double underline marks material to be set in small capitals.

≡≡≡   *Capital letters.* A triple underline marks material to be set in capital letters.

ℬ   *Lower case.* A slashed capital letter means it is in lower case.

(7 ml.)   *Spell out.* A circled number or abbreviation should be spelled out, e.g., seven, milliliter.

(seven milliliter)   *Use numerals or abbreviations.* Seven becomes 7; milliliter becomes ml.

℮   *Delete.* For small deletions, draw a line through the letter or words to be deleted and use the deletion mark to conclude the line. For large deletions, circle the material to be deleted and apply the mark to the circle or write the word "omit" alongside the circle.

Stet   *Let it stand.* Write and circle "stet" in the margin near the material that was erroneously marked for deletion.

∧   *Insert.* Place this mark at the point in the line where insert is to be made.

∿ or tr   *Transpose.* Use for transpositions of letters in a word or words in a line. For larger transpositions, circle the material that is to be moved and run a line with arrowheads from it to the point where it is to be inserted. Also write and circle "tr" in the margin alongside the material to be moved.

Figure 14.2 (continued)

- Printing the report
- Reports written to government specifications

Questions for Discussion

1. Recall your past written assignments, both in this course and other courses. How did you revise and edit your written work? Can you describe step by step how you went about the revision and editing process to generate the final document you turned in? Share your methods with the class, and discuss alternative editing procedures. Use the insights you gain from this class discussion to revise and edit your major report, following the assignments as given for this chapter.

Assignments

1. Write your instructor a memo addressing the following matters in the report you are writing in this course:

   a. On an average, how many words per sentence and how many syllables per hundred words do you have in your report?

   b. Are half of the sentences in your report complex sentences? Simple sentences comprise what percentage of your paper? Compound? Other types of sentences?

   c. On your rough draft, underline the topic sentences in your report. Encircle transition devices and structural paragraphs.

   d. Test each of your paragraphs by these criteria:

      What is the central idea (topic sentence)?

      What must the reader know to support it or explain it?

      Is there anything in it not related to the topic sentence?

      Are the sentences organized in a sequence which is sufficiently logical to support or explain the topic sentence clearly?

   e. Check your report for the common editing problems noted in this chapter.

   f. Check your report for the format considerations noted in this chapter.

   g. Does your report have sufficient variety in sentence length and structure to promote interest and avoid monotony? Does your report have transitional paragraphs, sentences, phrases?

   h. Do subjects and verbs of your sentences have grammatical agreement?

   i. Does each "it," "who," "which," and "that" clause refer to a definite word?

   j. Does the punctuation help your reader reach the exact meaning you want him to reach?

   k. Are your words and phrases accurate and precise?

   l. Is all spelling accurate?

   m. Do your headings accurately and logically reflect the content matter they describe?

   n. Does your report have adequate graphics to describe and present its data efficiently?

   o. If you are using tables in your report, do they conform to the checklist on page 377?

2. Using the information in this memo and your instructor's comments, edit and rewrite the draft of the report text you prepared in Chapter 13.

3. Go to your library. Examine the documentation in a representative group of scientific/technical periodicals. What variations do you find among periodicals in the same field? In different fields? Are there variations in details of punctuation and capitalization or in content and arrangement? Report your findings to your class.

4. Write to a professional organization in your field or discipline to obtain a copy of the guidelines or manual for its documentation practices. As a separate assignment, prepare the list of references/bibliography/footnotes required in your term assignment in accordance with those practiced in your declared professional field.

5. Prepare documentation, references, and bibliography for the report you are writing in accordance with the style and procedures recommended in this chapter. Turn these in to your instructor for comment.

6. Prepare a table of contents for your report.

References

1. *CBE Style Manual, Fifth Edition.* Bethesda, Maryland: Council of Biology Editors, 1983.

15
Graphic Presentation in Technical Writing

Chapter Objective

Provide understanding of the principles of the various types of graphic aids and their capabilities, and explain where and how to make use of them in technical writing.

Chapter Focus

- Role of graphics
- Planning graphic aids
- Types of graphic aids
  Lists
  Tables
  Photographs
  Drawings and diagrams
  Cartoons
  Exploded views
  Maps
  Graphs, charts, and curves
- Computer assisted graphics

Pictorial representation was our earliest means of nonverbal communication. It served the limited needs of primitive people well. Leonardo da Vinci was one of the first writers of science and technology who recognized the importance of integrating pictorial representation with verbal description as a way to communicate factual information clearly and efficiently. His notebooks established a useful tradition in technical writing because his graphic materials supplement his words and sometimes supplant them. The axiom that "one picture is worth a thousand words" is not the propaganda of the technical illustrator but a generalization that has much experience to substantiate it. It is also true that an improperly conceived, poorly integrated, or badly executed pictorial element may create irreparable confusion for a reader.

We will be concerned in this chapter with the types of visual aids that can help the technical writer to describe, define, explain, analyze, compare, classify, and narrate (for example, see Figure 15.1). This chapter is not meant to be a text on graphic aids. Its purpose is to identify and outline uses of graphic presentation of information in technical writing. Authoritative texts on techniques of graphic aids will be found in the bibliography. We will examine the uses of lists and tables, photographs, drawings, diagrams, graphs, charts, and flow sheets.

In technical literature, illustrations are seldom used for embellishment. Even though illustrations do perk up a manuscript, add to its attractiveness,

Figure 15.1
This picture on a 12th century B.C. Egyptian tomb shows the use of siphons to draw off Nile River water purified by sedimentation. This is one of the earliest descriptions of a technical process.[1]

[1]Figure 15.1 is taken from *Nuclear Energy for Desalting*. Oak Ridge, TN: U.S. Atomic Energy Commission, Division of Technical Information, Sept. 1966, p. 2 [17]. In the present chapter, I have chosen to use footnotes for acknowledgments because frequently several illustrations are referred to on the same text page and it would be difficult or confusing for the reader to follow the referrals. In other chapters, as for example in Figures 7.4 and 7.5, I have used the documentation method I advocate, because there is only one illustration referred to in the text page. All further superscript notations in this chapter will be placed at the initial text reference to the indicated figure.

and enhance its readability, their sole purpose is functional. Graphic presentations are functional because they can reveal more clearly, concretely, and accurately than words matters of a statistical or complex nature. They are functional and efficient because they promote easier reader comprehension by their ability to organize, confirm, and underscore data and interrelationships that might otherwise be obscure.

Graphic presentation helps the reader better to digest, ruminate, and analyze facts and ideas. Illustrations not only help the reader but also help the writer to organize and understand his subject. You will recall how in Chapters 7 and 8 illustrations of devices and processes serve as a point of departure in presenting descriptions and explanations. An illustration, like an image reflected in a mirror, presents a vivid, organized structure for the communication of required information.

As we cannot appropriate the words or ideas of others as our own in papers and reports we write without providing full credit to the originator, so we cannot appropriate graphic creations of others as our own. If we do, we are guilty of plagiarism or theft. Very few of us have the talents of a Leonardo da Vinci. Often we need help in the design of a graph, drawing, or chart, or in the formation of a table. Photographs of specialized subjects or situations often are beyond our resources and capabilities.

The best source for students of graphic aids are published materials, which usually are copyrighted. Whether the material we want to use is copyrighted or not, acknowledgment of the originator is necessary. Use of exact copyrighted text of 250 words or more or exact reproduction of a graphic aid, like text material, may be "paraphrased" or modified, but the user of the idea or process being communicated must provide credit to the original author. The documentation practices used in acknowledging text sources are used to credit borrowings of graphic material.

Planning the Graphic Aid

After all data are compiled, their organization, analysis, and interpretation frequently can best be accomplished through graphic means. Certain relationships between quantities can be shown most clearly by graphs. The first step is to examine those data that are statistical, involved, and/or technically complex to determine which might more clearly and concretely be communicated by graphic presentation. The nature of the material and the reader's knowledge level and requirements determine which type of graphic aid best meets the necessities of efficient transmission of the information. For example, statistical data in a report to a technical person might most accurately be presented in a table, while the same data for presentation in a semitechnical article might best be presented in a bar graph. The table offers more precise graphical data; the bar graph offers only rounded numerical data. The type of graphic aid selected depends on the reader's ability to understand and use it.

Where to Place Graphic Aids

Visual aid material should be placed near the text matter it illustrates. This is not always possible in typewritten papers and reports where several visual aids may be part of a discussion within a paragraph. If a visual aid is relatively small, it can be integrated into the appropriate paragraph of text. A full-page visual aid might be placed either facing or immediately after the page where it is discussed. Graphic material should not precede its discussion in the text because it will then only confuse the reader. In typed reports, illustrations that cannot be integrated within the text page are placed for convenience in a separate section, usually in the appendix. All visual aids, whether integrated within the text page or collected at the end of the article or report in an appendix, must be referred to in the text itself. The reader should be directed by explanatory sentences to the specific illustration or table for better understanding, interpretation, and correlation of the data being communicated. If an illustration or table supplements, clarifies, confirms, analyzes, or reveals conclusions, such explanations must be clearly stated in the text. The more complex the illustration, the more explanatory material is necessary. Directions for interpretation of illustrations may be included in the text or in captions below the illustrations. Illustrations need to be as simple as possible. Text matter and lengthy explanations are kept out of the illustration proper.

Lists

A list is a series of names, words, numbers, sentences, or similar data arranged, enumerated, or catalogued in a sequence demanded by a logic encompassing the grouping. Each column of a table could be considered a list. Lists identify data in some sort of sequence. While one item of a list might be compared with another, the purpose for the listing is identification rather than immediate comparison with equivalent data. Occasions for using lists are numerous; among the more frequent are catalogs, directories, tables of contents, indexes, parts lists, specifications, directions, and summaries. Often, within a paper or report, it is convenient to structure a series of parallel statements in a numbered, indented or tabulated form. Lists are a good display device for securing emphasis.

I have found it convenient to use lists throughout this text. The process of listing is an efficient device for summarizing major points of a subject. In Chapter 11, for example, after discussing the "problem concept" and its significance to scientific investigation, I listed four guides in defining and formulating a problem. In Chapter 13, before explaining the elements of a formal report, I listed them. The use of an illustration to list is a creative way to add interest to a mere catalog of items, as is demonstrated in Figure 15.2.[2]

---

[2]Figure 15.2 is from *Consumers Should Know: How to Buy a Personal Computer*. Washington, DC: Consumer Electronic Group, Electronic Industries Association, 1986, pp. 26–27. [4]

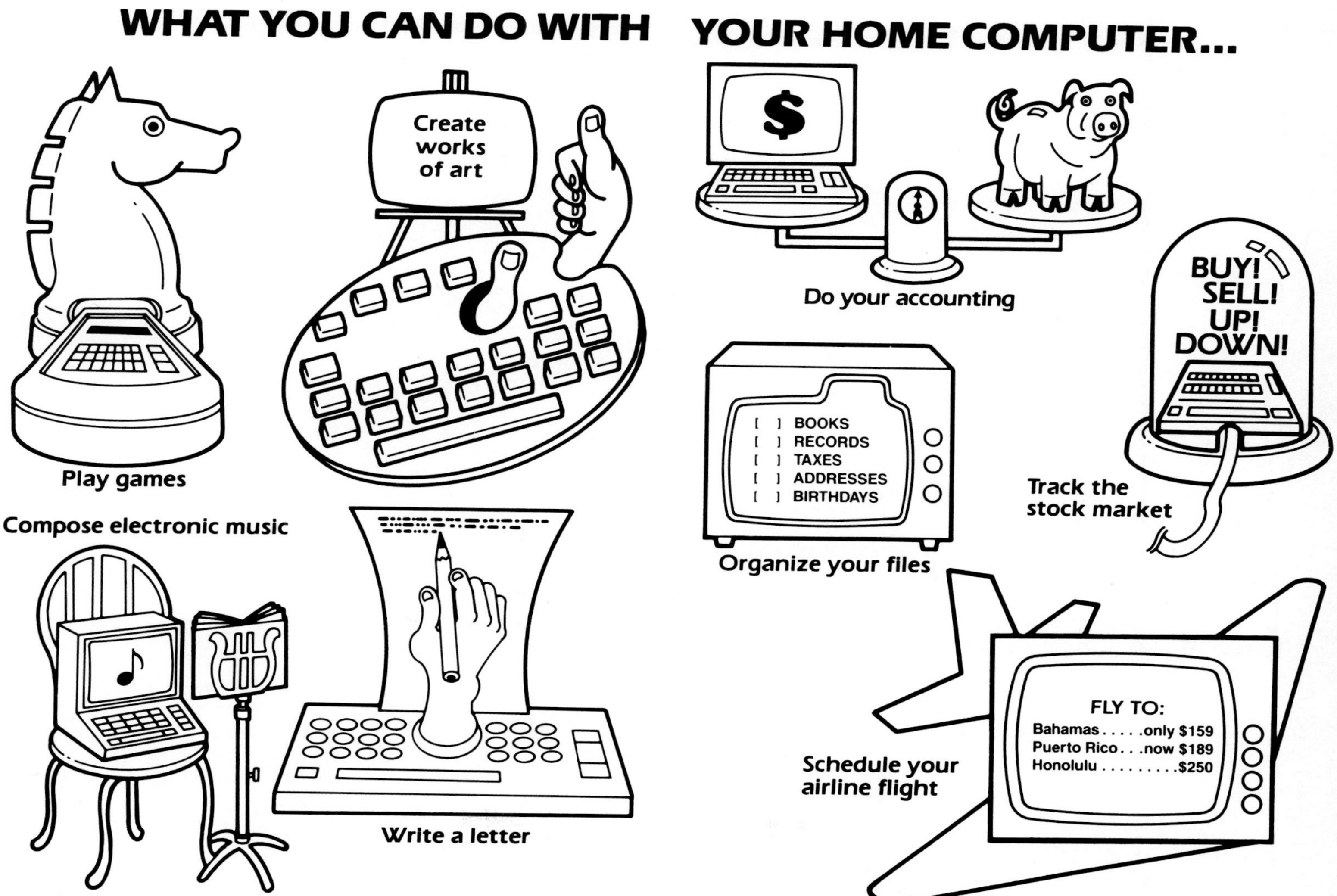

WHAT YOU CAN DO WITH YOUR HOME COMPUTER...

Create works of art

Play games

Compose electronic music

Write a letter

Do your accounting

Track the stock market

Organize your files

BOOKS
RECORDS
TAXES
ADDRESSES
BIRTHDAYS

FLY TO:
Bahamas.....only $159
Puerto Rico...now $189
Honolulu.........$250

Schedule your airline flight

BUY! SELL! UP! DOWN!

$

Figure 15.2
Example of a List in an Illustration Format

When and How to Use Tables

Consider a paragraph of text containing the following data:

> As of December 31, the Consolidated Electric Company had a total of 7,863 employees. Of these, 1,989 (or 25 percent) had less than five years' service; 1,590 (or 20 percent) had five to ten years' service; 1,275 (or 16 percent) had ten to fifteen years' service; 784 (or 10 percent) had fifteen to twenty years' service; 931 (or 12 percent) had ten to thirty years' service; 1,294 (or 17 percent) of the employees had thirty or more years' service with the Consolidated Electric Company.

The data, while fairly brief, are rather formidable for a reader to digest. Not that the information is complex or difficult, but it is full of numbers arranged in horizontal lines of text, which makes the information difficult for the reader to assimilate and interpret. Let us rearrange this information by tabulating it. All of the data in the paragraph concern the Consolidated Electric Company employees' length of service. That subject, then, could become the title of the table. The rest of the information could fit neatly, accurately, and in easily readable form into three columns. The first column might have the heading, Years of Service; the second column, Number of Employees; the third column, Percentage of Total. These figures would be listed under Years of Service: under 5, 5 to 10, 10 to 15, 15 to 20, 20 to 30, 30 or more. Under Number of Employees would be listed these figures: 1,989, 1,590, 1,275, 784, 931, and 1,294. Under Percentage of Total, 25, 20, 16, 10, 12, and 17 would appear. At the bottom of the Years of Service column would be Total as of December 31; under Number of Employees would be 7,863; and under Percentage of Total, 100 percent. The result is shown here as Table 15.1.

The table permits the reader to see clearly at a glance all of the numerical information and permits her to interpret and generalize. She can see that, while numerically a fourth of the employees have less than five years of service, more than half have been with the company for more than ten to as many as thirty years. The table lends itself very easily to comprehension and permits the reader to compare the various data groupings.

Table 15.1
Consolidated Electric Employees' Length of Service

Years of Service	Number of Employees	Percentage of Total
Under 5	1,989	25
5 to 10	1,590	20
10 to 15	1,275	16
15 to 20	784	10
20 to 30	931	12
30 or more	1,294	17
Total as of December 31	7,863	100 percent

Tables offer a convenient means for presenting characteristics of things, processes, and concepts. Tables offer the most precise way to present experimental data in a compact arrangement of related facts, figures, and values in orderly sequence, usually in lines and columns for convenient reference.

Tables are boxed or framed within a page when the data are self-sufficient and self-explanatory. Where significance and meaning of a table are dependent upon explanatory material preceding and succeeding the information in the columns, the table is usually not boxed, as in the following example:

The average surface wind velocities and direction at the Seattle Weather Station during the swimming season (for 20-year and 38-year averages) are as follows:

	June	*July*	*August*	*September*
Average wind velocity m.p.h.	4.2	3.9	3.7	3.9
Predominant direction	S	N	N	S

Thus, in summer, with light northerly or southerly winds, we could expect that the water in the swimming areas would be exchanged by surrounding lake water in from two to four hours.

Tables such as the one immediately preceding do not have numbers and titles. The amount of data they present is limited. They are called **dependent tables**.

Independent tables are those that are self-sufficient and have numbers and titles that often answer the question Who, What, Where, When, and How. For example:

TABLE XXI

U.S. Population Change by Size of County, 1985 to 1995

TABLE 102

Mortality From Leading Types of Accidents
in Canada by Sex and Age
1985–1995

TABLE V

Laboratory Z Analytical Data of Three Preparations
of Heparin Isolated from Rat Skin

Independent tables within a paper or report are numbered consecutively. The title states where the material was obtained and what it means. The number and title of the table are placed above the table. Subtitles and/or headnotes are frequently used. **Headnotes** give additional details such as condition under which data were recorded, limitations of the data, accuracy of source, and other explanatory matters. For example, the title of a table might be as follows: Absences during 1995 of Consolidated Electric Company Employees. The headnote under the title, in parentheses, might be (Figures represent work days lost). The first column at the left of a table is known as the stub. It serves to identify the horizontal line of data. The column headings to the right sometimes have a second tier of heads for subclassification. The table itself should be arranged as simply as possible. Footnotes should be used wherever necessary for explanation to avoid complicating the title or mixing the words and figures in the table itself. It is usually desirable to use symbols (asterisks and daggers, for example) for footnote references rather than numerals, so that the reader will not confuse the references with the tabular data. Double lines, or heavier lines, are used in the table to indicate division. Totals are placed at the bottom or at the right or at the top and left. Blank spaces within the table should be avoided. Either writing *0* in the space or indicating by a dash that no data were taken for that item is preferable. Tables, like any other visual aid, should be integrated with the text matter. The text discussion should consider the table, even though the table's information may be independent of the text (Figure 15.3).

Checklist for Constructing Tables

1. When four or more items of statistical information or data are to be presented, the material will be clearer in tabular form.
2. Quantitative, descriptive, and comparative data are more readily comprehensible in table form.
3. The data of a table should be crystallized into a logical unit. Extraneous data should be excluded; the table should be self-explanatory. Though self-contained, the table should be integrated with text matter for fuller explanation and interpretation.
4. The table should have both a number and title. Tables may be numbered in Roman numerals or in Arabic. The title should be concise, yet clearly identify the contents. A subtitle may be used for providing precise details.
5. Each vertical column and, as necessary, each horizontal line should have an identifying head.
6. Standard terms, symbols, and abbreviations should be used for all unit descriptions. The same unit system of measurement should be used for

Figure 15.3
Elements of an Independent Table

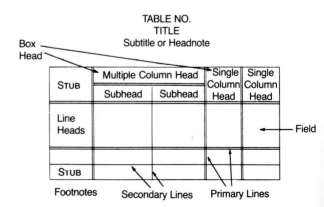

comparable properties or dimensions; for example, in linear measurements, feet and meters should not be intermingled.

7. If all the numbers in the table are measurements in the same units, then the unit is stated in the title.

8. Data to be compared should be placed in a horizontal plane.

9. If headings are not self-explanatory, footnotes should be used. If an item is repeated several times in a table, it should be removed from the data in the table and placed in the title, in a footnote, or in a column or line head.

10. Footnotes should be numbered or identified in sequence, line by line, from left to right, across the table.

11. Figures in columns are aligned to similar digits—ordinarily the right digit. However, when the data set up in a column are composed of different units, they should be centered in the column or aligned on the left. For example, in a table providing data of performance characteristics of an instrument, succeeding line heads would call for different units of data, as shown in Table 15.2.

Table 15.2
Data Units for Measuring Instrument
Performance Characteristics

Peak Volts	+10 kV
Peak Current	200 amps
Output Impedance	50 ohms
Risetime (10–90%)	<10 nS
Fall-time	150 nS
Throughput Delay	55 nS
Internal Repetition Rate	1, 5, 10 Hz

12. Fractions should be expressed in decimals. Decimal points are aligned in a column. When the first number of a column is wholly a decimal, a cipher is added to the left of the decimal point, e.g., 0.192.

13. Column or line headings should be used to group related data.

14. Tables containing similar data should be set up in the same manner within a given report.

15. Whenever possible, a table should be designed and structured so that it can be typed on one page. If the data cannot be made to fit one page, a continuation page should be used. The word *continued* should be placed at the bottom of the first page, as well as at the top of the second page, to indicate that the table has not been completed. Column heads must also be shown on the second page. Subtotals, when appropriate, should be shown at the bottom of the first page and at the top of the second page. Subtotals should always be clearly identified as such.

16. Only significant or summary tables should be placed in the main body of a paper or report. Supporting tables or those of record interest are placed in the appendix.

Examples of Types of Tables

Tables 15.3–15.6 and Figure 15.4 (pp. 400–405) are examples of five types of tables. Table 15.3 illustrates a table displaying experimental data; Table 15.4, statistical data; Table 15.5, comparison data; and Table 15.6 shows a process in table form; Figure 15.4 uses graphics to illustrate its data.

Use and Types of Photographs

Photographs offer not only realistic and accurate representations but also dramatic and artistic effects. Effective photographic illustrations require thoughtful planning, so that all desired details are shown at the most favorable angle. Poor photographs have such faults as cluttered backgrounds, inadequate lighting, poor camera angle, and improper focus. Good photographs are sharply detailed and do not contain distortions of any essential elements. Proper lighting eliminates or reduces strong shadows, which destroy fine details.

Table 15.3
Results of Colorado State University Laboratory Testing of Potential Sealants for the Coachella Canal

Sample No.	Material	Grit Content %	Colloidal Yield %	Wall Building Filter Loss (cc)	Cake (in)	Viscosity (centipoises)
S1-1	Coyote Well	1.3	53.5*	40	3/32	3
S1-2	Ackins Claim	12.1	42.9*	189	8/32	2
S1-3	Thermo Claim	2.5	48.9*	88.5	1/8	2
S1-4	Burslem Claim	20.7	28.9	69	3/16	2
S1-5	Armaseal	4.2	65.2	41	1/16	<4
S1-6	Maas Clay	5.7	60.1	38	1/16	1
S1-7	Western Clay (Utah)	17.5	55.5	28.5	1/16	6
S1-8	Western Clay (Utah) reserves	5.0	41.3	33.3	1/8	3
S1-9	Bent. Corp. (Utah)	4.1	84.6	14.5	5/64	8
S1-10	Baroid (Wyo) crushed	4.8	89.4	16.5	1/8	22
S1-11	Baroid (Wyo) 200 mesh	2.9	88.2*	16	3/32	23

*Dispersant (sodium tripolyphosphate—0.75 gms) added where tendency for flocculation noted.

Table 15.4
Bathing Beach Bacteriological Data* — King County, June 25–Sept 2, 19___

	Green Lake		Lake Washington		Puget Sound Golden Gardens	Lake Sammamish State Park
	West Beach	East Beach	Seward Park	Juanita		
No. of values	113	105	112	113	112	113
Median value +	23	230	230	60	230	60
Range of values	0 to 2400	0 to 24,000	0 to 13,000	0 to 7000	0 to 7000	0 to 2400
Avg. no. bathers	130	104	110	110	43	114
Avg. water temp.	70.3	71.1	69.0	70.5	58.2	71.9
Coliform per bather‡	0.18	2.2	2.1	0.54	5.4	0.53

*Most probable number of coliform bacteria per 100 ml sample, computed from Seattle-King Co. Health Dept. data
+Less than 45 proportioned equally among 0, 15, 30, and 45.
‡Coliform per bathers = $\dfrac{\text{Median}}{\text{Avg. no. bathers}}$

Table 15.5
Comparison of Common Nondestructive Evaluation Methods

Method	Characteristics Detected	Advantages	Limitations	Sample
Ulltrasonics	Changes in acoustic imped-ance caused by cracks, non-bonds, inclusions, or inter-faces.	Can penetrate thick materials; excellent for crack detection; can be automated.	Requires coupling to material either by contact to surface or immersion in a fluid such as water.	Adhesive as-semblies for bond integrity.
Radiography	Changes in density from voids, inclusions, material variations; placement of inter-nal parts.	Can be used to inspect wide range of materials and thick-nesses; versatile; film provides record of inspection.	Radiation safety requires pre-cautions; expensive; detection of cracks can be difficult.	Pipeline welds for penetra-tion, voids.
Visual-Optical	Surface characteristics such as finish, scratches, cracks, or color; strain in transparent materials.	Often convenient; can be auto-mated.	Can be applied only to sur-faces, through surface open-ings, or to transparent mate-rial.	Paper for sur-face finish.
Eddy Currents	Changes in electrical conduc-tivity caused by material vari-ations, cracks, voids, or inclu-sions.	Readily automated; moderate cost.	Limited to electrically conduct-ing materials; limited penetra-tion depth.	Heat exchange tubes for wall thinning and cracks.
Liquid Pene-trant	Surface openings due to cracks, porosity, seams, or folds.	Inexpensive, easy to use, readily portable, sensitive to small surface flaws.	Flaw must be open to surface. Not useful on porous mate-rials.	Turbine blades for surface cracks or po-rosity.
Magnetic Par-ticles	Leakage magnetic flux caused by surface or near-surface cracks, voids, inclusions, ma-terial or geometry changes.	Inexpensive, sensitive both to surface and near-surface flaws.	Limited to ferro-magnetic ma-terial; surface preparation and post-inspection demagnetiza-tion may be required.	Railroad wheels for cracks.

Table 15.6
A Process in Table Form

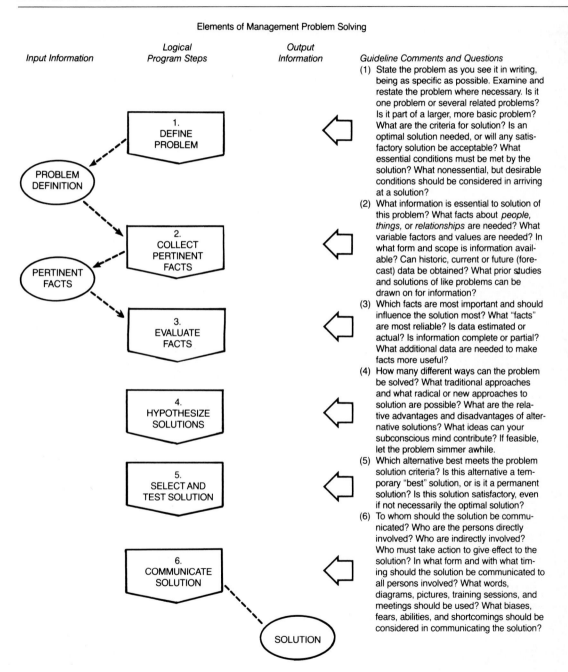

Elements of Management Problem Solving

Input Information	Logical Program Steps	Output Information	Guideline Comments and Questions

(1) State the problem as you see it in writing, being as specific as possible. Examine and restate the problem where necessary. Is it one problem or several related problems? Is it part of a larger, more basic problem? What are the criteria for solution? Is an optimal solution needed, or will any satisfactory solution be acceptable? What essential conditions must be met by the solution? What nonessential, but desirable conditions should be considered in arriving at a solution?

(2) What information is essential to solution of this problem? What facts about *people, things,* or *relationships* are needed? What variable factors and values are needed? In what form and scope is information available? Can historic, current or future (forecast) data be obtained? What prior studies and solutions of like problems can be drawn on for information?

(3) Which facts are most important and should influence the solution most? What "facts" are most reliable? Is data estimated or actual? Is information complete or partial? What additional data are needed to make facts more useful?

(4) How many different ways can the problem be solved? What traditional approaches and what radical or new approaches to solution are possible? What are the relative advantages and disadvantages of alternative solutions? What ideas can your subconscious mind contribute? If feasible, let the problem simmer awhile.

(5) Which alternative best meets the problem solution criteria? Is this alternative a temporary "best" solution, or is it a permanent solution? Is this solution satisfactory, even if not necessarily the optimal solution?

(6) To whom should the solution be communicated? Who are the persons directly involved? Who are indirectly involved? Who must take action to give effect to the solution? In what form and with what timing should the solution be communicated to all persons involved? What words, diagrams, pictures, training sessions, and meetings should be used? What biases, fears, abilities, and shortcomings should be considered in communicating the solution?

Diagram elements:
- PROBLEM DEFINITION → 1. DEFINE PROBLEM
- PERTINENT FACTS → 2. COLLECT PERTINENT FACTS
- 3. EVALUATE FACTS
- 4. HYPOTHESIZE SOLUTIONS
- 5. SELECT AND TEST SOLUTION
- 6. COMMUNICATE SOLUTION → SOLUTION

SOURCE: William N. McNairn, "Don't Blame Your Judgment on Your Genes," *Price Waterhouse Review*, vol. 22, no. 1, 1977, p. 31. [13].

THE MODERNIZED metric system

The International System of Units-SI

is a modernized version of the metric system established by international agreement. It provides a logical and interconnected framework for all measurements in science, industry, and commerce. Officially abbreviated SI, the system is built upon a foundation of seven base units, plus two supplementary units, which appear on this chart along with their definitions. All other SI units are derived from these units. Multiples and submultiples are expressed in a decimal system. Use of metric weights and measures was legalized in the United States in 1866, and since 1893 the yard and pound have been defined in terms of the meter and the kilogram. The base units for time, electric current, amount of substance, and luminous intensity are the same in both the customary and metric systems.

Symbol	When You Know	Multiply by	To Find	Symbol
in	inches	[A]25.4	[B]millimeters	mm
ft	feet	[A]0.3048	meters	m
yd	yards	[A]0.9144	meters	m
mi	miles	1.609 34	kilometers	km
yd²	square yards	0.836 127	square meters	m²
	acres	0.404 686	[C]hectares	ha
yd³	cubic yards	0.764 555	cubic meters	m³
qt	quarts (lq)	0.946 353	[D]liters	l
oz	ounces (avdp)	28.349 5	grams	g
lb	pounds (avdp)	0.453 592	kilograms	kg
°F	Fahrenheit temperature	5/9 (after subtracting 32)	Celsius temperature	°C
mm	millimeters	0.039 370 1	inches	in
m	meters	3.280 84	feet	ft
m	meters	1.093 61	yards	yd
km	kilometers	0.621 371	miles	mi
m²	square meters	1.195 99	square yards	yd²
ha	hectares	2.471 05	acres	
m³	cubic meters	1.307 95	cubic yards	yd³
l	liters	1.056 69	quarts (lq)	qt
g	grams	0.035 274 0	ounces (avdp)	oz
kg	kilograms	2.204 62	pounds (avdp)	lb
°C	Celsius temperature	9/5 (then add 32)	Fahrenheit temperature	°F

[A]exact

[B]for example, 1 in = 25.4 mm, so 3 inches would be (3 in) (25.4 $\frac{mm}{in}$) = 76.2 mm

[C]hectare is a common name for 10 000 square meters

[D]liter is a common name for fluid volume of 0.001 cubic meter

Note: Most symbols are written with lower case letters; exceptions are units named after persons for which the symbols are capitalized. Periods are not used with any symbols.

Multiples and Submultiples	Prefixes	Symbols
1 000 000 000 000 = 10^{12}	tera (ter a)	T
1 000 000 000 = 10^{9}	giga (ji ga)	G
1 000 000 = 10^{6}	mega (meg a)	M
1 000 = 10^{3}	kilo (kil o)	k
100 = 10^{2}	hecto (hek to)	h
10 = 10^{1}	deka (dek a)	da
Base Unit 1 = 10^{0}		
0.1 = 10^{-1}	deci (des i)	d
0.01 = 10^{-2}	centi (sen ti)	c
0.001 = 10^{-3}	milli (mil i)	m
0.000 001 = 10^{-6}	micro (mi kro)	µ
0.000 000 001 = 10^{-9}	nano (nan o)	n
0.000 000 000 001 = 10^{-12}	pico (pe ko)	p
0.000 000 000 000 001 = 10^{-15}	femto (fem to)	f
0.000 000 000 000 000 001 = 10^{-18}	atto (at to)	a

SEVEN BASE UNITS

meter-m
LENGTH

The meter (common international spelling, metre) is defined as 1 650 763.73 wavelengths in vacuum of the orange-red line of the spectrum of krypton-86.

1 METER — 1 650 763.73 WAVELENGTHS — ONE WAVELENGTH — "Kr ATOM

An interferometer is used to measure length by means of light waves.

The SI unit of area is the **square meter** (m²).

The SI unit of volume is the **cubic meter** (m³). The liter (0.001 cubic meter), although not an SI unit, is commonly used to measure fluid volume.

kilogram-kg
MASS

The standard for the unit of mass, the kilogram, is a cylinder of platinum-iridium alloy kept by the International Bureau of Weights and Measures at Paris. A duplicate in the custody of the National Bureau of Standards serves as the mass standard for the United States. This is the only base unit still defined by an artifact.

U.S. PROTOTYPE KILOGRAM NO. 20

The SI unit of force is the **newton (N)**. One newton is the force which, when applied to a 1 kilogram mass, will give the kilogram mass an acceleration of 1 (meter per second) per second.

$$1 N = 1 kg \cdot m/s^{2}$$

1 kg — 1N — ACCELERATION of 1m/s²

second-s
TIME

The second is defined as the duration of 9 192 631 770 cycles of the radiation associated with a specified transition of the cesium-133 atom. It is realized by tuning an oscillator to the resonance frequency of cesium-133 atoms as they pass through a system of magnets and a resonant cavity into a detector.

Schematic diagram of an atomic beam spectrometer or "clock." Only those atoms whose magnetic moments are "flipped" in the transition region reach the detector. When 9 192 631 770 oscillations have occurred, the clock indicates one second has passed.

CESIUM SOURCE — DEFLECTION MAGNET — TRANSITION REGION (CAVITY) OSCILLATING FIELD — DETECTOR — DEFLECTION MAGNET — OSCILLATOR — NBS ATOMIC TIME SCALE SYSTEM

The number of periods or cycles per second is called frequency. The SI unit for frequency is the **hertz (Hz)**. One hertz equals one cycle per second.

The SI unit for speed is the **meter per second** (m/s).

The SI unit for acceleration is the **(meter per second) per second** (m/s²).

The SI unit for pressure is the **pascal (Pa).**

$$1 Pa = 1N/m^{2}$$

The SI unit for work and energy of any kind is the **joule (J).**

$$1 J = 1 N \cdot m$$

The SI unit for power of any kind is the **watt (W).**

$$1 W = 1 J/s$$

Standard frequencies and correct time are broadcast from WWV, WWVB, and WWVH, and stations of the U.S. Navy. Many short-wave receivers pick up WWV and WWVH, on frequencies of 2.5, 5, 10, 15, and 20 megahertz.

Figure 15.4
Figure Using Graphics to Illustrate Its Data

ampere -A
ELECTRIC CURRENT

The ampere is defined as that current which, if maintained in each of two long parallel wires separated by one meter in free space, would produce a force between the two wires (due to their magnetic fields) of 2×10^{-7} newton for each meter of length.

FORCE 2×10^{-7}N — 1m — 1m — 1A — 1A

The SI unit of voltage is the **volt** (V).
$$1V = 1W/A$$

The SI unit of electric resistance is the **ohm** (Ω).
$$1\Omega = 1V/A$$

kelvin -K
TEMPERATURE

The kelvin is defined as the fraction 1/273.16 of the thermodynamic temperature of the triple point of water. The temperature 0 K is called "absolute zero."

On the commonly used Celsius temperature scale, water freezes at about 0°C and boils at about 100°C. The °C is defined as an interval of 1 K, and the Celsius temperature 0°C is defined as 273.15K.

The Fahrenheit degree is an interval of 5/9°C or 5/9K; the Fahrenheit scale uses 32°F as a temperature corresponding to 0°C.

The standard temperature at the triple point of water is provided by a special cell, an evacuated glass cylinder containing pure water. When the cell is cooled until a mantle of ice forms around the reentrant well, the temperature at the interface of solid, liquid, and vapor is 273.16K. Thermometers to be calibrated are placed in the reentrant well.

TEMPERATURE MEASUREMENT SYSTEMS

°F 212 / °C 100 / Water Boils
98.6 / 37 / Body Temperature
32 / 0 Water Freezes — 273.15
-40 / -40
2045 Platinum Freezes
Absolute Zero — 0
FAHRENHEIT °F — °C CELSIUS — KELVIN

THERMOMETER (ELECTRICAL RESISTANCE TYPE); WATER VAPOR; ICE; WATER; REENTRANT WELL; REFRIGERATING BATH; TRIPLE POINT CELL

mole -mol
AMOUNT OF SUBSTANCE

The mole is the amount of substance of a system that contains as many elementary entities as there are atoms in 0.012 kilogram of carbon 12.

When the mole is used, the elementary entities must be specified and may be atoms, molecules, ions, electrons, other particles, or specified groups of such particles.

The SI unit of concentration (of amount of substance) is the **mole per cubic meter** (mol/m³).

candela -cd
LUMINOUS INTENSITY

The candela is defined as the luminous intensity of 1/600 000 of a square meter of a blackbody at the temperature of freezing platinum (2045K).

CAVITY; FREEZING PLATINUM; INSULATING MATERIAL

The SI unit of light flux is the **lumen** (lm). A source having an intensity of 1 candela in all directions radiates a light flux of 4π lumens.

A 100-watt light bulb emits about 1700 lumens.

TWO SUPPLEMENTARY UNITS

radian -rad
PLANE ANGLE

The radian is the plane angle with its vertex at the center of a circle that is subtended by an arc equal in length to the radius.

ONE RADIAN

steradian -sr
SOLID ANGLE

The steradian is the solid angle with its vertex at the center of a sphere that is subtended by an area of the spherical surface equal to that of a square with sides equal in length to the radius.

Area r^2 — ONE STERADIAN

INCHES / CENTIMETERS / YARD / METER

Figure 15.4 (continued)

405

To give the reader an idea of equipment size or to lend human interest to a picture, include in the photograph, as is suitable, human beings who might be logically associated with the equipment. A woman in a bikini or in a sexy evening gown shown operating scientific equipment is not appropriate. The magnitude in size of the Department of Energy's Lawrence Berkeley Laboratory neutral beam injection apparatus is shown by the two men in Figure 15.5.[3] Sometimes smallness can best be illustrated by inclusion of a common article for comparison as is shown by the penny in the photograph of the experimental sprinkler heads in Figure 15.6.[4]

In addition to the techniques of the ordinary still camera, color photographs, aerial photographs, photographs taken by means of a Klydonograph

Figure 15.5
This neutral beam injection apparatus was developed by DOE's Lawrence Berkeley Laboratory in California for fueling and heating the Tokamak Fusion Test Reactor. This machine will produce high-energy deuterium ions for fusion experiments.

---

[3]Photo courtesy of the U.S. Energy Research & Development Administration. [20]
[4]Photo from "Estimating Lightning Performance," *Westinghouse Engineer*, January 1951, p. 33. [19]

(Figure 15.7),[5] X-ray photographs, photographs taken through an electron microscope (see Figure 15.9)[6] or through a telescope, and photo micrographs can lend impressive visual aid. The airbrush is frequently used on photographs to provide highlights and emphasis, to aid in sharpness of detail, or to allow deemphasis of unimportant details. Callouts are frequently added to the photograph to help provide distinctive details and identification. **Callouts** are labels with leaders that point to individual items in an illustration; the labels identify or give information about the item. When there are many elements to be identified, numbers or letters replace the callouts. Each letter, then, is referenced to the identifying caption. The photo cutaway view of the lawnmower (Figure 15.8)[7] has callouts for identifying parts, as does Figure 15.9.

Drawings

The drawing is one of the oldest forms of symbolic representation. Writing and drawing were identical in prehistoric Egypt and in early Greece. The Greek word *graphein* means both writing and drawing. Drawings are made with pencil,

Figure 15.6
Scaling down the cost of residential sprinkler protection. These closeup views provide a contrast between FMRC's six prototype residential sprinklers and a standard commercial sprinkler.

---

[5]Photo courtesy of the Factory Mutual *Record*, May–June 1978, p. 10. [18]
[6]Electron micrographs courtesy of Harlan F. Weisman, M.D. and Michelle K. Leppo, Division of Cardiology, Johns Hopkins Medical Institutions, 1990. [22]
[7]Photo courtesy of Ken Cook Co. Technical illustrator: Marvin Van Den Heuvel. [12]

To those schooled in the art of lightning measurement, this interesting figure is a film record of the electric field intensity developed by a thundercloud. The measurement was made by a specially developed Klydonograph capable of recording voltages about three times those obtainable with conventional Klydonographs. In essence, the record is produced on a sheet of photographic film that is placed between a probe electrode and a grounded plate—the greater the field intensity, the larger the figure. In this example, the record shows that an electric field intensity of approximately 2500 volts per inch existed just prior to the lightning stroke.

Figure 15.7

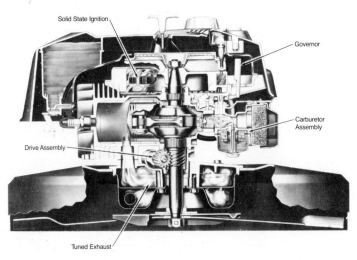

Figure 15.8
Callouts identify the lawnmower parts in this cutaway view.

Figure 15.9
Transmission electron micrograph of normal control canine heart tissue. The nucleus (N) shows the fine, evenly distributed granular appearance of chromatin. The contractile units or sarcomeres are separated by mitochondria (m). Original magnification = 8,000 times.

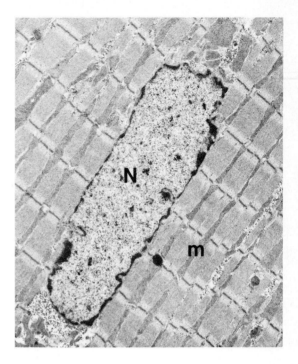

pen, crayon, brush, or computer software. There are many kinds, from a simple freehand sketch to detailed engineering and architectural drawings with elaborate minutiae of detail. Drawings offer more flexibility than photographs for showing the inner movements of equipment, cross-sections, and relationships. The more complex the subject matter of a drawing is, the greater the necessity for identifying callouts. Letter symbols with keys to identify the symbols are required.

The explanation of the process of genetic inheritance offered by the text in Figure 15.10 is made clearer and vividly lucid by the accompanying drawings.[8] The three examples in Figure 15.11 show how drawings or diagrams may be integrated with text to help explain technical matter.[9] Here they illustrate the design of the handcrank, its components, and the way they work. These drawings permit cutaway and cross-sectional views of schematics, which clarify the explanation in the text.

---

[8]Figure 15.10 is from the National Institute of General Medical Sciences, *What Are the Facts about Genetic Disease?* Washington D.C.: U.S. Government Printing Office, n.d. [15]

[9]Figure 15.11 is from *Weapons Systems Fundamentals, Basic Weapons Systems Components* Washington D.C.: U.S. Government Printing Office, July 15, 1960. [21]

How Dominant Inheritance Works

One affected parent has a single faulty gene (D) which dominates its normal counterpart (d).

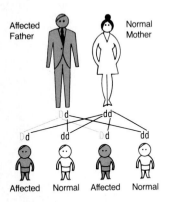

Affected Father — Normal Mother

Affected   Normal   Affected   Normal

Each child's chance of inheriting either the D or the d from the affected parent is 50%.

How X-Linked Inheritance Works

In the most common form, the female sex chromosome of an unaffected mother carries one faulty gene (X) and one normal one (x). The father has normal male x and y chromosome complement.

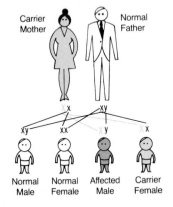

Carrier Mother — Normal Father

Normal Male   Normal Female   Affected Male   Carrier Female

The odds for each male child are 50/50:
1. 50% risk of inheriting the faulty X and the disorder
2. 50% chance of inheriting normal x and y chromosomes

For each female child, the odds are:
1. 50% risk of inheriting one faulty X, to be a carrier like mother
2. 50% chance of inheriting no faulty gene

How Recessive Inheritance Works

Both parents, usually unaffected, carry a normal gene (N) which takes precedence over its faulty recessive counterpart (n).

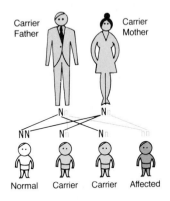

Carrier Father — Carrier Mother

Normal   Carrier   Carrier   Affected

The odds for each child are:
1. a 25% risk of inheriting a "double dose" of n genes which may cause a serious birth defect
2. a 25% chance of inheriting two Ns, thus being free of the recessive gene
3. a 50% chance of being a carrier like both parents

Figure 15.10
The Process of Genetic Inheritance

Use of Diagrams

Diagrams are a subclassification of drawings. Diagrams are representations of abstractions. They are symbolic configurations. Sometimes they are charts or graphs explaining or illustrating ideas or statistics. Some texts use the words *drawing* and *diagram* interchangeably. Drawings attempt to represent the likeness of their subject; diagrams attempt to show the operation of the subject. Diagrams present an analysis by means of a symbolic or conventional representation of an actuality. The verbal equivalent of a diagram is an outline. Therefore, the diagram outlines or sketches, rather than represents an actuality. But diagrams help to clarify, exemplify, delineate, analyze, emphasize, or summarize. Types are numerous and are limited only by the creativity of the technical writer and his or her illustrator. Among the most frequently used diagrams are these:

ADJUSTMENT DEVICES

FRICTION RELIEF HANDCRANK
The handcrank can be provided with an adjustable friction drive, by using a cup spring (or a helical spring) to apply pressure upon a gear bearing against a wood or composition disk. Adjustment of a clamp nut provides the means of varying the pressure to obtain the degree of friction required. If pressure becomes greater than the friction imposed upon it, the gear will slip and protect associated gearing from strain or damage.

ADJUSTABLE HOLDING FRICTION
The same handcrank may be provided with cork disks, a collar, and a bushing. This assembly puts a drag on the handcrank, keeps it positioned, and prevents motion from backing out through the handcrank.

POSITIONING PLUNGER
We can carry the design of the handcrank still further and add a plunger for the purpose of holding the shaft in either of two positions: an IN position and OUT position. In changing position, the shaft and the drive gear move in relation to the adapter housing. The plunger is pulled out and the handcrank pushed or pulled to its new position. When released, the plunger is returned by a spring and enters a hole in the bushing, locking the assembly in a particular position.

Moving the handcrank to the in or out position will cause it to engage or become disengaged, or this arrangement can be used to drive one or the other mechanism. By using a wide gear, this drive can be kept in engagement all the time, the in and out position being to control the drive of another gear. Thus, it is possible to drive one gear at all times, alone, or in conjunction with another. Further, we would include a switch actuated by the in or out position of the shaft.

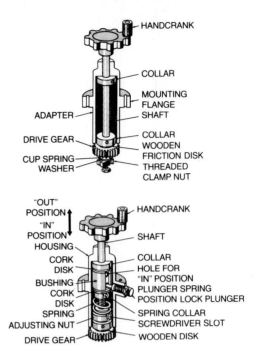

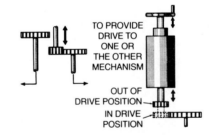

Figure 15.11
The excellent integration of text with illustrations enhances reader comprehension in this cross-sectional view.

1. *Schematic diagrams*
   a. One type is a cutaway (cross-sectional view) drawing of, for example, the interior of a device, organism, or geologic formation (Figures 15.12 and 15.13)[10,11]

---

[10]Figure 15.12 is courtesy of National Institute of Arthritis, Metabolism, and Digestive Diseases, N.I.H. [5]

[11]Figure 15.13 is from *Enhanced Recovery of Oil and Gas*. Washington, D.C.: U.S. Department of Energy, 1977, p. 10. [7]

Figure 15.12
A Schematic Diagram of the Digestive Tract

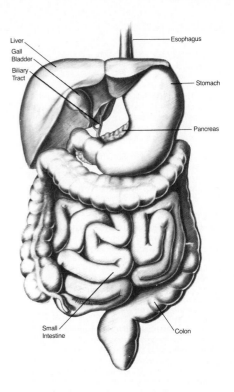

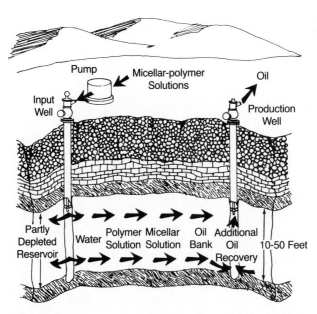

Figure 15.13
This cutaway diagram of a geologic formation containing oil illustrates the micellar-polymer process for recovering oil from "spent" wells.

b. The second type is a graphic representation in which symbols and lines are used to depict various components and their connections. This type of schematic is the equivalent of the blueprint. Of this second type of schematic diagram there are two subtypes:

(1) *Complete:* This type shows the arrangements of the components of an entire equipment or a major unit.

(2) *Stage:* This type is sometimes called a partial or functional schematic. It is a functional segment of usually one stage of a complete schematic diagram (Figure 15.14).

2. *Layout diagrams*

   a. *Wiring diagrams:* These are graphic representations of the relative physical location of all component parts of an electronic equipment. This type of diagram identifies all wires or interconnections between components. Leads are identified by a colored code; cable numbers and type of wire used are charted. Graphic symbols are not used in wiring diagrams. Parts are shown either pictorially or by simple rectangular boxes. The layout of the components follows the actual chassis arrangement.

   b. *Engineering and Architectural:* These present designs of structures and mechanical systems, providing details with dimensions of construction.

3. *Pictorial schematic diagrams:* These visuals are graphic representations that show components of equipment or systems as they appear normally to the eye. They may be drawn orthographically, isometrically, or angularly, through projection or perspective. Schematic drawings often are used to depict the design of an apparatus or process (Figures 15.15 and 15.16).[12] There are pictorial schematic diagrams of complex systems. By use of pictographs—representational symbols—concepts and/or operations are represented in the pictorial scheme of the diagram (Figure 15.17, p. 416).[13]

Figure 15.14
The Basic dc Wheatstone Bridge Circuit, Commonly Used to Measure the Microwave Power in a Bolometer (An Example of a Stage or Partial Schematic Diagram)

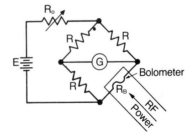

[12]Figure 15.15 courtesy of the Westinghouse Electric Corporation Research and Development Center. [11] Figure 15.16 from *Aerospace Science: The Science of Flight.* Maxwell Air Force Base, AL: Air Force Curriculum Division, 1988. [1]

[13]Figure 15.17 courtesy of the National Science Foundation, from "New Trails to Chemical Productivity," *New Frontiers of Science*, 1977. [16]

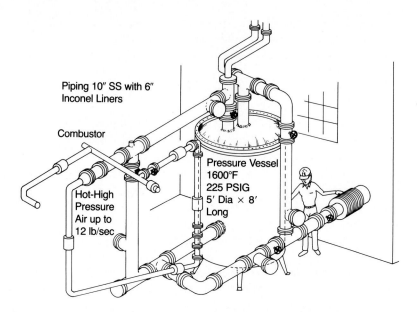

Piping 10″ SS with 6″
Inconel Liners

Combustor

Pressure Vessel
1600°F
225 PSIG
5′ Dia × 8′
Long

Hot-High
Pressure
Air up to
12 lb/sec

Figure 15.15
A pictorial diagram, such as this dimetric projection of the Westinghouse high temperature and pressure particulate removal test facility, imparts more information and helps the reader visualize the actuality of an apparatus better than does a block diagram or a straight line orthographic plan view.

4. *Block diagrams:* These diagrams show the function of an entire unit. All stages or functions are represented by rectangular boxes or similar symbols and are arranged in the order of activity flow. Each block is labeled clearly. Notes are frequently used to aid the explanation of the diagram (Figure 15.18 and Figure 15.19, pp. 417–418).[14]

Use of Cartoons

Cartoons can be appropriate and effective in technical presentations. They draw the reader's attention, hold her interest, and can provide relief from solid text

---

[14]Figure 15.18 from AGI/NAGT. *Laboratory Manual In Physical Geology*, 2d Ed. Columbus, OH: Merrill, 1990, p. 26. [2] Figure 15.19 from *Dimensions* Washington D.C.: National Bureau of Standards, April 1978. [6]

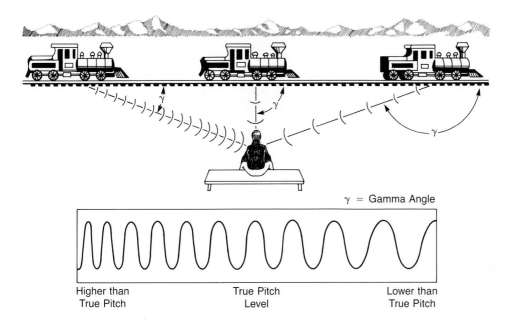

γ = Gamma Angle

Higher than True Pitch   True Pitch Level   Lower than True Pitch

The operation of the Doppler effect. The Doppler process is named after Christian Johan Doppler, a German mathematician who discovered the principle. The Doppler effect can be observed by listening to the whistle of a passing train. As the train approaches, its whistle, as heard by a stationary observer, has a fairly steady pitch which is higher than true pitch. The speed of the train is being added to the speed of the sound. When the train becomes parallel with the observer, the pitch drops quickly to a frequency below the true pitch and remains at the lower frequency as the train moves away from the observer; the train speed is being subtracted from the speed of sound.

Figure 15.16
A Pictorial Explaining a Process [1]

material. Notice how the flow chart of flat glass manufacturing (Figure 15.20, p. 418) is enlivened by the cartoon characters.[15]

Cartoons are excellent devices for emphasis and for attracting attention to important or highly technical matter. Young readers or users at lower reading skill levels are better reached with the use of cartoons. Note in Figure 15.21 (p. 419), a sequence from an army manual directed to soldiers with no more than a high school education, how the cartoon aids the reader to learn to measure the distance between two points on a curved, winding road.[16]

---

[15]Figure 15.20 from *Westinghouse Engineer*, March 1960, pp. 38–39. [23]

[16]Figure 15.21 from Human Resources Research Organization, *Guidebook for the Development of Army Training Literature.* Arlington, VA.: U.S. Army Research Institute, November 1975. [11].

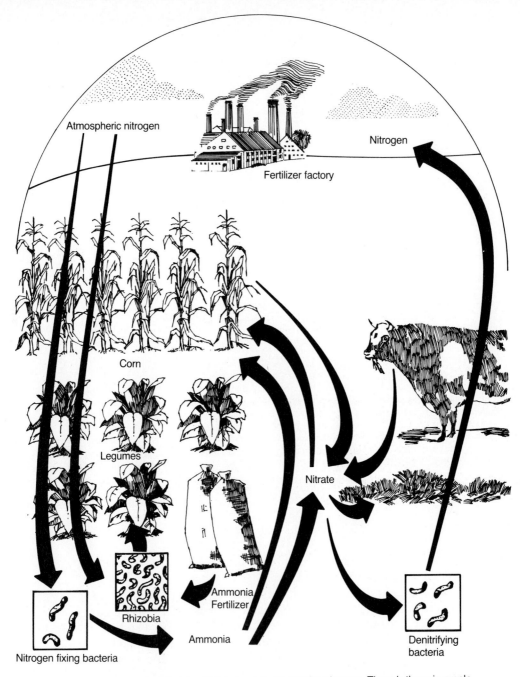

Atmospheric nitrogen

Nitrogen

Fertilizer factory

Corn

Legumes

Rhizobia

Nitrate

Ammonia
Fertilizer

Ammonia

Nitrogen fixing bacteria

Denitrifying
bacteria

Nitrogen is a key component of protein. All living creatures require nitrogen. Though there is ample nitrogen in the atmosphere, all living creatures must ingest it from the foods they eat. Bacteria at the roots of some plants assist in fixing atmospheric nitrogen. Man has helped the process by manufacturing synthetic nitrogen fertilizer.

Figure 15.17
A Schematic Pictorial Drawing of a Complex System

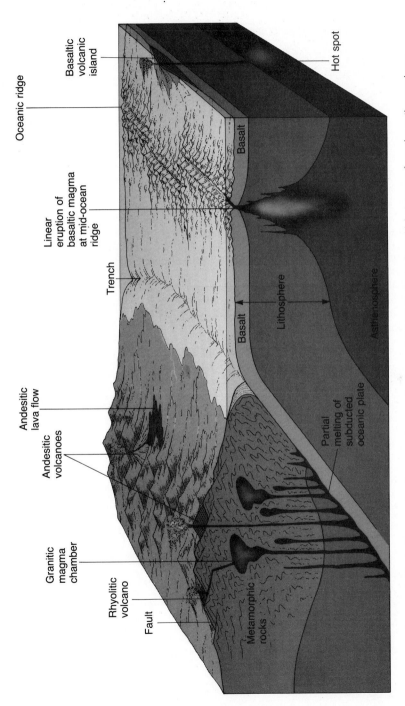

Oceanic ridge

Basaltic volcanic island

Hot spot

Linear eruption of basaltic magma at mid-ocean ridge

Trench

Basalt

Basalt

Lithosphere

Asthenosphere

Andesitic lava flow

Andesitic volcanoes

Partial melting of subducted oceanic plate

Granitic magma chamber

Rhyolitic volcano

Fault

Metamorphic rocks

Schematic block diagram showing some igneous features of the Earth. Mafic (basaltic) magmas from the asthenosphere commonly erupt to the seafloor along mid-ocean ridges, and above isolated *hot spots* such as Hawaii. Basaltic eruptions at mid-ocean ridges force the older basalt (rock) to move laterally into trenches, where it subsides beneath the continent. Partial melting of the subducted basalt commonly produces intermediate (dioritic) to felsic (granitic) magmas, plus andesitic and rhyolitic eruptions. (Mount Saint Helens in Washington State is such an andesitic volcano.) This model of *seafloor spreading* may explain the phenomena known as *continental drift* and *plate tectonics*.

Figure 15.18
Schematic Block Diagram

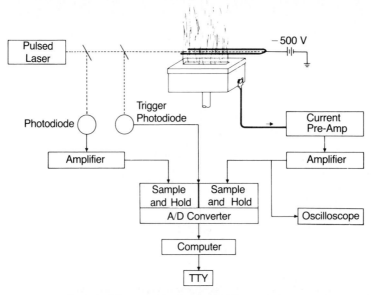

Analytical apparatus for trace metal detection in water utilizing the optogalvanic effect. A Commercial dye laser is directed into an analytical burner normally used in atomic absorption work. Electrodes are placed on either side of the flame and held at a potential of −500 volts. The burner head is allowed to float above ground potential and serves as the other electrode. When metal species in the flame are irradiated with laser light at a frequency corresponding to an atomic transition, increased thermal ionization results. This yields a measurable current change which is recorded. The current is proportional to the concentration of metal atoms present in the flame. With this apparatus, several metallic species have been detected at the sub parts-per-billion by mass level in water.

Figure 15.19
Block Diagram

Figure 15.20
Simplified "Flow Diagram" of Flat Glass Manufacture

418

Figure 15.21
Emphasizing Information for Lower Reading Levels

The Exploded View

Exploded views can be either photographs or drawings of a device or equipment in which the parts are shown in a disassembled state but arranged to simulate a perspective view (see Figure 15.22).[17] The parts are arranged in sequence in their respective axis of assembly. Each part is called out by name or by a reference designation.

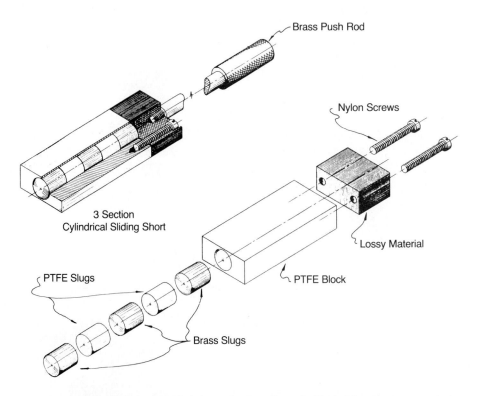

Exploded and assembled views of a 3-section cylindrical sliding-short, a new design which permits greater ease in making precision phase-shift measurements at microwave frequencies. PTFE block prevents brass slugs from contacting inside wall of waveguide and provides a stable mechanism to slide within the waveguide. Brass slugs maintain a light press fit within PTFE block; PTFE slugs maintain optimum spacing between brass slugs.

Figure 15.22
Exploded View

[17]Figure 15.22 from *National Bureau of Standards Technical News Bulletin*, January 1971, p. 17. [14]

Maps

The map is a symbolic, conventionalized representation of reality. Maps are most frequently used to show geographic or spatial distributions or relationships, representing areas of land, sea, or sky. Among the types used in technical writing are contour, profile, historical, linguistic, political, or demographic.

Maps make use of color and shading, as well as conventionalized symbols, to depict details of information and to indicate relationships. The maps should include a key to help explain the symbols and to enable the reader to read or interpret the maps satisfactorily (Figure 15.23). Scales are included where geographical distances are important.

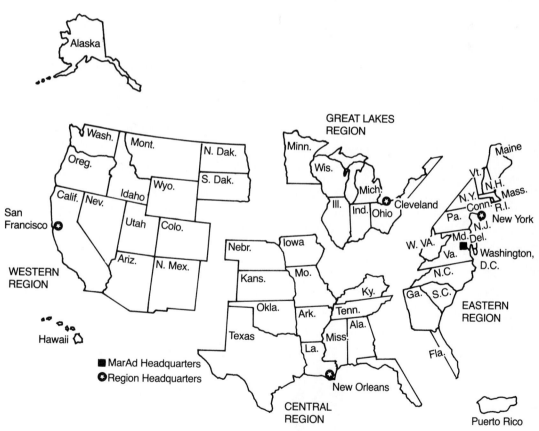

Figure 15.23
Map

Use of Graphs, Charts, and Curves

Graphs, *charts*, and *curves* are terms that are used interchangeably for the diagramming, mapping, and presenting of statistical information. These devices compare pictorially changes in value or interrelationships of variable quantities. Graphs, charts, and curves simplify statistical aspects of information and aid in their interpretation. Some types of visual aids within these groupings are the curve or line chart, the pie chart, the surface chart, the flow chart, the pictorial chart, the organization chart, the bar graph, and the line graph.

The line chart or graph is the simplest and most commonly used visual aid. Generally, the independent variable is plotted on the abscissa—the horizontal axis or scale—and the dependent variable is plotted on the ordinate—the vertical axis or scale. The most common independent variables include time, distance, voltage, stress, load; they are plotted on the horizontal or abscissa scale. Temperature, money, current, and strain—common dependent variables—are plotted on the vertical or ordinate scale.

Choosing the Proper Scale

Theoretically, each scale begins at 0. The zero is placed where the two scales intersect. However, this placement is not always possible. Occasionally, the abscissa or horizontal line needs to represent a variable where zero is meaningless. Examples of such a variable would be hours, months, or years during which an event occurred. The ordinate or vertical axis may also need to represent values wherein the zero is not appropriate—for example, where costs may begin at millions or more dollars. This problem is met by putting some value higher than zero where the scales intersect by putting a break in the scale. This is also frequently done by starting the scale at zero and then depicting the fracture that represents values between zero and the first value drawn to follow it on the scale. Draftsmen may or may not picture a broken scale with a jagged or wavy line. Care must be taken when broken grids are used because distortions of data can occur, as is shown in Figure 15.24.

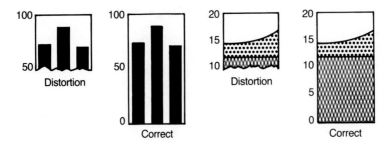

Figure 15.24

Further, choosing the proper scale values is important. Identical data presented on two graphs each of which has a different scale may give the viewer two entirely different impressions (Figure 15.25).

Types of Charts

Various types of charts are used to present information to the reader for convenient and easy comprehension:

1. *A line chart or line graph* is used to show comparative trends or values over a long period of time. Line graphs may be used in combination with bar charts or may be used to represent the profile curve of continuous or changing data. Examine the combination line chart and line graph in Figure 15.26.

2. The *bar chart* is used to show relative quantities by vertical or horizontal bars of varying lengths. Causal relationships cannot be shown by a bar chart. Multiple factors can be shown by bar charts by changing the appearance of the bars through crosshatching, utilizing different colors, and filling in the bars in different ways. A logarithmic horizontal bar graph is illustrated in Figure 15.27.

Figure 15.25
Effect of Choice of Scale

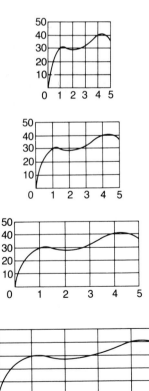

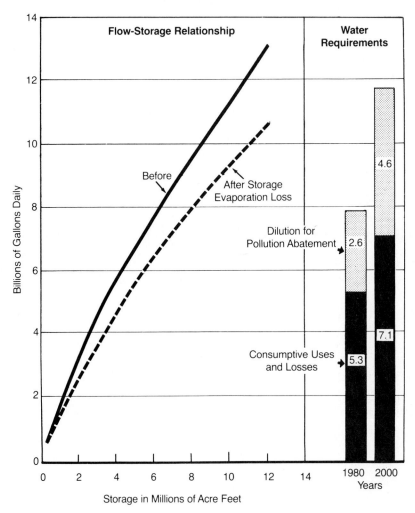

**UPPER ARKANSAS
RED RIVERS REGION**

Figure 15.26
Line/Bar Chart

3. A *pie chart* or circle graph is an excellent device to partition or classify a whole into various parts or elements. See the typical pie chart in Figure 15.28.
4. The *pictorial chart* is a visual aid in which units are represented by a picture symbol. Each symbol may stand for a stated number of units. A pictorial chart is illustrated in Figure 15.29.[18]

---

[18]Figure 15.29 from *Energy Conservation, Gas Heat Pumps: More Heat from Natural Gas*. Washington, D.C.: Office of Public Affairs, Energy Research and Development Administration, 1977. [8]

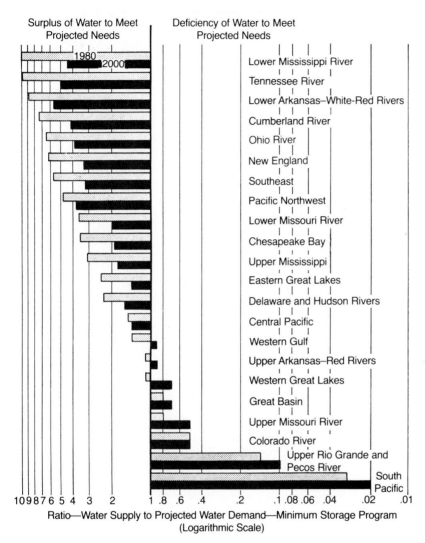

**Ratio of Maximum Obtainable Supply to
Minimum Flow Requirement
1980 and 2000**

Figure 15.27
A Bar Graph

5. A *surface chart* is a variation of the line chart. It can be used to show very clearly cumulative totals of two or more components; relative sizes of various components are dramatically, although not always accurately, shown through this charting device. The surface chart is made by shading or cross-

hatching the areas between index lines, so that the areas are differentiated. The areas at the bottom are shaded more darkly; the upper areas become progressively lighter. Surface charts are not practical when index lines cross each other. See the surface chart in Figure 15.30 for an example.[19]

6. The *organizational chart*, like the block diagram, is a directional chart. Blocks are arranged to simulate the sequence of the line of authority in an organization. Instead of functions of components, the titles of persons or the names

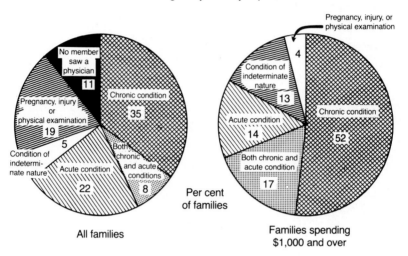

Figure 15.28
A Circle Graph or Pie Chart

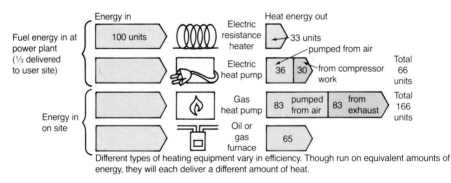

Figure 15.29
A Pictorial Chart

---

[19]Figure 15.30 from *Enhanced Recovery of Oil and Gas*. Washington, D.C.: U.S. Department of Energy, 1977. [7]

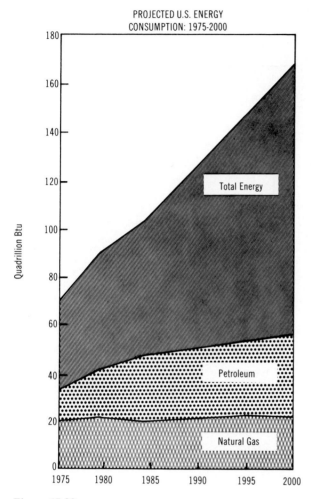

PROJECTED U.S. ENERGY CONSUMPTION: 1975-2000

Figure 15.30
Example of a Surface Chart. This graph projects the relative decline of oil and gas as United States sources of energy from about 75% in 1975 to 13% by the year 2000. During this period, the estimated growth in total energy consumption is expected to be sustained primarily by increasing reliance on coal and nuclear energy.

of departments are printed in the boxes. Reading is from the top to bottom. Equal functions are drawn from left to right. Direct relationships are indicated by solid lines connecting the boxes, indirect relationships by broken lines. A typical organizational chart is illustrated in Figure 15.31.[20]

---

[20]Figure 15.31 from Robert Colburn, *Fire Protection & Suppression*. New York: McGraw-Hill, 1976, p. 70. [3]

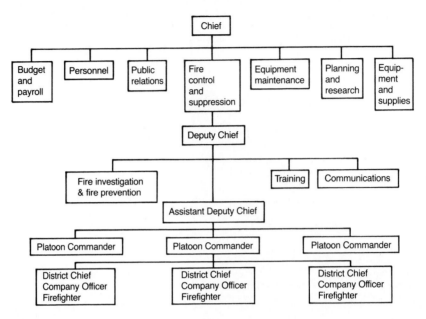

Figure 15.31
A Model Fire Department Line-and-Staff Organization

7. The *flow chart* traces pictorially the movement of a process from beginning to completion of an action, or from the raw material to a finished product. Processes may be pictured by simplified drawings or conventional symbols to represent the operations. Arrows are used to indicate flow of movement (refer again to the flow chart in Figure 15.20).

Computer Assisted Graphics

Most word processing programs have a line drawing feature that uses characters from the keyboard to draw lines, boxes, graphs, and simple illustrations such as flow charts and organization charts. Computer graphic programs are available to allow you to create, edit, display, and print a full spectrum of graphic images. These graphic program packages can range from those limited to drawing simple curves and bar charts to ones that create complex designs and pictures. *Paintbrush and Drawing Programs* allow the user to become an artist without paint, canvas, or brush. The creativity comes from the user, not the program. With input devices such as a mouse or light pen, the user sizes and shapes an image, draws and fills in shapes, and tries out colors and textures. *Design Programs* make drafting and design tasks easier and more efficient. Designers, using such programs, can create three-dimensional objects that can be rotated and viewed from every angle. Design programs are primarily for architects and engineers, but can benefit the efforts of the technical writer (Figure 15.32).

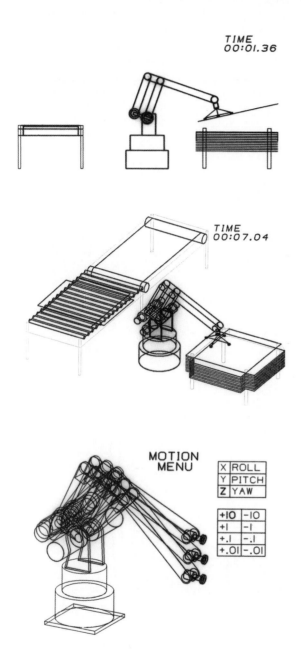

Figure 15.32
Computer-aided design software can be used to analyze and simulate how a robot will move in its factory-floor environment. The operator can check work-cell design elements, such as interference in the motion of the robot or "hand," with other work-cell components. (Reprinted with permission from Calma Co., a wholly-owned subsidiary of the General Electric Co., U.S.A.)

Business and Analysis Programs can transform numerical data into graphic representations—various types of charts and graphs (Figure 15.33).[21] Some of these programs are very user friendly. Simply by keying in your numerical data, you can convert them on screen with one or two strokes into sophisticated charts or graphs in several colors to provide clear analyses, comparisons, and insights.

Graphic Presentation Principles

In summary, the most effective graphic aids are those that serve the reader's needs, supplementing and clarifying the sense of the text. Illustrations for embellishment alone are a distraction. Poorly executed drawings or those illogically planned or placed confuse and frustrate the reader. Graphic aids should be functionally integrated into the discussion. Illustrations and tables should not be cluttered with excessive details or notations. They should bear proper identification by means of a table or figure number, title, and caption. Figures are numbered independently of tables and in sequence. Titles are identification labels; captions are brief but complete explanatory annotations.

Illustrations should be expertly rendered. Poor execution will distort the facts they are required to communicate. If you, the writer, are inexpert, make freehand sketches with ample explanatory notes and instructions for submission

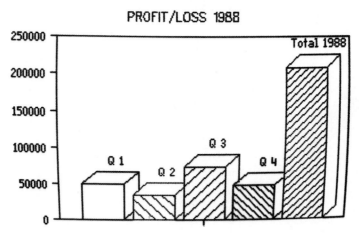

Figure 15.33
A business and analytical graph created with a spreadsheet package

---

[21]Figures 15.32 and 15.33 from Robert A. Szymanski et al., *Introduction to Computers and Information Systems.* Columbus, OH: Merrill, 1988. [19]

to an experienced technical illustrator. In your discussions with the illustrator, keep in mind reproduction requirements. Illustrations should be planned to have a uniform reduction ratio for reproduction. Recommended reductions are 1.5 to 1 for photographs and 2 to 1 for line drawings. Uniformity and consistency in such details as direction of arrowheads in callouts, in nomenclature, symbols, weight of lines, lettering, and similar matters give a professional quality to the printed paper or report.

Finally, you will find that an open mind and a willingness to accept advice and suggestions from the professional illustrator will frequently enhance the clarity and effectiveness of your ultimate product—the reproduced or published report.

Chapter in Brief

In this chapter we examined the types of visual aids that help a writer best to describe, define, explain, analyze, compare, classify, and narrate technical messages to the intended reader. We stated that the most effective graphic presentation is the one that best serves the reader by its ability to clarify and supplement the text. Computer aided graphics were also discussed.

Chapter Focal Points

- Lists
- Tables
- Photographs
- Drawings
- Diagrams
- Cartoons
- Exploded views
- Maps
- Graphs, charts, curves
- Computer assisted graphics

Questions for Discussion

1. Refer to the four different reports you located under Discussion Question 5 for Chapter 11. Examine also a representative number of technical periodicals. Select what you feel to be effective and ineffective examples of graphic presentations. What relationships do you find between subject matter and types of graphic aids? Are there any new or unusual combinations or uses of graphic aids? Discuss the effectiveness or lack of effectiveness of the visual aids used in four representative items of technical writing.

2. As an extra credit assignment, visit your university Computer Center or a commercial computer store. Ask to see a demonstration of graphics software packages for the

Macintosh computer and for the IBM (or its compatibles). Bring to class your description of the graphics packages, and if possible, sample printouts of the images they create. Tell the class which graphics package you would purchase, and justify your choice.

Assignments

1. Construct a table to present numerical data you have accumulated in your research project. If your research project does not lend itself to accumulation of numerical data, construct a table analyzing or comparing a number of factors within your problematic situation. Or find a newspaper or magazine article which includes information which could be advantageously shown in a table. Prepare a table for such data.

2. Prepare a bar or pie chart to show how you have allocated your time on the four phases of writing your report.

3. Check the report you are writing. Determine where graphic presentation can effectively promote the communication of your data. On the basis of considerations discussed in this chapter, determine which types of graphic aids you will need. First, make rough sketches. As you deem necessary, obtain advice and help from expert friends or teachers in the execution of your drawings and photographs. Construct the tables required. Lastly, rewrite your text to integrate the graphic presentations in your report.

4. Prepare a List of Tables and Illustrations you will include in your report.

References

1. *Aerospace Science: The Science of Flight.* Maxwell Air Force Base, AL: Air Force Curriculum Division, 1988.

2. American Geologic Institute and National Association of Geology Teachers. *Laboratory Manual in Physical Geology*, 2nd ed. Columbus, OH: Merrill, 1990.

3. Colburn, Robert E. *Fire Protection & Suppression.* New York: McGraw-Hill, 1976.

4. *Consumers Should Know: How to Buy a Personal Computer.* Washington, DC: Consumer Electronics Group, Electronic Industries Association, 1986.

5. *DDIC Bi-Monthly Memo.* Bethesda, MD: National Institute of Arthritis, Metabolism, and Digestive Diseases, March-April, 1979.

6. *Dimensions.* Washington, DC: National Bureau of Standards, April 1978.

7. *Enhanced Recovery of Oil and Gas.* Washington, DC: U.S. Department of Energy, 1977.

8. *Energy Conservation, Gas Heat Pumps: More Heat from Natural Gas.* Washington, DC: Office of Public Affairs, Energy Research and Development Administration, 1977.

9. "Estimating Lightning Performance." *Westinghouse Engineer*, Westinghouse Electric Research & Development Center, January 1951.

10. Higgins, Stanley A. Westinghouse Electric Research and Development Center, n.d.

11. Human Resources Research Organization. *Guidebook for Development of Army Training Literature.* Arlington, VA: U.S. Army Research Institute, November 1975.

12. Ken Cook Co., Milwaukee, WI, 1978.

13. McNairn, William N. "Don't Blame Your Judgment on Your Genes." *Price Waterhouse Review*, Vol. 22, No. 1, 1977.

14. *National Bureau of Standards Technical News Bulletin*, January 1971.

15. The National Institute of General Medical Sciences. *What Are the Facts about Genetic Diseases?* Washington, DC: U.S. Government Printing Office, n.d.

16. "New Trails to Chemical Productivity." *New Frontiers of Science*. Washington, DC: National Science Foundation, 1977.

17. *Nuclear Energy for Desalting*. Oak Ridge, TN: U.S. Atomic Energy Commission, Division of Technical Information, September 1966.

18. "Residential Sprinkler Systems." *Record, The Magazine of Property Conservation*, Factory Mutual System, May-June, 1978.

19. Szymanski, Robert A. et al. *Introduction to Computers and Information Systems*. Columbus, OH: Merrill, 1988.

20. U.S. Energy Research and Development Administration, Washington, DC, n.d.

21. *Weapons Systems Fundamentals, Basic Weapons Systems Components*, NAVWEPS OP 3000, Vol. 1. Washington, DC: U.S. Government Printing Office, July 15, 1960.

22. Weisman, Harlan F., M.D., and Michelle K. Leppo. Division of Cardiology, Johns Hopkins Medical Institutions, 1990.

23. *Westinghouse Engineer*. Westinghouse Electric Research and Development Center, March 1960.

PART FIVE
Short Technical Writing Forms

16
Short Reports

Chapter Objective

Provide understanding of principles and techniques, as well as proficiency in writing short reports.

Chapter Focus

- The Letter report
- The Memo report
- Recommendation reports
- Progress reports

*I*n Chapters 10–15, we covered the principles, forms, and elements of report writing. Emphasis was given to the more lengthy formal report because principles and techniques applying to it also apply to reports covering simpler situations. Reports, as we previously stated, are designed for the particular use a particular reader requires. Generally, its use forms the basis of a decision the reader must make. The report "form is the package in which you wrap your facts and analysis. Choose or design a package that is suitable for your material, your purpose, and your reader." In this chapter we shall examine some common short report forms.

The Letter Report

Frequently, in a reporting situation, the information to be transmitted is concerned with a relatively simple situation. The report's purpose is to provide the necessary information clearly, directly, and concisely. A letter of one to several pages best serves this purpose, if the information is being sent outside one's organization. If the information is being sent within one's organization, the memo form is used. (Review Chapter 2 for correspondence principles and forms.)

The letter report's appearance is similar to the business letter, having many of its conventional format elements. The printed letterhead of the originating organization is used. Sometimes, a subject line and internal headings within the letter appear. The letter report does not exhibit a "you psychology" tone. The information conveyed is the most important factor, not the personal tone; the tone reflects objectivity. However, the writer is free to use personal pronouns such as "I," "we," and "you."

The letter report is composed of the following elements, in the order stated here:

Purpose and scope
Findings
Conclusions
Recommendations, if appropriate

Letter reports of more than four pages may begin with a summary. Figure 16.1 is an example of a letter report.

The Memo Report

The memo report resembles the letter report, except that it is intended for members of the originating organization. There is a current tendency, however, to circulate reports in memorandum form to persons outside the originating organization. Like the letter report, it is used in those situations that are informal, of immediate interest, and of lesser scope. Because its material is intended

SMALL BUSINESS ADMINISTRATION
Washington, D.C. 20416
July 14, 19__

Mr. James Jacobi, President
Jacobi Paint Company
3200 East Monument Street
Baltimore, MD 21205
Subject: <u>Market Aids Newsletter</u>

Dear Mr. Jacobi:

Last February we wrote you as well as other subscribers of our <u>Market Aids Newsletter</u> about our publishing program. We asked whether you found our publication valuable enough to justify the costs of continuing to publish it. We also invited your comments on what uses you made of its information, what features you found useful, and what other subjects you wanted covered. Here are some of the findings:

     1. First of all, you told us that you do read the <u>Market Aids Newsletters</u> and that you analyze what they say in terms of your operations. Many people spoke of routing each new issue to key associates for similar study. A number mentioned keeping the back-numbers together in a notebook for handy reference. Still another large group said they use the <u>Market Aids</u> in customer- or employee-relations work, passing on certain issues of special interest. Some use them in discussions and training sessions. Quite a few re-read old <u>Aids</u> as "refreshers" when related problems come up.

     2. You liked our short format with a summary at the start and frequent subheadings in boldface type. You asked us to continue listing a few references for further reading at the end of each <u>Market Aids</u>. Some of you hoped for "more elementary material"; others requested just the opposite. (We'll try to strike a balance and issue some of both.) You urged us to keep the language and examples down to everyday businessmen's levels. (We'll work extra hard to do so.)

     3. On suggested future subjects you came up with scores of interesting ideas. The major areas of concern are human relations and communications, techniques of control, selling, money management, and computer programs for small businesses.

Purpose and scope

Numbered paragraphs 1–4 are findings of the survey.

Figure 16.1
Example of a Letter Report

The last two paragraphs are the terminal section of this letter report, offering the conclusion of the survey. They also serve as a public relations device to express appreciation for the replies received and they recommend that there is still time for readers who have not responded to the survey to do so.

4. More of you proposed potential authors than we had expected. This is all to the good. Many different fields of specialization are represented. You may be sure that all proposals will be carefully considered; some you will see in print later this year.

Based on your enthusiastic endorsement, we shall continue to publish <u>Market Aids Newsletter</u>. We are grateful for the thoughtful help of all who wrote us. We wish we could have replied to each one of you individually but the numbers were too great. Your interest, cooperation, and voiced support for the work we are trying to do on behalf of small businesses are very welcome.

To those subscribers who, for one reason or another, did not reply to our survey letter, let me say: It's never too late. We are always eager to get your reactions and ideas.

Sincerely yours,

Chief, Management Services Division

Figure 16.1 (continued)

for one's own organization, its tone is less formal and less personal than that of the letter report. Its information is presented in the memo format with "From," "To," and "Subject" lines.

Memo reports are used mostly for communication between administrative levels and between members of different departments within an organization. Often the recipient is familiar with much of the background of the reported situation. Therefore, more so than in the letter report, you should come directly to the point of your message.

Organizational Elements of a Memo Report

The opening paragraph states the purpose of the memorandum. The paragraph or paragraphs immediately following may present the conclusions reached, findings, or results obtained. The middle paragraphs provide whatever facts or explanations that are needed to substantiate the conclusions or results. Tabular or graphic data may be used as needed toward that end. The final paragraph presents, as appropriate, recommendations for the recipient. Figure 16.2 illustrates a memo report.

Recommendation Reports

Recommendation reports are analytical reports based on the examination or investigation of a problematic situation. They are written for the purpose of decision making and action. For data, the investigator relies on:

1. examination of the actual situation
2. reading the literature about the problem
3. interviews and consultations
4. testing/experimentation
5. his/her own experience

Because the client is interested in what to do about the problem, the major portion of the report is devoted to analytical considerations that lead to specific recommendations on actions to be taken. Very little space is devoted to the problem's background or the method of investigation. Most of the contents is devoted to an explanation of the data of the situation so as to lead the reader to the writer's conclusions and to acceptance of the writer's opinions and the recommended actions to solve the situation. Depending on the situation, recommendation reports can be in letter, memorandum, or formal report format.

The memorandum report in Figure 16.2 may also be considered a memorandum recommendation report. An example of a letter recommendation report is shown in Figure 16.3.

WESTINGHOUSE

TO: J. W. Wagner
    Section Manager, Lighting Division
    Cleveland Works

FROM: W. J. Robertson
       Materials Engineering
       East Pittsburgh Works

DATE: February 15, 19__

SUBJECT: Plaster Molding

The first paragraph presents the background and purpose of the memo.

In your letter of February 5, you asked several questions regarding plaster molding as a method of producing one-of-a-kind castings to suit the needs of your Industrial Designer. Specifically you asked about the type of plaster, processing of molds, and types of alloys that could be handled. The castings to be made are required to have good surface finish and be as accurate as possible.

In the second paragraph, the writer gets immediately to the point and presents to the recipient the factors causing the problem.

First of all, I want to point out that surface finish and accuracy (precise control of dimensions) are direct functions of pattern equipment regardless of the process used. Inherently, of course, different molding materials will have an effect but they cannot overcome defects in pattern equipment.

As far as molding processes are concerned, I believe there are several that can be considered for this application. Plaster molding is one, as you mentioned, but you might also consider the CO_2 molding technique, good dry sand molding such as applicable to critical core work, and investment type molds. Depending on the degree of surface smoothness and precision you need any one of these processes may be applicable.

In the middle paragraphs, three to six, the writer explains the methods, procedures, and factors involved in their successful use. Organizationally, these paragraphs represent findings based on experiential analysis.

There are two basic plaster molding processes—one using a permeable, foamed mixture, the other using a straight plaster mix, which is of low permeability. Plaster molding techniques do require considerable control in order to produce good molds and castings. Most commonly used are gypsum plasters, which work well on aluminum and copper base alloys except those copper alloys that are poured about 1100°C. In this temperature range, the plaster tends to break down giving very poor surface finish. Also, plaster molds are usually poured hot to assure that all moisture is driven off the mold. This is less critical with the foamed, permeable plasters but drying cycles are important and cannot be short-cut without trouble resulting. I am not familiar enough with plaster molding techniques to be able to spell out specific procedures for you to follow, but I suggest you contact one or more of the gypsum plaster manufacturers who can give you the details you want.

Figure 16.2
An Example of a Memo Report

You may also want to consider the investment type molds for this application but I suspect set-up and operation of this type of procedure for just occasional castings would not be justified. You would, however, get excellent finishes and dimensional control.

Sand molding procedures can also produce the type of quality you indicate if they are tailored to this need. Either the CO_2 process or dry sand procedures would be suitable. Surface finish would be directly related to sand composition in either case. The grain size and distribution of sizes of the sand would be important. I think Jim Drylie or one of his foundry sand experts could give you a good run-down on this possibility, although I don't know how familiar the foundry is with CO_2 molding. One big advantage to using a "precision" sand process lies in the greater size range that can be handled. If your interest in these special castings may range to large castings, i.e., larger than would fit in about a $12'' \times 12'' \times 12''$ space, then plaster may have limitations. These larger items could be handled more readily in a sand process.

Sperry Gyroscope Company has used a carefully controlled sand process for producing a variety of "precision" microwave components. They claim very good dimensional control and surface finish.Unfortunately, I don't know the details of their sand mixes, but they use both dry and green sand practice.

I suggest you contact one or more of the following for detailed processing information:

Investment Molds "Curacast"	Kerr Manufacturing Company 6081 Twelfth Street Detroit, Michigan 48208
Plaster	U.S. Gypsum Company
Sand	Archer-Daniels-Midland Company Federal Foundry Supply Division 2191 West 110th Street Cleveland, Ohio 44102
	B.F. Goodrich Chemical Company 3135 Euclid Avenue Cleveland, Ohio 44115 ("Good-Rite CB-40" binder)

For the CO_2 process possibilities, I think DuPont or Linde Air Products or any good foundry supply house can give you detailed information. I know what I have written doesn't exactly answer your question concerning details of the plaster molding process, but I think the other processes are worth considering, especially since sand procedures would not be foreign to your present foundry operations.

If you want me to pursue this in more detail for you, please let me know.

Paragraphs seven to nine represent conclusions and recommendations. Because of the many variables involved in the successful use of plaster cast molding, the writer suggests several knowledgeable sources for more specific information on the several approaches he has suggested to solve the foundry operation problem.

In the final two paragraphs the writer reaffirms the approaches he suggested and offers to be of further assistance.

Figure 16.2 (continued)

Mr. Darnell Winters
Production Superintendent
Palm Bay Boat Company
23719 Valley Road
San Diego, CA 92112

Dear Mr. Winters:

In your letter of July 17 of this year, you commissioned me to recommend means and procedures to reduce premature corrosion at or near weld areas of ships you are building. After some thought and study, I believe I can suggest an effective and inexpensive way to eliminate this troublesome and expensive problem.

When protective coatings break down in weld areas, it can usually be traced to the fact that harmful deposits formed during welding have not been fully removed before the coating process. Harmful deposits commonly found near weld seams are:

1.   Alkaline slag from the weld flux, which reduces the adhesion and durability of the coating film;
2.   Condensed flux fumes, which produce similar undesirable alkaline conditions;
3.   Oxides produced by the heat of welding;
4.   Weld metal spatter.

Beads of weld spatter may be as large as 6mm (¼ inches) in diameter, and their peaks are normally too high to allow adequate coverage by an average film thickness of coating. Spatter, therefore, presents vulnerable points for early rust formation.

A simple three-step process will eliminate problems caused by all four types of deposits:

1.   Treat the weld with 10% phosphoric or 10% hydrochloric acid to neutralize alkalinity. Scrub the acid into the weld area with a stiff brush. Commercial ready-made preparations (e.g., Rust-Oleum or Surfa-Etch) are also available for neutralizing the deposits, and they are easier to store and use. Be sure your workers wear protective rubber gloves, aprons, and goggles.
2.   After the acid preheatment, rinse the entire area thoroughly with fresh warm water: While the surface is still wet, remove any rust spots or oxides near the weld by rubbing with fine steel wool. Then dry very carefully.
3.   You can then remove the weld spatter by sand-blasting or grinding with power tools.

Figure 16.3
Example of a Recommendation Report

Surface preparation is, of course, only the first stage of weld protection; the primer coatings do the continuing rust prevention job. For optimum performance, use only those primer coatings specifically formulated to provide maximum anti-corrosive protection for steel surfaces, including welded areas. Examples include:

a. A lead-free red metal primer (X-60), which dries to the touch in 4-6 hours and may be exposed to the weather up to nine months before application of the finish coat.

b. A fast-drying (touch-dry in 30 minutes formulation C678).

Use an intermediate coat of 960 zinc chromate primer; it will help assure long term freedom from rust. Because of its light color, zinc chromate serves as an excellent undercoat when the finish coat is also light.

Durability of the coating system will depend on film thickness. Each coat, when dry, should be 25 (0.025mm; 0.001 in.) thick but not more than 50 (0.050mm; 0.002 in.). You must follow the instructions of the coating manufacturer for mixing, thinning, and application of the specific coating.

In case you do not have convenient referral to sources of primer and coating preparations, I am enclosing some catalog sheets on such preparations. If you have a need for any further information on steel weld corrosion problems, do please call on me again.

Sincerely yours,

Thomas F. Hood
Corrosion Consultant

Enclosures

Figure 16.3 (continued)

Progress Reports

All of us make progress reports; sometimes several times a day.

"What's new with your car, Fred?" a friend will greet you.

"Well, I got rid of that old clunker last week," you may reply. "It ate gas, burnt too much oil, and then the transmission went out. I bought a compact Saturday."

Or, your boss might say, "Jim, how are you doing with the problem in the Newby Paper Company plant?"

"It's not the transformers, according to my inspection yesterday," you report. "Though the transformers need a good, new coat of protection paint. I'm looking at the synchronous motors that drive the drying rolls. They seem to be in good condition. But I need to give a closer look at the motors driving the rack and suction rolls at the wet end of the machines. They're covered with pulp and gook and don't operate well at all. I'm going to look at them today and let you know by tomorrow afternoon what's wrong and what should be done."

The main purpose of a progress report is to give an accounting. In a formal situation, progress reports are issued at specified intervals to show what has been done, what is being done, and what is expected to be done. Funding agencies, managers, and supervisors require such reports as a necessary communications link for control purposes and intelligent management decisions, often on whether a project should be maintained, expanded, reoriented, or abandoned. Progress reports also are helpful to the performer. They enable him or her to focus on and assess periodically the work done and the work remaining in relation to allocated resources, time, and effort.

Reporting periods are specified by the funding agency, client, or administrator. The form of the progress report varies. It may be a letter, memorandum (if in-house), or a bound document with formal trappings like title page, table of contents, abstract, separate text sections, and appendix material. What is covered in progress reports depends on the subject matter and what the reader wants and needs to know. The following elements are usually included:

Introduction
    Identification of grant, contract, or work order
    Purpose of grant, contract, or work order
    Project description
    Summary of earlier progress
Details of progress during report period (include dates)
    Procedures
    Problems
    Results
    Discussion (include conclusions and recommendations if appropriate)
Work planned for next reporting period

Work planned for periods thereafter

Overall appraisal of progress to date

Most progress reports lend themselves to a chronological arrangement. In some cases, it may be more appropriate to arrange the progress by tasks or by subject matter such as equipment, materials, personnel, and costs in the case of production or construction work. Some progress reports are in letter format. A sample progress report follows (Figure 16.4).

Chapter in Brief

In this chapter we examined various types of short reports, their organization, formats, and techniques of composition.

Chapter Focal Points

• The Letter report
• The Memo report
• Recommendation reports
• Progress reports

Question for Discussion

Your institution, anticipating the future, is considering computer literacy as a requirement toward graduation. You are a member of the student council and a computer buff. You, personally, think the requirement is a good idea, and long overdue, but as a student representative you want to check the pulse of your community. Each council member of the seven divisions of your institution—Humanities, Biological Sciences, Physical Sciences and Engineering, Home Economics, Social Sciences, and Computing Center—surveys the students in her/his division. When the results are tabulated, the majority, except in Humanities and Home Economics, favor the requirement. As a classroom discussion project, how would you go about preparing a memorandum report to the president? What would its contents be? What recommendations should the memorandum report have?

Assignments

1. The president of your university is making a fund drive to renovate a number of buildings in your institution. He has named you to a committee to make a preliminary survey to identify the most appropriate candidate buildings to be included in his appeal for renovation funds. Among the buildings surveyed were:

Old Main—the Administration Building

Library

Gymnasium

Fourth Quarterly
Progress Report on
National Survey on Fires in Households
in the United States

Prepared for the U.S. Fire Administration
by XYZ Associates
1020 Connecticut Avenue NW
Washington, DC 20035

Prepared under Contract No. 7915 for the period
October 1–December 31, 19__

Figure 16.4
A Sample Progress Report

Introduction

This is the fourth quarterly progress report on the work being done under Contract No. 7915 in the fulfillment of the performance requirements listed in paragraph 8e, covering the period of October 1 to December 31, 19__. The purpose of the contract is to conduct a national survey of a stratified, representative sample of U.S. households to help answer the following questions:

1. How many household fires occur per year in the U.S.?
2. Where do fires start?
3. When do fires start?
4. What starts the fire and what catches fire first?
5. Who gets hurt or killed?
6. What is the dollar fire loss?

Work Previously Completed

Following approval of the scientific design, 33,000 statistically selected households were surveyed. The survey was completed in the last reported period and computer tabulation and analysis of responses were begun.

Present Work

Tabulation was completed and analysis is continuing. There were 2,463 fire incidents reported during the calendar year of concern in the survey. Analyses of responses to the second and fourth questions being investigated provide the following indications:

Where do Fires Start?

Kitchen (cooking)	40%	
Kitchen (no cooking)	25%	} 65%
Living Room	12%	
Bedroom	8%	
Basement	4%	
Utility Room	2%	
Bathroom	1%	
Other	8%	
Total	100%	

What Starts Fires?

Appliances (involving grease or food)	34%	
Other Appliances	28%	} 62%
Wiring	8%	
Smoking	7%	
Matches and the like	4%	
Other, or do not know	19%	
Total	100%	

Figure 16.4 (continued)

<u>What Catches Fire?</u>

Grease, food	41%
Appliances	26%
Wall, Floor covering	6%
Furnishings	5%
Clothing	2%
Other, or do not know	<u>20%</u>
Total	100%

<u>Problems</u>

Despite a careful attempt to define for respondents what to classify as a fire incident—both the observation of flames and the start of smoke coming from wiring or appliances—there is an evident confusion in the responders on what constitutes a fire. The number of fire incidents may be more than what was reported.

<u>Discussion</u>

Preliminary analysis indicates that accidental fires are usually started by appliances. Nearly half the time, fire incidents are associated with cooking. Though analysis of dollar losses has not been completed, present evidence indicates the average "cooking" fire does comparatively little damage because someone is usually there to discover it. Fires that appear to do the most damage are those where either the wall or floor covering catches fire or where the start of the fire is unknown. These preliminary conclusions will be verified in the next reporting period.

<u>Work Planned for the Next Reporting Period</u>

1. Average dollar loss per fire will be determined (answer to Question 6) and extrapolated nationally.

2. Activities connected with the start of a fire will be determined. (Completion of answer to Question 4.)

3. The answer to Question 3 will be determined.

4. Annual casualties in deaths and injuries will be determined and extrapolated nationally (Question 5).

<u>Completion of Project</u>

Work in the performance of this contract is proceeding on schedule. We anticipate completing analysis of survey returns and delivery of the Final Report as required at the end of the quarterly period following the next.

Figure 16.4 (continued)

School of Home Economics
Social Science/Humanities Building
Computer Center
Biology Building

Your committee will consider some of the following factors:

Need for air conditioning
Fire and other safety considerations
Class rooms and seating capacity
Comfort facilities
Heating system
Office space and student space
Parking
Laboratory and equipment resources

Write a memorandum recommendation report, identifying satisfactory and unsatisfactory building and space resources with recommendations to your university president.

2. After reading your report and discussing its recommendations with the university's board of trustees, your president decides it would be most practical to limit the fund drive to renovating just three buildings—Old Main, the Computer Center, and one more. He can't decide whether the third building should be the gymnasium or the library. He asks your committee to investigate and make a recommendation he can justify to both the alumni and the faculty. Write a report to meet your president's needs.

3. Your university president has appointed you to make a survey of traffic patterns in and around your campus with the object of determining whether student driving to or on campus should be banned. Write a memo report with recommendations for action the president should take. Your report might include tables and graphs to substantiate your conclusions and recommendations.

4. Write your instructor a progress report on your progress to date in writing the research report assigned for this course.

17

How to Prepare Proposals— From Concept to Document

Chapter Objective

Provide both understanding of the purpose and techniques of the proposal document and proficiency in its preparation.

Chapter Focus

- Definition of proposals
- The RFP (Request for proposal)
- Informal letter and memo proposals
- Formal proposals
- Solicited and unsolicited proposals
- The Proposal abstract
- Elements of a proposal
- Organizing and writing the proposal
  The Storyboard
- Proposal evaluation criteria

A communication, written or oral, that attempts to sell an idea, a concept, a service, a piece of equipment, a complex system, or anything else is a proposal. A proposition for research, a sketch of a carving for sculpture, an outline of a novel—all framed within a substantiated request to a foundation, a government agency, or a private enterprise for financial support is a proposal. The memo you write to your boss justifying a request to purchase a new pencil sharpener or suggesting a more effective system for fabricating the gizmo your company is producing is a proposal, as is the 2500-page document in ten volumes that discusses developing a space vehicle to land a team of astronauts on Mars and return them to Earth safely.

A proposal delineates a problem and lays out the essential groundwork or directions for its solution. The "solution" is the statement of the proposed work: *how* it will be done; *who* will do it; *where* and *when* it will be done, and *how much* it will cost. The successful proposal is one that convinces the prospective client or funding agency to invest money in your ideas, products, or services. The proposal, then, is a document that offers a plan to solve a consequential problem or provides a product or service that meets a stated need.

Solicited and Unsolicited Proposals

There are two types of proposals—solicited and unsolicited. A *solicited proposal* is developed in response to an announced, advertised, or invited request, or in response to a directive from a superior. The *unsolicited proposal* is initiated by an individual or organization seeking support, frequently financial, for the solution of an identified problem of interest. When an organization requires work from outside its own structure or requires noncatalog, specialized equipment, material, systems, or services, it needs to locate a qualified source for the required work or equipment. The medium for seeking a qualified source is the RFB (request for bid). When there is a problem for which an agency would like a bidder to propose a solution, the request medium is the RFP (request for proposal).

In solicited proposals, the RFB and RFP serve as instructions stipulating the exact specifications for the equipment, service, or work that is required, how that work is to be performed, and when it is to be delivered. An RFP helps a bidder know not only what is wanted, but also instructs the bidder how to prepare the proposal. It specifies what information is to be included, at which point in the proposal that information is to appear, how many copies of the proposal are due, and it frequently explains the point scale for the evaluation of the proposal. The proposal is judged by how well the bidder fulfills the terms of the RFP.

In unsolicited proposals, you, the presenter, initiate the offering. You identify a problem that needs to be solved, or an idea for a product or service of potential interest to a prospective client. The unsolicited proposal has to convince this prospective client that, first, there is in fact a problem or that the suggested idea, product, or service is practical and of definite benefit; second, that your suggested solution is both practical and effective; and third, you have the capability to fulfill what you promise in the proposal.

It is considered good business practice to write a letter of inquiry to a prospective client before you send your unsolicited proposal. This letter is to determine if the prospect is interested enough in your idea to pursue it in detail. This step is designed to save both you and the prospective client time and expense should the idea be deemed unsuitable. It also serves to increase your chances of submitting a proposal successfully by allowing you to target your submission to those prospects you know to be interested in your idea. A successful proposal generally results in what is called a *sole-source procurement*, which means that you will be contracted as the only provider of the product or service in question.

Not all proposals are directed to external sources. Internal proposals, prepared in a memo to management, are aimed to sell an idea, such as how a new item of equipment can improve procedures in the production line; how flextime can improve efficiency; or how a daycare program can reduce absenteeism.

Proposals can also be classified as informal and formal. Their chief differences are size and format. The *informal proposal* takes the form of a letter or memo of one to several pages. The organizational elements include an introduction outlining the problem that needs solution, a discussion of proposed actions and procedures for solving the problem, costs involved, and qualifications of the proposer. It may have an appendix for illustrations and for those calculations that could clutter up the letter. Letter and memo proposals may be solicited or unsolicited.

The *formal proposal*, whether prepared in response to an RFP or developed as an unsolicited document, is required for projects of sizable scope. The term "formal" refers to how "dressed up" the proposal appears. Its text is organized into standard elements: cover, title page, table of contents, list of illustrations, text in discrete sections, appendices, glossary, and bibliography—elements similar to that of a formal report.

Informal Proposals—Letters and Memos

Proposals on relatively simple matters, problems, equipment, or services require simple documentation and can be efficiently clothed in a letter or memo format. These informal proposals contain the following elements:

The *Introduction* answers why the proposal is written. Within this background information is presented a statement of the problem, the purpose and significance of the situation, the historical background of the problem, and the scope of its boundaries. Any definitions governing the approach and work to be done are indicated. The introductory section is important because it indicates to the supporting agency whether the proposer understands the problem to be investigated.

The *Technical Presentation* is the "guts" of the proposal. It describes the proposal plan for doing the work. In the technical presentation you—

1. Explain clearly what you propose to do. Indicate what specifications you will meet, the scientific or technical work you will do, and the equipment, system, product, or report you will deliver at what specified time.
2. Respond to the RFP point by point, item by item.
3. Explain if you have any exceptions to any of the requirements in the RFP, why those exceptions are necessary or why it will be advantageous to the supporting agency to make the exceptions.

The *Technical Description* is the creative contribution by the proposer for solving the problem. The method of attack, based on theory, state-of-the-art, or fundamental principles, is explained, justified, and substantiated. Diagrams to supplement the text should be used, as appropriate.

The proposer's *capabilities* in personnel, facilities, and experience are presented. Biographical data on personnel who will do the work are included.

Programming of the work should be carefully blocked out or charted. To be included are scheduling of work, personnel, facilities, and time.

Cost Schedules should include wages and salaries, equipment, and miscellaneous costs such as travel, communications, and outside service. Pertinent overhead costs, other administrative costs, and profit or fees should be clearly indicated and justified. An example of a letter proposal responding to a relatively simple RFP is shown in Figure 17.1.

The Formal Proposal

The usual structure of a proposal is a formal document, which may materialize into a multibillion dollar source of new or continued activities. The proposal represents the combined efforts of personnel in management and sales, as well as people from the scientific/technical, legal, accounting, and publications areas.

Proposals, whether generated by an RFP or initiated by an unsolicited request, must begin with a clearly identified problem. As John Dewey observed, a problem clearly identified and lucidly articulated is half solved. Once you have identified a problem and thought it through clearly enough to state succinctly, you can hypothesize a solution.

The first step, then, in preparing a proposal is to *state the problem clearly*. The second step is to write out the objectives to solve the problem. An **objective** is an action, usually measurable, to be accomplished for a purpose. If the objectives are well thought out, carefully developed, and clearly stated, the rest of the proposal should follow nicely. It might be helpful for you to review the discussion in Chapter 10, pages 158–164, on the application of the scientific method to problem solving. A clearly defined, delineated problem will have clearly defined, specific, and practical objectives.

Department of Mechanical Engineering
COLORADO POLYTECHNIC UNIVERSITY
Fort Collins, CO 80521
December 8, 19__

Mr. Robert Dolan
Manager, Wire Products Sales
The Colorado Fuel and Iron Corp.
Box 1920
Denver, CO 80201

Dear Mr. Dolan:

I am responding to your request of November 30, 19__ for a proposal to test CF&IC nails for withdrawal resistance in accordance with the requirements of paragraph 8 of "Gypsum Dry Wall Contractors International, Recommended Performance Standards for Nails for Application of Gypsum Wallboard." All nails will be furnished by CF&IC.

We could begin the tests immediately on the signing of a contractual agreement between your firm and the Colorado Polytechnic University. The tests will be conducted by me, a licensed professional mechanical engineer. A firm date of completion of the contract and delivery of the test reports with a final report would be six work weeks following contract signing.

The cost breakdown is as follows:

1. Salaries:
    Professional Engineer, 0.6 month $2500.00
    Clerical, 8 hrs. @ $9.00 ....   72.00
2. Equipment and supplies
    Strain gage elements 100.00
    Lumber and wallboard 250.00
3. Report duplication
    50 copies, ten pages each 200.00
4. Annuities, 6% of salaries 154.32
5. Overhead, 35% of salaries 900.20
    TOTAL 4176.52

Tests on additional types or sizes of nails can be made for 40% of the cost of the first test, or $1670.61 for each test.

Colorado Polytechnic University would be glad to have the opportunity to be of service to your firm and we look forward to hearing from you.

Yours truly,

Everet Norman
Head of Department

EN:wa

Figure 17.1
Example of a Letter Proposal

Preparing an Abstract

The abstract is a first step in writing a project proposal for funding. The abstract, as we learned earlier in Chapter 13, is an abbreviated representation of the contents of a document. Proposal preparation can be a costly, time-consuming activity. Large corporations spend millions of dollars competing for funds, which may not always be granted or contracted. The abstract has become a practical instrument to test a funding source's potential interest in a concept aimed to provide a solution to a problem in which it may have an interest. As stated previously, for unsolicited proposals, funding sources prefer to receive a letter of inquiry and an abstract rather than having to study a long document that may or may not be of interest.

In preparing a proposal, you, the proposal writer, like the report writer who prepares a report, must analyze the problem being undertaken. You must ask yourself a number of questions: What is the purpose of the work or actions to be taken? Who needs the answer to the problem? Why? How will the answer be used? What needs to be done to find the answer? Who will do the researching and/or developing? What resources are needed? How long will it take? What is the intended significance of the results?

After you have ruminated on these questions, you have research to do—background reading similar to the literature searching discussed in Chapter 11. If available to you, search pertinent on-line computer databases in addition to the library. Also, search your company's files for relevant information and data. Brainstorm and hold discussions with persons in and outside your organization who may have applicable information. After your background research, again relate the questions to the proposal situation and prepare a short abstract summarizing your thinking. Direct your abstract to the following points, as they are appropriate:

1. The problem or situation
2. The objectives of the proposal
3. Brief set of procedures to implement the objectives
4. Estimated costs and resources required
5. Personnel needed to do the work
6. Significance of the results

The abstract helps not only the proposal writer to see clearly and succinctly the problem, procedures, resources necessary, and the significance the results could provide, but it also serves as a prospectus for discussion with potential support agencies. The initial discussion usually leads to a revised abstract containing the particulars of interest to the potential supporting agency. Figure 17.2 is a sample proposal abstract.

Once the abstract has passed the first evaluation, either further discussion may follow to design the proposal's objectives, scope, and available funds in line with the requirements of the funding agency, or the preparation of the final proposal is in order.

ABSTRACT

TITLE: Fire Service Model Program for Information and Technology Transfer

THE PROBLEM: There is currently no centralized means for collecting, evaluating, and disseminating the many successful performance-improving and cost effective developments in the fire service. The last decade has produced numerous successful efforts in fire service planning, management, and operations that could and should be utilized by other fire departments but which receive limited or no dissemination. How can a concerned fire chief or local administrator find out about successful new techniques? Where can he find time to sift through the hundreds of variations of fire service practices? How can he know which ones actually work? And how can he become aware of superior programs that have not been publicized?

OBJECTIVE: The purpose of this program is to transfer and make available nationally the most efficient and cost-effective procedures being used successfully in some of the more progressive fire departments or in appropriately related areas.

PROCEDURES: Each year the International Federation of Fire Chiefs will invite state, local, paid, volunteer, private, and Federal fire protection departments to submit existing successful programs for consideration as "Exemplary." The invitations will be made by solicitation and public announcement. All candidate projects would be required to meet established qualitative and quantitative criteria of being significantly effective in improving efficiency and cost-effectiveness of fire departments, or reducing losses, injuries, and deaths, or be innovative in use of procedures, equipment, gear, etc.

The International Federation of Fire Chiefs with the concurrence of the U.S. Fire Administration will appoint a Review Board composed of leaders from a wide range of the fire service, city management, and the research community. The U.S. Fire Administration, as the support agency, will exercise final decision based on recommendations of the Review Board.

Projects designated as Model would be documented, made available in the form of manuals, and, if appropriate, developed into training courses by the U.S. Fire Academy. Localities submitting projects declared "Model" would receive plaques at special recognition ceremonies. Some Model projects may receive U.S. Fire Administration support as demonstration projects.

SIGNIFICANCE OF RESULTS: By creating an awareness and availability of innovations and cost-effective procedures, the fire service will be able to meet greater requirements more effectively and efficiently and help reduce fire losses.

ESTIMATED COSTS: $70,000.00.

PERSONNEL REQUIREMENTS: Chief James Shaw, Vice President IFFC, Principal Investigator, and Program Analyst, Clerical personnel, and ten Review Board members.

Figure 17.2
A Sample Proposal Abstract

The Storyboard—An Aid to Proposal Preparation[1]

You, as an entry-level employee, may not be responsible for preparing responses to RFPs. You may be involved, however, in proposal writing as a member of a team assigned to prepare elements of a proposal. Many companies who do extensive contract research and development proposal preparations assign technical personnel and writers to a team headed by an experienced editor. Team members often are persons who are knowledgeable in the technical content matter required in the RFP but lack communication skills, or who are entry-level technical writers. To help persons inexperienced in proposal preparation, many companies have borrowed a technique from film script preparation called the **storyboard**.

In film production the storyboard is a series of drawings arranged in order of the progression of shots in a film sequence. The drawings are captioned with appropriate descriptions, thus plotting and presenting the story of that sequence of film.

Much in the same way, the senior editor responsible for a proposal uses the storyboard approach to divide the technical content of the proposal into modules. A storyboard format is devised to help writers originate, organize, and present facts and ideas for the section or module for which they are responsible. Figure 17.3 shows a typical storyboard template.

Specifically, the storyboard organization design requires the writer to:

Draw a picture
List significant things about the picture in order of importance
Identify clearly features and benefits described
Write a one-sentence summary of what is in the storyboard.

The Proposal Document

Government agencies, foundations, and other funding sources have special proposal forms and formats to be followed. No matter what the form, all require certain essential elements. Though no single pattern of organization is appropriate for all proposals, the following elements identify essentials that may be modified to meet specific situations. The sequence is not sacred; other elements may be added and some combined.

The elements in outline form are:

1. Letter of Transmittal
2. Cover/Title Page
3. Executive/Summary

---

[1]How a large technical company makes use of the storyboard concept may be found in Jerome K. Clauser, "PODS: Pump Primers for Proposal Writers," *Technical Communication*, April, 1989.

Storyboard Organization Design for Proposals

RFP No.: <u>xyz</u>
Section No.: <u>IV</u>
Section Title: <u>Methods and Procedures</u>

Author: _____
Page: _____

1.   Draw a picture of what you are proposing, e.g., flow chart, organization chart, sample output, rack or shape projection, system configuration, table of data, plots, etc.

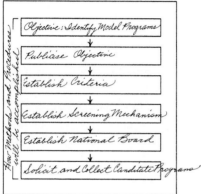

Figure No. 4
Write caption for this figure: <u>How Model Program will be initiated</u>
Indicate size:
□ quarter; □ third; ☑ half;
□ whole page; □ foldout

2.   List main points you want to make in this section. Number points in order of importance.
 1.   Both professional and volunteer fire service depts. will be approached.
 2.   National Board will be appointed from fire services; academia; professional societies; unions; media; insurance industry.
 3.   Criteria will be based on effectiveness, transferability, adaptability to both professional and volunteer depts.
 4.   Cooperation from media.

3.   List features and benefits

Features	Benefits
1. Competent administrative staff.	National Board relieved of tedious administrative tasks.
2. Fire services contacted by media and mail.	Insures everyone reached.
3. Criteria ID's Exemplary programs.	Exemplary programs incorporated into training.

4.   Write one sentence to summarize key items above:
 This project will improve national fire prevention and protection and identify those which are adaptable for transfer to both volunteer and professional departments.

Figure 17.3
The first step in using a storyboard is to draw a representation of the "story" that is to be told in a section of the proposal.

4. Table of Contents
5. Body of Proposal
    a. Introduction
        (1) Nature of the problem or situation requiring proposed work
        (2) Magnitude and significance of problem
        (3) Review of relevant literature (as appropriate)
        (4) How the results of the proposed work would solve/alleviate the problem
        (5) Who would benefit (How the results would be used)
        (6) Justification for addressing the problem at this time
        (7) Relation of the proposed program to the mission of the funding agency
        (8) Scope—what will and what will not be done
    b. Technical Work
        (1) Objectives to be accomplished (Sometimes these are translated into tasks to be accomplished)
        (2) Procedures or methods to achieve the objectives (Includes theory behind methodology)
        (3) Evaluation of techniques (to check validity of work performed)
        (4) Expected results
        (5) Recommended dissemination procedures
    c. Management
        (1) Staffing and resources
            (a) Résumés of proposed personnel
            (b) Statement of facilities, capability, and financial status
            (c) Management plan
            (d) Similar experience
        (2) Costs
            (a) Direct
            (b) Indirect
            (c) Other
        (3) Project time requirements (timeline chart)
6. Concluding statement (as appropriate)
7. Appendix (as appropriate)
    Backup materials, e.g., company brochures, financial statements, documentation of related work

The sample proposal in Figure 17.4 has been developed from the sample proposal abstract in Figure 17.2. Note how succinctly it meets the discussed requirements.

To prepare a proposal document, follow this outline, filling in the details as described here.

An Experimental Fire Service Model Program
For Information and Technology Transfer

I. The Background

In these stringent economic times there is increasing pressure on every public service to improve performance and become more cost effective. The fire service has not escaped this pressure and it may actually be one of the organizations most frequently singled out for criticism.

In the last decade progressive elements of the fire service have begun to overcome the image of a nonproductive agency mired in traditionalism. Out of this period has come a variety of research, innovative concepts, technology and techniques that deserve wider consideration and practice.

II. The Problem

There is currently no centralized means for collecting, evaluating, and disseminating the many successful performance-improving and cost-effective developments in the fire service. The last decade has produced numerous successful efforts in fire service planning, management, and operations, which could and should be utilized by other fire departments but which receive limited or no dissemination. How can a concerned chief or local administrator find out about successful new techniques? Where can he find time to sift through the hundreds of variations of fire service practices? How can he know which ones actually work? And how can he become aware of superior programs that have not been publicized?

III. The Objective

The objective of this project is to identify fire service programs that through the experience of implementation have proved productive, efficient, and cost effective. When any of these also meet the criterion of transferability to other departments or jurisdictions they will become a U.S. Fire Administration Model Program.

IV. Methods and Procedures

The International Federation of Fire Chiefs Foundation proposes to:

1. Publicize the purpose of the Fire Service Model Program by means of (a) direct mail to the constituencies of the Joint Council of Fire Service Organizations; (b) releases to the fire media; (c) invitations to the major public service organizations.
2. Establish criteria for candidate projects for the fire service model program.
3. Establish screening mechanisms for entries.
4. Establish a National Board, ten (10) members representing recognized authorities in the fire service, city management, research and development, education and training, and elected officials. (Items 2, 3, & 4 above will be done in consultation with the U.S. Fire Administration.)
5. Solicit, collect, log and acknowledge submission of candidate items for consideration.
6. Administer the activities under IV above.

1

Figure 17.4
A Sample Proposal

V.   National Review Board

The National Review Board will evaluate submissions in accordance with the established criteria. It will make three categories of selection:

1.   Exemplary Model Project
     A Model Project is one that significantly improves productivity, cost effectiveness and/or reduces loss of life, injury or property, and is readily transferable nationally. (Only a small portion of the candidate projects will be able to meet the standards.) The present budget is based on an estimate of costs for five (5) such projects.

2.   Programs of Excellence and Transferability
     These are programs meeting most but not all of the criteria of exemplary projects. Some may require further refinement for transferability. At this point in time, costs for programs requiring further refinement or testing are not possible to estimate and will be determined and negotiated in a succeeding phase of this program, as may be appropriate.

3.   Demonstration Projects
     These are innovative projects generated by candidate programs in 1 and 2 above but may require demonstration in other localities to promote transfer or require certain resources or some means of financial support to accomplish transfer or any required further testing. Costs for demonstration projects will vary with the individual character of those so identified and with specific requirements for their transfer and utilization by other communities and Fire departments. The present budget does not include such costs. Candidate demonstration projects will be identified and upon the decision of the U.S. Fire Administration to implement demonstration, costs will be determined and negotiated in a succeeding phase of this program.

VI. Expected Results

Candidate programs will provide documentation in the form of manuals, instructional handbooks, and exhibit material for transfer and utilization. In certain instances a "host" type program will ensue in which personnel from an exemplary model project or from a program of excellence will spend a period of time in another community to teach the innovative procedure.

It is also anticipated that documentation from these projects will become the basis of courses to be developed by the Fire Academy.

VII. Evaluation Procedure

Evaluation will address how effectively the objectives of this program were met. Evaluation instruments will be developed, e.g., questionnaires, check lists, observation, feedback. Documentation packets will meet criteria set. Target utilization audiences will be surveyed for utilization experience as to its use, cost effectiveness, results, and satisfaction.

VIII.   Significance of Results to Fire Service

By creating an awareness and availability of innovations and cost-effectiveness procedures, the fire service will be able to meet greater requirements more cost effectively and efficiently, and most important, fire losses will be reduced.

2

Figure 17.4 (continued)

IX. Staffing Requirements

Principal Investigator

Program Analyst

Clerical

Review Board (Ten in number)

X. Budget Requirements

A. Personnel

(1)	Principal Investigator 2 months @ 4,500 per month	$ 9,000
(2)	Program Analyst 4 months @ 3,000 per month	12,000
(3)	Clerical 6 months @ 1,500 per month	9,000

B.	Materials	1,000
C.	Communications and Mailings	2,500
D.	Final Report and Documentation Packets	3,500
E.	Plaques and Award Ceremonies	2,400
F.	Review Board (ten members; expenses only)	
	6 trips and per diem @ $500 average	51,000
G.	Personnel Benefits (75% of Item A)	22,500
	TOTAL	$112,900

XI. Time Requirements

The Fire Service Model Program for Information and Technology Transfer will be completed in 12 months. Schedule of Tasks and time required for each are as follows:

Months from Start

Task	1	2	3	4	5	6	7	8	9	10	11	12	13
1. Design	—												
a) Criteria	—												
b) Screening Mechanism	—												
2. Establish National Board		—		—		—		—		—			
Board Meetings		—											
3. Solicit Projects		—	—										
4. Progress Reports		—			—			—					
5. Evaluate Projects			—	—	—	—							
6. Awards											—		

3

Figure 17.4 (continued)

Months from Start

Task	1	2	3	4	5	6	7	8	9	10	11	12	13

7. Final Report   ——————
    a) Program   —
        Review
    b) Draft Report   —
    c) Draft Review   —
    d) Revision &   —
        Reproduction
8.   Documentation   —
      Packets

XII. <u>Deliverables</u>

    1. Quarterly Reports (3)

    2. Final Report (One Reproduction and ten copies)

    3. Documentation Packets

XIII. <u>Post Project Utilization</u>

Once a project has been designated "Exemplary," brochures will be prepared* to summarize its operations, highlighting results and particularly innovative elements. Instruction manuals* are also developed, detailing operating methods, budget, staffing, potential problems, and measures of effectiveness. Particular attention is focused on evaluation methods so that other locations adapting the program can gauge their own success and shortcomings.

———————

     *These items are not included in the present budget, since they belong to a second phase of the program, if results of the present experimental activity so warrant.

4

Figure 17.4 (continued)

Letter of Transmittal

The Letter of Transmittal is the formality for presenting the proposal to the recipient. Its duties are similar to those of the transmittal letter of a report. It reminds the funding source of the circumstances that called for the proposal (the specific RFP or the interest expressed by the recipient in previous interactions such as receipt of an abstract). The letter may call attention to certain aspects of the proposed work or the responder's past experience in the problem involved; however, the transmittal letter's major duty is to present the proposal. That document should speak for itself.

Cover/Title Page

Proposals of ten or more pages are bound to ensure that pages do not become loose. Identifying information is listed on the cover. The title of the proposal, the organization and address to whom the proposal is submitted and, if appropriate, the RFP or announcement that the proposal addresses. The date of submission is included because RFPs set time limits for receipt of proposed work, and costs for the proposed work reflect a specified time period. The full name and address of the presenting organization is also identified on the cover.

The title page includes all of the information listed on the cover. If the title of the proposal was abbreviated on the cover, it is stated completely. The title page may have a proposal number assigned to it by the organization forwarding the document. It may identify the division and individual responsible for the proposal, as well as the specific recipient.

Executive Summary

The executive summary of formal proposals prepared in response to RFPs serve a purpose similar to the executive summary element in formal research reports, as discussed in Chapter 13. It synthesizes the important issues of the proposal; its purpose is to recompose the proposal into an abbreviation of the document so that busy decision makers, who often are unfamiliar with the technically formidable content of the text, need not wade through the entire document to understand its purpose, procedures, results, and expected significance. The executive summary identifies the problem to be investigated or the work to be performed, the methodology to be used, and the proposed solution or hoped-for results.

Placed strategically first, the executive summary often serves as a persuasive sales instrument because it calls attention to the key points that may solve the problem of concern. The *abstract* prepared for an unsolicited proposal, together with a letter querying the interest of a funding source in a proposed project in that instance serves the same purpose as an executive summary.

Table of Contents

A proposal of six or more pages should have a table of contents to list its divisions and parts.

Body of Proposal

Headings should be used to make the document easier to read and follow. The items of the outline on pages 460–462 become contents to list its divisions and parts.

Similar to the introduction of a formal report, the proposal's *introduction* identifies the problem being addressed, its purpose, and its scope. In both the solicited and unsolicited proposals the identification and significance of the problem or work are strategic. Your analysis indicates that you understand the issues involved. In unsolicited proposals particularly you need to convince the funding source of the importance of the proposed work and its ramifications to the organization's concerns. This is best accomplished by highlighting the current state of knowledge and the history of prior related investigations. Previous work is described, along with results, shortcomings, or new opportunities. You should clearly define the proposed scope of your activity, what you will do and what you will not do, to avoid any misunderstandings later. Your aim is to convince the funding source that the problem needs to be solved, that with support your solution is feasible, that you can solve it, and that your solution will bring positive benefits.

The section(s) describing the *technical work* is critical. It is the basis for evaluating your understanding of the problem and your ability to find a working solution. It should answer such questions as:

How will the work be done?

What material, methods, and personnel will be used?

What resources and facilities are needed and does the proposer have them?

When will the work be started and how long will it take? (time schedule)

How much will the work cost? Why?

What will be the results?

Will the results solve the problem?

Are the results practical and reproducible?

Your proposal will compete with others; it should, therefore, stipulate clear-cut objectives with a concrete work plan that will accomplish them. The work plan should demonstrate your mastery of the state of the art so as to lead the planned work from the problem to a logical and practical solution. Ambiguously and vaguely conceived work plans are not competitive. Depending on complexity, the technical section may be written as a single unit or in several sections.

The *management* section forms the basis for the evaluation of the competence of your organization to fulfill the promise of your technical work statement. The quality of your organization is measured not only by its resources— financial, material, and personnel—and by its experience and reputation, but also by its ability to document these factors in terms of a management plan to accomplish the proposed work. Elements to be included are time allocations and responsibilities of the principal investigator and other project staff members, facilities, equipment, and resources available. Resumes of the principal investigator and of the project staff, summarizing their experience, education, and their publications related to the proposed work should be included. The budget should be itemized under headings of salaries and wages, equipment, expendable materials, other costs, and administrative costs. Budgets prepared under RFPs are usually separated from the proposal itself, so costs will not influence technical evaluations. Special budget sheets with instructions on allowances are usually furnished by funding agencies. The proposal should also include plans for dissemination of project results.

Concluding Statement

A formal conclusion section may not be found in many proposals, especially those that are lengthy and complex. However, you should conclude your document, as some highly successful proposals do, with a re-emphasis of the need for the project and with a restatement of the benefits that would accrue. Sometimes such concluding remarks are placed in the executive summary and in the transmittal letter.

The Appendix

This section includes such backup materials as: company brochures, annual reports, reprints of related work, highly technical data that could clutter the text but would be relevant to technical experts reading the proposal, and oversized illustrations.

Proposal Evaluation Criteria

Proposal preparation calls for the investment of much time, serious effort by key professionals, and, of course, money. Unfortunately, the expensive effort is not always rewarded with a grant. Many are rejected for any or all of the following reasons:

1. The problem is not within the scope of interest of the potential sponsor or it has a low priority based on the funds available.
2. The objectives are vague.
3. The solution to the problem isn't feasible or seems unfeasible from the approach suggested.
4. The proposed program is too complex.
5. The written document is poorly presented and disorganized; the language ambiguous and unclear.
6. The overall project is not adequately detailed.
7. The proposal contains errors and is slovenly in appearance.

Table 17.1
Point Values for Proposal Evaluation Criteria

Evaluation Item	Point Value
Understanding of Requirements	10
Soundness of Approach	
Analysis	5
Viability	5
Success probability	5
Qualifications	
Staff	20
Bidding organization	10
(Adequacy of facilities; past record on similar projects)	
Management Competence	
Project organization	10
Methods and procedures	15
Costs	20
TOTAL	100

Funding agencies often include their evaluation criteria in their RFP, showing the point rating of each item in the criteria. The rating scale reflects the importance the funding source attaches to the separate items in their criteria. While criteria may vary from agency to agency and from proposal project to proposal project, the ratings listed in Table 17.1 are general for most proposal evaluations.

Proposal Discussion Overview

1. A proposal is a document that addresses a problem and attempts to convince a funding source to support the plan for its solution.
2. A solicited proposal is developed in response to an announced or advertised request.
3. An unsolicited proposal is initiated by someone seeking support for solving an identified problem of interest to a funding source.
4. A careful delineation of a problem is basic to a proposal; without a recognized problem there is no need for a proposal document.
5. Proposals on relatively simple problems can be presented in letter format.
6. More complex situations require extensive analysis, research, and documentation, similar in approach to a formal research report; the proposal document includes many of the elements of a formal report.
7. Because of the time-consuming effort and expense necessary to prepare a formal document, the abstract has evolved as part of a querying instrument

for gauging the interest of a funding source in the problematic situation depicted.

8. The storyboard technique borrowed from film scriptwriting has evolved as a method for developing parts of the technical sections of a proposal. It is a productive technique for team effort, and can be used just as productively by an individual.

9. The technical section of a proposal presents the methodology and solution to the problem.

10. The management section describes the personnel involved in the work, as well as how the management will ensure that the work will be productive and delivered on time.

11. The cost section (often bound and evaluated separately so that the technical persons at the funding source are not influenced by cost considerations in their evaluation) conveys the price of doing the work.

12. The appendix contains all relevant support material.

The Convenience and Advantages of a Computer Prepared Proposal

As with any document that goes through several drafts or is prepared piece-meal, the proposal lends itself to the convenience of being prepared on a computer. Needed information and data can be obtained readily from computerized databases. Hypertext technology can contribute portions of the proposal text. Sections and graphics can be prepared separately by different personnel and later merged into a single document. Deletions, additions, and format and style changes can be made gracefully and quickly. Certain universal sections, called **boilerplate** material, as for example, resumes of staff assigned to work on the project, facility information, and budget and financial data, can be prepared ahead of time and used again and again. Word processing lets you update, edit, and revise the proposal text with timesaving advantages. Appearance, so important to the sales ability of your proposal, can be enhanced through the formatting, typography, graphics, and desktop publishing capability of the technology.

Chapter in Brief

In this chapter we discussed how to prepare proposals. We first examined informal memo and letter proposals, and then concentrated on formal solicited and unsolicited proposals. Prewriting preparations and techniques, and guidelines for organizing and writing the proposal elements were discussed, as was using the storyboard when preparing the proposal's technical section. We looked at the criteria that funding sources use to evaluate proposals. Finally we reviewed the convenience of using a word processor to prepare proposals.

Chapter Focal Points

- Definition of a proposal
- Informal letter and memo proposals
- Requests for proposals (RFPs)
- Role of the abstract in the proposal process
- The storyboard concept
- Evaluation criteria
- Guidelines for preparing proposal elements

Assignments

1. Write a letter proposal to your instructor soliciting her or his approval to investigate and report the problem or situation of your formal report. Remember the proposal is a sales document. To receive your instructor's approval you must be convincing that—

   The problem to be investigated is significant to the effort required;

   You have an understanding of the problem;

   It is amenable to investigation;

   It can be done within the time and scope of your course period;

   The result will be beneficial (to whom and to what extent);

   You have or can achieve the technical competence required;

   Your approach to the investigation and solution is sound; and

   That this work will advance your development both in the course and professionally.

2. Assume you have been asked by your university or college Student Council to identify your institution's most urgent problem, for example:

Student housing	Student counseling
Drug/Alcohol addiction	Lack of adequate health services
Campus traffic/Parking	Grading system
Overemphasis on athletics	Library resources
Laboratory resources	Computer resources
Lack of student employment opportunities	Relations between town and campus
	Increasing tuition costs

   Select one of these or a comparable topic (with your instructor's approval) and develop a proposal abstract.

3. Assume the Student Council has approved the work in your abstract for further development. Follow through in developing a proposal, containing the elements identified in this chapter:

   Background of the problem or situation

   Significance of the problem or situation

   Objectives of the proposal

   Work statement (methods and procedures to implement objectives)

   Resources and costs (if any) required

   Personnel needed to do the work

   Significance of results

18
Oral Reports

Chapter Objective

To provide guidelines for oral presentations of reports.

Chapter Focus

- Informal, impromptu reports
- Semiformal, extemporaneous reports
- Formal oral presentations
- Preparation strategies
- Audience analysis
- Gathering information
- Organizing the information
- Practicing the delivery
- Techniques for delivering the oral report
- Audiovisual aids

W hy should a text on a technical writing be concerned with oral delivery of information? There are three principal reasons:

1. All of us daily are called upon to share and present information—to the class, to the boss, to colleagues, or, informally, to a friend or family.
2. Speaking, like writing, is a medium of communication; principles of the process of communication, of language, of semantics, of exposition, of analysis, and of organization apply to both speaking and writing.
3. Principles of report preparation, organization, composition, and preparation of graphic aids are equally applicative.

As you advance professionally, your ability both to write and speak becomes more and more important. At conferences, at meetings, you will be called upon to discuss your work, to propose solutions to problems, to analyze trends, to justify budgets and activities. Many written reports often require an oral presentation.

The purpose of this chapter is to provide guidelines for making oral reports. It is not a substitute for a course or a text on speech. Like its counterpart, the oral report must be well prepared, factual, informative, clear, systematic, well organized, technically valid, and appropriate for its audience. There are significant differences, which translate into advantages and disadvantages, between the spoken and the written report.

Advantages

1. Immediate feedback enables you to adjust the content, style, and delivery of your information to make your report more relevant. (You can see immediately if your audience is with you.)
2. Your personality can help more directly to influence your audience's reception of your information.
3. Your information can be received simultaneously by many more people.
4. You have an opportunity to clarify obscure or complex points either in a question and answer period or during delivery.

Disadvantages

1. An oral report has limitations of time and complexity. Highly technical information must often be simplified and telescoped to fit a specific time-frame and yet be valid. In a written report, readers can study complex data as long as they need to.
2. Nervousness on your part, poor delivery, or problems with projection of your visuals can pervert and becloud your carefully prepared information.
3. Your listener has only one opportunity to grasp your information and cannot go back and review and study it.
4. Your reader sets his or her own pace for comprehension of the written report, skimming some parts and pausing to study other elements. In an oral report,

you establish the pace; if you misinterpret or ignore feedback, your listener may be lost.

5. Your audience cannot see headings between sections and subsections. (However, there is no reason why you cannot project these on a screen or note these on a blackboard or felt board.)

The differences between the two media not only identify the problems that exist in the speaker-listener relationship but also call attention to advantages of the oral communication situation: You can use your personality, voice, and gestures as well as eye contact, visuals, and feedback to engage your receiver's attention effectively and elicit direct response. On the other hand, your written report is easier to organize and refine, can be more complex, can be studied by your reader at his or her own pace, and can be more readily reviewed by decision makers. Keep these differences in mind when you plan any oral presentation.

Types of Oral Reporting Situations

Despite the previously identified distinctions, types of and forms for the oral report are similar to the written report. You should review Chapters 11, 12, and 13.[1] Depending on the situation, your oral reports will vary in style, extent, complexity, and formality.

Informal, Impromptu Oral Reports

These are often casual, one-on-one presentations of information about a situation or problem. They can be as informal as telling a family member what you did at work today or bringing back to your boss information on why the gizmo you were assigned to design will not be finished today as you promised.

When you are unexpectedly called upon to comment on some matter, your response is impromptu, unprepared. How you respond can mark you as a professional or an amateur. Not all of us are walking, articulate encyclopedias; all of us, however, can avoid becoming embarrassed. With experience and practice we can learn to stay poised. You are not expected to give a formal speech in response to an impromptu question. So, pause; take a deep breath; think while you're breathing about the subject you were asked about; take another breath; as items of recall start coming to mind, speak directly on what you know or what your opinion is on the matter. Say simply, "This is what I know." Or, "This is what my opinion is." Don't rattle on and on or fumble and fuss.

Listen to yourself as you speak; listen to others who express themselves well and clearly. Notice how they look and how they speak. Clear speaking is the result of clear thinking. Analyze what these articulate persons do and how

---

[1] You should also review Chapter 3, as well as the chapters in Part 2, "Modality and Media," and Chapter 15, "Graphic Presentation in Technical Writing."

they talk. Can you emulate them in the future? Try. Practice. Articulate expression comes with experience; experience comes with practice. While you can't practice to respond to matters you know little about, you can practice to respond clearly and with poise on what you know, then offer to learn more for a future discussion.

Extemporaneous, Semiformal Reports

These types of oral presentations follow certain conventional formalities. One example of the semiformal report would be your presenting before your class a synopsis of your written report. In a work situation, you may be called upon to brief your colleagues or members of your work team on what you found out about the problem, task, or situation at hand. In staff meetings, including teleconferencing, we are often called to review progress, analyze work problems, predict trends, or present briefings on policies. A **briefing** is a short, factual summary of the details of a current or projected situation, or a concise rationale of a condition. For a briefing, you might make one- or two-word notes as reminders and to ensure the adequacy and accuracy of your information. These notes need be no more than a listing of topics in an order that provides focus, coherence, and unity.

Formal Reports

Like its written equivalent, the formal oral report follows the constructs of the reporting situation. In this category are talks at professional meetings, video presentations at teleconferences, reports at national conventions, and speeches to civic groups and to similar formal assemblies.

The effectiveness of your oral report depends on factors similar to the effectiveness of your written report. In an oral report, there are six major steps:

1. Prespeaking—preliminary analysis and planning
2. Gathering the information
3. Organizing the information
4. Composing the presentation
5. Practicing the delivery
6. Delivering the oral report

Prespeaking—Preliminary Analysis and Planning

Of course, the medium of communication for the oral report is different from that of the written. The principles, however, are the same. To develop the presentation, you go through similar analytical processes. You begin by establishing answers to a number of key questions:

1. Why am I speaking on this subject (problem)?
2. To whom am I speaking (lay person, executive, expert)?

3. What does the listener (audience) want or need to know?
4. Am I supposed to offer a solution to the problem?
5. Do I expect the listener to take any action? If so, what action?
6. Do I want questions, suggestions, or comments from the audience?
7. What is the purpose of my oral report? Is it to communicate information? To motivate my audience to accept my conclusions and recommendations? To stimulate them to take action?

Just as you did in the written report, formulate your purpose into a clear, concise statement. Once you have determined the purpose and identified the target audience, you can establish the scope of the information you need.

Gathering the Information

Procedures for gathering the information required to reach your purpose have been covered in Chapter 11.

Organizing the Information

The approaches to organization in Chapter 12 apply to the oral report as well as to the written. Factors governing organization are directed to (1) purpose, (2) needs of the audience, and (3) nature of the subject (problem).

Your organization should be geared to the following criteria:

1. Answers the question: Why am I making this presentation?
2. States the thesis of the presentation.
3. Presents substantiation for the thesis.
4. Accomplishes the desired audience response to the presentation.

Composing the Oral Presentation

Because of the factors of time available and the diverse mix of the usual audience, the oral report is, in effect, a summary of the written report. Plan on a fifteen to twenty minute delivery period. The approach taken in the composition of the Executive Summary to a formal report, discussed in Chapter 13, is appropriate. The oral report has three major divisions—the *Beginning* or *Introduction*, *Body (Middle)*, and *Conclusion (Close)* of the presentation.

The Beginning or Introduction

The *Beginning*, *Opening*, or *Introduction* serves to capture the attention of your audience, introduce the subject or purpose of your report, and establish your credibility for your conclusions and thesis. Your first few sentences must interest your listeners or they will mentally drift away and never return. To catch your listeners' immediate attention, begin with a rhetorical device appropriate for your topic—a startling statement, an extraordinary statistic, a personal appeal, a striking figure of speech, or a humorous story relevant to your subject. Humor is universally appealing but it can be a two-edged sword if it is ill conceived or inappropriate to the thesis. Beginning with a visual aid or a sound effect can set the tone, establish the thesis, create audience identification with the subject, be dramatic, or result in an agglomeration of all these responses. Any one of the introductory approaches might be appropriate under one set of circumstances and inappropriate under a different set of circumstances. The job of the introduction is to get the attention of your audience, introduce your subject, and prepare your listeners for your thesis.

Body/Middle

Your subject, once introduced, needs to be explained in whatever details are necessary to accomplish the objective of your presentation. That is the purpose of the body of your speech. It develops the thesis by presenting the material to substantiate it. Remember your time is limited. To develop your thesis, you should consider what is the minimum information necessary to bring your ideas across and what the most effective logic for their sequences should be. Depending on the speech circumstance, you should also decide whether to handle questions from the audience during your presentation, at specific points, or at the end.

Methods for developing the body include these:

1. Examples illustrating points in spoken text or projections of visuals.
2. Repetition of the major points leading to the thesis in order to drive the points home and to ensure audience comprehension.
3. Statistics sparingly used and, more effectively, projected as visuals.
4. Comparisons and contrasts to touch the experience of the audience.
5. Testimony of experts or of participants of or witnesses to an event.

Conclusion/Close

Good salespeople know they cannot "peter out" at the end of their sales pitch. Their final words must convince the potential customer and reinforce the points they have made. Similarly, in the speech situation, the last thing you say is as important as the first in which you gained your audience's attention. Your whole presentation must have a point—the thesis of your report. It is the "payload" you are delivering. To be sure your audience receives the payload, in your conclusion, review the purpose of your presentation, summarize and underscore the main points, and, if appropriate, appeal for the action you want your audience to take.

Your conclusion, or close, should not be lengthy, but it should be vivid and address the points you want your listeners to carry away with them. Sometimes an illustrative example of the effects of your thesis is a productive way to conclude your presentation.

To summarize the organization of your oral report, the Introduction states the idea of your presentation, the Body develops the idea, and the Conclusion restates and reinforces it.

In composing your oral report follow the encoding principles of communication[2] and the elements of the written report.[3] Mode of delivery also plays a role. In memorized delivery, you write out your report and then memorize it completely. Though it has the benefit of cohesive structure and careful use of language, unless you are experienced and can afford the time to practice, it often sounds mechanical. There is also the danger that you might forget a word, a phrase, a line, or even an important section. The result can be disastrous.

You may be tempted to read your report. You have the same benefit of careful structure, but you also face the hazard of boring and losing your listeners. Very few people can read a speech, maintain eye contact with the audience, and provide the necessary supplementary body language and tonal qualities that convey a live personality.

The **extemporaneous** delivery allows naturalness and spontaneity. Delivery is carefully planned, practiced, and based on notes, organized to keep you on track. The notes, either topical or in sentence outline form, enable you to stay in control of your report material.

Practicing the Delivery

Practice means rehearsal. The best way to rehearse your speech is to practice it as closely as possible to the actual speech situation. If you will stand when you speak, practice standing up; if you will not have a speaker's stand to hold your notes, practice holding your notes; if you will use a blackboard, felt board, have an actual model, or, if you are going to project slides or transparencies, practice with these visual aids. If you can practice in the actual room where you will make your presentation, it will be well to do so. If not, try to visualize the room's environment as you practice. Be sure to number your notes. Type or print them triple spaced to make their reading easier.

Try to practice at least once before friends. If that is not possible, use a full length mirror and a tape recorder. Revise any part of your presentation that your feedback indicates is unclear. Check not only your delivery but also your organization to ensure your message is actually delivered.

---

[2]See Chapter 3.
[3]See Chapter 14.

Delivering the Oral Report

Many of us would like to think that what we have to say is more important than how we say it. Unfortunately, that is not the case, according to researchers in public speaking. They have found that effective, fluent delivery increases a speaker's credibility and improves audience comprehension and retention of a speaker's message.

"All right, so delivery is important!" you may say. "I have practiced my oral report, but delivering it in practice is not the same as giving it before an audience. I am frightened stiff. What shall I do?"

Your feelings of apprehension are not different from those of a quarterback before a big game or those of an actor at opening night. Your feelings are normal and actually healthy. You need an increase in adrenalin to perform what you have practiced. Winston Churchill, probably the greatest speaker in our century, was asked, "You are always so poised. Didn't you ever suffer from stage fright?" "When I was much younger, I did," said Sir Winston. "The day I had to make my first speech in Parliament I told a colleague how frightened I was. He was an experienced, wise and kindly man, a brilliant orator. With a twinkle in his eye, he said, 'You will do very well—as you rise, imagine that all of those distinguished old fogies before you are stark naked. That's what I do.' And I did just that. No audience stark naked is a frightening spectacle."

I am not suggesting necessarily that you imagine your audience as being stark naked. I am suggesting that even the most accomplished speakers have a certain amount of tension before a speech situation. Some tension is essential to effective speaking. Have you ever watched a race horse at the starting gate? The horse is nervous and ready to go. The animal has been trained to perform at the sound of the gun and is straining to run. Its adrenalin is flowing. We behave similarly before a speech situation. So, know you will experience tension, but the tension will help you deliver the report.

Communication, as you will recall from Chapter 3, is a two-way process. The advantage to the sender (you) is that you can obtain immediate feedback from the receiver (audience). Feedback, of course, can be negative as well as positive. Use the cues you receive to help you adjust and meet the requirements of the speech situation. If you have carefully taken the five previous steps before delivery, you will speak with confidence even though within you are full of nervous tension. A confident speaker has poise; with confidence, you will find your audience reacting favorably. Be natural, be yourself. Your audience should not be a blur of faces. On the other hand, you cannot look directly into the eyes of every member unless your group is ten or less. Select some individuals with whom to establish firm eye contact so you can observe their facial reactions.

You will find the following guidelines helpful in your speech delivery.

Breathing

Breathing is the basis of speech. In order to make normal speech sounds, you must first inhale. Speech is manufactured when you exhale with pressure; the vibratory motion of your larynx (Adam's apple) sends sound waves out through your mouth and nose. Patterns of individual sound are formed by the tongue, hard and soft palates, the gum ridges, and teeth. You have been breathing and speaking all your life. Just continue doing what comes naturally. But before you speak, be sure you have a good supply of air in your lungs. Inhale. Pause and take several good breaths. As your lungs fill, you can feel them push your diaphragm out. The pause, you will find, helps you to get your audience's attention. Look at your audience and inhale softly. Breathe frequently during your presentation. Break your speech material into small segments, so that you don't run out of air before your next breath. But be sure your breath-group phrases are cohesive.

Pitch

Pitch is the level of sound or tone of your voice. It is based on the physics of the number of vibrations of your vocal chords. Variations in tone are called inflections. An interesting voice has varied inflections, but uses a conversational tone. A pitch different from your natural speaking voice is distracting.

Voice Quality

Resonance, as you know from listening to stereo, adds richness to sound. In practicing your speech, experiment sending the sound waves through the resonating chambers of your speech mechanism—chest, mouth, head, and nasal cavities.

Intensity

Intensity is the force or loudness with which you project. Depending upon the size of your audience and the room arrangement, you should speak slightly louder than you would in normal conversation. Your volume should be loud enough for everyone to hear, but not so loud as to overpower your listeners.

Rate

Rate or speed is the tempo of your speech. It is a factor in being understood. If you speak too fast or too slowly, you will irritate and lose your audience.

Pause

Related to rate, the pause can be effective in drawing attention to important points.

Pronunciation

Mispronunciations undermine an audience's confidence in the speaker. If you are unsure how a word is pronounced, look it up in the dictionary.

Enunciation

Enunciation is the way a speaker articulates words. It is the clear, precise formation of sounds of a word. Again, faulty enunciation is distracting and undermines an audience's confidence.

Body Language

Whether you use gestures or not, your body as well as your voice speaks for you. A poised speaker is one who appears self-confident, relaxed, and capable of doing what the speech situation calls for. Gestures do not come naturally to the inexperienced speaker, but they can be effective in emphasizing a point. Do what comes naturally. Gestures should call attention to an idea but not to the gesture. Your face is important. It reflects you in action. Your confidence, conviction, and sincerity will be communicated by your facial expression. Some people have distracting mannerisms—scratching their nose, tugging at an ear, twisting a forelock, or looking down or away from the audience. Check your delivery before a mirror or friends to catch any distracting mannerisms. Always try to be natural. Be comfortable with yourself and what you do, and your audience will be comfortable with you.

Audiovisual Aids In Oral Presentations

Visual aids serve three functions.[4]

1. They help keep the attention of your audience (keeping them interested).
2. They present information in clearer fashion by means of the visual channel.
3. They help your audience to retain the information.

The visuals most commonly used are models, graphs, maps, charts, photos, drawings, and short printed passages. They may be slides and transparencies projected on a screen; they may be drawn or lettered on a blackboard or displayed on flannel boards; or held. They can be on sheets of paper, reproduced and distributed to the audience. Visuals have many advantages. You can often show an audience something more effectively than you can tell them about it. Showing *and* telling are often more successful than using either technique by itself.

Using Audiovisual Aids

1. If you are using a microphone, be sure it is adjusted for volume and your height before you begin your presentation.

---

[4]Review Chapter 15, "Graphic Presentations in Technical Writing."

2. Check your audiovisual equipment: slides, transparencies, felt, chalk, and poster boards *before*, not after, your presentation begins. Be sure your slides and transparencies are in proper order.

3. Be sure that your audience will be able to see illustrations and read the captions. The size of lettering and images should be determined by size of audience and viewing room.

4. Firmly anchor illustrative materials on their felt boards so they don't fall during your presentation.

5. Illustrate only one concept on a visual; two or more become confusing.

6. Don't block your audience's view by standing in front of the visuals you are explaining.

7. If you are using an overhead or slide projector, don't leave the visual on screen after you are through using it; your audience will shift its attention from you to the projection. Also, turn off the projector when you finish. Its noise and lights are distracting.

8. If you have prepared handouts, be judicious about when to distribute the material; your audience will read the material rather than tune in on you. Handouts are useful when you want your audience to follow specific or intricate details you wish to emphasize as you talk. If you want your audience to take the information with them, distribute the handout at the end of your presentation.

Guidelines for Oral Report Presentation

An oral report requires similar preparation to a written report. It is based on careful and thorough research and objective analysis. Like the written report, the oral begins with an introduction, followed by the body, which discusses findings, and ends with conclusions and recommendations. Because of time limitations and greater diversity of audience members, the oral report often follows the psychological organization scheme, so that conclusions and recommendations are presented to follow the brief introduction.[5] The body, offering substantiating details, follows. The concluding section in this organizational scheme summarizes the entire report in order to reinforce the thesis of the presentation.

Steps in oral reporting are these:

1. Establish the purpose of your presentation.
2. Analyze your audience in terms of their knowledge and attitudes.
3. Research the problem or situation to be reported.

---

[5]See Chapter 12, pages 299–300.

4. Analyze the data in relation to the purpose.

5. Prepare an organizational scheme, concentrating on major points you want your audience to receive if your purpose is to be met.

6. Compose your presentation, formulating it for your intended audience and placing particular effort into a strong introduction and conclusion.

7. Practice your presentation—more than once in order to gain confidence and poise.

8. Deliver your presentation, using audience feedback to make adjustments if necessary.

9. Summarize and reinforce your ideas to make your thesis comprehensive and acceptable.

10. Conclude in time to allow questions from your audience.

After the presentation, evaluate your experience from audience feedback and reaction (from both classmates and instructor). The more you understand your successes or failures, the more likely you will be successful in your next oral presentation.

Though this chapter is not a substitute for a good course in speech, you will find, if you follow the guidelines, that your oral report was not a frightening experience, and that you may even look forward to the next opportunity. The more experience you gain in speaking, the more confident, poised, and successful you will become. The attribute management looks for in professionals is ability in both written and oral communication.

Chapter in Brief

Similar to the written technical report, the oral report must be well prepared, factual, informative, clear, systematic, well organized, technically valid, and appropriate for its audience. Various types of oral reports were discussed in this chapter, as well as how to prepare, organize, and deliver them. Techniques for oral delivery and guidelines for using audiovisual aids were presented.

Chapter Focal Points

- Informal, impromptu oral reports
- Semiformal extemporaneous oral reports
- Formal oral reports
- Prespeaking preparation
- Organizing the report
- Delivering the report
- Vocal techniques
- Audiovisual aids

Assignments

1. Observe a lecture or technical presentation at your college. Evaluate it according to the principles and guidelines in this chapter.

2. Each of us is an expert in some matter or idea or belief about which we feel strongly. Assume you have been asked to speak on that subject matter, belief, or idea. Prepare a four to six minute presentation to inform or persuade your audience.

3. Prepare a short presentation of four to five minutes describing a device/organism or one explaining a process. Include visuals.

4. Assume the instructor in one of the courses of your major field of study needs to be away from class and has asked you to take the class for that class period. Prepare a ten minute lecture on the subject so as to stimulate class questions and discussion. Your success in this assignment may be measured on the amount of discussion stimulated among your class members.

5. Prepare an oral report based on the formal report you have written for your technical writing class, following the suggestions in this chapter. Prepare visuals, outline your presentation, and practice it. Present it to the class in ten to twenty minutes (as your instructor may assign). A question and answer period will follow your presentation.

6. Write an evaluation of your presentation and turn it in to your instructor for comments.

19
The Technical Paper and Article

Chapter Objective

To explain the purpose and role of professional scientific papers and technical articles and to provide guidelines on how to write them.

Chapter Focus

- Communications to the editor
- Technical articles for primary and technical journals
- Review and semitechnical articles
- Popular science articles for the general public
- Query letters
- How to write the article

A giddy hostess once asked Albert Einstein to explain his theory of relativity in "a few well-chosen words."

"Let me tell you a story instead," said the scientist. "I was once walking with a blind man. The day was hot. I remarked that I would like a glass of milk."

"What is milk?" asked my blind friend.

"A white liquid," I replied.

"Liquid I know, but what is white?"

"The color of a swan's feathers."

"Feathers I know, but what is a swan?"

"A bird with a crooked neck."

"Neck I know, but what is crooked?"

"Thereupon, I lost patience," said Dr. Einstein. I seized his arm and straightened it. "That's straight," I said. And then I bent it at the elbow. "That's crooked."

"Ah!" said the blind man. "Now I know what you mean by milk!"

This little story illustrates the difficulties in communicating about the advances of science and technology.

Writings about science and technology, other than reports, can be classified into three major types:

1. Technical papers for peers in primary and technical publications
2. Semitechnical and review articles for other scientists
3. Popular pieces for the general public

Technical Papers

It is traditional for scientists to write and publish their original findings and results in the primary journals of their fields. Scientific periodicals are the forum where peers evaluate their work. Through publication, scientists establish their reputations. It has become a practice for authors to obtain reprints of their papers and circulate them among other workers in their fields. This practice has reached the point where it has been humorously said that eminent scientists obtain the bulk of their knowledge of others' research from reprints that are sent to them.

Nevertheless, many scientists and engineers are loath to write a paper, report, or article, claiming that they dislike the idea of standing up before an audience to read the results of their work. Scientists and engineers are reluctant to write for the same reason other professionals are—fear of being unable to express their thoughts in speech or on paper. They have a wealth of information and knowledge but not the writing aptitude, or, more accurately, the writing *experience*. Yet, by necessity and tradition writing is part and parcel of the scientific pursuit and a required, "normal" activity.

Research is not complete until the data have been recorded, disseminated, and published. In our era of the computer, research data are sometimes disseminated by disk or by computer network, but such dissemination is not a final refereed paper but raw data. This is not yet considered **publication** of the work, which must still be written up and presented according to the conventions of the discipline. It is surprising how much work is carried to completion with great expenditure of funds and time and then left in notebooks or in computer files until they are too stale to be of use. This state of affairs may have been permissible in the days when scientists financed their own research. It is hardly acceptable when the money comes from others. Publication should be prompt and complete.

Research and significant advances may be reported in a number of ways. They may be set forth in report form, not for publication but for use within a given organization or for an organization's clients. Some scientific and engineering research organizations confine themselves solely to research. Their only tangible product is the report. In such an instance, the report may be circulated within the organization and/or to outside organizations that have a need for, or are entitled to the research information developed. At other times, the research may be published as a short communication to an editor, as a full-length paper, as a longer article for one of the journals which specializes in the field of research, and/or as a book.

Communications to the Editor

The communication or letter to the editor is a form that permits quick publication and dissemination of short summaries of significant research on topics of high current interest. Some important work is never published in any other form; details of problems, techniques, data, and results are exchanged among a small group of interested investigators and are never widely disseminated. The communication is written to contain only that information which other investigators require to carry on similar work.

Format conventions of the communication to the editor vary from publication to publication. *Science* titles its section "Reports"; *Nature* calls them "Letters" (and also has a section entitled, "Scientific Correspondence" for comments on scientific matters). The *Journal of Chromatography* carries the communication in a section entitled, "Short Communications"; the *Journal of Abnormal Psychology* as "Short Reports"; the *Journal of the Oil Chemists Society* as "Letters to the Editor"; the *Journal of Physical Chemistry* as "Letters." The American Physical Society, which publishes many periodicals covering the wide spectrum of physics, publishes several periodicals devoted strictly to letters: *Physical Review Letters*, *Optics Letters*, *Applied Physics Letters*, *Astrophysical Letters*, *JEPT Letters* (translations from a Soviet publication).

A number of other professional societies as well as commercial publishers have instituted periodicals devoted solely to this form: *Environmental Letters*, *Polymer Letters*, *Information Processing Letters*, *Europhysics Letters*, *Chemical Physics Letters*, *Analytical Letters*, *IEEE Electronic Devices Letters*, and many others.

Since communications to the editor permit quick publication and dissemination, this format has become popular with research scientists in all fields, particularly those who wish to establish priority of discoveries or wish to obtain feedback on their research.

In their published form, these short communications to the editor bear little resemblance to a conventional letter. Though signed by the author, they are actually short reports. Their size varies from about 200 words to almost 2000. Their organizational structure comprises an introduction (which explains the problem and its means of investigation), body (which discusses the resultant data), and terminal section (which offers generalizations or conclusions and, if appropriate, recommendations). Some contain sectional headings, abstracts, and documentation; many have illustrations, tables, and charts to clarify significant points.

Here is an example of a typical communication appearing in the "Reports" Section of *Science* [5:106–7].

Radiotelemetry of Physiological Responses in the Laboratory Animal

Introduction

The measurement of physiological reactions in laboratory animals has long presented problems. Recording is usually accomplished under conditions varying from complete restraint to the partial restraint of extended wires. The animal may never grow entirely accustomed to the attachments, and this factor may constitute a source of stress. Restraint is reported in the literature as being extensively used as a stressor, the work of Selye on this topic being widely known. It is apparent, therefore, that measurement of physiological variables under conditions of even minimal restraint may yield a distorted picture of such responses. In addition, the cues inherently present in attachments of any kind, as well as the transfer of an animal to the study environment, are major—if not, at times, fatal—methodologic obstacles to classical conditioning. The development of the transistor offers new possibilities for classical conditioning.

Method of Experiment

With this in mind we set out to develop a system of radiotelemetry for use as an adjunct to studies in classical conditioning. This methodological advance will permit the monitoring and recording of selected physiological reactions in intact, unanesthetized laboratory animals during their normal daily routines in a simulated normal environment uncontaminated by the intervention of the experimenter and experimental procedures, except for planned changes in the controlled environmental chambers. Transistorized, miniaturized, battery-power packages are being developed in our laboratories and permanently implanted in laboratory animals. The modulation of the radiofrequency carrier with biological information permits short-distance propagation of selected physiological activities, with the possibility also of providing remote control of selected stimuli (1). Continuous measurement of physical conditions within the environmental chambers—ambient temperature, humidity, air ionization, barometric pressure, air velocity, light intensity, chemical composition of the air within the chambers, and such other physical parameters as may be shown to be significant—may be recorded along with the physiological activity specific to the animals. The behavior of the animals may be observed by a remote visual system. Classical conditioning experiments with animals may be conducted over periods of

Figure 19.1

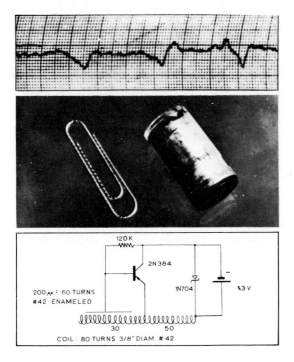

weeks and months, with observation periods before, during, and after experimental manipulation. We feel this innovation, growing out of the technique of telemetering, to be important and an improvement over techniques more familiar to those engaged in classical conditioning investigation.

[Figure 19.1] (top) shows a signal output of respiration from a laboratory rat, obtained by means of the accelerometer principle incorporated into a small capsule [Figure 19.1, center] very much like that reported by Mackay (2). A brass pellet mounted on a rubber diaphragm near the oscillator coil modulates the radiofrequency carrier. We have found that we can obtain a radiofrequency signal of 6.8 Mcy/sec and of about 250 μV at the antenna terminals of the receiver, with circuit shown in [Figure 19.1] and battery content of 200 μa. This by no means represents a lower limit to power requirements, but an arbitrary stopping point for the moment. The characteristics of the transistor are such that an increase or decrease in power is affected by changing the collector voltage and adjusting the emitter bias to increase or decrease the collector current. There is no reason that the current cannot be of the order of 50 μa, if this is permitted by the radiofrequency noise level within the environmental chamber. A transistorized, miniature, radiofrequency amplifier can be constructed within the antenna probe and supplied with low-voltage d-c power through the coaxial cable. We have found that, by exciting the capsule from an external source of radiofrequency power of about 3.9 Mcy/sec from a 100-watt exciter, about 2 ma of reverse current can be realized in the battery circuit within the capsule, in which we have included the Zener diode rectifier shown in the circuit diagram [Figure 19.1]. A rechargeable dry cell is being tested for the planned indefinite im-

Discussion

plantation of the capsule. After the transmitter has been assembled and dipped in toughened paraffin, it is inserted by surgical procedures into the abdominal cavity of the rat, where ballistic movements are sensed by the accelerometer. The signal is picked up by a "ferriloop-stick" antenna mounted within the environmental chamber and conducted through a coaxial cable to a communications receiver. The incoming frequency-modulated signal is mixed with a beat-frequency oscillator within the receiver, or with another radiofrequency signal from a frequency meter. The audio output of the receiving system is recorded on magnetic tape for subsequent playback and data analysis.

Results

While the base line of the trace [Figure 19.1, top] is an accurate (0.2° per 1000 cy/sec of audio output) and reproducible (± 0.5 percent) measure of the core temperature of the animal, a transmitter alone will not detect rapid changes of temperature. The response time constant of the radiosonde used at this writing is of the order of 100 seconds. It is therefore necessary to introduce a thermistor transducer into the circuit to effect this measurement.

Conclusions and Recommendations

The improvement in techniques inherent in microminiaturization and telemetry permit the coupling of classical conditioning experiments into an on-line digital computer to form a closed-loop systems approach to experimentation. This, in conjunction with automatic data processing and reduction, would seem to lead to qualitatively different testing of old and new hypotheses with multiple independent and dependent variables, under conditions of experimental control previously impossible.

Samuel J. M. England

Benjamin Passaminick

*Columbus Psychiatric Institute and Hospital,
Columbus, Ohio*

References:

1. M. Verzeano, R.C. Webb, Jr., M. Kelly, *Science* 128, 1003 (1958).
2. R.S. Mackay, *IRE Trans. on Med. Electronics* 6, 100 (1959).

Papers for Primary Journals

The major form of scientific publication is the full-length paper appearing in a primary journal. Despite special interests, scientists want to know—must know—what their colleagues elsewhere in the country and in the world are doing, and they find out by reading the primary journals and the technical or trade publications of their special fields.[1]

---

[1]In the present context, technical and trade publications are considered synonymous since the technical articles in each are by "experts" for "experts." It is true that distinction can be and is made frequently. Some associate the term *horizontal* with the technical journal and the term *vertical* with the trade journal. The implication is that the latter covers a very narrow industry or technical specialty and the former covers a more general area or group of related industries.

The chief difference between the primary journal and the technical journal is usually the source of publication. A primary journal generally is published by a professional organization for its members and the profession at large. Commercial publishers have entered the primary journal field either in collaboration with a professional society or on their own with an editorial board of prestigious scientists in the field. The technical journal is a periodical covering a specific technical area for workers in that field. The Institute of Electrical and Electronic Engineers (IEEE) publishes more than 60 primary periodicals covering the diverse but specialized interests of its large membership. On the other hand, *Electronics* is published by McGraw-Hill Publishing Company and deals with a similar general electronics area. The periodicals of the IEEE are more scholarly, often archival, and are esoteric in the way they treat their subject matter. Another distinction is the proportion of material devoted to basic rather than developmental research; *Electronics* concerns itself with applied research and practical matters that have a direct bearing on how line engineer-readers can do their jobs better. Some professional societies publish both journals on the state of the art of their profession and a technical periodical for the practitioners of the industrial activity served. Thus the Society of Mechanical Engineers publishes its *Transaction* series to cover the former and publishes *Mechanical Engineering* to cover the latter, and the IEEE publishes *Spectrum*, directed to line engineers.

The article for a scientific primary journal documents a subject researched for the purpose of extending knowledge and exchanging information, facts, and ideas that are current and important to the field. The subject may be a new theoretical concept, a rethinking and invalidation of a previously accepted theory or approach, a new application, or a new discovery—device, technique, process—that changes existing conditions within the field.

In content, organization, and structure the article resembles the formal report. The basic difference is not in subject matter or in organizational structure, but rather in motive. The article is written to test and share information; the report is written to obtain action. The article is written for a reader who may choose whether to read it; the report is written for a "captive" reader who has a stake in the information it conveys.

The most frequently included elements in the primary journal article are almost always found in the formal report:

1. An abstract
2. An introduction stating the purpose, thesis, background information, or background theory
3. Methods and materials
4. A discussion of results
5. An analysis of results and conclusions
6. Generalizations about the significance of the subject
7. References cited (Bibliography)

Publications serving different fields and disciplines have their own peculiar characteristics based on individual requirements and situations. In general, however, papers within various journals bear many similarities, varying usually in detail only.

Each primary publication has its own requirements both as to content and style. Before beginning to write, you might survey for the publication most suited for your subject and purpose. Check with the editor for interest in the content of your article and in your approach and for stylistic and mechanical requirements. Most publications and societies have style manuals or style sheets, which are helpful to contributors. If a publication does not have a style sheet, the editor will be happy to indicate by letter the requirements. A study of the periodicals is the best way for a writer to learn firsthand of the content material most suitable, structural aspects, and stylistic characteristics. An editor looks more favorably upon an article which is written specifically to the journal's requirements.

Articles for a primary journal are **refereed**; that is, the editor of the publication sends the manuscript for peer review to specialists within the subject area of the article for evaluation of the validity of its content, its written expression, and its publication worthiness. Many papers receive several reiterations before they become acceptable for publication. It is advisable to show a draft of one's article to technical experts within one's own organization for critique and suggestions before sending the article off to a journal.

Articles for Technical Magazines

Technical or trade periodicals are geared for readers who are not esoteric specialists but workers in the field. They may be professionals, administrators, or high-level technicians. Each field of activity has its own publications. The subject matter will be of an applied nature. Since these are commercial publications, their financial support comes from advertisers within that area of activity. Consequently, articles are interspersed within many pages of advertising. The advertising itself is of great interest to the readers. Many new products, processes, and services are of professional interest to them. New knowledge of the field is frequently disseminated in the form of announcements of new products, instruments, and services. Readers of the technical publications read the advertising with as much interest as they read articles.

Whereas papers in primary journals are stylistically formal, scholarly, and composed rigorously to reflect the scientific validity of the work described, articles in the technical and trade publications are streamlined for quick impact because their readers can give them only minimum, often cursory, attention. Language is less formal than in the primary journal. Long subtitles frequently are used to give a bird's-eye view of the article. Paragraphs are shorter, as are sentences and words. There is greater use of heads and subheads.

The whole approach in technical trade magazines is to make the article easy to read and understand. The writing, layout, and illustrations are designed to that end. Illustrations—diagrams, charts, tables, and photographs—play an important part in telling the story. Tight editing is done on each article submitted. Unnecessary paragraphs, sentences, and words are eliminated. Mathematics are converted to simple English words or to readable graphics. With these editorial techniques the magazine's pages are made inviting, comfortable, and understandable to busy readers. Such editorial practices are inimicable to the primary journal, but the aims of each type of publication are different and each serves an important need. Today, with many publications and other demands competing for a reader's time, many articles receive only a glance or a quick scanning. Mere recital of facts interesting in themselves will not hold the reader because the reading set is "Why should I read this article? What can I do with this knowledge?" Purpose or value to the reader is primary in the semitechnical article. In planning the article, you, the writer, must ask yourself, Why should the reader of the publication be interested in this information? How can I present it so that its significance reaches him quickly and easily? Many editors refer to articles as "stories." Each article tells a "story." Like a story, the article's narrative or sequence of information is influenced by the purpose and the thesis. The basic objective in the semitechnical piece is communication of useful information. Interest is maintained through emphasis on the significance of the information, through relation of the subject matter to the reader's experience, and through illustrations. Photography, flow charts, diagrams, and color also are used to advantage.

In some semitechnical articles, interest may be created through reader identification with the writer who may unfold a series of obstacles or developments and in the writing tell how they were overcome or achieved. The semitechnical article, unlike the primary article, will make use of the narrative structure that will lead to a climax or to a high point—reaching the objective in which the reader has an interest. The objective is the satisfaction readers achieve through their desire to learn about the subject.

Review Articles

Scientists write articles for the purpose of exchanging the knowledge of their fields with scientists in related and other fields. Such articles commonly present a review of a field or the broad implication of a new theory, process, instrumentation, or trend. A review summarizes and interprets the state of knowledge about a field or the current status of research on a significant problem. In the article, particular processes, techniques, and descriptions of apparatus and tests are generalized; conclusions and significance rather than methods are emphasized. The review article uses less-specialized language, fewer mathematical data, and fewer symbols since its readers are not always familiar with the technical diction of the specialized field. Typical media are: *Science, Scientific American, American Scientist*, and *Psychology Today*. Readership is intelligent, alert, and has a variety of interests.

Stylistically, the review article may be written from an objective (third-person) or from a very personal point of view, with the author playing a prominent part in the telling. The objective style is used more frequently. The basic content is that of the technical report—purpose, dicussion and analysis of the situation, and conclusion—though not necessarily in that arrangement. Generally, the writer will not begin with a statement of purpose, since a reader's attention must first be caught. Beginning with a statement of the significance of the topic may accomplish this. Writers of reviews in semitechnical publications have borrowed an approach professional writers in general magazines use, that of starting their piece with a rhetorical device—an interesting example, an extraordinary statistic, a startling statement, a striking figure of speech, an analogy, a comparison, a definition, or an appeal to personal interest.

For example, in his article. "High Stakes Molecular Chess," Storad, writing about the human immune system, carries the striking analogy of the title into the opening paragraph:

> Under normal circumstances, the immune system functions as a personal physician that cures and protects us against an array of disease-causing opponents. It is also the body's representative in a never ending match of high stakes molecular chess. The opponents—bacteria, viruses, and other microbes—look to exploit every opportunity. Their gambits take the form of infection and disease. The stakes often are life or death. [8:22]

The use of an extraordinary statistic to capture attention is illustrated by the opening paragraph in the article by Wessel and Dominski entitled "Our Children's Daily Lead," appearing in the *American Scientist*:

> Lead poisoning is a preventable disease, yet each year 200 children die of this condition; 12,000 to 16,000 are treated and survive, but at least one quarter of these suffer permanent injury to the nervous system. Recent studies suggest that children with moderate increase in body burden of lead, although clinically asymptomatic, experience serious interference with important body processes. If one includes this asymptomatic group, at least 400,000 children in the U.S. suffer annually from deleterious effects of lead. [12:294]

An example of a startling statement is the opening used by Kluger in his article, "Space Plane," appearing in *Discover*:

> If a machine is judged by the simplicity of its design, then the space shuttle is a downright mess. The ship may look uncluttered enough when it's in orbit, but to get there it needs the help of a mammoth steel and concrete launching pad, a 347-foot gantry, a blimplike external tank, and two tempermental solid-fuel boosters—hardly the most elegant arrangements. [6:81]

Three lines of a well-known children's song are skillfully used by a psychologist and linguist to introduce the theme of their semitechnical article. "The First Two R's," an article by Tzeng and Wang, appearing in *American Scientist*, shows that the way different languages reduce speech to script affects how visual information is processed in the brain:

"School days, school days, dear old golden rule days;
'Readin' and 'ritin' and 'rithmetic,
Taught to the tune of a hickory stick." [10:238]

The last line of this children's song calls to mind the old-fashioned classroom with its stern discipline, which has by now all but vanished from the American scene. It also highlights an interesting fact—that reading and writing are skills that do not come naturally, the way speech does. Typically, by the time he has reached school age, a child has effortlessly soaked up from his environment all the basic structures of the language spoken around him, whether it be English, Chinese, or Telugu. Learning the written language, however, is frequently an arduous process. ". . . This contrast between the two forms of language—speech versus script—is all the more striking given that written language is invariably based on spoken language."

Tulving, in his article, "Remembering and Knowing the Past," published in *American Scientist*, begins with an interesting definition, that of memory:

> In popular thought, a powerful association exists between memory and information storage: memory is a means of storing information. To have memory, in this view, means to be able to produce or receive information and then to keep it in the system over long periods of time. Thus trees with their rings, card files, and phonograph records can be said to have memory, along with brains and computers. [9:361]

In his *Scientific American* article, "Antisense RNA and DNA," Weintraub describes how certain molecules called antisense molecules can bind with specific messenger RNA's to selectively turn off genes. The author believes these antisense molecules may eventually be used to treat certain genetic diseases. Weintraub leads the reader comfortably into this highly technical subject by beginning with a simple statistic and following it with two questions whose answer is the theme of his article:

> It takes 100,000 genes to make a human being. What exactly do they do, and how do they do it? To answer these questions, biologists must tinker with individual genes—in effect, remove or turn off the genes—and observe the effects on organisms or individual cells. Studies of mutations have always afforded this information, but mutations are random by nature, which has made systematic study of individual genes difficult (or, in the case of human beings and other complex and long-lived organisms, impossible). [11:40]

A reference to a specific occurrence is used by Sapolsky in his article, "Stress in the Wild," which appeared in *Scientific American*:

> The year was 1936. Hans Selye, a young physician just starting off in research at McGill University in Montreal, had a major problem. He had been injecting rats daily with a chemical extract to determine the extracts's effects and had identified consistent changes in the animals: peptic ulcers, atrophy of the immune system tissues and enlargement of the adrenal glands. To his surprise, however, the rats in the control group, which had been injected with saline solution alone, showed identical changes. [7:116]

These examples of beginnings illustrate only a few of the rhetorical devices which can be used by the writer of the semitechnical article. The purpose of the beginning is to entice readers into the depth of the article. This emphasis on literary devices, however, should not mislead you into thinking that the primary purpose of the semitechnical article is to entertain. Its purpose is to inform— bring to readers definite information of practical or intellectual value. Because readers of the semitechnical piece are not specialists in the field written about, they may either be bored or frightened off by the specialized character of that discipline or the details and terminology before they reach the matter of interest. Hence the interest-creating devices are called for not only in the introduction but also in the body of the article and in the terminal section. The terminal section in the technical review piece more often than not summarizes the author's perception of the state of knowledge and offers recommendations on how challenges are to be met, or offers predictions on the direction the specialty will progress in the foreseeable future.

The Popular Science Article

Trade periodicals such as *Writers' Digest* and *Writer* note that currently the best market opportunities for writers are primarily in nonfiction; they foresee no end in the decline of markets for fiction. Many prominent magazines exclude fiction entirely from their pages. The magazines publishing both fiction and nonfiction are devoting more pages to the latter.

The reason for the situation, these trade publications say, is that the changing world we live in is so absorbing that to most readers, articles explaining the changes carry much more interest than fiction. Humans landing and walking on the moon have made laymen aware that they are on the threshold of a great age of scientific and technological progress. Fiction can no longer compete with the events of our space age. People want to be ready for a new, miracle age, which they feel is but a door away. They read voraciously and want to understand.

There is a great gulf separating scientists from laymen. A century ago, there was no such gulf. A contradiction of our times is that we can be cultivated in the humanities and still be ignoramuses. The cultured, educated person of 100 years ago was up not only on the Spenserian stanza but also on the theory of evolution, the laws of thermodynamics, the principles of telegraphy. Knowledge since has grown to the extent that no one person can encompass its boundaries.[2]

The basic content ingredients of popular science articles are those of technical reports. The writers treat the material more generally. They soft-pedal or eliminate all reference to complex data or theory. They simplify methods and

---

[2]This is true not only in the sciences but also in the humanities. Witness the number of Ph.D. dissertations on such esoterics as "Non-conforming Tercet Rhymes in the Terza Rima."

give only general or metaphoric descriptions of equipment; for example, they might describe an oscilloscope as a TV-type gizmo no bigger than a bread box.

The organization varies in the popular article even more than in the semi-technical piece, depending on the readership and the technical knowledge level of the audience of the particular magazine or newspaper. A major problem in writing about science is that the simplest technology is incomprehensible to the literate and educated person not trained in the symbols, language, methods, and customary thought processes of the specialist in the field.

The technical writer writing for the general reader should remember Pope's admonition, "The proper study of mankind is man." The more the subject matter deviates from the reader's interests and experiences, the greater the skill the writer must exercise to gain and hold the reader's attention.

Popular articles on science and technology differ in the degree to which they must be popularized because magazines and newspapers vary in the average educational level of their readers. Newspapers with mass circulations require articles written in the simplest terms. Yet, some newspapers, for example the *New York Times* and the *Christian Science Monitor*, have a readership level above that of *Cosmopolitan* or *Reader's Digest*.

The following publications are ranked in order of increasing difficulty:

1. *Parade*
2. *Prevention*
3. *Cosmopolitan*
4. *Reader's Digest*
5. *Popular Science Monthly*
6. *Playboy*
7. *DISCOVER The World of Science*
8. *Smithsonian*
9. *Atlantic Monthly*
10. *Harper's Magazine*
11. *Science*[3]
12. Primary journals such as *Journal of Immunology, Journal of Experimental Psychology, Physical Review,* and others.

Establishing contact with the reader at the very start is all-important. If the reader's interest is lost after a few sentences, it is never regained. The reader has gone to another article, another magazine, or another diversion.

The types of beginnings noted in our previous examination of the semi-technical article are even more appropriate for the popular piece. These approaches are based on human psychology. They include appealing to the read-

---

[3]*Science*, the journal of the American Association for the Advancement of Science, covers primary advances and state-of-the-art reviews of the broad spectrum of biological, physical, and social sciences.

ers' self-interest and curiosity, their interest in other people, and their interest in the concrete or tangible matters in everyday experience. Here are some examples, not previously mentioned, of types of openings based on these major appeals:

1. Appeal to a basic interest
2. Direct question
3. Anecdote
4. Historical background
5. Direct statement of thesis
6. A narrative episode
7. Reference to a specific occasion
8. Metaphors or similes

Steps in Writing the Popular Science Article

The first step is selection of the subject. Choose one with which you are intimately familiar or have researched enough to be technically competent to know what you want to say about it and who would want to read about the subject.[4] Then survey which magazine readers primarily compose the audience you want to reach. If more than one magazine has the audience you want, aim for the one with the highest circulation (you are making a market survey).

Obtain a number of issues of the periodical over a period of about two years. Study style, structure, and layout of their feature articles. Locate issues with articles parallel in scope to the one you have in mind. Analyze and reduce several to outline form to give you a better idea of structure and presentation. In studying your target periodical, analyze the literary style and physical format of the articles. Are they written in the third person or first person? How are difficult, technical words handled? What devices are used to explain theory or abstract concepts? To what extent are illustrations used? How are they used as expository devices? How long is the average paragraph, sentence, word? What types of openings are used in the article? What rhetorical devices? Does the article have advance organizers, prompting cues, inserted questions, overviews, topic sentences? How are heads used? Do heads aid in explanation? Do heads spell out the organization of articles?

---

[4]Writing about science and technology for the general reader is one of the most difficult challenges for the freelance writer. There are freelancers whose specialty is writing the popular science piece. Though some may command from $500 to more than $5,000 per article, researching the piece costs money out of their own pocket and can involve considerable personal time and travel. Most magazine editors have their favorite science writers (who have proved themselves) whom they commission to write on subjects of current interest. Many newspapers, such as the *New York Times* and the *Washington Post,* employ a staff of science writers and illustrators to write stories about the latest advances in science and technology. Some professional technical writers, engineers, scientists, and teachers do occasional freelance writing, utilizing their knowledge of specialized technical subjects. They are usually most successful, however, in contributing to trade publications.

Study the ads carefully. Advertisers are the major support of the general magazine. Illustrations and text of ads should be revealing about the periodical's audience and its level of education, intelligence, and technical understanding.

After such a study, you are ready to rough out an outline for your article. List topics and ideas you will cover. Set down the approach or point of view from which you will present the information.

Then, if necessary, obtain permission to write about the subject from the person, company, or organization involved.

Next, contact the periodical selected to publish the article, inquiring about its interest in the subject matter. A brief outline accompanies or is included in your letter of inquiry. Your letter should explain why the information of your proposed article would be of interest and value to the magazine's readers. Indicate, also, the extent and type of illustrations you are planning to have.

The Query Letter

Here is what an editor of a general magazine has recommended as the ingredients of a good query letter:

> First of all, it should reflect some knowledge on the part of the writer of the general editorial content of the [magazine]. It should also indicate whether the topic being suggested has been covered elsewhere previously, and, if so, when and where. The author's sources of information for this story material should also be given.
>
> An outline of the story line of the proposed article should be included (this need not be more than 150–200 words). Perhaps most important, the specific angle, or peg for the piece should be described. The editor should be told why the writer feels the particular topic would be pertinent and interesting for the current issue of the magazine. Some indication should also be given as to how many words the writer feels he would need to develop the subject properly (our single-page stories run between 850–1000 words, and our double-page stories between 1500–2000). The writer trying us for the first time might also list some of his previous magazine and/or newspaper credits.
>
> We are impressed by succinctness in queries. I might also say that we prefer to have writers confine themselves to not more than two or three suggestions at one time. More than that is a little too much to digest at one sitting. [13]

An experienced editor and publisher considers the following example as a particularly fine query letter:

> Dear Sir:
>
> The second largest telescope in the world is now nearing completion at Lick Observatory, Mt. Hamilton, California. Under the direction of Mr. Donald Shane, this 120-inch mirror has been specifically designed for study of outer stellar galaxies and mapping of the entire universe.
>
> In the 20 years since the building of the 200-inch telescope, Palomar, many advances in electronics and optics have made the most advanced and— according to the engineers—the most accurate telescope in the world.

I am co-ordinating public relations for Judson Pacific-Murphy Corporation, builder of the telescope, and the University of California Public Information Office to present, at last, the story of this fantastic astronomical instrument to the public.

This story, I believe, will interest your readers; it deals with the building of the telescope. Here a firm known for building bridges—with tolerances in inches—successfully completes this 145 ton instrument with tolerances of one-two thousandths of an inch or one-twentieth the thinness of the paper this is written on and using only bay area firms as subcontractors rather than European or Eastern specialists as has always been the case in the past.

Complete records, progress photos, engineering data, and the cooperation of the astronomer at Lick Observatory are available to me for a feature story geared to the format you require on this number one astronomical event in this International Geophysical Year. [13]

Writing the Article

After you have received word from the editor of interest in your proposal, outline the article more fully (preferably in sentence form) before you begin the actual writing.

Keep Readers in Mind

Start the rough draft. The opening is important. It must capture the readers' attention, interest them to read further, and lead them to the point or thesis of the article.

To maintain reader interest, the writing must remain within the range of the readers' intelligence and experience, structured within an organizational plan that will maintain the interest. Useful expository techniques to bring very complex or technical aspects of a subject within the realm of the readers' experience are such rhetorical devices as anecdotes, analogies, and literary allusions. The anecdote is one of the most frequently used devices in popular writing because it illustrates points more effectively than will long explanations. References to prominent people in history, to episodes, or to situations in literature can sometimes aid in establishing a common bond between the writer and readers and make the point the writer wants to make more graphically. Sigmund Freud (a superior writer of science) incisively used literary allusions to Greek myths and the characters in Shakespeare's plays.

A widely used and effective device for bringing unfamiliar ideas within the readers' experience is the figurative analogy. The **analogy** is a comparison between two unlike things which, nevertheless, have certain essentials in common. Among commonplace analogies are the comparison of the earth to a ball, the heart to a pump, the camera to the eye, or the computer to a superbrain. There is danger in the use of analogies in that readers often understand only the analogical counterpart but not the matter compared.

Related to the analogy and equally as effective in explaining scientific concepts are the use of similes and metaphors. A **simile** is a direct comparison between two unlike things. Such comparisons are introduced by the prepositions *like* or *as*. A **metaphor** does not state a comparison but proceeds indirectly to indicate the likeness of the two things involved. Poets use similes and metaphors to help create sensual images in their readers' minds. Science writers use these two figures of speech to help define and explain to readers difficult things, processes, and concepts by comparing them directly (using a simile) or implicitly (using a metaphor) with more familiar matters.

In a *Washington Post* article timed to the spring growing season, Booth uses the device of a metaphor to help explain the mystery of how leaves develop the Washington area trees. He likens plant life to that of the human community. A flower is only a cloying, sexually overactive tart, but a leaf, on the other hand, promotes the growth and development of the plant through its factory-like activity of photosynthesis:

> Every year Washingtonians go gaga over the cherry blossoms. But, of course, blossoms are not very serious things. Hypersexual and cloying, flowers are the showy tarts of the plant community. It is the leaves, now appearing on a tree near you, that do the real work. Leaves are the business end of a plant.
>
> It is the leaves, after all, that show up for work each spring and arrange themselves into so many solar panels to collect the energy of the sun and convert it into sugar and starch, the hard currency of the plant kingdom. The leaves are the factories, humming with the engines of photosynthesis, that sustain and nourish the growing tree. [4]

Handling Scientific Terms

Science is constantly at work exploring new fields or modifying old techniques; it keeps reaching out for new words and phrases to describe its innovations. These descriptive terms are usually based on or coined from Latin and Greek, though occasionally borrowed from other languages—German in the case of nuclear and rocket research. Many new terms are created by the addition of prefixes or suffixes or word roots. *Transistor* is an example (transfer and resistor). New, strange, and sometimes unpronounceable words, together with complicated sentence structure, create difficulty for the reader—lay or otherwise.

When it is necessary to use new terms, immediate definition or explanation should follow without interrupting the article. This can best be accomplished by defining or explaining the term in an appositional phrase set off by commas. Definitions should start with what readers know and then add no more than they are able to comprehend. Never define a hard word by a harder word. For example, do not define *calorie* to the general reader as do some dictionaries by explaining it as the quantity of heat necessary to effect a rise in temperature of 1° C. of a gram of water. The general reader would understand calories better if you say that "100 calories of energy can be derived from three cubes of sugar or from a small pat of butter" or that "a man uses up 100 calories an hour to keep his body running and 160 calories when he is working hard."

Acker and Hartsel, in their article "Fleming's Lysozyme" in *Scientific American*, very effectively explain difficult words and terms by offering the more common meaning first and then following it by the more difficult term:

> Bacterial anatomists are indebted to the late Sir Alexander Fleming for a sensitive chemical tool with which they have been studying bacteria, dissolving away the cell wall, and exposing the *cell body* or *cytoplasm* within. In 1922 at St. Mary's Hospital in London, six years before his epochal discovery of penicillin, Fleming found "a substance present in the tissues and secretions of the body, which is capable of rapidly dissolving certain bacteria." Because of its resemblance to enzymes and its capacity to *dissolve*, or *lyse*, the cells, he called it "lysozyme." [1]

The writers, instead of using first the difficult terms *cytoplasm* and *lyse*, used first the more commonly known terms *cell body* and *dissolve*. This very effective device gives the readers confidence when they go from the common term to the more uncommon term. They have no difficulty understanding what the writers have written. Psychologically this is a very effective device in that it builds confidence in readers, making them feel that they understand the more difficult terms because they have first been exposed to their meanings.

Avoid shoptalk. Shoptalk is the vocabulary used by a community of workers in their everyday work. Be sure you define every specialized term you use as you would highly technical ones.

Structure and Style

Use personal pronouns; they create empathy between writer and reader and add vitality and interest to the article. Use clearly worded sentences, expressing one idea at a time. Cement the sentences of your paragraph with a topic sentence. Organize paragraphs into divisions clearly identified by heads that point to the direction in which your article is going.

A plan of organization is a must to keep reader interest. The plan should be simple enough for readers to grasp and follow. Wherever drama, conflict, and suspense are inherent in the subject, the narrative elements delineating these will hold interest. However, not all matters in science and technology are suitable to the narrative form. Cause and effect, analysis of contributing factors, evidence and conclusions, predictions, and other methods may be more appropriate. (Review the various patterns of organization in Chapter 12.)

The ending of the popular science article is as important as the beginning. It must clinch the readers' understanding of the thesis. Sometimes a summary is appropriate, sometimes a typical application of the technical points developed in the body, and sometimes a final astonishing fact. The ending should demonstrate the significance of the article so that readers will leave it feeling it has been worth their while to read.

Figure 19.2 is a succinct summary of the preceding discussion, using four published articles as examples. Figure 19.3 (pp. 509–511) is an example of a student-level popular science article. Figure 19.4 (pp. 512–513) is an example of a professional article written for a primary journal, and Figure 19.5 (pp. 514–516) is a professional-level popular science article. Study these articles to see how they incorporate the elements covered in this chapter, and how the elements are adapted to each article's chosen style, audience, and writing level.

A Student's Comparative Analysis of Four Different
Forms of Technical Writing

The approach to any piece of writing depends upon its reading audience. Different audiences require a different style, organization, and language.

In comparing a report, a professional paper, a semitechnical article, and a popular article, all on the same subject, many differences as well as some similarities become apparent.

The subject of monomolecular films for preventing water evaporation from reservoirs is relatively new. Means for making these films practical have initiated much research.

An article appearing in the Sunday <u>Denver Post,</u> entitled "Chemical Film May Save Western Slope Water" by Roscoe Fleming was used as an example of a popular article.

"Reducing Evaporation from Small Reservoirs" by Nedavia Bethlahmy was selected as an example of a semitechnical article. This article appeared in <u>Northwest Science,</u> Vol. 33, No. 3.

A reprint from the <u>Australian Journal of Applied Science</u>, Vol. 10, No. 1, entitled "The Influence of Monolayers on Evaporation from Water Storages" by W. W. Mansfield was used as an example of a professional paper.

Geological Survey Professional Paper 269, entitled "Water-Loss Investigations: Lake Hefner Studies, Technical Report," was used as an example of a technical report.

It may be generalized from the titles alone, that as the material becomes more technical, the title tends to become more specific concerning the material contained and therefore often less appealing to the lay reader. As the writing becomes more technical, the terminology used in the title also tends to become more technical.

The differences in length of the various types of writing are very noticeable. The popular article was a little shorter than the semitechnical article, which was four pages long, about 1600 words. The professional paper was nineteen pages long, while the report was 158 pages long. It is interesting to note that the popular article was printed in a Sunday edition of the newspaper, when more space is available.

The organization differed greatly among the various pieces of writing. The popular article first created interest and aroused the curiosity of the reader. The concept of the "chemical films" was explained, some of the experiments being carried on were cited, and the potential importance of these films was presented. Next, after tracing the history of the interest in the films, the article explained how the chemical films function. Simple terminology as well as analogy was used. The last three paragraphs mentioned some of the organizations that were working on the experimental projects. No references were cited. There did not appear to be any superimposed pattern of organization for the article. Rather, its text was organized for maintaining reader interest.

1

Figure 19.2
Student Analysis of Technical Writing Forms

The semitechnical article also began with a section designed to create interest. Some of the historical aspects of the research on chemical, evaporation-reducing films were mentioned, and some of the details concerning what the films are and how they work were explained. An experiment carried out in Oregon was described. A section discussing the practical applications of the chemical films concluded the article. A bibliography of eight items followed the body of the article. In general, the semitechnical article was designed to be of informative value to an educated reader who lacked technical knowledge on this particular subject. It provided some background on the subject, cited some experiments being carried out, and discussed the usefulness and practicality of the hexadecanol.

The professional paper presented a sharp contrast to the popular and semitechnical articles. The paper was divided into two sections; the first was concerned with evaporation and seepage from water storages, and the second with the action of wind, waves, and dust upon monolayers. Each of the two sections was similarly organized and written. A short summary of the entire article preceded the introduction, which went immediately into the subject without any devices for creating reader interest. The various aspects of the problems involved were each systematically discussed; experiments were cited and results given. A discussion section followed; then came a section of acknowledgments, and finally, a list of references. The organizational pattern of the paper was definitely evident. It was designed to communicate specific information to those who might be involved in the same problem, or to those who are technically competent to receive this specialized information.

The technical report began with preliminary sections; a foreword, a preface, a table of contents, an abstract. A section stating the conclusions and recommendations followed. The formal introduction included an historical review, the statement of the problem, and a listing of personnel and their qualifications. This was followed by a section giving the reason for the choice of Lake Hefner for the experiment.

The body of the report contained a number of individual sections, each of which dealt with a certain aspect or component of the study. Each of these sections contained a long description of the instrumentation and methods used, followed by a discussion of the results obtained, conclusions, and a list of references. The report ended with a summary and recommendations section, appendix, and an index. The entire report was carefully, systematically organized. Each of the individual sections in the middle was written by a different person, but all of the sections were organized and presented in the same organizational structure. The report's data were systematically and concisely presented, aimed toward a technically trained reader.

The writing style of the different types varied greatly. In the popular article, the author was concerned in making his information as interesting as possible. He used devices to pique curiosity and he succeeded with the devices to maintain reader interest. Though the writer succeeded in sticking to facts for the most part, he tended at times to generalize and make sweeping statements, designed to create interest, but the statements were not always entirely accurate. The author thereby reduced highly technical information to the level of understanding of the general reader. The article was created by involving the reader in the matters explained.

2

Figure 19.2 (continued)

The semitechnical article made an attempt to gain reader interest in the opening, but did not follow through with interest-creating devices in the rest of the paper, whereas the popular article made successful use of interest-creating devices throughout. Many technical concepts were explained in simple terms. There were many statistics offered, with references. Facts were related in simple language. A short summary of the present state of knowledge concerning evaporation-retarding films was directly and effectively presented.

In the professional paper, there was no attempt to gain or hold reader interest. The introduction presented the background of the problem. The rest of the paper was devoted strictly to analyzing the problem and sharing its present state of knowledge with interested, technically trained persons. The analysis progressed in a systematic, scientific manner. Highly technical concepts were used without explanation.

In the report, the background material necessary for the full understanding of the investigation was provided. There was no attempt to gain or hold the interest of the reader. The report directly related the investigations and experiments being carried out, explained why and how this was being done, and what the results were. Many technical terms and concepts were used without definition or explanation. The methods and instruments used were described in detail. Both positive and negative results were reported. A difference in style of writing was discernible in each of the authors who contributed to the various sections of the report.

As the technical level of publications increased, the language used became more technical:

In the popular article, only three technical terms were used, and two were defined. The term "molecule" was not defined. The term "hexadecanol" was defined, but its abbreviated, popular form, "hexy" was thereafter used.

In the semitechnical report, five technical terms were used; only two of these were defined. The language was relatively simple. Technical terminology was used only when necessary.

There were thirty-seven technical terms in the professional paper. None of these were defined or explained. Exceptions were factors in equations, which were defined by the use of another equation.

The report had innumerable technical terms included within it, none of which were defined or explained. As in the professional paper, factors in equations were defined by the use of another equation.

In the popular article, analogy was used three times to explain concepts and terms; comparison was used once, example was used once, and a simple explanation was used once.

Simple explanation was the only method used to explain technical concepts or terms in the semitechnical article.

In both the professional paper and the report, no definitions or explanations were given. There were no analogies, anecdotes, or other special rhetorical devices used.

3

Figure 19.2 (continued)

No special technical background in monomolecular films is needed to understand the popular article. Technical background is not necessary, although it would be of help to understand the semitechnical article. A background in advanced mathematics and physics is especially necessary for the professional paper and the technical report.

As the level of writing became more technical, the more precise the information became; graphs, tables, and equations communicated many important data in the writings.

The popular article had no graphs, tables, or equations. There was one photograph of a simple, but interesting process. This picture was fully explained in the text and in the caption.

The semitechnical article employed one minor table, used for comparison. There were no graphs, equations, or pictures.

There were fifty-one equations, eight graphs, and one table appearing in the professional paper. Also used were many statistics that, along with all the equations, rendered this paper extremely technical.

The technical report contained twenty pictures, forty-four tables, three maps, eighty-nine graphs, and one hundred eight equations. Through the use of the graphs, tables, and equations, very precise information was easily recorded and compared.

As the writing became more technical, more references were cited. There were no direct references cited in the popular article. The semitechnical article cited eight references. The professional paper cited twenty, and the technical report cited a total of one hundred thirty-four. It is interesting to note that many of the references cited in a technical report were written in foreign languages, primarily German, indicating a deeper study of the literature in the technical report than in any of the other writings.

In summary, it might be generalized that the similarities in the four types of writing here compared were few. Other than a common subject, each was consciously directed to a specific audience. Each had an informative purpose. Elements to evoke emotion or controversy were deliberately avoided.

On the basis of the present comparison, one could easily plot a linear curve to show that the greater the intended reading audience, the less technical the information becomes, the less formal the organizational structure, the less objective the style of writing, the less precise the language use, the less specific the data and details, and the greater the use of rhetorical elements and literary devices, and the greater emphasis on personal elements to involve the reader.

Conversely, as the intended reading audience becomes more specialized, the more technical and specialized is the information communicated, the more formal the organizational structure, the more objective the style, the more specific the details and data, the more precise the use of language, entailing more and more graphic presentational devices (use of diagrams, drawings, charts, tables, etc.) and the greater the use of specialized terminology, symbols, and processes.

Roger M. Hoffer

4

Figure 19.2 (continued)

HOW AIRPLANES FLY

by

David M. Graham

Today's airplanes vary in weight from a few hundred pounds to many thousands of pounds. It is truly amazing, and seemingly a miracle, that these heavy machines are able to stay up in the air and even travel through it at tremendous speeds without falling from the sky. However, the fact that a large airplane can fly at a height of thousands of feet is due to a fairly simple principle. Jet planes are held up by the same principle, but they are capable of forward flight for a different reason than propeller planes. We will learn in this article how the planes that are propeller-driven operate.

First, how does an airplane keep from falling down once it gets up into the sky? The airplane wings' main function is to hold the airplane up. These wings are not just slabs of metal, all flat and stiff. They have a very definite shape which helps them lift the airplane.

Figure A

Notice in Figure A that the bottom of the wing is fairly flat while the top is curved. The front of the wing is thick, and the rear, or trailing edge, is very thin. When the airplane is flying, air rushes over the wings at high speed. Because the top of the wing is curved, the air passes over the top of the wing much faster than under the bottom of the wing. The air passing over the top, shown by the lines in Figure B, causes a vacuum, or suction, which tends to lift the wing. The air going under the wing doesn't create such a vacuum. So, the airplane, which is attached to the wing, is lifted instead of pushed down. If both surfaces of the wing were equally flat, the air rushing over the surfaces would push on both sides with an equal pressure. This wouldn't help the airplane stay up in the air. That's why the top is curved—so the air will help lift the airplane while it is flying. This vacuum, which lifts the plane, works by trying to suck the wing up to fill the space where there is less air just above the wing. The suction, like the suction from a vacuum cleaner, is called lift.

Figure B

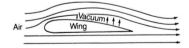

The airplane propeller pulls the aircraft through the air. The blades of the propeller are just like small wings, the front of the propeller blade is curved and the trailing edge is thin and flat. The side of the blade that corresponds to the top of the wing is facing towards the front of the airplane. The rear side of the blade corresponds to the under-side of the wing. As

1

Figure 19.3

Example of Student Popular Science Article

(Written as a student assignment for a publication like *Boy's Life*.)

the propeller turns, the air rushing past the curved part of the blade creates a vacuum just as it did on the wing. This vacuum pulls the propeller forward and the airplane with it (Figure C). The action of the propeller is very much like the action of a screw being driven into a piece of wood. As the screw turns, it pulls itself into the wood like the propeller pulls the airplane through the air. The propeller, with the engine, is really the most important part of the airplane. When the airplane is just taking off from the ground, the propeller has to pull the airplane forward until it is going fast enough to give lift to the wings.

Figure C

Front

(Exploded
View)

Direction of
Rotation

Once the airplane is up in the air, how is it steered so it won't just fly in a straight line and smack into a mountain or something? Don't worry, the airplane can be steered. There are three different controls that enable the airplane to go up and down, and turn to the right or the left. These are called control surfaces and their specific names and locations are: the ailerons, which are located on the trailing edges of the wings; the elevators, which are part of the small wings on the rear of the airplane; and the rudder, which is located on the tail fin above the little wings (Figure D).

Figure D

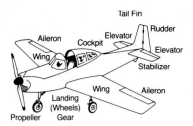

The control surfaces are moved by means of a steering wheel in the cockpit. This wheel turns just like the one in a car, but it can also be pushed forward and backward. Turning the wheel to the right makes the airplane turn, or bank, to the right. A left bank is made by turning the wheel in that direction. Pushing the wheel forward makes the airplane point down towards the ground. Pulling it out makes the airplane climb. Some of the older airplanes and most of the big airliners have a different system of moving the airplane around. The controls are the same but, instead of having a single wheel that does everything, these airplanes have a small steering wheel on a "stick" which operates just the elevators and the rudder. On the cockpit floor, there are two big pedals, like the ones in an automobile, which move only the ailerons. Now let us find out what happens when the airplane controls are moved.

2

Figure 19.3 (continued)

The ailerons are long, thin, door-like pieces. They fit on the trailing edge of the wings and can be moved up or down. If the aileron lies straight out, and even with the rest of the wing, the air rushing by is not disturbed and the airplane flies straight. Now, suppose the aileron pointed down. The air that normally whizzed under the wing would strike the aileron with a lot of push. This causes the wing to be forced up. The same thing happens when the aileron points up instead of down. The air hits the aileron and forces the wing down. The ailerons are so connected that when one of them is pointed down, the other points up. Then the rushing air pushes one wing up and pushes the other down.

On the rear of the airplane, called the tail section, are what look like three small wings. Two are level like the large wings in front, and one points straight up. This one is called the tail fin, while the level ones are called stabilizers. The elevators on the stabilizers operate just like the ailerons with one exception—both elevators move in the same direction, instead of opposite directions as with the ailerons. These control surfaces make the airplane either climb or dive. When the elevators point up, the air pressure forces the tail down, like with the ailerons, and points the airplane up so it can climb higher. Just the opposite happens when the elevators point down.

The remaining control surface, the rudder, steers the airplane just like the rudder on a boat. The air does the job instead of water. If the rudder points to the right, the airplane turns to the right, and to the left if the rudder points to the left.

Now, let's climb in and make an actual flight. First, the engine is turned on. A knob on the dashboard controls the gas just like the accelerator pedal in a car. We push the knob in to make the propeller turn faster and pull the airplane forward. Out on the runway, we push the knob all the way in and we speed down the runway faster and faster. The rushing wind creates lift on the wings and the airplane leaves the ground. We're flying! We climb high into the air so we can test our controls. You may have noticed, we had to pull back on the wheel to get the plane into the air, and we kept the wheel pulled back until we were far enough above the ground. Now that we're high enough, let's make a right turn. We slowly turn the wheel to the right and the airplane seems to tip over to the right. The right aileron points up and the left one points down. Pulling back on the wheel just slightly helps the airplane go through the turn by raising the elevators which pushes the tail down and to the left. We do the opposite movements when we want to make a left turn. But, we still have to pull back slightly on the wheel to help push the tail around where we want it. Since we've already seen how the airplane climbs, let's put it into a dive. We push the wheel forward slowly and the plane flies toward the ground. If we were to push the wheel forward too fast or too far, the airplane would point down too far and get out of control. Then we might crash. Now, as the wheel goes forward, the elevators are pointing down and pushing the tail up because of the air pressure. To pull out of the dive, we just pull back on the wheel.

It's time to take the airplane back for a landing. We turn the airplane so we can see the runway straight ahead. We push the wheel in slowly to fly the airplane closer to the ground. We keep down until we are very close by the time we're over the end of the runway. As soon as we can feel the wheels touch the ground, we pull out the gas knob to slow the propeller down. But, we keep the wheel pushed in slightly. In slowing down, the tail section will settle by itself as the air can no longer hold it up. And, we've landed safely.

That is how an airplane works. With a little practice, flying becomes easier than driving a car. But we have to start out in the airplane with much more caution than we use when driving a car. After all, an airplane doesn't travel as close to the ground as does the car.

3

Figure 19.3 (continued)

Freeze Avoidance in a Mammal: Body Temperatures Below 0°C in an Arctic Hibernator

BRIAN M. BARNES

Hibernating arctic ground squirrels, *Spermophilus parryii*, were able to adopt and spontaneously arouse from core body temperatures as low as −2.9°C without freezing. Abdominal body temperatures of ground squirrels hibernating in outdoor burrows were recorded with temperature-sensitive radiotransmitter implants. Body temperatures and soil temperatures at hibernaculum depth reached average minima during February of −1.9° and −6°C, respectively. Laboratory-housed ground squirrels hibernating in ambient temperatures of −4.3°C maintained above 0°C thoracic temperatures but decreased colonic temperatures to as low as −1.3°C. Plasma sampled from animals with below 0°C body temperatures had normal solute concentrations and showed no evidence of containing antifreeze molecules.

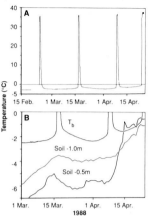

Fig. 1. Body temperature (**A**) of a hibernating female arctic ground squirrel as indicated by an abdominal temperature-sensitive radiotransmitter; (**B**) with an expanded scale, abdominal temperature during the last three arousals from torpor and concurrent adjacent soil temperature. Break in curve indicates missing data.

HIBERNATION IN MAMMALS IS EXpressed by a fall in body temperature (T_b) to near the ambient temperature of the hibernaculum. Torpid animals maintain low T_b's for up to several weeks until a brief (<24 hours) spontaneous arousal to high T_b occurs, after which animals recool. The lowest T_b's previously reported for natural hibernation in a variety of mammalian hibernators are between 0.5° and 2°C (*1*) and in ambient conditions of 0° to 3°C (*1*). In experimental conditions, slowly lowering ambient temperatures below 0°C leads either to an increase in an animal's metabolism and stabilization of T_b or an "alarm arousal" after which the animal, upon returning to torpor, will actively regulate T_b at 2° to 3°C (*2*). Some ectothermic vertebrates can endure subzero T_b's either by avoiding or tolerating freezing. For example, many species of polar and north temperate fish, through use of blood antifreeze proteins or glycoproteins, live at temperatures of −1.9°C (*3*), and painted turtles and four species of frogs can pass the winter frozen at temperatures of −3° to −7°C (*4*). Accounts of endotherms surviving subzero T_b's are either anecdotal (*5*) or describe the artificial induction of subzero body temperatures, a condition from which the animal could not independently arouse (*6*). I report telemetric and direct evidence of the regular, prolonged, and spontaneously reversible adoption of core T_b of as low as −2.9°C in the arctic ground squirrel, *Spermophilus parryii*, hibernating in outdoor enclosures.

Arctic ground squirrels were captured during late August 1987 in the northern foothills of the Brooks Range, Alaska, near the Toolik Field Station of the University of Alaska Fairbanks (68°38'N, 149°38'W; elevation 809 m) and transported to Fairbanks. Animals were implanted abdominally with miniature temperature-sensitive radiotransmitters that had been previously calibrated (*7*). On 19 September 1987, seven males and five females were released in Fairbanks into outdoor wire cages (0.9 by 0.9 by 1.8 m, buried to 1.3 m) where they dug burrows and remained for the next 8 months (*8*). Each cage was fitted with copper wire loop antennas (two or four each) housed in plastic pipe and connected to coaxial leads.

Each lead was connected to a radio receiver with an interface to a computerized data acquisition system (*9*). Bandpass filters were used to overcome radio interference from a local AM radio station, and data collection began in mid-February 1988. In spring, after each animal emerged from the hibernaculum, transmitters were recovered and recalibrated (*10*). Soil and air temperatures at the site were recorded with thermocouples and a thermocouple thermometer. To determine the temperature regimes arctic ground squirrels experience during hibernation in the environment at which they were collected, soil temperatures at a depth of 1.0 m at two natural burrow sites near the Toolik Field Station were recorded over winter on automated remote recorders (*11*).

Minimum T_b's of six hibernating ground squirrels occurred in February and March and averaged −1.9° ± 0.3°C (range −2.9° to −1.1°C). The T_b of the individual that reached the lowest T_b (−2.9°C) is shown during the last 2 months of hibernation in Fig. 1A; an expanded scale for the last three arousals shows T_b and adjacent soil temperatures (Fig. 1B). The pattern of change in T_b shown in Fig. 1 is typical for hibernating ground squirrels: prolonged bouts of continuous torpor interspersed by short spontaneous arousals. What is unusual is for T_b to fall below 0°C. As animals entered hibernation, the cooling rate of T_b slowed significantly after reaching −1.5°C, which indicates that the animal either increased its insulation or more likely began to actively produce heat in order to prevent cooling below some further minimum T_b. During deep torpor, T_b in the region of the transmitter did not vary more than 1°C, and it remained between 1° and 3°C above the temperature of the soil until mid-April when T_b approximated the temperature of the warming soil just before the final arousal (Fig. 1B). Several days before each spontaneous arousal, T_b began to slowly rise, increasing by approximately 0.5°C before rapid arousal ensued. Early indications of arousals have also been shown by a rise in the hypothalamic set point of T_b in hibernating marmots (*12*). These patterns were similar in all six animals studied.

Soil temperatures at a depth of 1.0 m at natural burrow locations on the North Slope of Alaska recorded over the winter of 1987–88 reached a minimum of −18°C, whereas temperatures at the same depth in the caged burrows in Fairbanks never fell lower than −7°C. Nest chambers where animals hibernated in the experimental burrows were excavated: the spherical nests were constructed of straw, approximately 30 cm in diameter, and usually located in a corner of the cage at 1.2 ± 0.05 m depth. Depth of natural hibernacula are limited by the permafrost table; ground squirrels appear not to dig into frozen ground (*13*). The permafrost table lies between 25 and 100 cm deep over most of Northern Alaska (*14*).

To reproduce conditions of freeze avoidance under laboratory conditions, arctic ground squirrels were housed in an environmental chamber whose temperature was gradually reduced in fall 1988 from 5° to −4.3°C during 1 month. Body temperatures of hibernating animals were measured at several locations, and blood was sampled by cardiac puncture from individuals that exhibited subzero rectal temperatures. Plasma was separated from blood cells, measured for solute concentration (*15*), and screened for the presence of antifreeze properties by testing for thermal hysteresis of melting and freezing points (*16*).

In ambient temperatures of −4.3°C arctic ground squirrels adopted colonic, foot, and subcutaneous temperatures that ranged from −1.3° to 0°C, and maintained oral and thoracic temperatures of −0.70° to 0.7°C (Fig. 2). Thus, under these conditions hibernating ground squirrels had heterogeneous T_b's and typically sustained across body temperature gradients of 1° to 2°C. Subzero body parts seemed fully perfused as

Institute of Arctic Biology, University of Alaska Fairbanks, Fairbanks, AK 99775–0180.

Figure 19.4
Example of Scientific Primary Journal Article [2]

subdermal wounds inflicted on toes and abdominal skin bled promptly. In six animals with colonic temperatures averaging $-0.63°C$, concentrations of plasma solutes were normal (302 ± 4.4 mmol/kg), and freezing and melting points of plasma were not different ($-0.59° \pm 0.02°C$ and $-0.56° \pm 0.01°C$, respectively) and were similar to equilibrium freezing points of blood in nonhibernating mammals (17).

Animals withstand body temperatures below the freezing point of water by being freeze tolerant (4), by solute-dependent freezing point depression (18), by using antifreeze molecules (3), or by supercooling (19). There was no evidence of an exotherm (thermal heat of fusion) at subzero T_b's, which indicates that body water did not freeze. Plasma solute concentrations measured in ground squirrels with subzero deep body temperatures would have offered protection from freezing to temperature of approximately $-0.6°C$, but for the core temperatures measured of $-1.3°$ to $-2.9°C$ a further mechanism of freeze avoidance must be offered. Antifreeze molecules depress freezing points relative to melting points by providing resistance to the growth of ice crystals (3). Freezing and melting points of plasma taken from ground squirrels hibernating at ambient temperatures of $-4.3°C$ were equal, indicating that, under these conditions, antifreeze substances are not present. By exclusion this leaves supercooling, which is a metastable state of below freezing temperatures that persists in the absence of a nucleator which would readily instigate crystallization (20). Rats, hamsters, and other small mammals can be artificially supercooled to colonic temperatures of $-2.5°$ to $-5.5°C$, with up to 100% survivorship after they are artificially rewarmed and resuscitated (6). However, the tenure of subzero T_b in such supercooled animals must be brief (<60 min); if it is prolonged, spontaneous crystallization occurs and partially frozen animals usually (but not always) cannot be revived. Arctic ground squirrels in this study maintained subzero T_b's for more than 3 weeks.

The ability of arctic ground squirrels to undergo deep and prolonged supercooling is a new finding, despite several decades of measuring T_b's in diverse species of hibernating birds and mammals (21). This ability may relate to the prolonged and extreme conditions under which arctic ground squirrels must overwinter: 8 to 10 months within the hibernaculum with soil temperatures at nest depth declining to $-18°C$. Dormant seasons for other species of hibernators are usually shorter and recorded hibernaculum temperatures remain above freezing (22).

Supercooling to near $-3°C$ should offer energetic advantages over maintaining greater than $0°C$ T_b's to ground squirrels hibernating at ambient temperatures substantially below $-0°C$. Few metabolic measurements have been made of hibernators maintained in subzero conditions and none

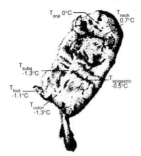

Fig. 2. Regional body temperatures of a hibernating arctic ground squirrel (scale 1:6) housed in an ambient temperature of $-4.3°C$. Average temperatures ($\pm$SE) and depth of temperature probe for 6 to 11 animals also at $-4.3°C$ were: colon $-0.62° \pm 0.11°C$, 6 cm; foot $-0.65° \pm 0.15°C$, 3 mm; abdominal $-0.59° \pm 0.13°C$, 2 cm; thoracic $0.49° \pm 0.12°C$, 1 cm; oral $-0.16° \pm 0.16°C$. Colon temperatures were measured with a thermocouple thermometer (BAT-12) and a RET-2 thermocouple (Sensortek, Clifton, New Jersey); other temperatures were measured with 30-gauge thermocouple wire housed in a 19-gauge needle. Thermocouple temperature readings were calibrated with a glass thermometer (10). Ground squirrels normally hibernate while curled in a ball.

have been made for animals with below $0°C$ T_b. However, extrapolating from existing data on the greatly elevated metabolic costs of hibernators that maintain above $0°C$ T_b at ambient temperatures of $0°$ and $-2°C$ (23) suggests that supercooling to $-3°C$ might save ten times the energy expended by maintaining above $0°C$ T_b (24). Any metabolic savings accrued over the hibernation season would be advantageous to ground squirrels—presumably in the forms of increased overwinter survivorship and of energy stores left after hibernation for use during the short but frenetic reproductive season that begins at emergence from hibernation.

REFERENCES AND NOTES

1. O. B. Reite and W. H. Davis, *Proc. Soc. Exp. Biol. Med.* **121**, 1212 (1966); H. T. Hammel, T. J. Dawson, R. M. Abrams, H. T. Anderson, *Physiol. Zool.* **41**, 341 (1968); L. C. H. Wang, *Am. J. Physiol.* **224**, 673 (1973); C. P. Lyman and R. C. O'Brien, *ibid.* **227**, 218 (1974); H. C. Heller and S. F. Glotzbach, *Int. Rev. Physiol. Envir. Physiol. II* **15**, 147 (1977).
2. C. P. Lyman, *J. Exp. Zool.* **109**, 55 (1948).
3. A. L. DeVries, *Comp. Biochem. Physiol.* **90B**, 611 (1988).
4. W. D. Schmid, *Science* **215**, 697 (1982); K. B. Storey and J. M. Storey, *Physiol. Rev.* **68**, 27 (1988); ———, S. P. J. Brooks, T. A. Churchill, R. J. Brooks, *Proc. Natl. Acad. Sci. U.S.A.* **85**, 8350 (1988).
5. R. J. Hock, in *Cold Injury*, M. E. Ferrer, Ed. (Josiah Macy, Jr. Foundation, New York, 1958), pp. 61–133; A. Svihla, *J. Mammal.* **39**, 296 (1958).
6. R. K. Anjus, *J. Physiol.* **128**, 547 (1955); N. I. Kalabukhov, in *Recent Research in Freezing and Drying*, A. S. Parkes and A. U. Smith, Eds. (Blackwell, Oxford, 1960), pp. 101–118; A. U. Smith, *Biological Effects of Freezing and Supercooling* (Williams & Wilkins, Baltimore, 1961), pp. 304–368; P. Popovic and V. Popovic, *Am. J. Physiol.* **204**, 949 (1963).
7. Transmitters (model VM-FH disk, Mini-Mitter Co, Inc., Sunriver, OR) were implanted in animals anesthetized with methoxyflurane (Metofane, Pit-

man-Moore, Washington Crossing, NJ). Animal care was in accordance with University of Alaska Animal Use and Care Committee guidelines.
8. Ten of 12 animals survived to emergence in spring, but signals from only six transmitters were consistently received. Absent signals were due to battery failures in three cases or nest locations that were not close enough to an antenna for signal reception in three cases.
9. Dataquest III, Data Sciences, Inc., Minneapolis, and Mini-Mitter Co., Inc., Sunriver, OR.
10. Transmitters were recalibrated in a refrigerated alcohol-water bath over the range of $-4°$ to $+2°C$ within 24 hours of recovery from each animal. Computed temperatures based on initial calibration values were monitored from each animal's own receiver simultaneously with temperature on a mercury thermometer calibrated at ice point as described in T. E. Osterkamp ["Calibration and field use of Hg-in-glass thermometers for precise temperature measurements near 0°C" (Geophysical Institute, Fairbanks, AK, 1977)]. The total uncertainty for the thermometer relative to International Practical Temperature Scale was $\pm 0.025°C$ (Brooklyn Thermometer Co. Test No. 221378). Computed temperatures averaged 0.09°C high (range $-0.39°$ to 0.60°C, $n = 6$). Data presented are corrected to reflect the recalibration values.
11. Datapod, Omnidata International, Inc., Logan, UT.
12. G. L. Florant and H. C. Heller, *Am. J. Physiol.* **232**, R203 (1977).
13. E. A. Carl, *Ecology* **52**, 395 (1971); W. V. Mayer, *Anat. Rec.* **122**, 437 (1955).
14. J. V. Drew, J. C. F. Tedrow, R. E. Shanks, J. J. Koranda, *Trans. Am. Geophys. Union* **39**, 697 (1958); K. A. Linell and J. C. F. Tedrow, *Soil and Permafrost Surveys in the Arctic* (Clarendon Press, Oxford, 1981).
15. Plasma volumes of 10 μl were measured for solute concentration with a Wescor 5500 Vapor Pressure Osmometer (Logan, UT).
16. Ice crystal melting temperatures and ice crystal growing temperatures were measured in plasma volumes of 20 μl after a small seed crystal was introduced by spray freezing. Procedures are described by A. L. DeVries [*Methods Enzymol.* **127**, 293 (1986)].
17. D. S. Dittmer, *Biological Handbooks: Blood and Other Body Fluids* (Federation of American Societies for Experimental Biology, Washington, DC, 1961).
18. M. S. Gordon, *Animal Physiology: Principles and Adaptations* (Macmillan, New York, ed. 3, 1977).
19. P. F. Scholander, L. Van Dam, J. W. Kanwisher, H. T. Hammel, M. S. Gordon, *J. Cell. Comp. Physiol.* **49**, 5 (1957).
20. M. J. Taylor, in *The Effects of Low Temperatures on Biological Systems*, B. W. W. Grout and G. J. Morris, Eds. (Arnold, London, 1987).
21. C. P. Lyman, J. S. Willis, A. Malan, L. C. H. Wang, *Hibernation and Torpor in Birds and Mammals* (Academic Press, New York, 1982).
22. L. C. H. Wang, *Can. J. Zool.* **57**, 149 (1979); G. J. Kenagy and B. M. Barnes, *J. Mammal.* **69**, 274 (1988).
23. F. Geiser and G. J. Kenagy, *Physiol. Zool.* **61**, 442 (1988).
24. This estimate results from extrapolating metabolic costs of torpor shown from Geiser and Kenagy (23, figure 2) to an ambient temperature of $-10°C$ and a T_b of either $0°$ or $-3°C$. At ambient temperatures below T_b greater than Q_{10} effects [see K. Schmidt-Nielson, *Animal Physiology: Adaptation and Environment* (Cambridge Univ. Press, Cambridge, 1979), p. 207] on metabolism are seen as animals must produce heat to maintain a gradient between body and ambient temperatures. The extent of energy savings due to supercooling would depend on the proportion of metabolically active tissue that attains the supercooled state. Since measurements of body temperatures in hibernators at an ambient temperature of $-4.3°C$ suggest that only posterior regions supercool, and since the most metabolically active tissues during torpor likely reside in the anterior of the body (heart, brain, brown adipose tissue), then energetic advantages of partial supercooling over maintaining above 0°C temperatures throughout the body may be significantly less than this estimate.
25. Supported by NIH grant HD 23383. I thank H. A. Maier, A. S. Porchet, and A. D. York for assistance, D. Borchert for the drawing, and A. D. York, G. J. Kenagy, L. K. Miller, and R. Elsner for reading the manuscript.

8 Feburary 1989; accepted 26 April 1989

Figure 19.4 (continued)

How Animals Survive the Big Chill

By Brian M. Barnes [3]

BECAUSE THE Earth is slightly tilted on its axis, the northern and southern hemispheres alternately face closer to the sun during the year. We recognize the consequences of this tilt as the passage of seasons, an effect exaggerated the further one moves toward the poles. This mechanical accident of rotation has had a profound effect on the evolution of animals that live in arctic and boreal zones. Every part of their biology is affected, most notably in the ways they cope with winter cold and diminished food availability.

Our knowledge of how animals endure the cold has improved dramatically in the past decade. We now know that some species begin to prepare well ahead of the first snows with changes in their reproductive systems, insulation and fat metabolism. And recently we have learned that some animals go into extreme states of suspended animation during which they either freeze nearly solid or prevent their bodies from freezing. In both cases, the process occurs reversibly and without injury as a natural part of overwintering.

Three Strategies

There are three basic strategies animals use to cope with winter: Migration, acclimatization and hibernation. Acclimatization involves ceasing energy-expensive activities such as reproduction and growth, and developing heftier insulation and an increased ability for producing heat (thermogenesis), triggered by exposure to decreasing daylengths and temperatures. In hares, voles, mice and hamsters, thermogenetic ability derives primarily from an increase in brown fat. Brown fat contains a specialized protein called thermogenin which uncouples fat metabolism from energy production so that potential energy is released directly as heat in a process called non-shivering thermogenesis. Adult humans have little or no brown fat, but in human infants and winter-acclimatized small mammals, thick deposits can develop inside the cardiac cavity and line the major arteries leaving the heart. During cold stress these tissues act as a central furnace to heat the blood.

Hibernation—spending the winter in a lethargic state—typically entails a profound decrease in body temperature extending from several days to several weeks, causing a low metabolic rate and reduced need for food. Hibernating animals can subsist 50 to 100 times longer on a given amount of energy than they can at high body temperatures. Ground squirrels, marmots, woodchucks and chipmunks retreat into underground hibernacula for five to seven months and cool their body temperatures by 30 to 40°C; in contrast, hibernating bears drop by only 5 to 7°C.

To save energy, it might be expected that hibernating animals would attempt to become as cold as possible. Until recently, the freezing point of water was considered the absolute lower limit, since freezing and thawing of biological tissues are usually quite lethal. When body water crystallizes into ice, cells become severely dehydrated; this effect can worsen as tissues slowly thaw and tiny ice crystals recrystallize and spread. In the process, subcellular structures become deformed and the lipid and protein membranes that surround each cell shrink and fold. If this continues, cells and tissues die, as anyone who has had severe frostbite can tell you.

Frozen Frogs, Sweet Dreams

Yet recent research has shown that certain frog species, a turtle and even a mammal—the arctic ground squirrel—can readily survive core body temperatures well below freezing. They manage this in one of two ways: freeze tolerance, the ability to organize freezing within tissues so that it does not harm cells; and freeze resistance, the ability to prevent ice formation in the first place.

Wood, gray tree, chorus and spring peeper frogs, as well as painted turtles and undoubtedly several other kinds of northern amphibians and reptiles, burrow into the mud and duff of ponds and forest floors each autumn. When surrounding temperatures drop below freezing, most of the water in their bodies becomes ice. In the spring, they thaw and revive. The phenomenon has been seen in a variety of insects and marine invertebrates, but was not reported in a vertebrate until 1982.

William Schmid, a biologist at the University of Minnesota, had collected a number of wood frogs from nearby forests to display to students. He accidently left them in the trunk of his car overnight and was dismayed to find them frozen and apparently dead the next morning. He took them inside to his lab anyway, thinking that they would still be good for dissection, and after being away for an hour giving a lecture, returned to be surprised by their revitalization. Further experimentation showed that a wood frog readily survives about two-thirds of its body water becoming ice. If cut, it will not bleed; its limbs will break if bent. Its heart does not beat and masses of ice fill the abdominal cavity. But it is alive.

Figure 19.5
Example of a Popular Science Article

How? Ken and Janet Storey at Carleton University in Ottawa determined that, when ice first begins to form, the frog becomes excessively sweet. Large stores of glycogen in its liver are transformed into glucose which is circulated throughout the body, reaching 10 times normal levels in the liver, heart, kidney and brain; in blood, glucose concentrations can rise a hundredfold, to one part in 10. The high glucose levels act to keep water inside cells from freezing and protect cell structures from injury due to desiccation. When water outside tissue cells begins to crystallize into ice, the concentration of dissolved molecules or solutes left behind becomes greater. This leads to a movement of water from inside the cells to the outside, as a result of osmosis (the tendency of water to move across a cell membrane when it separates solutions of different concentrations). Water leaving the inside of the cell results in less water available there to freeze, and it also produces an increase in the intracellular solute concentration. This leads to a decrease in the freezing point of that intracellular water, as putting salt to ice on sidewalks causes ice to melt at mildly freezing temperatures. Glucose may bind intracellular water, making it unavailable for ice formation. When spring comes, all of these changes are completely reversible.

The frogs exhibit freeze-tolerance. Freeze-resistance can take one of three forms. One is synthesis of antifreeze molecules; another is solute-dependent freezing-point depression. The third is "supercooling."

Supercooling or undercooling is when liquids stay fluid below their freezing points. Ice crystals require a "nucleator" around which to form; without it, pure water can be cooled to −38°C before it crystallizes spontaneously. Usually water is contaminated with particles that act to nucleate ice at far warmer temperatures, although crystallization is always a random event dependent on numerous variables. Once an ice embryo forms, ice will spread quickly through supercooled water unless antifreeze is present.

The Supercooled Squirrel

The discovery of mammals that adopt body temperature below freezing resulted from my studies of hibernation and reproductive biology in arctic ground squirrels, which weigh up to 2.5 pounds and abound on the Alaskan northern tundra wherever there are dry soils that can be easily dug.

Arctic ground squirrels breed seasonally, which means they have reversible puberty. Each year they undergo reproductive maturation around the time that they emerge from hibernation. The mating season lasts only two weeks and by early summer their gonads have reverted to an undeveloped state. Each year this cycle repeats itself.

To study how hibernation influences the ability of the endocrine system to control puberty, we brought back squirrels from the Brooks Range to Fairbanks. We first needed to determine the exact pattern of normal body temperature change during hibernation. To do this under as natural conditions as possible, we monitored hibernating ground squirrels in outdoor burrows where nest temperatures would decline naturally. We also wanted to determine how soon after ending hibernation and resuming normal body temperatures ground squirrels tunnel from their burrows and emerge from their hibernacula, and exactly when relative to these events they undergo puberty.

We surgically implanted miniature radio-transmitters into the abdominal cavity of each animal. These AM units, about the size of a stack of five nickels, transmit pulses with a frequency that varies with temperature. In mid-September we released the implanted animals into individual outdoor pens in the woods. Buried in the pens were large copper wire loop antennas that connected to radio receivers tuned to pick up the signals from the transmitters. A computer was programmed to translate transmitter frequencies into body temperatures every half hour. Within a few days, each squirrel had honeycombed its pen with tunnels and built a strawlined nest in the deepest part to serve as its hibernaculum. By late October, they had sealed their burrows and would not emerge again for six months. Meanwhile, we would be able to eavesdrop on their body temperatures. (When we first listened in, however, we heard country and western music. There are only four AM stations in Fairbanks, but one was transmitting near our frequency from a tower less than a mile away. Two months of adjustments later, we finally got clear signals.)

That winter, air temperatures above the pens averaged below −18°C November through February and often remained below −30°C for weeks. Soil temperatures near the hibernacula dropped to −10°C.

By spring several results were apparent. Male ground squirrels stop hibernating and return to high body temperatures weeks before emerging above ground, and it is during this time that they sexually mature. Females emerge within a day of rewarming and become ready to mate in that short interval. Throughout winter, the animals showed patterns of body temperature typical for hibernators: two- to three-week bouts of continuous torpor broken by day-long resumptions or arousals to normal levels.

But the most surprising finding was that our squirrels were cooling to core body temperatures as low as −3°C. We then spent the next winter confirming our findings by directly measuring subfreezing body temperatures in animals hibernating in the lab at temperatures similar to those outdoors.

Next we investigated how these animals can survive. We knew that they weren't actually freezing because,

Figure 19.5 (continued)

had ice formed, we should have measured a temporary increase in their body temperature as heat was released during crystallization. (Called latent heat of fusion, this occurs every time a liquid turns into a solid.) Yet as their temperatures passed below freezing, we observed nothing but a steady downward trend.

This implied that the ground squirrels were freeze-resistant. The question was: Which kind of freeze-resistance? We ruled out excessive freezing-point depression due to blood solutes, because when we tested blood from subfreezing animals its concentration was normal for mammals and worth only a half a degree of protection.

We tested their blood for antifreeze by placing a small volume into a glass tube and slowly lowering its temperature while watching through a microscope. At 0°C we introduced a tiny ice crystal into the blood by quickly freezing the side of the tube and watched to see when the crystal began to grow; then we reversed the process to determine the melting point at which the crystal disappeared. Were antifreeze molecules present, the freezing point would be lower than the melting point, indicating molecular inhibition of ice growth. But freezing

and melting temperatures were identical, indicating that, at least at the body temperatures we tested, our animals had no antifreeze.

The remaining hypothesis was supercooling: avoidance of freezing by keeping ice crystals from forming in the first place. Before entering hibernation, arctic ground squirrels may cleanse their blood and tissues of naturally occurring nucleators and during hibernation guard carefully against exposing themselves to the frozen soil surrounding their nests. Supercooled yet still breathing and with beating hearts, these animals would seem to live a precarious existence. Should an internal nucleator become active, or an ice crystal penetrate the tissues, ice would immediately spread through their bodies.

So far the arctic ground squirrel is the only mammalian hibernator known to naturally adopt and spontaneously arouse from body temperatures several degrees below freezing. Evolutionary selection might have favored this trait: Metabolic studies of other hibernators suggest that supercooling could save arctic ground squirrels up to 10 times the amount of energy they would have to spend if they were not freeze-resistant. It will be interesting to discover whether the one other hiber-

nating species that shares the northern tundra with the arctic ground squirrel—the Alaskan marmot—also supercools.

Determining how arctic ground squirrels are able to supercool when other mammals cannot and how wood frogs and other freeze-tolerant animals withstand internal ice formation promises interesting future research. For example, how do wood frogs tolerate such high levels of glucose? What changes occur in the blood and tissues of arctic ground squirrels before they are able to supercool? Are there specific naturally occurring nucleators that can be eliminated or otherwise neutralized? Would similar changes in other species allow them to supercool too?

Conceivably, such research might even have future indications for human clinical medicine. If we could understand how wood frogs control the massive release of glucose, and how their bodies tolerate what should be toxically sweet blood, we might have new weapons against diabetes. Research on supercooling in ground squirrels may give clues to better treatment of frostbite, new techniques in cryosurgery or longer storage time for donor transplant organs. Or, who knows—maybe someday supercooled astronauts will circle the solar system.

THE FREEZE FACTOR-COPING WITH WINTER

Until very recently, no mammal had ever been known to naturally adopt body temperatures below freezing. New research however, shows that the body temperature of the Artic Ground Squirrel routinely drops well below the freezing point of water. Yet ice crystals do not form in their blood.

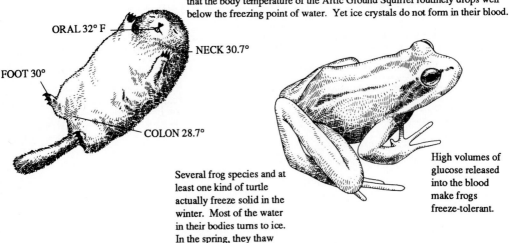

ORAL 32° F

NECK 30.7°

FOOT 30°

COLON 28.7°

Several frog species and at least one kind of turtle actually freeze solid in the winter. Most of the water in their bodies turns to ice. In the spring, they thaw and revive.

High volumes of glucose released into the blood make frogs freeze-tolerant.

Figure 19.5 (continued)

Chapter in Brief

In this chapter we examined the various types of papers and articles scientific and technical persons write in their professional activity. Included were communications to the editor, papers for primary journals, articles for technical and trade magazines, semitechnical articles, and articles for general readers.

Chapter Focal Points

- Communications to the editor
- Papers for primary journals
- Articles for the technical and trade press
- Review articles
- Semitechnical articles
- Popular science articles
- Query letters
- Handling scientific terms

Class Project

Professor Brian M. Barnes has written two articles on the same topic, dealing with his research on how some animal hibernators survive temperatures below the freezing point. One article appeared in *Science*, a primary scientific journal of the American Association for the Advancement of Science, and the other in the pages of a newspaper, *The Washington Post*. By studying both pieces (Figures 19.4 and 19.5), you can gain insight into the approach, technique, style, and language used in writing for two vastly different readerships. After reading both articles, consider the following questions:

1. What is the thesis of each article? Is the thesis the same?

2. Compare the title of each article.

3. Compare the length of each article.

4. What about their illustrations? Any similarities? Any differences? Compare the captions. What role do the illustrations play?

5. The unlabeled abstract at the beginning of the *Science* article serves as an "advance organizer" or an "overview." (See Chapter 4, pages 134–135). Dr. Barnes omits this reader aid in *The Washington Post* article. Why? The language of the abstract is less technical than of its article. Why?

6. Do language, word choice, style, and voice differ in the two articles?

7. The opening of the *Science* article is direct and specific to the research performed by Dr. Barnes. The opening in the popular science article in *The Washington Post* is generalized and provides an explanatory background as well as data from other researchers before discussing the research of the author. Why?

8. Are there differences in qualities of style? (See Chapter 4.) What are the differences? Is there any use of personal pronouns? Do both articles show objectivity? Any differences in length of sentences? How much biology or zoology course work do you need to follow the technical data in the *Science* article? How much in *The Washington*

Post? Did the abstract help you with the specialized technical terminology of the *Science* article?

9. What purpose do the references and notes serve at the end of the *Science* article?

10. *The Washington Post* article in its terminal section suggests hibernation research "might even have future implications for human clinical medicine," but that is not mentioned in the *Science* article. Why?

11. Which article took you longer to read? To understand? Were the illustrations and captions helpful to your understanding?

12. Do the articles stimulate you to want to know more about the subject of hibernation?

Questions for Discussion

1. Locate a report, a professional paper, a semitechnical article, and a popular article— all dealing with the same subject. Discuss the organization, style of writing, language, opening, and viewpoint of the four types of writing. Comment on the writer's success in transmitting his information to his intended readers of each of these types.

2. Choose a technical journal and a popular magazine, each carrying advertisements. Compare copy, illustrations, and layouts of the ads in each periodical. Are there more differences than similarities? How much specialized background does the reader need to have to follow the text or copy of the ads in the technical journal? What inferencs can be drawn about writing articles for a particular publication from studying its ads?

3. Choose a technical concept that you feel sufficiently informed about. Structure an analogy that will make this concept more intelligible to the uninformed reader.

4. Choose a technical subject with which you are familiar (for example, how an automobile carburetor works, how a plant grows from seed to flower, how a rock is formed, how immunization works, how an electronic blackboard works) and write two explanations: one to a friend with a scientific background and one to a ten-year-old child.

5. Read a number of articles listed in the literature guides about the particular subject of your interest. Study them as to organization, style, language, illustrations, viewpoint, literary devices. Compare them according to reading levels and note their special interests.

6. Go to the *Reader's Guide and Index to Periodical Literature* and other literature guides; look up a subject of your own special field of interest. Find an article in one of the professional journals and one in a popular magazine such as *McCall's Magazine* or a Sunday supplement such as *Parade*. Study and compare the two articles. Note particularly:

   Opening paragraphs

   How many technical terms are used and how many are defined by each?

   What literary devices were used to help explanations of concepts—analogies, contrast, repetition, anecdotes, humor, etc.?

   What other special devices were used—formulas, tabulations, illustrations?

   Is there a marked difference in level of specialized knowledge necessary to understand these special devices? In the number of illustrations?

7. Establishing contact with the reader at the beginning of an article is particularly important, especially in popular magazines. How do you do this? Does capturing your reader's attention at the beginning guarantee success of your article? How do you continue to keep your reader's interest throughout the article?

Assignments

1. Choose a subject for a 1,000-word popular article. Plan the article for a particular publication, taking note of the suggestions in Problem 6, above. Prepare a sentence outline and develop suitable illustrations and any pertinent tables. Write two or three opening paragraphs using any of the openings suggested in your text. Write a query letter to the editor of the magazine in which you wish to have your article published. Write the article using the techniques you learned in this chapter, and submit it for publication. Good luck with your submission.

2. Select a topic for the professional journal of your field. (It could be the subject matter of your research report.) Write the editor of that professional journal for format and style requirements. Plan your paper in accordance with the principles in this chapter and this text. Prepare a sentence outline.

References

1. Acker, Robert F., and S. F. Hartsel, "Fleming's Lysozyme." *Scientific American*, June, 1960, pp. 132–42.

2. Barnes, Brian M., "Freeze Avoidance in a Mammal: Body Temperatures Below 0° C in an Arctic Hibernator." *Science*, June 30, 1989, pp. 1593–95.

3. Barnes, Brian M., "How Animals Survive the Big Chill." *The Washington Post*, March 4, 1990, p. C3.

4. Booth, William, "Plants' Workhouses Mysteriously Answer Springtime Siren." *The Washington Post*, April 23, 1990, p. A3.

5. England, Samuel J. M., and Benjamin Passaminick, "Radiotelemetry of Physiological Responses in the Laboratory Animal." *Science*, January 15, 1961, pp. 106–7.

6. Kluger, Jeffrey, "Space Plane." *DISCOVER The World of Science*, November, 1989, pp. 81–84.

7. Sapolsky, Robert M., "Stress in the Wild." *Scientific American*, January, 1990, pp. 116–23.

8. Storad, Conrad J., "High Stakes Molecular Chess." *Arizona State University Research*, Fall, 1989, pp. 22–25.

9. Tulving, Endel, "Remembering and Knowing the Past." *American Scientist*, July-August, 1989, pp. 361–67.

10. Tzeng, Ovid J. L., and William S. Y. Wang, "The First Two R's." *American Scientist*, May-June, 1983, pp. 238–43.

11. Weintraub, Harold M., "Antisense RNA and DNA." *Scientific American*, January, 1990, pp. 40–46.

12. Wessell, Morris A., and Anthony Dominski, "Our Children's Daily Lead." *American Scientist*, May-June, 1977, pp. 294–99.

13. *Writer's Digest*, March, 1958.

PART SIX
Reference Index and Guide to Grammar, Punctuation, Style, and Usage

$\mathcal{S}$ome writers take refuge in the popular notion that grammar is not important. However, language is a code, and grammatical writing is easier to read and understand than ungrammatical writing. Unless people signal to each other in the same code, they cannot exchange intelligible messages. Their agreement to use the same code is the basis of correctness of grammar. Incorrect or ungrammatical means "not customary to the accepted code." You may not change the customary code without risking the possibility of your message being garbled.

This Reference Index is not meant to be a text on grammar and usage. Its purpose is to provide a ready and convenient guide to some of the more common problems the technical writer will meet in writing requirements. Fuller treatment of some of these problems will be found in the Selected Bibliography at the end of this book.

a, an. *a* is used before words beginning with a consonant (except *h* when it is silent), before words beginning with *eu* and *u*, pronounced *yu*, and before *o*, pronounced as in *one*. *An* is used before all words beginning with a vowel sound and words beginning with a silent *h*.

a	*a*	*a*	*an*	*an*
beacon	hard disk	eugenicist	alloy	herb
drift	helix	one-hour rating	eel	*H*
shunt circuit	hydrate	unit	*N*	homage
tentacle	hysteresis	uropod	impulse	hour

abbreviations. Abbreviations are used frequently for convenience in technical writing. However, they are acceptable only if they convey meaning to the reader. You should explain nonstandard abbreviations fully when you introduce them in a report. Abbreviations are appropriate in compilations, tables, graphs, and illustrations where space is limited. Their use in the text should be limited except when preceded by numerals, as in giving dimensions or ratings. Unfortunately, the spelling of abbreviations is not universal. The American National Standards Institute has a list which has been approved by many scientific and engineering societies. However, other professional societies, such as the Institute of Electric and Electronic Engineers, and government agencies have their own abbreviations. Dictionaries give frequently used abbreviations in their main listing of words; some also compile a list in a back section. You should consider including a glossary of abbreviations if you use a large number of them in a report. Always write the word out if you are unsure of the proper abbreviation. Use periods with abbreviations except for names of well-known organizations or government agencies, for instance: NOW, NSF, DOD, DOE, UNESCO, NATO, IEEE, PTSA, NAACP. To avoid confusion, use periods when the abbreviation spells out an actual word: W.H.O., C.A.B., A.I.D.; however, usage has ruled AIDS for acquired immunodeficiency syndrome. In a company or corporation name, omit the period when the abbreviation has become the official name: the IBM Corporation. When letters within a single word are used as an abbreviation, they are capitalized, but do not take periods: DDT, DNA, RNA, TB, TV. V.D. requires periods because it stands for two words. Stylistic convention within a field often determines whether an abbreviation is capitalized. We frequently see AI (artificial intelligence), PE (potential energy), and OD (outside diameter), but bp (boiling point), qid (4 times a day), emf (electromotive force), and both nmr and NMR (nuclear magnetic resonance). This is true of ac and dc, frequently seen as AC and DC. It is true also of both ram and RAM (random-access memory) and rom and ROM (read-only memory). What should you do about the capitalization of abbreviations? Be consistent in their use. Incidentally, when ac and dc are used as nouns they are usually spelled out (alternating current and direct current), but as adjectives they are abbreviated. Through usage, abbreviations of many business terms are acceptable in formal writing: COD (cash/collect on delivery), EOM (end of the month), FOB (free on board), GNP (gross national product), LCL (less than carload lots), R&D (research and development).

above. Grammatically, this word occurs most frequently as an adverb (the equipment identified *above* [*above* modifies the verb *identified*]) or a preposition (*above* the lithosphere). *Above* is sometimes used as an adjective (the *above* diagram) or as a noun (the *above* is an equation of merit). Many competent writers avoid using *above* except as a preposition because the sentence has a stilted construction and the thought transmitted tends to be ambiguous since the reference of *above*, especially as a noun, is vague. *Above* as an adjective or adverb, in the context previously indicated, should be used with discretion and specificity. I would avoid its use as a noun entirely, unless you don't mind sounding stuffy.

accept, except. *Accept* (verb) means to take when offered, to receive with favor, to agree to. *Except* (verb) means to exclude or omit, *except* (preposition) means with the exclusion of, but. The customer *accepted* delivery of four of the equipments; of the five equipments built, only one was *excepted* from delivery. All *except* model 5 were delivered.

access, excess. *Access* means approach, admittance, admission (to gain *access* to the laboratory). *Excess* means that which exceeds what is usual, proper, or specified (there was an *excess* of ten liters of liquid in the tank.)

acronym. An *acronym* is a pronounceable word formed with the first letter or letters of each of a series of words: *laser* from light amplification by stimulated emission of radiation; *radar* from radio detecting and ranging. *ABM* is not an acronym, but an abbreviation for antiballistic missile; NSF is the abbreviation for National Science Foundation.

activate, actuate. *Activate*, an old English word that was once considered obsolete, has been reintroduced as a special term in chemistry and physics to denote the process of making active or more active as: to make molecules reactive or more reactive; it is used especially in the sense of promoting the growth of bacteria in sewage and of making substances radioactive. The military have adopted *activate* to mean: to set up or formally institute a military unit with the necessary personnel and equipment. *Activate*, however, should not replace the word *actuate*, which means to set a machine in motion or to prompt a person to action.

adapt, adopt. These two words are often confused, even though their meanings are entirely different. *Adopt* means to take by choice into some sort of relationship (he *adopted* the older engineer's design). *Adapt* means to modify, to adjust, or to change for a special purpose (he *adapted* the cam to provide a slower movement).

adjective. An *adjective* is the part of speech that modifies or limits nouns and pronouns. Its purpose is to clarify for the reader the meaning of the word it modifies. The adjectives in the expression *electrostatic* reaction, *inverted-V* antenna, *igneous* magna, *alpha* particle, *standstill* torque, and *heat-resisting* steel provide more meaning to the reader than the nouns by themselves. Adjectives are classified as demonstrative, descriptive, limiting, and proper. A demonstrative adjective points out the word it modifies; *this* computer, *these* cells, *that* alloy, and *those* theories. A descriptive adjective denotes a quality or condition of the word it modifies: an *automatic* titration; a *progressive* disease, a *bent* pipe; a *dedicated* word processing system, and a *slower flow* rate. A limiting adjective designates the number or amount of the word it modifies: a *5 percent* potassium chloride solution; a *600,000 candela* intensity; 5 grams of ascorbic acid; and a *10 millimeter-per-second* flow. The articles *a*, *an*, and *the* are limiting adjectives. A proper adjective is one derived originally from a proper name: *Josephson* effect; *Schick* test; a *Hollerith* code; *Gibb's* function; *Californian* jade; and *Brownian* movement. Proper adjectives that have lost their sense of origin are written in lower case: *india* ink; *macadamized* road; and *angstrom* unit.

adverbs. An *adverb* is a word, phrase, or clause that modifies a verb (the experiment went *smoothly*); an adjective (the *very* slow titration took longer than I expected); another adverb (*almost* immediately the change occurred); or an entire sentence (*Fortunately*, the power had already been turned off). Most adverbs are adjectives plus the ending *ly*: quick*ly*, complete*ly*, correct*ly*, simp*ly*, final*ly*. Some adverbs derived from Old English have no special adverbial sign: *now, quite, since, below, much,* and *soon*. Some adverbs have the same form as adjectives: *best, early, fast, slow, straight, well,* and *wrong*. Some of these adverbs also have an ly form: *slowly, wrongly*. When the adverb is used with a compound verb, it should *normally* be placed between the elements of the verb as this sentence and the following example illustrate: The cerebal cortical neuron of the cat will *usually* show a characteristic bursting firing pattern on injection of PTZ.

advice, advise. *Advice* is a noun meaning guidance, counsel, or a recommended opinion. The mechanic's *advice* was to install a new alternator. Statements without recommendations or opinions are not *advice*, as for example: The present *advice* is aspirin should not be used. In this instance, the use of *advice* is wrong. The sentence should read: Present knowledge indicates that aspirin should not be used in this situation. *Advise* is a verb meaning to give information, to recommend, to caution, to give counsel. The physician *advised* the patient to take two aspirins. The policeman *advised* the motorist to drive in first gear across the icy bridge.

adverse, averse. *Adverse* means opposed to or being unfavorable to. *Averse* means unwilling to or reluctant to. The investigating committee was *averse* to making an *adverse* recommendation.

adviser, advisor. Both spellings are acceptable. What adds to the confusion is that *adviser* is used more frequently, but the spelling of *advisory* is the correct one.

affect, effect. These words are frequently confused because many people pronounce them alike. *Affect* is almost always a verb, meaning to influence or to make a show of: The new economic measures *affected* the industry's recovery; he *affected* an English accent. *Affect* as a noun is used as a term in psychology, pertaining to feeling, emotion, and desire as factors in determining thought and conduct. *Effect* is used infrequently as a verb. As a noun it means result or consequence. The *effect* of cooling on the system increased its efficiency; the *effect* of the design change was a saving in time and money. *Effect* as a verb means to bring about: The change in design *effected* a great saving.

agenda. *Agenda* is the plural of the Latin *agendum*. In today's usage *agenda* is singular and *agendas*, plural.

agree to, agree with, agree on, agree about. These are idiomatic expressions. You *agree to* things, *agree with* people, and people *agree on* something by mutual consent: He *agreed to* the contract; he *agreed with* the client; they *agreed on* the details of the contract; they *agreed about* the details.

all right. *Alright* is never *all right*. Do not use it.

allude, refer. To *allude* to something is to speak of it without direct mention; to *refer* to something is to mention it directly.

alphabetizing names. When dealing with *Mc* or *Mac*, alphabetize by the second letter: *Maberry, MacTavish; Mallard; McNeil.* In a listing, if family names are written before given names, the expression *Jr.* (or *Sr.* or *III*, etc.) comes last: *Rockefeller, John D., Jr.,* not *Rockefeller, Jr., John D.*

alternate(ly), alternative(ly). As adjectives and adverbs, these two words are frequently confused. *Alternate* means by turn; *alternative* means offering a choice: The problem had two *alternative* solutions. The interstate highway is the shortest route to the city, but since it becomes crowded very early, there are several *alternative* routes to take. The weather *alternated* between rain and sunshine. The wall had *alternating* layers of brick and stone. *Alternate* as a noun means one that takes the place of or alternates with another: If you cannot attend, send an *alternate*.

although, though. Both words are often used interchangeably to connect an adverbial clause with the main clause of a sentence to provide a statement in opposition to the main statement, but one that does not contradict it. *Although* is preferable to introduce a clause that precedes the main clause; *though* for a clause that follows the main clause: *Although* we were short of test equipment, we managed to check out the system. We managed to check out the system, *though* we did not have all the test equipment.

A.M. and P.M., also a.m. and p.m. These are abbreviations for *ante meridiem* (before noon) and *post meridiem* (after noon). Some persons are confused about how to write "12 noon" and "12 midnight." Though *M* is the abbreviation for noon, writing *12 noon* is used more commonly; midnight is written as *12 P.M.* or *12 p.m.*

among, between. *Among* denotes a mingling of more than two objects or persons; *between*, derived from an Old English word meaning "by two," denotes a mingling of two objects or persons: *Among* several designs, his was the most practical. The instrument's accuracy was *between* ±1 and ±2 percent. *Among*, however, expresses a collective relationship of things and *between* seems to be the only word available to express the relation of a thing to many and surrounding things, severally and individually: disagreement *among* bidders; to choose *between* courses; the space lying *between* three points.

amount, number. *Amount* refers to things or substances considered in bulk; *number* refers to countable items as individual units: the *amount* of gas in the container (but the *number* of containers of gas); the *amount* of corn in the silo (but the *number* of bushels of corn).

and/or. This expression, usually frowned upon in literary writing, is commonly used in agreements, contracts, business and legal writing, and technical writing to show there are three possibilities to be considered: He offered his house *and/or* automobile as collateral.

ampersand. *Ampersand* is the word for the symbol &. Its primary use is to save space. Its use as a substitute for the word *and* is frowned upon in formal writing. Many firms use the *ampersand* as a formal part of their official or incorporated name as in J.C. Mason & Co., Inc. In addressing firms, use the form indicated in their formal letterhead.

apostrophe. The most common use of the *apostrophe* is to show possession and ownership: Dr. *Stanton's* equation; the *agency's* director; the *formula's* derivation; the *book's* index. The *apostrophe* is used to indicate omission of one or more letters in a contracted word or figure: *'95* for 1995; *didn't* for did not; *they're* for they are. The *apostrophe* is used in the plural form of numbers, letters, and words: The *1990's*; *ABC's*; Programming is spelled with two *m's*; The second of the two *which's* is not necessary in the sentence. There is some tendency to omit the *apostrophe* in plurals of numbers and letters, such as 1920s, four Ws, or second of three thats. The *apostrophe* is not used with the possessive pronouns: his, hers, ours, theirs, yours, its. Sometimes a singular idea is expressed in words that are technically plural; in such a case, the plural form of the possessive is used: *United States'* scientists; *General Motors'* earnings. Almost all singular words ending in *s* require another *s* as well as the apostrophe to form the possessive: The *Times's* story, *Gibbs's* paper. But the *s* after the apostrophe is dropped when two or more sibilant sounds precede the apostrophe: Kansas' wheat. However, when a name ends with a sibilant letter that is silent, the possessive is formed with *'s*: *Arkansas's* boundary; *Malraux's* writings. Plurals of names ending in *s*, such as Jameses', Charleses', Joneses' also have the apostrophe following the final *s*.

appendix. The plural most commonly used is *appendixes*. Purists prefer *appendices*.

appositives and their antecedents. An *appositive* is a term that modifies a noun or other expression by placing immediately after it an equivalent expression that repeats its meaning: the resistor, V101; videotext, an interactive electronic information system; our director, Dr. Roberts. *Appositives* should have clear antecedents. Vague appositive reference can be corrected by placing the appositive immediately after the word or phrase it modifies or by rewording the sentence.

Faulty: Maxwell formulated a new theory while experimenting with Faraday's concept, a major contribution to the study of electricity.

Revised: While experimenting with Faraday's concept, Maxwell formulated a new theory, a major contribution to the study of electricity.

or: Maxwell contributed greatly to the study of electricity when he formulated a new theory based on Faraday's concept.

appraise, apprise, apprize. *Appraise* means to evaluate, *apprise* means to inform; *apprize* is the British spelling.

Wrong: Employees were *appraised* of the new company policy.

Right: Employees were *apprised* of the new company policy.

Wrong: Employees were *apprised* only on their performance.

Right: Employees were *appraised* only on their performance.

Note the prepositions: *appraised on, apprised of*

articles. There are three articles in the English language: *a, an,* and *the. A* and *an* are indefinite articles: *a* dog; *an* apple; dog and apple refer to typical or unidentifiable things. *The* dog, *the* apple refer to a particular or identifiable dog and apple.

assay, essay. These two words are sometimes confused. *Assay* means to test; *essay* means to attempt. *Assay* also has the meaning of analyzing or appraising critically: We shall *assay* the sample to determine its properties. Having obtained a more precise instrument, the scientist again *essayed* the experiment.

as to whether. This phrase is used by persons who would rather use several words when one would serve, as in the sentence, *As to whether* the new software program would work was open to question. Are the words *as to* needed? Not really. *Whether the new software program would work was open to question* is all that is needed.

awhile, a while. *Awhile* is an adverb; *a while* is a noun phrase, often used as the object of a preposition: the odor lasts *awhile*. For *a while*, the odor remained in the room.

back of, in back of. Though both are grammatically correct, *behind* is less wordy.

backward, not *backwards.*

because, for, since, as. *Because* is used to introduce a subordinate phrase or clause, providing the reason for the statement in the main clause: *Because* of its low boiling point, the substance could not be used in space applications. When a sentence begins with *The reason is* or *The reason why . . .* is, the clause containing the reason should not begin with *because* but with the word *that. Because* in this usage is an adverb; *that* is a pronoun. The linking verb *is* requires a noun rather than an adverbial clause. *For, since,* and *as* can be used in similar constructions, but these conjunctions are less formal and less emphatic.

before is better than the wordy and stilted *prior to.*

beside, besides. *Beside* means near, close by, by the side of. *Besides* means in addition to, moreover, also, aside from: The replacement assembly was placed *beside* the faulty instrument. *Besides* corrosion resistance, the new alloy had many other desirable properties.

biannual, biennial. *Biannual* means twice a year (as does *semiannual* and *semiyearly*); *biennial* means every two years.

biweekly, semiweekly. *Biweekly* means every two weeks; *semiweekly* means twice a week.

building names. Capitalize the names of governmental buildings, churches, office buildings, hotels, specially designated rooms, etc.: the Capitol (state or national), Department of Commerce Building, Washington Cathedral, Beth El, Kennedy Center, Oak Room.

brackets []. *Brackets* are used whenever it is necessary to insert parenthetical material within parenthetical material. *Brackets* are also used to make corrections or explanations within quoted material. In quoting material, use *sic* in brackets [*sic*] to indicate that an error was in the original quoted material: "He lives in New Haven, Conneticut [*sic*]."

but. *But* is the coordinating conjunction used to connect two contrasting statements of equal grammatical rank. It is less formal than the conjunction *however* or *yet* and is more emphatic than *although*: Not an atom *but* a molecule. The signal was short *but* distinct. The crew worked industriously under his supervision, *but* the minute he turned his back, they sloughed off.

can, may. *Can* expresses the power (physical or mental) to act; *may* expresses permission or sanction to act. In informal or colloquial usage, *can* is frequently substituted for *may* in the sense of permission, but this substitution is frowned upon in formal writing. *Could* and *might* are the original past tenses of *can* and *may*. They are now used to convey a shade of doubt or a smaller degree of possibility: We *can* meet your stringent specifications. Under the FDA regulations, products with 3 percent hexachlorophene solutions *may* be sold by prescription only. The absorbed hydrogen *might* be removed by pumping out the system. The use of *could* suggests doubt or a qualified possibility: The possibility is remote that under those conditions the contractor *could* meet his delivery schedule.

cannot, can not. Both forms are used. *Can not* is more formal, but *cannot* is used more often.

capital, Capitol. Use lowercase *capital* for the city that is the seat of national or state government. Washington D.C. is the *capital* of the U.S.; Sacramento is the *capital* of California. The U.S. Senate meets in the *Capitol*.

capitalization. The rules for capitalization in technical writing are the same as in other formal writing. Use capitals for:

1. Titles, geographical places, and trade names:
   Origin of Species
   Atlantic Ocean, Antarctica, Ohio
   Freon, Polaroid, Teflon
2. Proper names and adjectives derived from proper names, but not words used with them:
   Dundee sandstone, Pliocene period, Gaussian, Coulombic, Boyle's law, Huntington's chorea, Einstein's theory of relativity
   Words derived from proper nouns and which, through long periods of usage, have achieved an identity within themselves are no longer capitalized, for example:
   bunsen burner, galvanic cell, ohm, ohmic drop, petri dish
3. Scientific names of phyla, orders, classes, families, genera: *Decapoda, Megaloptera, Colymbiformes, Urochorda*
4. The first word of a sentence is capitalized.
5. In quotations, the first word of a quoted sentence or the first word of a part of a sentence is capitalized. When the quotation is interrupted, the second part is not capitalized unless it is a complete sentence:

   > "Press the move key," the instructor told the word processing class, "the control F-4 key." He then added, "The status line at the bottom of the screen shows you have five choices."

6. Heads, subheads, and legends require capitalization.

7. Names of computer services and systems, software programs, and databases should be capitalized. Names of computer languages are regularly styled either with initial capitals or all capitals:

> DIALOG; NEXT; COBOL or Cobol; BASIC or Basic; WordPerfect; FORTRAN or Fortran; PL/1; Lotus; MS-DOS; Ventura Publisher.

catalog, catalogue. If you prefer the second spelling, you are a traditionalist fighting a losing battle.

cement, concrete. Cement and concrete are not interchangeable. *Cement* by definition is an adhesive. *Concrete* is produced by mixing Portland cement, water, sand, and gravel.

center about, center around, center on, center upon. In formal writing, the idiom is *center on* or *center upon*; in less formal writing, *center around* or *about* is the idiom found in use, but *revolve around* is better.

centigrade, Celsius. *Celsius* is the name of the Swede who invented the centigrade system. The metric system has preferred to use his name for designating the centigrade system: The boiling point of water at sea level is 100 degrees *Celsius*. In technical writing, use *Celsius*.

chairman, chairperson. *Chairperson* is preferable regardless of the presiding individual's gender.

chemical elements and formulas. Names of elements are not capitalized: *carbon, hydrogen, strontium 90*. In chemical formulas, the first letter of the abbreviation of the element is capitalized: H_2O; SO_2Na; Cl_2O; $KMnO_4$.

circumlocution. A long word meaning wordiness, excessive verbiage, dead wood. Perhaps because they have a subconscious belief that they are paid according to the number of words they write, technical writers often overwrite and pad. Excessive verbiage spreads ideas thin.

Wordy: Caution must be observed . . .

Revised: Be careful, don't . . . .

Wordy: We are about to enter an area of activity.

Revised: We are ready to begin.

Wordy: It is not desirable to leave filters in a system after resistence has increased to the point where there is a substantial decrease in the flow of air.

Revised: Filters should not be left in a system after resistance has caused a substantial decrease in the flow of air.

Wordy: On the basis of the foregoing discussion it is apparent that . . . .

Revised: This discussion shows . . . .

Wordy: The pursuit and capture of winged, air-breathing arthropods is more easily effected when a sweet, as opposed to sour, substance is, for purposes of beguilement, made use of: For example, the viscid fluid derived from the saccharin secretion of a plant and produced by hymenopterous insects of the super family *Apoidea* has proved to be more successful in this endeavor than has dilute and impure acetic acid.

Revised: More flies are caught with honey than with vinegar.

clause (subordinate). A *subordinate* or *dependent clause* is an element of complex and compound sentences (see pages 124–126); it usually has a subject and a verb and is grammatically equivalent to a noun, adjective, or adverb. The dependent clause is introduced by a subordinating conjunction, e.g., *as, because, since, when*; or by a relative pronoun, e.g., *that, who, which*.

> *That the waveguide overheated* was not a surprise to him (subject or noun clause).
>
> *Because the expert knows where to look for data on free-radical reactions in solution,* he seems to have little sympathy for the nonkineticist who needs such data (adjective clause).
>
> The administrator *who received the report* will not read it (adjective clause). The endogenous RNA was removed *when the preparation was subjected to alkaline hydrolosis* (adverbial clause).
>
> Few experiments have been carried out *because nematodes have little or no capacity for regeneration* (adverbial clause).

collective noun. A *collective noun* is one whose singular form carries the idea of more than one person, act, or object: army, class, crowd, dozen, flock, group, majority, personnel, public, remainder, and team. When the collective noun refers to the group as a whole, the verb and pronoun used with it should be singular. When the individuals of the group are intended, the noun takes a plural verb or pronoun: The committee *is* here; The committee *were* unanimous in disapproval; The corporation *has given* proof of its intention. The plural of a collective noun signifies different groups: *Herds* of deer graze in the upper valley.

colon (:). The colon is used as a mark of introduction to a word, phrase, tabulation, sentence, or passage to be quoted. It is also used in giving clock time. The first word after a colon is not capitalized if what follows is not a complete sentence. Use the colon:

1. After the salutation of a letter:
   Dear Dr. Reimann:
   Gentlemen:
2. In memoranda following TO, FROM, and SUBJECT lines:
   TO:
   FROM:
   SUBJECT:
3. To introduce enumerations, usually *as follows, for example, the following,* etc.
   The components are *as follows*:
4. To introduce quotations:
   The statement read:
   We the undersigned members of the corporation believe:
5. To separate hours and minutes:
   The meeting will begin at 10:30 a.m.

comma. The modern tendency is to avoid *commas* except where they are needed for clarity. Use commas:

1. To set off nonrestrictive modifiers or clauses. A word, phrase, or clause that follows the word it modifies and that restricts the meaning of that word is called a restrictive modifier or restrictive clause. Restrictive clauses are *not* set off by commas. When such a modifier merely adds a descriptive detail, gives further information, it is called *nonrestrictive* and is set off by commas. If the modifier is omitted, will the sentence still tell the truth or offer the meaning you intend? If it does, then the clause or modifier is *nonrestrictive* and should be set off by commas. If it does not tell the truth and if it does not give the meaning intended, you must *not* use commas. Example: Our chief chemist, who got her Ph.D. degree from Stanford University, is now in Europe. *Who got her degree from Stanford University* is not necessary to the sense intended, so that clause is set off by commas. However: Our chief chemist who checked the computations of the experiment does not agree with the conclusions reached. The clause *who checked the computations of the experiment* is necessary to the sense of the sentence. It is restrictive and therefore should not be set off by commas.

2. With explanatory words and phrases or those used in apposition: To meet the deadline, Mr. Simon, our president, supervised the experiment.

3. With introductory and parenthetic words or phrases, such as *therefore, however, of course,* and *as we see it*; As you are aware, the legislation did not pass, this action, nevertheless, is necessary.

4. In inverted construction where a word, phrase, or clause is out of its natural order: That the result was unexpected, it was soon apparent.

5. To avoid confusion by separating two words or figures that might otherwise be misread: By 1990, 30,000 students are expected to enroll.

6. To avoid confusion by separating words in a series: The client manufactures furniture, pottery items, electric fixtures, and steel garden implements.

7. To make the meaning clear when a verb has been omitted: I covered the door, John covered the hallway, Bill covered the windows, and Tom, the remaining exit.

8. To separate two independent long clauses joined by a coordinating conjunction (*and, but, for, yet, neither, therefore,* or *so*): Analog recorders of .05 percent accuracy are available, but frequent maintenance is often required to hold this tolerance. (A comma is not used when the coordinate clauses are short and are closely related in meaning: A panel of experts might be chosen from particular areas of specialization and their report would be published).

9. To separate a dependent clause or a long phrase from its independent clause: For complex mixtures of acids such as those found in physiological fluids, a five-chamber concave gradient is generated. Although good recoveries of both methyl pyruvate and methyl lactate could be obtained at column temperatures below 100° Celsius, there was no indication that such recoveries could be achieved when either of the acids was in excess.

10. To separate the day of the month from the year:
December 21, 1916
When the day of the month is not given, the usual practice today is to omit the comma: December 1916. In construction in which the day precedes the month, no comma is used: 21 December 1916.

11. To separate town from state or country when they are written on the same line:

Bethesda, Maryland; Washington, D.C.; Stockholm, Sweden; Fort Collins, Larimer County, Colorado.

12. In figures to separate thousands and millions: 528,121; 10,894,082.

13. To introduce a quotation: Professor Thomas said, "The new alloy will work."

14. To introduce a paraphrase similar in form to a quotation but lacking quotation marks: The question may be asked, Will this new approach work?

compare, contrast. There is some confusion in the application of these two words because *compare* is used in two senses: (1) to point out similarities (used with the preposition to); and (2) to examine two or more items or persons, to find likenesses or differences (used with the preposition *with*). *Contrast* always points out differences.

complement, compliment. These two words, though pronounced alike, are completely different in meaning. *Complement* means to fill up, to complete a whole: The *complement* of this sixty-three degree angle is the one of twenty-seven degrees. *Compliment* means to praise: The professor *complimented* the student on her design.

compound predicate. A compound predicate consists of two or more verbs having the same subject. It is often used to avoid the awkward effect of repetition of the subject or the writing of another sentence: The Emperor Van de Graaff accelerator is precisely controllable. Another feature is that it is easily variable. It is also continuous. A more economical and smoother way of saying the previous sentences is to combine them with a compound predicate: The Emperor Van de Graaff accelerator is precisely *controllable, easily variable,* and *continuous.*

compound subject. Two or more elements that serve as the subject of one verb form: The *deposition time* and *amount of carbon available in the gases determine* (not determines) the resistance value in the film resistors.

computerese. (See also *jargon.*) *Computerese* is the special vocabulary or sublanguage evolved by people working with computers and information processing systems for communicating with one another. It is a user unfriendly language, virtually incomprehensible to the uninitiated. Here is an example:

> *1ON1* = 3 is a dBaseIII Plus work-alike. It is compatible with dBaseIII commands and functions in the dot prompt and programming mode; the assist mode, however, has been modified to make it easier to use. 1ON1 = 3 is compatible with most dBaseIII Plus files and can read and report on information in dBaseII files. It requires 512K of RAM.

To the uninitiated, this unfriendly description of a program, to use a computerese term, is "GIGO" (garbage) and meaningless.

conjunction. Conjunctions introduce and tie clauses together and join series of words and phrases. Types of conjunctions include:

> coordinating: *and, but, for*
> correlative: *either . . . or, not only . . . but*
> conjunctive adverbs: *however, therefore, consequently, accordingly, nonetheless*
> subordinating: *as, because, since, so that, when*

contractions. *Contractions* are words from which an unstressed syllable is dropped in speaking. Their use is acceptable in informal writing but not in formal writing.

> can't: use *cannot*
> didn't: use *did not*
> I'll: use *I shall, I will*

council, counsel, consul. These three words are sometimes confused. A *council* is an advisory group; *counsel* means advice and, in law, it means one who gives advice; a *consul* is an official representing his or her government in a foreign country.

dangling modifiers. The *dangling modifier*, the most prevalent fault in technical writing, is usually a verb form (often a participle) that is not supplied with a subject to modify, and which seems to claim a wrong word as its subject. It is said to dangle because it has no word to which it logically can be attached. Infinitive and prepositional phrases may also be dangling.

Wrong: Calibrating the thermistor through the temperature range of 17° to 19° Celsius, a value of 4.0 ± 0.1°C.1 molar was obtained.

Revised: Calibrating the thermistor through the range of 17° to 19° Celsius, we obtained a value of 4.0 ± 0.1°C.1 molar.

Wrong: After adjusting the valves, the engine developed more power.

Revised: The engine developed more power after the valves were adjusted.

or: After adjusting the valves, the mechanic found the engine developed more power.

Wrong: To write a program for our computer, it helps to know Fortran.

Revised: To write a program for our computer you would find it helpful to know Fortran.

Better: A knowledge of Fortran is helpful in writing a program for our computer.

Wrong: Near Kamchatka, Alaska, Figures 11 through 17 show the typical appearances in both the television pictures and the infrared data, of several stages in the life cycle of the cloud vortices of a cyclonic storm.

Revised: Figures 11 and 17 show typical appearances in both television pictures and infrared data of several stages in the life cycle of the cloud vortices of a cyclonic storm near Kamchatka, Alaska.

dash (—). The *dash* is used to indicate a change of thought or a change in sentence structure. It is also used for emphasis and to set off repetition or explanation. The *dash* may be used in place of parentheses when greater prominence to the subordinate expression is desired. Dashes should not be used with numbers because they might be mistaken for minus signs.

> The rectifier—that was the guilty component.
>
> I am suspicious of prognosticators—but I would not be surprised if tangible evidence of extraterrestrial, intelligent life will be found before the end of the present century.
>
> A traveling wave tube has been made with a helix fifteen-thousandths of an inch an inside diameter—about three times the diameter of a human hair.

These molecules—formic acid, acetic acid, succinic acid, and glycine—are the very ones from which living things are constructed.

data. The word *data* is the plural of the Latin word *datum*. The singular form, *datum* is rarely used in English. The word *data* is often defined as raw facts or observations. In science and technology, *data* are characterized by their tendency toward quantification. Because the singular form is rarely used, the word *data* is becoming more acceptable in all but the most formal English as either singular or plural. In the computer field, *data* has come to mean a collection or mass of information; in that sense, it has lost its plural connotation: *Data is stored for retrieval.* Historically, Latin plurals sometimes become singular English words (e.g., *agenda, stamina*). Some writers treat *data* as a collective noun and use the singular verb with the word. For instance: The experiment's *data* was made availble to three laboratories. However, there is a loud and armed camp of data pluralists who will split an infinitive with alacrity but will scream with outrage and scorn at any usage of a singular English verb with the Latin plural form. My advice is that it is always safe to use the plural unless you can afford a battle—and can afford to lose it, because you may well do so.

dates. The more common form for writing dates is: August 11, 1990. The form, 11 August 1990, originating with the military services, is seeing more use because it makes a comma unnecessary. When only the month and year are used, August, 1990 is becoming obsolete; in more and more communications, the comma is being eliminated: August 1990. If saving space is important, months having more than four letters can be abbreviated: Jan., Feb., Mar., Apr., Aug., Sept., Oct., Nov., and Dec. In informal writing, figures are often used: 8-11-90 (for August 11, 1990). In England and other European countries, the day usually comes first: 11-8-90 (for August 11, 1990).

deadwood. See *circumlocution.*

decimals. Use figures for all numbers that contain decimals: 1.4 *liters of fluid; 4.5 inches of rain.* A zero is placed to the left of the decimal point when the fraction is less than a whole number: 0.3. Do not mix fractions with decimal fractions:

> *Not:* 2½ lb., 2.2 oz.
> *Instead* write: 2.5 lb., 2.2 oz.

defect. Use the preposition *in* for a defect in a thing: The defect *in* the engine, caused the accident. Use *of* for a person's shortcomings. The astronaut's defect *of* perception caused the accident.

diagnosis, prognosis. *Diagnosis* is the process for determining by examination the nature of a circumstance for a disease. *Prognosis* follows diagnosis. It is the physician's educated guess as to the course and outcome of the disease. The words are not interchangeable.

different from, different than. *Different from* is the established usage. *From* is a preposition and *than* is a conjunction. Different things differ *from* each other. Many authorities recommended that different *than* be used (though *than* is a conjunction) if it avoids awkwardness or wasteful words. For example: *Computers use different programming languages now than they did ten years ago,* rather than *Computers use different programming languages today from those which they used ten years ago.*

dimensions, measurements, weights and proportion. When any of these consist of two or more elements, or when a decimal is used, they should be rendered in numbers, even those below ten: *6 by 18; 12 feet 6 inches by 22 feet 9 inches; 7 years 5 months 15 days (age); 8 pounds 4 ounces; 5 parts gin, 1 part vermouth.* However, when a single dimension or measurement below 10 is given, it should be spelled out when it contains an ordinary fraction: *two and a third* miles; *six* feet tall; *seven* pound baby. But when the fraction is given in decimal form, use figures: *21.6-inch* snowfall.

disc, disk. Dictionaries say each is a variant spelling of the other, but in the computer age, *disc* should be reserved for phonograph records, and *disk* for the computer's magnetic information storage device.

discreet, discrete. *Discreet* means prudent. *Discrete* means separate or distinct. To use one for the other advertises ignorance.

disinterested, uninterested. *Disinterested* means fair, impartial, unbiased; *uninterested* means lacking interest or without curiosity. Although in recent years the two words have been used interchangeably in informal speech, your use of *disinterest* to mean lacking interest may raise some readers' eyebrows.

ditto marks ("). Ditto marks are a convenience in tabulation and lists that repeat words from one line to the next, but ditto marks should not be used in formal writing.

divided into, composed of. *Divided into* applies to a thing once whole that has been made into separate parts; *composed of* applies to a whole created from several or many parts.

division of words. Break words only between syllables. When in doubt about syllabication, look up the word in the dictionary. Avoid breaking up a compound word that requires a hyphen in its spelling. Do not divide words of two syllables if the division comes after a single vowel: *among, along, atom, enough.*

dollars and cents. Sums of dollars and cents are usually written in figures: *10 cents; 75 cents; $12; $24.95; $11,914.* But *$1 million; $3.94 million; $4.6 billion* is acceptable. *$4 to $10 million* should be written as *$4 million to $10 million* to avoid confusion. With dollars and cents, the convention of spelling out numbers below 10 does not apply: *3 cents, $8.* Round amounts are written in words: One hundred dollars, three thousand dollars, a million dollars.

due to. Grammarians say that *due to* is properly used in the sense of *caused by* or *resulting from* when *due* is an adjective modifying a noun: The hissing sound was *due to* the malfunction of the number three pump. The modified noun is *sound.* But *due to* should not be used when there is no noun modified. The malfunction occurred *due to* the leak in the number three pump is grammatically unacceptable. *Due to* in the present instance modifies the verb *occurred.* You will solve your problem and please grammarians by using an adverbial construction like *because of:* The malfunction occurred *because of* the leak in the number three pump.

ecology, environment. These words are not synonymous. *Ecology* is the science or study of the relationship between an organism and its environment. *Environment* is the totality of conditions surrounding an organism or organisms, including human beings.

editorial "we." The substitution of *we* when *I* obviously is intended is considered by many as affected and pompous. In those circumstances where there are several authors of a report, the use of *we* is certainly called for. However, a writer may want to take the reader with him over intellectual territory, as for example:

> We have seen the effect of X on variable Y in our experiment; we will therefore proceed now to add Z into the mixture to see whether thus and thus will then occur.

In correspondence, the use of *we* may be desirable when the writer is expressing his organization's policy or desires.

either. *Either* means *one or the other of two.* It can be *either* an adjective or a pronoun: *Either* alternative is not sound. Nevertheless, I shall have to take *either* (one). To use *either* to refer to three of more objects is inaccurate:

Poor: *either* of three choices
Better: *any* of three choices

e.g. *e.g.,* an abbreviation of the Latin *exempli gratia,* meaning "for example," is used to introduce parenthetical examples.

electric/electrical, electronic. The words are not interchangeable. Use *electric* to describe anything that produces, carries, or is activated by an electric current: *electric* appliance, *electric* charge, *electric* circuit. Use *electrical* to describe things that pertain to but do not contain or carry electricity: *electrical* analog, *electrical* engineer. Use *electronic* to describe devices that are activated by the flow of electrons: computers, radio receivers, television sets are *electronic* devices.

ellipsis (. . .). A punctuation mark of three dots indicates that something has been omitted within quoted material. Four dots are used to indicate that the omission occurs at the end of a sentence. The *ellipsis* also is used to indicate that a statement has an unfinished quality.

enormity, enormousness. *Enormity* means horror or great wickedness; *enormousness* refers to size: *the enormity of crime, the enormousness of the national debt.* (Some might call the national debt an *enormity.*)

ensure, insure, assure. *Ensure* and *insure* are often pronounced the same and in some contexts provide a similar meaning: to make certain. In common usage, *insure* means to protect against financial loss, as of life and property. *Assure* means to impart trust, an act of making a person so confident as to set his mind at ease. In England and Canada, life insurance companies use the word *assurance* in their name.

> We took the stereo to an authorized dealer to *ensure* it would be repaired properly.
> At any rate, we were glad we had *insured* the stereo.
> The technician *assured* us the Dolby Noise Reduction unit was fixed.

etc. *etc.,* an abbreviation for the Latin *et cetera,* which means "and so forth," is sometimes used at the end of a list of items. Unless there is some reason for saving space, avoid the use of *etc.* An effective way to avoid this use is to introduce the list with *such as* or *for example.* In textual material, such as this book or a long formal report, use *etc.* only in parenthetical or tabular material. In the body of the text, use *and so forth.* The use of *and* with etc. is redundant.

every, everybody, everyone. *Every* is an adjective; *everybody* and *everyone* are pronouns. *Every* and its compounds are grammatically singular:

> Every instrument in the bench setup was working.
> Everybody is here.
> Everyone who saw the phenomenon took back his own impressions.

Every one becomes two words when used in the sense of every or each one of a group name. *Every one of the six recommendations was adopted.* Though *every one* seems all-inclusive, it requires a singular verb.

exclamation mark (!). An *exclamation mark (or point)* is used after an emphatic interjection or forceful command. It may follow a complete sentence, phrase, or individual word. It should be used sparingly in technical writing.

farther, further. In informal speech, there is little distinction between the two words. In formal writing, *farther* applies to physical distances; *further* refers to degree or quantity.

> We drove on sixty miles *farther.*
> He questioned me *further.*
> The more he read about the matter, the *further* confused he became.

Federal. Capitalize the word when it is part of a name or when used as an adjective synonymous with the United States: *Federal Reserve Board, Federal courts, Federal Government.*

fewer, less. Use *fewer* in referring to a number of individual persons or things. *Fewer than three resistors failed.* Use *less* in referring to quantity: *Less copper was used in this design.*

forego, forgo. To *forego* means to go before, to precede. To *forgo* means to give something up.

gender, the generic "he." Gender in language is a grammatical term that indicates the sex or sexlessness of the referent. Many languages have special endings for feminine, masculine, and neuter nouns and for adjectives modifying them. English abandoned this system several hundred years ago. Gender in English is determined by meaning. All nouns naming living creatures are feminine or masculine according to the sex of the individual, and all other nouns are neuter. In English, as in other languages, we attribute gender or personification to things, inanimate objects and to nouns generalizing persons:

> *The sun sent his warm rays down to earth.*
> *The moon cast her pale light . . .*

> *The ship made her way through the channel.*
> *A good teacher will not lie to his pupils.*

The expression of gender in these examples is based not on grammar, but on traditions of long usage. For most English words, gender is identifiable only by the choice of the meaningfully appropriate pronoun (*he, she, it*):

> *The speaker chose his words carefully.* (a male speaker)
> *The playwright was brilliant in her characterization of the hero.* (a female playwright)
> *The storm made its impact felt.*

Today, most writers reject as sexist the use of the generic "he" for words or terms in which the sex of the referent is ambiguous. For example, the above sentence, *A good teacher will not lie to his pupils* becomes *A good teacher will not lie to his or her pupils.* This simple revision is more applicable to the thought expressed. In other cases, however, a similar revision might result in awkward phrasing. For example, *Every chairman must accept his responsibility to meet with each of his club members* would become *Every chairperson accepted his or her responsibility to meet with each of his or her club members.* Although the word *chairperson* has come into common usage, the repetition of *his or her* is clumsy. Such sentences can almost always be restructured to avoid these conflicts altogether. The above sentence could be written: *Every chairperson (or presiding officer) must accept responsibility for meeting with each club member.* A common way to avoid the use of awkward *his or her* is simply to make the referent plural whenever possible. For instance, *The client is usually the best judge of the worth of his therapy* becomes *Clients are usually the best judges of the therapy they receive.* And *A good teacher will not lie to his pupils* becomes *Good teachers will not lie to their pupils.*

In other examples, *humanity* or *people* could be substituted for *mankind,* and *work force* or *personnel* for *manpower.*

genus, species. A *genus* is a taxonomic class of plants or animals that includes groups of closely related species. A *species* is a taxonomic category immediately below a genus. The name of the species is always preceded by the name of the genus, or larger category of which it is a subdivision. Only the genus is capitalized: *Homo sapiens; Myotis lucigigus.*

good, well. *Good* is an adjective; *well* is both an adverb and adjective. One can say:

> I feel good (adjective).
> I feel well (adjective).

However, the meanings are different. *Good,* in the first sentence implies actual body sensations. *Well* in the second sentence refers to a condition of health—being "not ill." In nonstandard spoken English, *good* is sometimes substituted for *well.* For example:

> The centrifuge runs *good.*

Most educated persons would say:

> The centrifuge runs *well.*

Government, government. Capitalize when referring to a specific national government: the *United States Government*, the *Israeli Government*. Use lower case for general state and local governments: the *city government*, etc.

guarantee, guaranty. Used as nouns, each form is proper. Used as verbs, *guarantee* is more common.

hanged, hung. A person is *hanged*; a picture is *hung*.

heretofore. *Heretofore* is stilted; replace it with *until now*.

hertz, cycle. *Hertz* is a unit of frequency equal to one cycle per second; it has been adopted by the scientific community to replace *cycle*. Thus, it is kilohertz *not* kilocycle. The abbreviation is kHz. Kilohertz is used for both singular and plural.

historic, historical. *Historic* is preferred in the sense that something was or is momentous in human events. *Historical* means related to history: *historic* battle, *historic* invention, but *historical* accuracy.

however. As a conjunction, *however* is a useful connective between sentences to show the relation of a succeeding thought to a previous one in the sense of "on the other hand" or "in spite of." Inexperienced writers tend to overuse *however*. *But* is often more appropriate because it is a more direct connective. Compare the following uses of *however*:

1. In an attempt to make certain that only the fragments of the minus strand were synthesized, the recommended precautions were followed; *however*, when the reactions were carried out, the results indicated a vast excess of the plus strand.
2. In an attempt to make certain that only the fragments of the minus strand were synthesized, the recommended precautions were followed; *but* when the reactions were carried out, the results indicated a vast excess of the plus strand.
3. In an attempt to make certain that only the fragments of the minus strand were synthesized, the recommended precautions were followed. When the reactions were carried out, the results indicated a vast excess of the plus strand, *however*.

The more experienced writer will select use 2 or 3 over the usage in 1 above. *However* is also an adverb; it may modify an adjective or another adverb:

However great the difficulty was, he still did his best.
However hard he tried, he could not finish in time.

hyphen (-). *Hyphens* are a controversial point in style. Modern tendency is to eliminate their use. Consult a good dictionary or the style manual of your organization. A *hyphen* is a symbol conveying the meaning that the end of a line has separated a word at an appropriate syllable and that the syllables composed at the end of one line and the beginning of the next are one word. The *hyphen* is also used to convey the meaning that two or more words are made into one. The union may be ad hoc—for that single occasion—or permanent. Light-yellow flame has a different meaning than light, yellow flame. *Hyphens* are, therefore, used to form compound adjectival descriptive phrases preceding a noun: a beta-ray spectrum, cell-like globule, 21-cm radiation, less-developed countries. Conventionally, *hyphens* are used to join parts of fractional and whole numbers written as words: thirty-three-, one-fourth.

Hyphens are used to set off prefixes and suffixes in differentiating between words spelled alike but having different meanings:

> Recover his composure, and re-cover his losses.
> Recount a story, re-count the proceeds.
> Fruitless endeavor, fruit-less meal.

Hyphens are used between a prefix and a proper name:

> pre-Sputnik, ex-professor
> Pro-relativistic quantum theory

Hyphens are used between a prefix ending in a vowel and a root word beginning with the same vowel:

> re-elected
> re-enter

Hyphenation has become more a publisher's worry than a writer's. Most organizations set a style policy in hyphenation. Some follow the style manual of the University of Chicago (the style used for this text), or the style manual of the U.S. Government Printing Office. Some professional societies have style manuals for the writing done in their fields (e.g., the American Institute of Physics and the Conference of Biological Editors).

Word processing systems have a hyphenation feature to take care of hyphenation requirements. There are three types: a hard hyphen; a soft hyphen; and a nonbreaking hyphen. A hard hyphen is inserted by the user with a hyphen key. The *hard hyphen* stays with the word even if it is wordwrapped to the middle of a line after editing. If the word no longer requires hyphenation, the user must delete it. *Soft hyphens* are inserted by an automatic feature. If the position of the word changes during revision rendering the hyphen unnecessary, the automatic hyphen feature removes the hyphen. A *nonbreaking hyphen* insures that the word will not be broken at the end of a line; as for example, in the phrase: a 1-liter beaker, the characters "1-" would not be separated at the end of the line from the word, "liter," which might otherwise appear at the beginning of the next line. The nonbreaking hyphen feature keeps the phrase "1-liter" together.

i.e. The use of *i.e.,* the abbreviation of the Latin *id est* (meaning *that is*), often saves time and space, but the English *that is* is preferable.

if, whether. *If* is used to introduce a condition; *whether* is used in expressions of doubt.

> *If* the weather holds good, the space launch will be made.
> I wondered *whether* the space launch would be made.
> I asked *whether* the space launch would be made.

imply, infer. These two words are often confused. *Imply* means to suggest by word or manner. *Infer* means to draw a conclusion about the unknown on the basis of known facts.

in, into, in to. *In* shows location; *into* shows direction.

> He remained *in* the laboratory.
> He came *into* the laboratory.

The construction *in to* is that of an adverb followed by a preposition:

> He went *in to* eat.

in-, un-. The prefixes, *in-* and *un-* and their variants, *il-* and *im-*, usually give words a negative meaning: *inadequate, inaudible, incomplete, illiterate, impractical, unacceptable, unnecessary, unresponsive.* Both *in-* and *un-* are used with some words: *in*decipherable, *un*decipherable; *in*distinguishable, *un*distinguishable, *in*supportable, *un*supportable. Problems arise because some related words use a different prefix: *indigestible, undigested; inadvisable, unadvised.* If you are not sure whether a word takes *in-, un-, im-,* or *il-,* consult a dictionary. Not all words beginning with *in* are negative: *inflammable* means the same as flammable. Some other *in* words that do not have negative meaning are: *incubate, indemnity,* and *invaluable.*

include. *Include* is used to introduce a number of items that do not constitute a complete listing: *The new generator included a mercury pump, a rotor, and a set of turbine wheels.*

incredible, incredulous. A statement, story, or situation is *incredible* (unbelievable); a person is *incredulous* (unbelieving).

indentation. *Indentation* in a manuscript or printed copy is beginning a line in from the left-hand margin. In typewritten or word processed copy, paragraphs are indented five spaces. In word processing, you indent by hitting the Tab key for a paragraph first line *indentation.* If you want the whole text material indented (for example, when you quote more than three lines) you hit the indent key.

irregardless. The use of this word, having a negative prefix, *ir,* and a negative suffix, *less,* is considered nonstandard in formal writing. Use *regardless* or *irrespective* instead.

its, it's. *Its* is a possessive form and means belonging to it. *It's* is the contraction for *it is.*

-ize, -ise. English has many verbs ending in the sound of *iz,* some of which are spelled *-ise,* some *-ize,* and some both ways.

   British spelling prefers *-ise*; American spelling, *-ize*: anesthe*tise,* anesthe*tize,* standar*dise,* standar*dize.* In some words, *-ise* is the usual spelling in American English: *advise, devise, revise, surmise.* Both *-ize* and *-ise* are used in a number of words: *advertise, advertize; analyze, analyse.* If you are writing for an American audience, follow American usage.

jargon. *Jargon* or shoptalk is the term for the specialized words or phrases used in a particular profession, trade, science, field, or occupation, as for example: *input, output;*

throughput; infrastructure; metalanguage; heuristic; peer group; dichotomy; complex; party (for *person* in law). Such words are the cliches of specialized fields and should be avoided or used with care to be sure the meaning is clear for the reader. (See *computerese, shoptalk.*)

kind, kinds; sort, sorts; type, types. As nouns, *kind, sort,* and *type* are singular nouns in form: This *kind* of illness. . . . This *sort* of equipment. . . . This *type* of system. . . . Informal usage, however, has added an *s* when the noun they stand before is plural: These *sorts* of problems. . . . These *kinds* of ideas. . . . These *types* of solutions. . . . In formal writing, the addition of an *s* should be avoided.

know-how. This term meaning technical skill has become acceptable in formal writing.

lay, lie. These verbs are often confused. *Lay* (principal parts: *lay, laid, have laid*) is a transitive verb meaning to put something down. *Lie* (principal parts: *lie, lay, have lain*) is an intransitive verb meaning to rest in reclining position.

> I *laid* the tool on the bench.
> The old man *lay* resting on the sofa.

lb., lbs. *lb.* is an abbreviation of the Latin *libra,* the plural of which is *librae.* Because there is no *s* in the Latin plural, some purists argue that *lb.* should be used for both singular and plural forms. Usage has brought acceptance of *lbs.* for the plural of the word pound.

leave, let. *Leave* means to go away or to part with; *let* means to allow or permit.

like, as. Despite common usage, *like* is not fully accepted as a conjunction. The preposition *like* is used correctly when it is followed by a noun or pronoun without a verb. In formal writing, avoid using *like* as a conjunction; *as* should be used instead.

> He writes *like* an ignoramus.
> He looks *like* me.
> Watch and do *as* I do.

literally. This word is often used when *figuratively* is meant: The strike has management *literally* walking a tightrope. The proper word in this sentence is *figuratively.*

madam, madame. *Madam* can be the keeper of a bordello.

material, materiel. *Material* is the term for the substance or substances of which a thing is made or composed. *Materiel* refers to apparatus, arms, and military equipment. The field of business has appropriated the term for the aggregate of things used or needed in an enterprise.

may be, maybe. These two forms are often confused. *May be* is a compound verb; *maybe* is an adverb meaning perhaps.

mean, median. In statistics, a *mean* is an average; a *median* is the figure that ranks midway in a list of numbers arranged in ascending or descending order. For example, in a discussion of the varying wages of 41 workers, the *mean* is the total of their pay divided by 41. The *median* is a wage that is higher than 20 of the wages and lower than the remaining 20.

Messrs. It is the abbreviation of the French word, *messieurs*. It is used as the plural of *Mr.*: *Messrs.* Campbell and Hamilton. In the salutation of a letter when a firm name is addressed, as for example, Campbell and Hamilton & Company, the word *Gentlemen:* should be used. *Messrs.* may be used in text material to avoid repetition of *Mr.* in a listing of three or more: *Messrs. Clark, Boehne, Markman, and Shaw attended the meeting.*

militate, mitigate. The meaning of these two words is sometimes confused. To *militate* (used with the preposition *against*) means to have weight or effect against; to *mitigate* means to ease or to soften a situation.

misplaced modifiers. Modifiers should be placed near the words they modify.

Wrong: Throw the horse over the fence some hay.

Revised: Throw some hay over the fence to the horse.

The accurate placing of the word *only* is critical to precise meaning in technical writing. Notice the change in meaning brought about by shifting the word *only* in the following sentences:

1. *Only* the physicist calculated the value of X in the equation.
2. The *only* physicist calculated the value of X in the equation.
3. The physicist *only* calculated the value of X in the equation.
4. The physicist calculated *only* the value of X in the equation.
5. The physicist calculated the *only* value of X in the equation.
6. The physicist calculated the value of *only* X in the equation.
7. The physicist calculated the value of X *only* in the equation.
8. The physicist calculated the value of X in the *only* equation.

Ms. This title may be used for either single or married women, analogous to Mr. for men; its plural is *Mss.*

namely. When you introduce a series of items, as in the following sentence: The apprentices were thoroughly grounded in the fundamental processes of the work, namely, planning, designing, and building, does the word *namely* add anything? Not really. Experienced writers would omit the word.

neither, neither . . . nor. *Neither* is a word used for two items or subjects and requires a singular verb: *Neither* of the two alternatives was desirable. After a *neither-nor* construction, if the subjects are singular, use a singular verb: *Neither the pump nor the exhaust was working.* If the subjects are both plural, use a plural verb: *Neither the personnel nor the systems were properly evaluated.* If one subject is singular and the other plural, use the number of the one after the *nor*: *Neither the director nor the scientists were surprised by the results.*

No. The abbreviation for number, *No.* is written with an initial capital.

nobody, nothing, nowhere. All three are written as single words. *Nobody* and *nothing* are singular grammatically. *Nowheres* is nonstandard for *nowhere;* do not use it.

none, no one. *None* is commonly used to refer to things (but not always); *no one* is used to refer to people. *None* may take either a singular or plural verb, depending on whether a singular or plural meaning is intended. *None* of the keypunchers are able to work Sundays. *No one* always takes a singular verb.

number (the noun). *Number* is a collective noun, taking a singular or plural verb according to total or individual units meant: A number of pages in the manual *were* missing; the number of missing pages *was* exasperating.

number (of nouns and verbs). The singular and plural aspect of nouns and verbs is termed *number.*

1. A verb agrees in number with its subject; a pronoun or pronominal adjective (my, our, your, his, her, its, their) agrees in number with its antecedent. Examples:

   Right: Our greatest need *is* (not are) modernized plant designs.

   Right: The remaining two tubes of the circuit are V-101A and V-101B. *They* (not *it*) control the power input.

   Right: The closing sentence of both paragraphs and sections often *summarizes* (not *summarize*) the contents and *shows* (not *show*) their significance to the whole.

   Right: Determination of the choice and of the placement of punctuation marks *is* (not *are*) governed by the author's intention of meaning to be conveyed.

2. A compound subject coordinated by *and* requires a plural verb, regardless of the individual number of the member subjects.

   Right: A hammer and a saw *are* on the bench.

   Exception: When the compound subject refers to a unity, a singular verb is used.

   Right: A brace and bit *is* on the bench. (However, a brace and a bit *are* on the bench.)

   Right: Johnson and Sons *is* a reliable distributor.

3. Nouns or pronouns appearing between a subject and a verb have no effect upon the number of the verb. The number of the verb is determined solely by the number of its subject.

   Right: The director of engineering, in addition to two senior engineers and an engineering aid, *was* in the laboratory.

   Right: He, as well as I, *is* to be assigned to the project.

4. Collective nouns (e.g., committee, crowd, majority, number) may be either singular or plural, depending on whether the whole or the individual membership is emphasized.

   Right: The committee *is* holding its first meeting.

   Right: The committee *are* in violent disagreement among themselves.

5. Expressions of aggregate quantity, even though plural in form, generally are construed as singular.

Right: Two times two *is* four (some people say two and two are four).

Right: Two-thirds of the corporation's income *has* been embezzled.

Right: Forty kilowatt-hours *was* registered by the meter.

6. When the expressions of quantity not only specify an aggregate amount but also stress the units composing the aggregate, they are construed as plural.

Right: Two hundred bags of Philippine Copra *were* piled on the dock.

Right: More than a billion pounds of Copra *was* purchased in the Philippines.

7. A relative pronoun (*who, which, that*) should not be taken as singular when its antecedent is a plural object of the preposition *of* following the word *one.*

Right: He was one of the ablest hydrographic engineers who *have* devoted *their* skill to the Coast and Geodetic Survey.

numerals. Practice varies in determining which numbers should be spelled out and which should be written as figures. Authorities frequently state rules that seem reasonable or preferable to them. A current trend is to use figures for all units of measure, such as meter, gram, liter, volt, hectare, and kelvin. Aggregate numbers are those resulting from the addition or enumeration of items. The tendency to differentiate between the two is rapidly disappearing. Most authorities recommend the spelling out of numbers under ten. Round numbers above a million are frequently written as a combination of figures and words: a 56 million dollar appropriation, an employment force of 100 million.

Most authorities recommend writing out a number at the beginning of the sentence. When two numbers are adjacent to each other, the first is spelled out: six 120-watt lamps.

The practice is to spell out small fractions when they are not part of a mathematical expression or when they are not combined with the unit of measure: three-fourths of the area, one-third of the laboratory, one-half of the test population However: use ½ mile, ¾ inch pipe, ⅟₅₀ horsepower, ¼ ton.

Decimals are always written as figures. When a number begins with a decimal point, precede the decimal point with a zero: an axis of 0.52.

Hours and minutes are written out except when A.M. or P.M. follow: 8:30 A.M. to 5:00 P.M.; six o'clock, nine-thirty, half past two; ten minutes for coffee breaks.

Street numbers always appear in numerals: 1600 Pennsylvania Avenue, N.W.

When several figures appear in a sentence or paragraph they are written as numerals: In Experiment 2, there was a total of 258 plants. These yielded 8,023 seeds—6,022 yellow, and 2,001 green.

Numbers indicating order—first, second, third, fourth, etc.—are called *ordinal* numbers. (Numbers used in counting—one, two, three, etc.—are *cardinal* numbers.) First, second, third, etc., can be both adjectives and adverbs. Therefore, the *ly* forms—firstly, secondly—are unneccessary and are now rarely used.

ongoing. This overworked adjective should be replaced by any of its many synonyms: *continuing, progressing, underway, growing.*

on, onto, on to. The distinction between these three words is similar to *in, into,* and *in to* (See entry). The first two words of each set are prepositions; the third term consists of an adverb and a preposition. *On* indicates position: the model on the bench. It also means "time when": *On September 13th,* or continued motion: the technicians worked *on through the night. Onto* suggests movement toward: He climbed *onto the roof* to adjust the antenna. When *on* is used as a separate adverb, the combination *on to* is written as two words: After visiting the library, she went *on to* the laboratory.

oral, verbal. The literate individual does not use these terms interchangeably. *Oral* conveys the idea of spoken words. *Verbal* has the general meaning of words used in any manner spoken, written, or printed. In everyday usage, the distinction is often blurred so that we say a *verbal* agreement for a spoken agreement rather than an *oral* agreement.

or. *Or* is a coordinating conjunction that connects words, phrases, or clauses of equal value.

page numbers. Unless beginning a sentence, use lowercase: page 1; pages 87–213. Abbreviations may be used: p. 5; 331 pp.; pp. 7, 18, 34, and 203; pp. 416–597.

parallel structure. Parallelism promotes balance, consistency, and understanding. Operationally, it means using similar grammatical structure in writing clauses, phrases, or words to express ideas or facts of equal value. Adjectives should be paralleled by adjectives, nouns, by nouns; a specific verb form should be continued in a similar structure; active or passive voice should be kept consistently in a sentence. Shifting from one construction to another confuses the reader and destroys the sense of the meaning. A failure to maintain parallelism results in incomplete thoughts and illogical comparisons.

1. *Faulty:*   Assembly lines poorly planned and which are not scheduled properly are inefficient.

   Revised: Poorly planned and improperly scheduled assembly lines are inefficient.

   Faulty: This is a group with technical training and acquainted with procedures.

   Revised: This group has technical training and a knowledge of procedures.

   Faulty: Before operating the boiler, the fireman should both check the water level and he should be sure about the draft.

   Revised: Before operating the boiler, the fireman should check both the water level and the draft.

   Faulty: U.S. Highway 40 extends from the Coastal Plain, runs across the Piedmont, and into the mountains.

   Revised: U.S. Highway 40 extends from the Coastal Plain, across the Piedmont, and into the mountains.

2. *Illogical shifts:*
   Shift in tense: These effects are usually concentrated in the blood system of the animal. The count of red and white blood cells is affected greatly. The percentage of hemoglobin was also reduced.

   Revised: These effects were usually concentrated in the blood of the animal. The count of red and white blood cells was affected. The percentage of hemoglobin was also reduced.

   Shift in person: First, a new filing system can be introduced. Second, you can train personnel to handle the present complicated system.

   Revised: First, a new filing system can be introduced. Second, personnel can be trained to handle the present complicated system.

Shift in voice:	The technicians began the tests on July 7, and the results were tabulated the following day.
Revised:	The technicians began the test on July 7 and tabulated the results on the following day.
Shift in mood:	Class members should take meaningful notes on the lectures. Do not cram for quizzes that are to follow.
Revised:	Class members should take meaningful notes on the lectures. They should not cram for the quizzes that are to follow.
Confused sentence structure:	It was because of a natural interest that made me choose the profession of technical writing.
Revised:	A natural interest in writing led me to choose the profession of technical writing.

parentheses ( ). *Parentheses* are used to enclose additional, explanatory, or supplementary matter to help the reader in understanding the thought being conveyed. These additions are likely to be definitions, illustrations, or further information added for good measure. *Parentheses* are used also to enclose numbers or letters to mark items in a listing or enumeration. When a clause in parentheses comes at the end of a sentence and is part of it, put the period outside the parentheses: *One of the many principles secreted into the blood stream by the anterior pituitary "master gland" is that which stimulates the secretion of the adrenal cortex (Adreno Cortico Tropic Hormone; ACTH).* If the parenthesized material is independent of the sentence, it requires a period. The period is placed inside the closing parenthesis mark: *When used for replacement or production of artificial hyperadrenatism, the corticosteroids are not curative. (They merely provide symptomatic relief.)* Do not place a comma before a parenthesis mark; if a comma is indicated after the parenthetical expression, the comma should be placed outside the closing parenthesis mark: *The laboratory (National Institute of Standards and Technology), he said, was not involved.*

people, persons. Use *people* for round numbers and groups (the larger the group, the more appropriate people sounds), and *persons* for precise or quite small numbers: *One million people were affected: The corporation notified 2967 persons. Only two persons were infected.*

per. *Per* is a preposition borrowed from the Latin, meaning *by, by means of, through,* and is used with Latin phrases that have found their way and use in English: *per diem, per capita, percent* (or, archaically, *per cent*), *per annum. Per* has established itself by long range usage in business English.

percent, per cent, percentage, proportion. Generally written as one word, *percent* often replaces the word *percentage* or even the word *proportion:* Only a small *percent* of the program was run. With numerals, the *percent* symbol, %, is ordinarily used: 66.67%. *Percent* is used with definite figures. *Percentage* needs a qualifying adjective: A large *percentage* of the population favor the measure. The number of the verb used depends on the object of the preposition *of:* Ten *percent* of the boxes were empty. Ninety *percent* of the book is dull.

period. *Periods* are used to mark the end of complete declarative sentences and abbreviations. Periods are also used in a request, phrased as a question out of courtesy: *Won't you let me know if I can be of further service.*

phenomenon, phenomena. The plural form *phenomena* is frequently misused for the singular *phenomenon*.

plurals of abbreviations, letters, and figures. These plurals are formed by adding *'s* as in *M.D.'s, C.P.A.'s, q's,* and *size 7's.*

plurals of compound nouns. Plurals of compound nouns are formed by the addition of *s* to the most important word; The plural for assistant attorney general is assistant attorneys general, since they are *attorneys* not *generals.* But the plural for deputy associate director is deputy associate *directors.* Similarly, it is *sisters*-in-law, *editors*-in-chief and *points* of view.

possessive (for two names). When you are writing about one thing owned by two persons, it is permissible to use the possessive only once:

Correct: Jack and Jill's hill

Correct: Jack's and Jill's hill

When more than one item and separate ownership is involved, you should show two possessives:

> We shall use Jack's and Jill's computers.

possible, probable. Many things may be *possible* but not *probable.* If your meaning is that the matter is in the realm of likelihood, then the proper word is *probable.* If your meaning is that the matter can be achieved despite difficulties, the proper word is *possible.*

practicable, practical. Any thing that can be achieved or done is *practicable,* but whether it is *practical* depends on other factors: It could be *practicable* to develop pollution-free gasoline, but not *practical* because of cost and political factors.

prepositions. A *preposition* is a word of relation. It connects a noun, pronoun, or noun phrase to another element of the sentence. Certain prepositions are used idiomatically with certain words. For example:

knowledge of

interest in

hindrance to

agrees with (a person), agrees to (a suggestion)

agrees in (principle to a suggestion)

agrees to (a plan)

obedience to

responsibility for

fear of (high places), fear for (her safety)

means of (winning)

connected with

information on, information about

If a person uses an unidiomatic preposition, the reason is probably unfamiliarity with the word or confusion resulting from divided usage of that expression. Usually, we learn

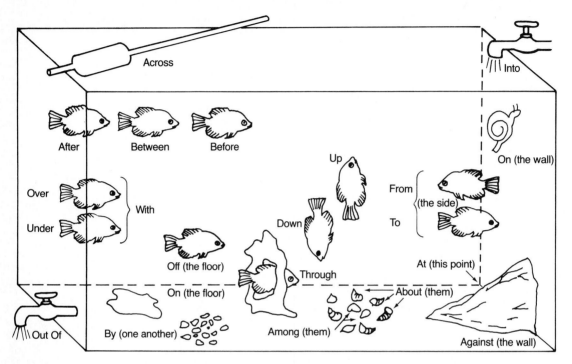

Figure A
From I. A. Richards, Design for Escape. *NY: Harcourt, Brace & World, 1968.*

the use of the proper preposition by hearing or seeing it in its usual construction. Most prepositions, however, are governed by the logic of the relationship they bring to the noun, phrase, or clause from some other element of the sentence. Figure A illustrates the logic of the relationship provided by prepositions.

A number of years ago, it was fashionable for grammarians to put a stigma upon prepositions standing at the end of sentences. Actually, it is a characteristic of English idiom to postpone the preposition. Many technical writers still feel the pain inflicted in years gone by when some die-hard high-school teacher slapped their wrists for ending a sentence with a preposition. Today, they still fall into very clumsy constructions to avoid it. Winston Churchill had the final word on final prepositions. Sir Winston, a famous stylist of his time, had little patience with rigid rules of grammar. When an assistant underlined a Churchillian sentence and noted solemnly in the margin, "Never end a sentence with a preposition," Sir Winston marginally noted back, "This is the sort of English up with which I will not put."

presume, assume. When you *assume* something, you suppose the matter to be a fact with or without a basis of belief. When you *presume* something, you regard the matter as true because there is a reason to do so and you have no evidence to the contrary.

principal, principle. *Principal* is used both as a noun and an adjective. As a noun it has two meanings: (1) the chief person or leader; and (2) a sum of money drawing interest. As an adjective, *principal* means "of main importance" or "of highest rank or authority."

Principle is always a noun and means fundamental truth or doctrine, or the basic ideas, motives, or morals inherent in a person, group, or philosophy.

prior to. This is a stilted expression. Replace it with *before.*

programed, programmed. Purists favor *programed* because they can cite the rule for doubling consonants: If a word is not accented on the last syllable, the consonant is not doubled. Scientists prefer the double consonant in *programmed,* and scientific publications use the double *M.* Your dictionary will tell you that both the single and double *M* are acceptable. Which form should you use? Use the form your community of readers feel most comfortable with, but be consistent in your usage.

provided. Use *provided* not *providing* in the sense of *if: the transducer will work, provided it is coupled to the amplifier.*

pronoun. A *pronoun* refers to something without naming it. Its meaning in the reader's mind must be completed by a clear reference to some other word or group of words called its antecedent.

1. *Pronouns* must have definite, clearly understood antecedents.

Antecedent implied but not expressed:	The generator was overloaded and could not carry it.
Revised:	The generator could not carry the overload.
or:	The overload was so large that the generator could not carry it.
Vague second-person reference:	The generator is not satisfactory when shunt-excited. You must have separate excitation.
Revised:	The generator is not satisfactory when shunt-excited. It must have separate excitation.

2. A *pronoun* agrees with its antecedent in number.

Plural:	Multi-stage amplifiers are the heart of radio, television, and almost all electronic equipment. *Their* circuits have (not *its* circuit has) features of great practical importance.
Singular:	The *antibody* is a protein and is produced as a response to the presence in the blood of foreign antigens; *it* passes (not *they*) into the blood through the lymphatic vessels.

3. Place relative *pronouns* (*who, that, which*) as close as possible to their antecedents.

Faulty:	As transistors and amplifiers have become available that have higher powers and less noise, we have succeeded in developing instruments of improved ranges of operation.
Revised:	As transistors that have higher powers and as amplifiers that have less noise become available, we have succeeded in developing instruments of improved ranges of operation.

punctuation. *Punctuation* is one means by which a writer can achieve clarity and exactness of meaning. Sloppy punctuation can distort meaning and confuse the reader. There are two principles governing the use of punctuation marks:

1. The choice and placement of punctuation marks are governed by the writer's intention of meaning.
2. A punctuation mark should be omitted if it does not clarify the thought.

There is a modern tendency to use open punctuation marks—that is, to omit all marks except those absolutely indispensable. A common mistake of the inexperienced writer is to overuse punctuation marks, especially the comma. A good working rule is to use only those marks for which there is a definite reason, either in making clear or in meeting some conventional demand of correspondence. To illustrate, notice the difference in meaning provided by the punctuation in the following two sets of sentences:

1. (a) The professor said the student is a fool.
   (b) "The professor," said the student, "is a fool!"
2. (a) All of your resistors, which were defective, have been returned.
   (b) All of your resistors which were defective have been returned.

The differences in meaning of sentences 1(a) and 1(b) because of the changes in punctuation are quite obvious. The use of commas in sentence 2(a) makes the sentence say that "all of your resistors were defective, and I have returned them all." Sentence 2(b) without the commas tells the reader that "only those resistors which were defective have been returned." The ultimate test for any punctuation mark is "Is the punctuation needed to make the meaning of my words clearer?"

question mark(?). This punctuation mark is used to denote the end of a question. The question mark is inside quotation marks only when the quoted matter itself is a question.

quotation marks ("). *Quotation marks* are used to enclose direct quotations, titles of articles and reports, and coined or special words or phrases. Quotations longer than three typewritten lines should be indented. No quotation marks are used, and in print, the size of type is usually reduced. In double-spaced typewritten papers, such quotations are single-spaced. The placement of punctuation marks that are not part of the quotation is controversial. Most American publishers place them inside the close-quote. The reason is that the quotes help fill the small spot of white that would be left if the punctuation marks came outside: *What is variously referred to as "inoculation," "homologous serum," "transfusion," or "syringe" jaundice is treatment-produced, resulting from injections of virus-contaminated substances or from injections given with virus-contaminated needles or syringes.* When quotation marks are to be used within a quotation, a single quotation mark is used around the inner quote. For example: *The Research Review Board concluded, "The evidence implicating oncogenes in the development of human cancers is still largely circumstantial. It is 'guilt by association' at the moment, but the association is provocative."*

Re. *Re* is derived from the Latin *in re,* meaning "thing." *Re* is used to mean "in the matter of" or "in reference to." *Re* is used in business letters or memos preceding the text to indicate the subject of the communication; it can be used also in the body of the letter or memo. Its use is standard in legal papers, but is considered pretentious in nonbusiness or nonlegal writing.

re (as a prefix). As a prefix meaning "again," *re* is sometimes followed by a hyphen, sometimes not. When the word to which it is attached begins with an *e,* a hyphen is used; re-employed, re-enlist. If the hyphen is omitted between the prefix and the word and the result is a word with an unintended meaning, the hyphen is needed to avoid confusion: *recover* from an illness; *re-cover* a sofa.

rebut, refute. *Rebut* means to respond to a statement or speaker, to take issue; *refute* means to prove the statement or speaker wrong or false.

recourse, resource, resort. These three nouns are sometimes confused because they are similar in sound and there are some instances in which one might be substituted for another. A *resource* is a reserve asset that is readily available to a person as needed. A *recourse* is the act of turning to someone or something for help, or it is the person or something you *resort* to for aid, giving rise to the expression, "a last *resort*."

recurrence, reocurrence. *Reocurrence* means a second happening; *recurrence* gives the meaning of happening repeatedly or periodically.

redundancies. The same thing said twice is a *redundancy,* a waste of words, and is an annoyance to readers: His plan was increasingly *more* justified. The finance office is *up* above the third floor. Enclosed *herewith/attached* is the contract. It is customary *practice* to charge a higher interest for an unsecured loan. The words in italics are extra verbiage. The expressions are improved without the repetition.

reparable, repairable, irreparable. Both *reparable* and *repairable* have the meaning of capable of being mended, repaired, or remedied, but their usage differs. *Repairable* is used with things: a broken piece of furniture is *repairable,* but a bad situation might be *reparable.* If it can't be *reparable,* then it is *irreparable.*

said. In legal documents, the use of *said* as a demonstrative pronoun (this, that, these, those) has a long tradition of use (*e.g.,* *said* Mr. Peterson, *said* dwelling). However, in business correspondence, it should be avoided as a cliché.

semicolon (;). The principal use of the *semicolon* is to separate independent clauses that are not joined by a conjunction or are joined by a conjunctive adverb or some other transitional term (e.g., therefore, however, for example, in other words). A *semicolon* is also used to separate clauses and phrases in a series when they already contain commas:

> Among the articles offered for sale were a harpsichord, which was at least 200 years old; a desk, which had the earmarks of beautiful craftsmanship; and a table and four chairs, which were also antique.

shoptalk. *Shoptalk* is the specialized vocabulary used by specialists in their everyday work activity. It is a jargon usually known only to the initiated and is both obscure and obstructive to those not "in the know." It is not limited to the shop occupations. Physicians, lawyers, and college professors have their own "talk" that obfuscates. *Shoptalk* is appropriate in informal channels of communication within particular occupations, but it is out of place in formal writing aimed at persons or groups outside of that specialized activity. (See also *computerese, jargon.*)

should, would. *Should* and *would* are used in statements that suggest some doubt or uncertainty about the statement that is being made. Many years ago, *should* was restricted to the first person, but usage now is so divided in the choice of these words that personal preference rather than rule is the guide today. However, consistency in usage should be followed:

> I would greatly appreciate your granting me an interview.
>
> or:
>
> I should greatly appreciate your granting me an interview.

sic. *Sic* is Latin for *thus, so.* Placed in brackets, *sic* is used to mark an error in quoted matter. It shows the reader that the error was in the quoted material and was not made by the quoter:

> The Mississippi River starts at Lake Itaska, Minesota [sic].

since, because. Both words can be used interchangeably to indicate a reason for an action, serving to introduce a subordinate clause.

> *Since* he did not try, he failed.
>
> *Because* he did not try, he failed.

Since, however, functions as several parts of speech:

> As an adverb: She was searching ever *since.*
>
> As a preposition. The victim has been missing *since* last Tuesday.
>
> As a conjunction: She improved her SATC scores *since* she took them last year.

slow, slowly. *Slow* is both an adjective and an adverb; *slowly* is an adverb. As adverbs, the two forms are interchangeable, but *slow* is more forceful than *slowly.*

some, somebody, someone, somewhat, somewhere, etc. *Some* is used as a pronoun and an adjective.

> *Some* think otherwise (pronoun).
>
> He has *some* ideas on the problem (adjective).

Somebody (pronoun), *someone* (pronoun), *something* (pronoun), *someday* (adverb), *somehow* (adverb) are written as one word. As pronouns, they require singular verbs.

spacial, spatial. *Spatial* is the preferred adjectival form of *space. Spatial* is derived directly from the Latin word for space, *spatium.* The word *space* came into English from the French word, *espace.* The nonstandard word *spacial* creeps into news stories and articles frequently.

species, specie. The word *species* is both singular and plural. It refers to a logical division of a genus:

> This *species* of turtle is nearly extinct. (singular)
>
> Many *species* of turtles are amphibious. (plural)

The word *specie* means a coin, usually of gold or silver. Its plural is *species*. To use the term *specie* to represent a single organism of a certain class is erroneous.

spilled. Not spilt.

split infinitives. Usage has won out. To split an infinitive is no longer viewed as a grammatical misdemeanor. Frequently the interpolation of a word between the parts of an infinitive adds clarity and emphasis. For example:

> To carefully examine the evidence
> To forcefully impeach
> To seriously doubt
> To better equip

structure. This is an overworked transitive verb, which can often be replaced by *build, construct, or organize.* Similarly, *restructure* can often be replaced by *rebuild, recast, reconstruct, reorganize, revamp,* or *revise.*

target. *Target* as a verb is military or governmental jargon. *Targeted,* for example, should be changed to *set a target, aimed at,* or *concentrated on.*

tautology. Needless repetition or meanings differently expressed is an extreme form of wordiness. Here are examples:

> true facts
> when gases combine together
> consensus of opinion
> first and foremost
> same identical
> basic fundamentals
> initial beginning

telephone numbers. Standard usage calls for placing parentheses around area code numbers: (301) 468-8119; (202) 634-7658. Current popular usage tends to forgo the parentheses and simply inserts a hyphen between the area code and the number: 301-468-8119. Either usage is acceptable.

than, then. *Then* is an adverb relating to time; *than* is a conjunction in clauses of comparison.

> *Then* came the dawn.
> Fortran is a more universal programming language *than* the other machine languages.

that, which. *That* is preferred in restrictive clauses: *The treatment that the physician selected was a diuretic to reduce intra-ocular pressure.* In nonrestrictive clauses, *which* is mandatory: *The diuretic, which was not expensive, helped reduce the intra-ocular pressure.*

there is, there are. These expressions delay the occurrence of the subject in a sentence. The verb in the expression must agree in number with the real subject.

> *There* is a difficult problem associated with the system.
> *There* are two solutions to the problem.

Often these expressions can be omitted with no loss to the sentence.

> A difficult problem is associated with the system. Two solutions are available for the problem.
>
> or:
>
> The problem has two solutions.

thousand, thousands, hundred, hundreds, million, millions, trillion, trillions. When the number of thousands, hundreds, millions, or trillions is given as in *ten thousand, ten hundred, ten million, a hundred trillion,* the form is singular. If the number is not given, the phrasing is *thousands* of *dollars . . . trillions* of *dollars.*

toward, towards. The two forms are interchangeable.

trademarks. Names of products and processes that are the exclusive property of an individual or company are capitalized: *Coca-Cola, Formica, Frigidaire, Kodak, Polaroid, Technicolor, Xerox,* etc.

underlining. In typewritten or handwritten copy *underlining* is used in place of italics:

1. To indicate titles of books and periodicals.
2. For emphasis: Words that would be heavily stressed when spoken are *underlined.*
3. To indicate foreign words.

unique. The word means single in kind or excellence, unequalled; therefore, it cannot be compared. It is illogical to speak of (or to write): a more *unique* or most *unique* thing.

upward. Not upwards.

use, utilize. Technical writers often overuse *utilize* for the simpler verb *use.*

very. *Very,* an intensive word, has become so overused that its force is slight. If you are tempted to use *very* in a sentence or in the complimentary close of a letter, ask yourself what the word adds to the meaning you wish to convey. If the answer is negative, do not use it.

via. *Via* is the Latin word for *way, road,* or *path.* In English, *via* is used as a preposition to show a relationship to a place such as a highway or a route that passes through a geographical area: You can drive to Minneapolis *via* Chicago. They flew to Japan *via* Hawaii. Purists frown on extending the use of *via* to mean through the agency of or by the means of, since *via* in this context does not refer to a means of transportation. Purists prefer the preposition *by* over *via:* The information reached us *by* telex.

viz. *Viz.* is the abbreviation of the Latin *videlicet* meaning to wit or namely. Since the word *namely* hardly ever adds anything to the thought of a sentence, its Latin abbreviated equivalent certainly does not. Omit its use. Lawyers, alas, will not obey this injunction.

we. *We* is frequently used as an indefinite pronoun in expressions like *we find, we believe.* It is preferable to the passive and impersonal construction: *it is found; it is believed.* The use of *we* is especially recommended in conclusions and recommendations. The reader wants a living, warm-blooded human being giving him findings and beliefs based on judgments. Of course, if the writer represents a single individual the pronoun *I* is used rather than the "editorial" *we.*

when, where. Despite frequent informal usage, it is poor style to define a term by using the phrase *is when* or *is where. When* and *where* are adverbs and so cannot properly introduce noun clauses.

Wrong: A simile is *when* two unlike things are compared.

Right: A simile is a figure of speech in which two unlike things are compared.

Wrong: An injunction *is where* the court orders you to do or not to do something.

Right: An injunction *is a court writ* requiring the doing or refraining from doing a specified act.

Frequently *where* is wrongly used for *when. Where* should introduce an adverbial clause of place; *when,* of time: *Where* experts disagree, we should not rush to quick judgment. *When* should be used in the previous sentence.

which. *Which* is a pronominal word, used in both singular and plural constructions. It refers to things and to groups of people regarded impersonally (The crowd, which was large . . .)

who, whom. *Who* is the form used in the nominative case (as the subject) and *whom,* even in formal usage, is used only as the object of a preposition (objective case).

> He is the one *who* is responsible.
>
> For *whom* the bell tolls.
>
> He struck *whoever* was in his way.
>
> To *whomever* the rule applies.

who's, whose. *Who's* is a contraction of *who is; whose* is the possessive form of the pronoun *who.*

-wise. Use this suffix with care. It is acceptable in traditional forms like *clockwise, lengthwise, otherwise, pennywise, slantwise.* Many people frown on the faddish usage such as *energywise, healthwise,* or *wagewise.*

Xerox. Many people will frown if you use this trademark as a verb. Use *reproduce* or *photocopy.* There are many other reproduction technologies that do not use the Xerox process.

your, you're. *Your* is possessive pronoun; *you're* is a contraction for *you are.*

Your company.
You're to leave before noon.

Discussion Questions and Assignment Exercises in Grammar and Usage

1. Which indefinite article is used with the following words?

   a ___ an ___ aardwolf   a ___ an ___ herbaceous perennial

   a ___ an ___ helium diving bell   a ___ an ___ hour circle

   a ___ an ___ historical event   a ___ an ___ eutherian mammal

   a ___ an ___ M   a ___ an ___ urine specimen

   a ___ an ___ Henle's membrane   a ___ an ___ urea resin

2. What is the difference between an abbreviation and an acronym? List five acronyms.

3. Your instructor will not (accept ___ except ___) late reports unless a student has been (accepted ___ excepted ___) from the previous assignment.

4. Though the engineer (adapted ___ adopted ___) his supervisor's suggestion, the modified circuit would not (adapt ___ adopt ___) properly.

5. The subject's attitude (affected ___ effected ___) the outcome of the experiment.

6. a. Turn the gear (slow ___ slowly ___).

   b. The (slow ___ slowly ___) gear would not move.

7. The (Advisery ___ Advisory ___) Board reached its decision.

8. Professor Jones (alluded ___ referred ___) to the first chapter of *An Essay on the Principle of Population* by T.R. Malthus to substantiate his thesis.

9. The discussions (among ___ between ___) the three corporations ended in an agreement to divide the markets (among ___ between ___) them.

10. The (amount ___ number ___) of barrels with sediment in the experiment was greater than the (amount ___ number ___) of receptacles available, but the total (amount ___ number ___) of sediment was not as great as expected.

11. Place an apostrophe where it belongs in the following.

    Lises first experiment
    The three electronic companies losses
    Everybodys mistakes
    Ours is not the reason why . . .
    The first of two thats
    United States crops during the 80s
    Sir James Jeans experiments
    O.K.d.
    Three cents worth
    Achilles heel

12. The apositive is misplaced in the following sentence. Revise the sentence to correct it.

In acute intermittent porphyria, the liver produces an excessive amount of porphobilinogen, an inborn error of metabolism.

13. Explain why the following sentence is grammatically wrong and then correct it.

    The reason why the experiment failed is because the wrong specimen was used.

14. What is wrong with the following sentence?

    The foreman sent his report in biweekly, every Monday and Thursday.

15. Rewrite the following sentences to eliminate the excessive verbiage and to make the thoughts clearer.

    a. A class A fire is the type of fire that occurs in ordinary everyday materials such as wood, paper, animal and vegetable fibers which are the organic chemical compounds that contain carbon along with some hydrogen, oxygen, and nitrogen and which are normally almost always controlled and extinguished by cooling or the removal of fuel.

    b. Asbestos which has been known to be implicated as a contributing agent involving cancer of the chest and of abdominal membranes has obvious beneficial uses in the manufacture of any number of useful products in our modern complex society such as cement pipe, fireproof clothing, brake linings in automobiles, materials for roofs of houses and buildings, and for floor tiles, to name only a few of these.

    c. A decision tree branches out just like a tree's branches in the form of a network to illustrate basically two things, namely, decisions and results or sometimes these are called events to show the potential consequences of a series of decisions.

    d. In this study a unique field of research relative to fire casualty, i.e., the toxicity of metals in a fire atmosphere by means of concentrated analysis of soot given off during fires by means of atomic absorption spectography.

    e. In this paper I explore and describe the itinerary of personal concerns and experiences that led me to designing a grade school science curriculum which used food as a medium, and from there to another science syllabus, also for fifth graders in which food has become the message.

    f. After being ingested, zinc is absorbed from the digestive tract where it acts like an emetic, partly arresting the chance of systemic poisoning, but, nevertheless, its corrosive properties frequently cause severe gastrointestinal irritation, leading to marked nausea and diarrhea, as well as an inflammation of the stomach and duodenum and to gastric necrosis, resulting in congestion in abdominal viscera, the kidneys and the liver.

    g. Also known as tertiary recovery, a micellar solution of detergents, other chemicals, and water is pumped down injection wells, a multistage pumping operation dependent on chemical collection of rock-held oil.

    h. When the graphite soaks up light, taking in about 90 percent of the visible light that reaches it, heat is transferred to the flowing gas, drawing the heated gas ultimately from a large network of mirrors and pipes by means of a gas turbine coupled to an electric generator, such being a solar heat system.

    i. An every day example of convection heating is the blast of heat experienced when opening a hot oven door.

    j. Instead, a detector known as a heat flux transducer is used whose surface temperature increases in proportion to the intensity or heating rate of the radiation; which is important because it ultimately determines how long a firefighter can withstand the heat from a fire.

16. In the following sentences make the verb and or pronoun agree with (its ___ their ___) subject or antecedent.

    a. Mr. Conway is one of those supervisors who (has ___ have ___) been very helpful with (his ___ their ___) suggestions.

    b. This laboratory is one of several that (require ___ requires ___) a lot of expensive upkeep. We do not believe (it ___ they ___) (is ___ are ___) worth the money.

    c. Many of our patients who (smokes ___ smoke ___) and (uses ___ use ___) saccharin (has ___ have ___) to be worried about the carcinogenic effects on (his ___ their ___) health.

    d. Not one of the students in the chemistry class finished (his ___ her ___ their ___) test questions.

    e. Perennial grasses offer year-round protection, even in a draught, if (it ___ they ___) are not overgrazed. However, grass provides no income for any of the region's farmers who (does ___ do ___) not have cattle.

17. If you think any of the following sentences are grammatically wrong, explain the error and correct the sentence.

    a. On reviewing the various compound gases, it was found there were examples of duplication of the volume of one of the constituents.

    b. To avoid, therefore, confusion and circumlocution, and for the sake of greater precision of expression than I can otherwise obtain, it will be necessary to garner succinct meaning into the framing of thoughts which reflect accurately and validly and not fallaciously the situation my fallible human frailty may allow.

    c. Visual observation was made that Ilford X-ray plates were used which were mounted in a plate-holder, the front of which was covered with opaque black paper.

    d. By the combustion of cyanic acid with copper oxide, it was obtained 2 volumes of carbon dioxide and 1 volume of nitrogen, but by the combustion of ammonium cyanate one must obtain equal volumes of these gases, which proportion also holds for urea.

    e. "d'Alembert's paradox" is when a uniformly translated body suffers no hydro-dynamic drag by virtue of its motion through an infinitely extended inviscid fluid.

    f. In their efforts to summarize and systematize the great proliferation of nuclear particles that were being produced by accelerators on the high-energy frontiers of the 1950's, the quark was discovered.

    g. Silvicultural research in western larch forests prior to the mid-1960's were directed primarily at growing trees for commercial timber production, being a natural outgrowth of demands placed on the forest for timber at that time.

    h. The primary objective of most stand improvement work is to grow bigger and better trees, have a faster growth rate, as well as being economical.

    i. The tape inserted at system initialization is initialized by rewinding the tape and then you write a header record to the tape.

    j. Home fires in the U.S.A. resulted in an estimated 3×10^8\$ of damage in 1976, and the total cost of fire related activities was 13×10^8\$.

    k. During operation of the system, it may be required to examine certain system parameters or initiating specific system functions.

    l. Two spectrographs are included among the space telescope instruments which split light up into its component colors or wavelengths, both measuring the intensity of radiation in each wavelength interval.

m. Within regularly coiled gastropods we can recognize two principal coiling types: isotrophic and anisostrophic, the former being bilaterally symmetrical and the latter which are not.

n. The electron is the most basic particle in chemistry and biology, being responsible for the initial light absorption that causes photochemistry and has been proved to initiate biological events such as photosynthesis as well as the visual transduction process.

o. Can techniques of in vitro fertilization and transplantation of the embryo damage the resulting fetus and lead to abnormal children was a question that is being considered.

18. The following sentences contain restrictive and nonrestrictive clauses or modifiers. Place the commas that are needed for clarity of meaning.

a. This study will determine the conditions under which damaging moisture problems occur in the attic.

b. This study which was to determine the conditions under which damaging moisture problems occur in attics was never funded.

c. The National Institute of Standards and Technology recently completed its first validation test of a tiny electronic device that will be used to protect computer data in transmission and storage based on the Federal Data Encryption Standard.

d. The National Institute of Standards and Technology recently completed its first validation test of a tiny electronic device that will be used to protect computer data in transmission and storage.

e. On another front, scientists in the dental clinical research program are studying improvements in composite resins which will give restorations a smoother surface and make them bond better to the tooth.

f. In the smog formation research project modeling consists of mathematical representation of an "airshed" which is the air over a certain region of the country.

19. The modern tendency is to avoid using commas except where they are needed for clarity of meaning. In the following sentences, place commas to ensure clarity. Some of these sentences will require full restructuring to achieve clarity:

a. As children's requests change with increasing sophistication their mothers switch from establishing the sincerity of a request to identifying the object wanted.

b. Characteristically less than 5 percent of the mother's responses to a child's requests before he is 17 months old have to do with agency or who is going to do or control something.

c. The concept of addiction once thought to be clearly delineated in both its meanings and its causes has become cloudy and confused.

d. Rather than deadening frustration and providing an excuse for aggressive and illegal acts the depression of inhibitory centers through alcohol lubricates cooperative social interactions at mealtimes and other structured social occasions.

e. Plaques develop most often in people who suffer from high blood pressure, diabetes, or obesity and in those who smoke heavily include a great deal of fat in the diet or have a family history of severe cardiovascular disease.

f. Were the top corporate tax rate cut to 42 percent and the top personal income tax rate to 50 percent pre-tax earnings need to net $300 of dividends to top-bracket stockholders would be reduced to $1034, equivalent to $3.45 for each $1 of the stockholder's after-tax dividends.

g. That all areas of temperature were properly monitored it soon became apparent was critical to the experiment.

h. Although checking the thermometer at the freezing level does not guarantee absolute accuracy at the working level say 100° it does assure with a little record keeping that it has not changed from the last "Ice Point" check.

i. The hot and cold water are fed separately into a mixing valve properly called a temperature control unit which in turn compensates for variations of temperature and pressure input thus providing output water of proper precision temperature to the processor.

j. When a pair of these lasers oriented at 90° was placed on a rotating table beat frequency excursions of about 275 kilohertz were observed associated entirely with laboratory-fixed environmental factors presumably the geomagnetic field.

k. In 1980 563 fossils from the Cambrian period 500–600 million years ago were discovered.

l. On the Coram Experimental Forest and elsewhere studies that have included site preparation as one of the variables have demonstrated that exposed mineral soil prepared either by prescribed burning Figure 21 or scarification Figure 22 provides the best environment for establishment of larch regeneration and also as shown in table 7 for other more shade-tolerant species.

m. You will find the data on page 12 Section III.

n. The project engineer wanted to continue the experiment the director of engineering decided to abort.

o. On the other hand that there are short-term fluctuations as measured by sun spots and solar-flare activity seems firmly established and there is mounting observational evidence that the so-called solar constant the amount of solar radiant energy striking a unit surface at the top of the atmosphere oriented perpendicular to the solar beam undergoes small but measurable and possibly regular changes.

20. Correct the dangling modifiers in the following sentences.

a. Combining with tin dental amalgams have been experimented with gold alloys.

b. This blight is known as urban or photo chemical smog interacting automobile emissions with the atmosphere and sunlight forming ozone.

c. Measured and updated by laboratories around the world they require accurate data on chemical reactions rates.

d. Charred, but still readable in the smoldering ruins of a burned out apartment the firefighters found a Smokey the Bear poster.

e. Terrestrial agriculture being seen as inadequate to supply the world's food needs the oceans are a major source of food for mankind and have been looked to with increasing attention.

f. To accept biorhythm it helps to know about circadian cycles.

21. Which of the following two sentences is correct? Explain why.

a. Western larch is well adapted to direct seeding due to its seeds being small so they are not subject to rodent depredation.

b. Failure of the spring seeding was due to moisture depletion.

22. Check the correct word to be used:

a. The metallurgist was able to recover (fewer ___, less ___) than one milligram of the substance.

b. (Fewer ___, Less ___) than half of the committee members voted for the recommendation.

c. Most shoppers bought (fewer ___, less ___) coffee last year.

d. The (council ___, counsel ___) presented its recommendations.

e. The students exhibited their (disinterest ___, uninterest ___) in the professor's lecture by yawning.

f. The first arm of the radio telescope antenna extends 11.8 miles, the second, a little (farther ___, further ___).

g. He went (farther ___, further ___) in his reply than he intended.

h. Two Columbia University seismologists are worried (if ___, whether ___) the three nuclear power plants built on the Ramapo fault up the Hudson River from New York will be (affected ___, effected ___) by a quake.

i. (If ___, Whether ___) there is a 5 to 11 percent chance for a severe quake to occur, the Nuclear Regulatory Commission, they say, made a mistake in approving the plants.

j. His metabolism drove (literally ___, figuratively ___) with the speed of light.

23. Check the correct verb, pronoun, or pronominal adjective in the following sentences.

a. "This pain from shingles (is ___, are ___) too much for me," the patient said.

b. The therapist, using many strategies, (manipulate ___, manipulates ___) the expectations of his subjects.

c. The patient population (has ___, have ___), in this setting, great faith in (its ___, their ___) therapists.

d. One times many (is ___, are ___) a lot.

e. The accelerated recovery and use of the most abundant fossil fuel on earth—coal (are ___, is ___) becoming increasingly complex.

f. The use of electric vehicles (has ___, have ___) (its ___, their ___) pros and cons.

g. Topics of the bilateral discussion (include ___, includes ___) the full spectrum of science.

h. The Ames Laboratory with (its ___, their ___) team of scientists of many disciplines (have ___, has ___) prepared a feasibility study for detecting signals from extraterrestrial intelligence.

i. Commercial reprocessing of these materials (is ___, are ___) among several answers to the dilemma facing this country.

j. Taylor and Smith (is ___, are ___) an expert consulting firm.

k. The president of the firm, as well as his two chief officers, (is ___, are ___) here to take charge.

l. The committee (is ___, are ___) heterogeneous, divided into three groupings.

m. Conscious attention to the comments and questions (provide ___, provides ___) awareness and (permit ___, permits ___) emphasis of elements of judgment that may otherwise be habitually overlooked.

n. From this New Jersey study, it is apparent that the acknowledgment of risk and previous awareness of the cancer issue are additive; those who knew about the cancer rates show less interest in obtaining information than those who did not, and the same basic relationship between appraised risk and information seeking (is ___, are ___) observed for both groups.

24. Which of these numbers should be spelled out and which written as figures?

    a. (one ___, 1 ___) gram
    b. (a million ___, 1,000,000 ___) members
    c. (eighty-three ___, 83 ___) liters
    d. (six seventy-five-watt ___, 6 75-watt ___, six 75-watt) lamps
    e. (⅓ ___, one-third ___) of the subjects in the test
    f. The Redskins scored in the (¼ ___, fourth ___) quarter.
    g. There were (33 ___, thirty-three ___) questions in the test.
    h. (Thirty-three ___, 33 ___) white mice were used.
    i. There was (3 ___, three ___) inches of snow on the ground.
    j. A (¹⁄₅₀ ___, one-fiftieth ___) horsepower unit

25. Correct the following sentences for lack of parallelism and other faults:

    a. Microscopic examination of the centrifuged sediment was accomplished so you could see the contaminants and chemical determinations showed ketone, some yeasts and parasites revealed by the microscope as well as quantification of bacteria.
    b. Four barn owls were used in these experiments which injected a light anesthesia intramuscularly, so we could begin to explore the influence of sound source location on the response projectories of the owl's central auditory neurons for which a movable speaker was used to deliver sound stimuli under free-field conditions we placed in an anachoic chamber.
    c. Clean water is a basic human need, together with diminishing sources of energy, comprising ever increasing this country's current problems.
    d. Once the astigmatism has been corrected, other aberrations still remain, being called spherical aberration, coma, and curvature or field and they arise from the shape of the mirror surface cross section in the tangential plane.
    e. Scientists use the micron to express small distances and wavelengths of light, being denoted by μ and is equal to 10^{-3} mm. in length.
    f. He estimated that the world population of tigers was about 100,000 in 1900 and says the population is only about 7,000 today.
    g. The system has been used in hospitals around the country. You can get personal information in a matter of seconds. We don't have to tell you how important speed and efficiency are in a hospital.
    h. There are two kinds of coronary bypass operations. In the first and most common, you take a length from the saphenous vein in the leg and transplant it in the chest. The second, but less common, uses the internal mammary artery, a small artery that runs alongside the breastbone.
    i. Methadone is still being promoted as a treatment for addiction because it blocks the negative effects of heroin.
    j. There is more to communication than simply indicating and requesting. You use it for social change.

26. Check the correct idiomatic preposition.

    a. abstain (from ___, to ___, with ___)

b. accede (on ___ , to ___)

c. adhere (on ___ , to ___)

d. agree (on ___ , to ___) a plan

e. agree (on ___ , to ___) a proposal

f. agreeable (to ___ , with ___)

g. concur (in ___ , on ___) an opinion

h. concur (with ___ , to ___) a person

i. consist (in ___ , of ___) material

j contend (against ___ , for ___) an obstacle

k. contend (for ___ , with ___) a principle

l. contend (against ___ , with ___) a person

m. correspond (in ___ , to ___ , with ___) a thing

n. correspond (to ___ , with ___) a person

o. differ (about ___ , from ___ , with ___) a question

p. differ (about ___ , from ___ , with ___) a quality

q. differ (about ___ , from ___ , with ___) a person

r. envious (about ___ , of ___)

s. expert (about ___ , at ___ , in ___)

t. identical (to ___ , with ___)

u. independent (of ___ , on ___)

v. infer (from ___ , about ___)

w. need (for ___ , of ___ , in ___)

x. part (from ___ , with ___) a person

y. part (from ___ , with ___) property

z. subscribe (in ___ , to ___)

27. Relative pronouns are misplaced in the following sentences. If there are other faults, revise to make the corrections.

    a. A theory when lightning storms converted such molecules as water methane and ammonia into a "primordial soup" of prebiotic molecules such as amino acids and carbohydrates explains that life originated on earth.

    b. The oryx has traditionally scrabbled its living from sparse desert plants amid the sand and rock of the Arabian peninsula and which is a large beautifully marked antelope.

    c. When atomic nuclei combine with each other to form larger nuclei, some of the mass of the original nuclei is converted into energy according to Einstein's famous formula $E = mc^2$, which is a process we call fusion.

28. The following expressions have circumlocutions or "deadwood," and redundancies (needless repetition). Revise.

    a. any and all

    b. advanced forward

    c. in this day and age

    d. give and convey

    e. vermilion in color

    f. rectangular in shape

    g. large in size

    h. return back

    i. 33 in number

29. For restrictive clauses the pronominal, *that* is preferred. For nonrestrictive clauses, *which* is the pronominal of preference. Check the proper pronominal in the following sentences. Use commas as are needed.

    a. We believe the decrease (that ___, which ___) is associated with the trajectory of resonant energetic electrons is due to the drift they experience in the afternoon sector as they move away from the earth.

    b. Among nonflying mammals (that ___, which ___) regularly visit flowers for food are marsupials of Australia.

    c. The viscous boundary layer (that ___, which ___) is where the flow region is closest to the hull is greatly affected by the ambient pressure field generated by the body's motion through the fluid.

    d. For high-speed memories those (that ___, which ___) require less than about 1-nanosec access time the company uses three-junction interferometers.

    e. The structure (that ___, which ___) took four years to complete is laid out in the shape of a C.

30. Check the proper interrogative pronoun:

    a. Among the ten (who ___, whom ___) were chosen for the award were three women.

    b. The ten (who ___, whom ___) the Board recommended for an award included three women.

    c. The student (who ___, whom ___) the instructor described as the cheater, was not guilty.

    d. (Who ___, whom ___) do you think you are fooling?

    e. (Who ___, whom ___) do you think is responsible?

Selected Bibliography

General References

Abbreviations for Use on Drawings and Text. New York: American Standards Institute, 1972.

Alred, Gerald J., Diana C. Reep, and Mohan R. Limage. *Business and Technical Writing: An Annotated Bibliography.* Metuchen, NJ: Scarecrow Press, 1981.

Barzun, Jacques. *On Writing, Editing and Publishing.* Chicago: University of Chicago Press, 1971.

Barzun, Jacques. *Simple and Direct.* New York: Harper & Row, 1975.

Belkin, Gary S. *Getting Published, A Guide for Business People and Other Professionals.* New York: John Wiley, 1984.

Bernstein, Theodore M. *The Careful Writer: A Guide to English Usage.* New York: Atheneum, 1971.

Boston, Bruce O., ed. *Stet! Tricks of the Trade for Writers and Editors.* Alexandria, VA: Editorial Experts, 1986.

Caernarven-Smith, Patricia. *Audience Analysis and Response.* Pembroke, MA: Firman Technical Publications, 1983.

Calvert, Patricia. *The Communicator's Handbook.* Gainsville, FL: Maupin House, 1990.

Chen, Ching-Chih. *Scientific and Technical Information Sources*, 2d ed. Cambridge, MA: MIT Press, 1986.

Council of Biology Editors, *Style Manual*, 5th ed. Washington, DC: American Institute of Biological Sciences, 1983.

Day, Robert A. *How to Write and Publish a Scientific Paper.* 3d ed. Philadelphia: ISI Press, 1987.

Dobrin, David. *Writing and Technique.* Urbana, IL: National Council of Teachers of English, 1989.

Dodd, Janet S., ed. *The ACS Style Guide for Authors and Editors.* Washington, DC: American Chemical Society, 1986.

Dohiny-Farina, Stephen, ed. *Effective Documentation: What We Have Learned from Research.* Cambridge, MA: MIT Press, 1988.

Ebbit, Wilma R., and David R. Ebbitt. *Writer's Guide and Index to English.* Glenview, IL: Scott, Foresman, 1983.

Evans, Bergan, and Cordelia Evans. *A Dictionary of Contemporary English Usage.* New York: Random House, 1967.

Felker, Daniel B. *Document Design: A Review of the Relevant Research.* Washington, DC: American Institutes for Research, 1980.

Felker, Daniel B. *Guidelines for Document Design.* Washington, DC: American Institutes for Research, 1980.

Firman, Anthony H. *An Introduction to Technical Publishing.* Pembroke, MA: Firman Technical Publications, 1983.

Flower, Linda. *Problem Solving Strategies for Writing.* New York: Harcourt Brace Jovanovich, 1981.

Fowler, H. W. *Modern English Usage.* (Rev. ed. by Ernest Gowers.) New York: Oxford University Press, 1965.

Frank, Francine Wattman, and Paula A. Treichler, eds. *Language, Gender, and Professional Writing: Theoretical Approaches and Guidelines for Nonsexist Usage.* New York: The Modern Language Association of America, 1989.

Goldfarb, Ronald L., and James C. Raymond. *A Guide to Legal Writing.* New York: Random House, 1982.

Gregg, Lee W., and Erwin R. Steinberg, eds. *Cognitive Processes in Writing.* Hillsdale, NJ: Lawrence Erlbaum, 1980.

Guin, Dorothy M., and Daniel Marder. *A Spectrum of Rhetoric.* Boston: Little, Brown, 1987.

Guidelines for the Preparation of Bibliographies. Beltsville, MD: National Agricultural Library, U.S. Department of Agriculture, 1982.

Harre, Rom, and Roger Lamb. *The Encyclopedic Dictionary of Psychology.* Cambridge, MA: MIT Press, 1983.

Harrison, Colin. *Readability in the Classroom.* Cambridge, England: Cambridge University Press, 1980.

Interrante, Charles G., and Frank J. Heymann, eds. *Standardization of Technical Terminology: Principles and Practices.* Philadelphia: American Society for Testing and Materials, 1983.

Karush, William. *Webster's New World Dictionary of Mathematics.* New York: Webster's New World, 1989.

Kiger, Anne Fox. *Acronyms and Initialisms in Health Care Administration.* Chicago: American Hospital Association, 1986.

Kirkman, John. *Good Style for Engineering and Scientific Writing.* London: Pitman Publishing, 1980.

Klare, George R. *A Manual for Readable Writing.* Glen Burnie, MD: REM, 1975.

Klare, George R. *The Measurement of Readability.* Ames, Iowa: Iowa State University Press, 1963.

Kronick, David A. *A History of Scientific Periodicals.* Metuchen, NJ: Scarecrow Press, 1962.

Lapedes, Daniel N., ed. *McGraw-Hill Dictionary of Scientific and Technical Terms.* New York: McGraw-Hill, 1974.

Lederer, Richard. *Anguished English: An Anthology of Accidental Assaults Upon Our Language.* New York: Dell, 1989.

Lippman, Thomas L., comp. *Washington Post Deskbook on Style,* 2d ed. New York: McGraw-Hill, 1989.

Longyear, Marie M. *The McGraw-Hill Style Manual, A Concise Guide for Writers and Editors.* New York: McGraw-Hill, 1983.

McCullo, Marion. *Proofreader's Manual.* New York: Richard Rosen Press, 1969.

McGraw Hill Encyclopedia of Science and Technology, 6th ed. New York: McGraw-Hill, 1987.

A Manual of Style, 13th ed. Chicago: Chicago University Press, 1982.

Meadows, A. J., and Gordon A. Singleton. *A Dictionary of New Information Technology.* London: Kogan Page & New York: Nichols, 1982.

Miles, Thomas H. *Critical Thinking and Writing for Science and Technology.* San Diego: Harcourt Brace Jovanovich, 1989.

Miller, Casey, and Kate Swift. *Handbook of Nonsexist Writing.* New York: Harper & Row, 1980.

Morris, William, and Mary Morris. *Harper Dictionary of Contemporary Usage.* New York: Harper & Row, 1975.

Morton, L. T., ed. *Use of Medical Literature.* 2d ed. London: Butterworths, 1977.

Mullins, Carolyn J. *A Guide to Writing and Publishing in the Social and Behavioral Sciences.* New York: John Wiley, 1977.

Odell, Lee, and Dixie Goswami, eds. *Writing in Nonacademic Settings.* New York: Guilford Press, 1985.

O'Connor, Maeve. *The Scientist as Editor: Guidelines for Editors of Books and Journals.* New York: John Wiley, 1979.

Parker, Sybil P., ed. *McGraw-Hill Dictionary of Scientific and Technical Terms.* New York: McGraw-Hill, 1984.

Quiller-Couch, Sir Arthur. *On the Art of Writing.* New York: G. P. Putnam's, 1916.

Roget's II, The New Thesaurus. Boston: Houghton Mifflin, 1988.

Scientific and Technical Reports—Organization, Preparation, and Production: ANSI Z39. New York: American National Standards Institute, 1987.

Sheehy, Eugene P., ed. *Guide to Reference Books.* Chicago: American Library Association, 1976.

Smith, Peggy. *Proofreading Manual and Reference Guide.* Alexandria, VA: Editorial Experts, 1981.

Strunk, William, Jr. *The Elements of Style.* (Revised by E. B. White). New York: Macmillan, 1959.

Timmons, Christine, and Frank Gibney, eds. *Brittanica Book of English Usage.* New York: Doubleday, 1980.

Toulman, Steven, et al. *An Introduction to Reasoning.* New York: Macmillan, 1984.

Tyrrell, William Blake. *Medical Terminology for Medical Students.* Springfield, IL: Charles C. Thomas, 1979.

U.S. Government Printing Office Style Manual. Washington, DC: U.S. Government Printing Office, 1984.

Walvoord, Barbara E. *Three Steps to Revising Your Writing for Style, Grammar, Punctuation, and Spelling.* Glenview, IL: Scott, Foresman, 1988.

Webster's Encyclopedic Unabridged Dictionary of the English Language. New York: Crown Publishers, 1989.

Webster's Guide to Abbreviations. Springfield, MA: Merriam-Webster, 1985.

Webster's Medical Desk Dictionary. Springfield, MA: Merriam-Webster, 1986.

Webster's Standard American Style Manual. Springfield, MA: Merriam-Webster, 1985.

Weisman, Herman M. *Information Systems, Services, and Centers.* New York: Becker & Hayes; John Wiley, 1972.

Williams, Joseph M. *Style: Ten Lessons in Clarity and Grace,* 2nd ed. Glenview, IL: Scott, Foresman, 1984.

Zinsser, William. *On Writing Well: An Informal Guide to Writing Nonfiction,* 2d ed. New York: Harper & Row, 1985.

Technical Writing and Editing

Alley, Michael. *The Craft of Scientific Writing.* Englewood Cliffs, NJ: Prentice-Hall, 1987.

Anderson, Paul V. *Technical Writing: A Reader-Centered Approach.* New York: Harcourt Brace Jovanovich, 1987.

Andrews, Deborah, and Margaret D. Blickle. *Technical Writing, Principles and Form.* New York: Macmillan, 1978.

Beene, L., and P. White, eds. *Solving Problems in Technical Writing.* New York: Oxford University Press, 1988.

Brunner, Ingrid, J. C. Mathes, and Dwight W. Stevenson. *The Technician as Writer: Preparing Technical Reports.* Indianapolis: IN: Bobbs-Merrill, 1980.

Burrell, Craig D. *Medical Journalism.* New York: Sandoz Corporation, 1986.

Cain, B. Edward. *The Basics of Technical Communicating.* Washington, DC: American Chemical Society, 1988.

Carter, Sylvester P. *Writing for Your Peers.* New York: Praeger, 1987.

Chandler, Harry E. *Technical Writer's Handbook.* Metals Park, OH: American Society for Metals, 1983.

Damerst, Wiliam A., and Arthur H. Bell. *Clear Technical Communication.* San Diego: Harcourt Brace Jovanovich, 1989.

Dirckx, John H. *Dx + Rx—A Physician's Guide to Medical Writing.* Boston: G.K. Hall, 1977.

Dragga, Sam, and Gwendolyn Gong. *Editing: The Design of Rhetoric.* Amityville, NY: Baywood, 1989.

Fearing, Bertie E. and W. Keats Sparrow, eds. *Technical Writing: Theory and Practice.* New York: Modern Language Association of America, 1990.

Helgeson, David V. *Handbook for Writing Proposals That Win Contracts.* Englewood Cliffs, NJ: Prentice-Hall, 1985.

Holtz, Herman. *The Consultant's Guide to Proposal Writing.* New York: John Wiley, 1986.

Houp, Kenneth W., and Thomas E. Pearsall. *Reporting Technical Information*, 6th ed. New York: Macmillan, 1988.

Jordan, Stello, ed. *Handbook of Technical Writing Practices.* New York: John Wiley, 1971.

Kapp, Reginald O. *Presentation of Technical Information.* New York: Macmillan, 1957.

Kolin, Philip C., and Janeen L. Kolin. *Professional Writing for Nurses.* St. Louis, MO: C.V. Mosby, 1980.

Kolin, Philip C. *Successful Writing at Work.* Lexington, MA: D.C. Heath, 1986.

Lannon, John M. *Technical Writing*, 4th ed. Glenview, IL: Scott, Foresman, 1988.

Loring, Ray, and Harold Kerzner. *Proposal Preparation and Management Handbook.* New York: Van Nostrand Reinhold, 1982.

Mancuso, Joseph C. *Mastering Technical Writing.* Reading, MA: Addison-Wesley, 1990.

Markel, Michael. *Technical Writing: Situations and Strategies.* New York: St. Martin's Press, 1988.

Mathes, J. C., and Dwight W. Stevenson. *Designing Technical Reports.* Indianapolis, IN: Bobbs-Merrill, 1976.

Michaelson, Herbert B. *How to Write and Publish Engineering Papers and Reports*, 3rd ed. Phoenix, AZ: The Oryx Press, 1990.

Mills, Gordon H., and John A. Walter. *Technical Writing*, 5th ed. New York: Holt, Rinehart & Winston, 1986.

Moran, Michael G., and Debra Journet, eds. *Research in Technical Communication, A Bibliographic Sourcebook.* Westport, CT: Greenwood Press, 1985.

Niederlander, Carol, David Kvernes, and Sam Sutherland. *Practical Writing.* New York: Holt, Rinehart & Winston, 1986.

Olsen, Leslie A., and Thomas N. Huckins. *Principles of Communication for Science and Technology.* New York: McGraw-Hill, 1983.

Pearsall, Thomas E., and Donald H. Cunningham. *How to Write for the World of Work.* New York: Holt, Rinehart & Winston, 1986.

Pfeiffer, William S. *Proposal Writing, The Art of Friendly Persuasion.* Columbus, OH: Merrill, 1989.

Rook, Fern. *How to Prepare a Science Project Report, Write a Research Paper, Format a Report.* Phoenix, AZ: Fern Rook, 1982.

Samuels, Marylyn Schauer. *The Technical Writing Process.* New York: Oxford University Press, 1989.

Sherman, Theodore A., and Simon A. Johnson. *Modern Technical Writing,* 4th ed. Englewood Cliffs, NJ: Prentice-Hall, 1983.

Sides, Charles H. *Technical and Business Communication.* Urbana, IL: National Council of Teachers of English, 1989.

Souther, James W., and Myron L. White. *Technical Report Writing.* New York: John Wiley, 1977.

Stuart, Ann. *The Technical Writer.* New York: Holt, Rinehart & Winston, 1988.

Warren, Thomas L. *Technical Writing.* Belmont, CA: Wadsworth, 1985.

Whalen, Timothy. *Writing and Managing Winning Proposals.* Boston: Artech House, 1987.

Young, Matt. *The Technical Writer's Handbook.* Mill Valley, CA: University Science Books, 1989.

Zimmerman, Donald E., and David G. Clark. *The Random House Guide to Technical and Scientific Communication.* New York: Random House, 1987.

Graphics and Production

Beach, Mark, Steve Shepro, and Ken Russon. *Getting it Printed: How to Work With Printers and Graphic Arts Services to Assure Quality, Stay on Schedule, and Control Costs.* Portland, OR: Coast-to-Coast Books, 1986.

Bethune, James D. *Technical Illustration.* New York: John Wiley, 1983.

Burden, William J. *Graphics Reproduction: Photography.* New York: Hastings House, 1980.

Caird, Ken. *Cameraready.* Pasadena, CA: Cameraready Corp., 1973.

Campbell, Alastair. *The Graphic Designer's Handbook.* Philadelphia: Running Press, 1987.

Cleveland, William S. *The Elements of Graphing Data.* Monterey, CA: Wadsworth Advanced Books and Software, 1985.

Demoney, Jerry, and Susan E. Meyer. *Pasteups and Mechanicals: A Step-by-Step Guide to Preparing Art for Reproduction.* New York: Watson-Guptill, 1982.

Hanks, Kurt, and Larry Belliston. *Draw! A Visual Approach to Learning, Thinking, and Communicating.* Los Altos, CA: William Kaufman, 1977.

Hartley, James. *Designing Instructional Texts.* New York: Nichols, 1978.

Heller, Steven, and Seymour Chwast. *Graphic Style from Victorian to Post-Modern.* New York: Harry N. Abrams, 1988.

Jastrzebski, Zbigniew. *Scientific Illustration.* Englewood Cliffs, NJ: Prentice-Hall, 1985.

Lojko, Grace R. *Typewriting Techniques for the Technical Secretary.* Englewood Cliffs, NJ: Prentice-Hall, 1972.

MacGregor, A. J. *Graphics Simplified: How to Plan and Prepare Effective Charts, Graphics, Illustrations, and Other Visual Aids.* Toronto: University of Toronto Press, 1979.

Matkowski, Betty S. *Steps to Effective Business Graphics.* San Diego, CA: Hewlett Packard, 1983.

Modley, Rudolph, et al. *Pictographs and Graphs: How to Make and Use Them.* New York: Harper, 1952.

Nelms, Hemming. *Thinking With a Pencil.* New York: Barnes & Noble, 1964.

Parker, Roger C. *Looking Good in Print: A Guide to Basic Design for Desktop Publishing.* Chapel Hill, NC: Ventura Press, 1988.

Pratt, Dan, and Lev Ropes. *35mm Slides: A Manual for Technical Presentations.* Tulsa, OK: American Association of Petroleum Geologists, 1978.

Quick, John. *Artists' and Illustrators' Encyclopedia*, 2d ed. New York: McGraw-Hill, 1977.

Rehe, Rolf F. *Typography: How to Make it Most Legible.* Indianapolis, IN: Design Research Publications, 1977.

Robinson, Artur H., and Barbara Petchnik. *The Nature of Maps.* Chicago: University of Chicago Press, 1976.

Sanders, Norman. *Photographing for Publication.* New York: R.R. Bowker, 1983.

Schenkman, Roger. *The Typing of Mathematics.* Santa Monica, CA: Repro Handbook, 1978.

Schmid, Calvin F., and Stanton E. Schmid. *Handbook of Graphic Presentation*, 2d ed. New York: John Wiley, 1979.

Swann, Alan. *How to Understand and Use Design and Layout.* Cincinnati, OH: North Light Books, 1987.

White, Jan V. *Editing by Design: A Guide to Effective Word-and-Picture Communication for Editors and Designers.* New York: R.R. Bowker, 1982.

White, Jan V. *Graphic Design for the Electronic Age.* New York: Watson-Guptill, 1988.

White, Jan V. *Mastering Graphics, Design and Production Made Easy.* New York: R.R. Bowker, 1983.

Willows, Dale M., and Harvey A. Houghton, eds. *The Psychology of Illustrations, Vols. 1 and 2.* New York: Springer-Verlag, 1978.

Woefle, Robert M. *A Guide to Better Technical Presentations.* New York: IEEE Press, 1975.

Business Communication

Anderson, Paul. *Business Communication.* New York: Harcourt Brace Jovanovich, 1989.

Backman, Lois J., Norman Sigband, and Theodore W. Hipple. *Successful Business English.* Glenview, IL: Scott, Foresman, 1987.

Berryman, Gregg. *Designing Creative Resumes.* Los Altos, CA: William Kaufman, 1985.

Bowman, Joel P., and Bernadine P. Branchaw. *Business Report Writing.* Chicago: Dryden Press, 1984.

Bradley, Patricia Hayes, and John E. Baird, Jr. *Communication for Business and the Professions*, 2d ed. Dubuque, Iowa: Wm. C. Brown, 1983.

Brusaw, Charles T., Gerald J. Alred, and Walter E. Oliu. *The Business Writer's Handbook,* 3d ed. New York: St. Martin's Press, 1987.

Croft, Barbara L. *Getting a Job, Resume Writing, Job Application Letters, and Interview Strategies.* Columbus, OH: Merrill, 1989.

Davis, Ken. *Better Business Writing, A Process Approach.* Columbus, OH: Merrill, 1983.

Dumont, Raymond M., and John M. Lannon. *Business Communication.* Boston: Little, Brown, 1987.

The Encyclopedia of Business Sources (EBIS). Detroit: Gale Research Inc., 1988.

Frank, Darlene. *Silicon English: Business Writing Tools for the Computer Age.* San Raphael, CA: Royall Press, 1985.

Harty, Kevin J., and John Keenan. *Writing for Business and Industry, Process and Product.* New York: Macmillan, 1987.

Hatch, Richard. *Business Writing.* Chicago: Science Research Associates, Inc., 1983.

Holtz, Herman. *Beyond the Resume: How to Land the Job You Want.* New York: McGraw-Hill Paperbacks, 1984.

Jacobi, Ernst and G. Jay Christenson. *On the Job Communication for Business, the Professions, Government and Industry.* Englewood Cliffs, NJ: Prentice-Hall, 1990.

Kogen, Myra, ed. *Writing in the Business Professions.* Urbana, IL: National Council of Teachers of English and the Association for Business Communication, 1989.

Leonard, Donald J., and Robert L. Shurter. *Effective Letters in Business,* 3d ed. New York: McGraw-Hill, 1984.

Murphy, Herta M., and E. W. Hildebrand. *Effective Business Communication,* 4th ed. New York: McGraw-Hill, 1984.

Munter, Mary. *Business Communication: Strategy and Skill.* Englewood Cliffs, NJ: Prentice-Hall, 1987.

Sigband, Norman, and Arthur H. Bell. *Communication for Management and Business.* Glenview, IL: Scott, Foresman, 1986.

Tibbetts, Arn. *Practical Business Writing.* Boston: Little, Brown, 1987.

Weisman, Herman M. *Technical Correspondence: A Handbook and Reference Source for the Technical Professional.* New York: John Wiley, 1968.

Wilkinson, C. W., Peter B. Clarke, and Dorothy C. Wilkinson. *Communicating Through Letters and Reports,* 8th ed. Homewood, IL: Richard D. Irwin, 1983.

Wilkinson, C. W., et al. *Writing and Speaking in Business,* 9th ed. Homewood, IL: Richard D. Irwin, 1986.

Yate, Martin John. *Resumes that Knock 'Em Dead.* Holbrook, MA: Bob Adams, 1988.

Communication

Barker, Larry. *Communication,* 4th ed. Englewood Cliffs, NJ: Prentice-Hall, 1987.

Berger, Charles R., and Steven Chaffee, eds. *Handbook of Communication Science.* New York: Sage, 1987.

Billner, John R. *Fundamentals of Communication,* 2d ed. Englewood Cliffs, NJ: Prentice-Hall, 1988.

Borden, George A. *An Introduction to Human-Communication Theory.* Dubuque, IA: Wm. C. Brown, 1971.

Bryson, Lyman, ed. *The Communication of Ideas*. New York: Institute of Religious and Social Studies, 1948.

Cherry, Colin. *On Human Communication*, 3d ed. New York: John Wiley, 1978.

Dance, Frank E. X., ed. *Human Communication Theory*. New York: Holt, Rinehart & Winston, 1967.

Dance, Frank E. X., and Carl Larson. *The Functions of Human Communication*. New York: Holt, Rinehart & Winston, 1976.

DeVito, Joseph A., ed. *Communication, Concepts and Processes*, Rev. & enl. Englewood Cliffs, NJ: Prentice-Hall, 1976.

Doubleday Pictorial Library of Communication and Language. New York: Doubleday, 1965.

Gordon, George N. *The Languages of Communication: A Logical and Psychological Examination*. New York: Hastings House, 1969.

Holland, A. *CADL, Communicative Abilities in Daily Living*. Baltimore, MD: University Park Press, 1980.

Linsey, Peter H., and Donald A. Norman. *Human Information Processing*. New York: Academic Press, 1972.

Myers, Gail E., and Michele T. Meyers. *The Dynamics of Human Communication*, 5th ed. New York: McGraw-Hill, 1988.

Probert, Walter. *Law, Language and Communication*. Springfield, IL: Charles C. Thomas, 1972.

Schramm, Wilbur, and Daniel F. Roberts, eds. *The Process and Effects of Mass Communication*, Rev. ed. Urbana, IL: University of Illinois Press, 1971.

Schramm, Wilbur. *The Story of Human Communication: Cave Painting to Microchip*. New York: Harper & Row, 1987.

Thayer, Lee. *On Communication: Essays in Understanding*. Norwood, NJ: Ablex, 1987.

Language and Semantics

Andrews, Edmund. *A History of Scientific English*. New York: Richard R. Smith, 1947.

Bates, E. L. *Language and Context; The Acquisition of Pragmatics*. New York: Academic Press, 1975.

Bernstein, Theodore M. *The Careful Writer: A Guide to English Usage*. New York: Atheneum, 1971.

Bernstein, Theodore M. *Reverse Dictionary*. New York: Quadrangle/New York Times, 1977.

Bloomfield, Leonard. *Language*. New York: Holt, Rinehart & Winston, 1933.

Bodmer, Frederick. *The Loom of Language*. New York: W.W. Norton, 1985.

Britton, Bruce K., and John Black. *Understanding Expository Text: A Theoretical and Practical Handbook for Analyzing Explanatory Texts*. Hillsdale, NJ: Lawrence Erlbaum, 1985.

Chase, Stuart. *Power of Words*. New York: Harcourt, Brace, 1954.

Chomsky, Noam. *Knowledge of Language: Its Origin and Use*. New York: Praeger, 1986.

Clark, Herbert H., and Eve V. Clark. *Psychology and Language*. New York: Harcourt Brace Jovanovich, 1977.

Condon, John C., Jr. *Semantics and Communication*, 3d ed. New York: Macmillan, 1985.

Dirckx, John H. *The Language of Medicine.* Hagerstown, MD: Harper & Row, 1976.

Fromkin, Victoria, and Robert Rodman. *An Introduction to Language,* 4th ed. New York: Holt, Rinehart & Winston, 1988.

Giles, Howard, and R. N. St. Clair. *Recent Advances in Language, Communications and Social Psychology.* Hillsdale, NJ: Lawrence Erlbaum, 1986.

Hayakawa, S. I. *Language in Thought and Action,* 4th ed. New York: Harcourt Brace Jovanovich, 1978.

Hughes, John P. *The Science of Language.* New York: Random House, 1962.

Huxley, Aldous. *Words and Their Meaning.* Los Angeles: Ward Ritchie Press, 1940.

Jesperson, Otto. *Language: Its Nature, Development and Origin.* New York: Macmillan, 1949.

Kess, Joseph F. *Psycholinguistics, Introductory Perspectives.* New York: Academic Press, 1976.

Lee, Irving J. *Language Habits in Human Affairs.* New York: Harper & Brothers, 1941.

Linsky, Leonard, ed. *Semantics and the Philosophy of Language.* Champaign, IL: University of Illinois Press, 1952.

Lyons, John. *Semantics,* Vols. 1 and 2. Cambridge: Cambridge University Press, 1977.

Mencken, H. L. *The American Language,* 4th ed. New York: Alfred A. Knopf, 1936. Supplement 1, 1945; Supplement 2, 1948.

Ogden, C. K., and I. A. Richards. *The Meaning of Meaning,* 8th ed. New York: Harcourt, Brace, 1947.

Richards, I. A. *Design for Escape.* New York: Harcourt, Brace & World, 1968.

Sapir, Edward. *Language.* New York: Harcourt, Brace, 1921.

Urban, Wilbur M. *Language and Reality.* New York: Macmillan, 1951.

Walpole, Hugh R. *Semantics: The Nature of Words and Their Meaning.* New York: W.W. Norton, 1941.

Whorf, Benjamin Lee. *Language, Thought and Reality: Selected Writings of Benjamin Whorf.* (Edited by John B. Carroll.) New York: John Wiley, and MIT Press, 1956.

Woodger, J. H. *Biology and Language.* Cambridge, England: Cambridge University Press, 1942.

Research and Its Methods

Altick, Richard D. *The Art of Literary Research.* New York: W.W. Norton, 1963.

Barzun, Jacques, and Henry F. Graff. *The Modern Researcher,* 4th ed. New York: Harcourt Brace Jovanovich, 1985.

Behling, John H. *Research Methods: Statistical Concepts & Research Practicum.* Washington, DC: University Press of America, 1977.

Berkman, Robert I. *Find It Fast: How to Uncover Expert Information on Any Subject.* New York: Harper & Row, 1987.

Bloom, Martin. *The Experience of Research.* New York: Macmillan, 1986.

Braithwaite, R. B. *Scientific Explanation.* New York: Harper Torchbooks, Harper & Brothers, 1960.

Cohen, Morris R., and Ernest Nagel. *An Introduction to Logic and the Scientific Method.* New York: Harcourt, Brace, 1934.

Cohen, Victor. *News & Numbers, A Guide to Reporting Statistical Claims and Controversies in Health and Related Fields.* Ames, Iowa: Iowa State University Press, 1989.

Conant, James B. *Science and Common Sense.* New Haven, CT: Yale University Press, 1951.

Cottam, Keith M., and Robert W. Pelton. *Writer's Research Handbook.* New York: A.S. Barnes, 1977.

Dewey, John. *Logic: The Theory of Inquiry.* New York: Henry Holt, 1938.

Dominowski, R. *Research Methods.* Englewood Cliffs, NJ: Prentice-Hall, 1980.

Emory, William C. *Business Research Methods*, Rev. ed. Homewood, IL: Richard D. Irwin, 1980.

Feibleman, James K. *Scientific Method: The Hypothetico-Experimental Laboratory Procedure of the Physical Sciences.* The Hague, Netherlands: Martinus Nijhoff, 1972.

Games, Paula A., and G. R. Klare. *Elementary Statistics: Data Analysis for the Behavioral Sciences.* New York: McGraw-Hill, 1967.

Grazians, Anthony M. *Research Methods: A Process of Inquiry.* New York: Harper & Row, 1988.

Hibbison, Eric A. *Handbook for Student Writers and Researchers.* Englewood Cliffs, NJ: Prentice-Hall, 1984.

Horowitz, Lois. *Knowing Where to Look.* Cincinnati, OH: Writers' Digest, 1988.

Huff, Darrell. *How to Lie With Statistics.* New York: W.W. Norton, 1954.

Keppel, Geoffrey. *Design and Analysis: A Researcher's Handbook.* Englewood Cliffs, NJ: Prentice-Hall, 1973.

Kish, Leslie. *Survey Sampling.* New York: John Wiley, 1965.

Leaver, R. H., and T. R. Thomas. *Analysis and Presentation of Experimental Results.* New York: Halstead Press, 1975.

Lesko, Matthew. *Lesko's New Tech Sourcebook: A Directory to Finding Answers in Today's Technology-Oriented World.* New York: Harper & Row, 1986.

McCormick, Mona. *The New York Times Guide to Reference Materials.* New York: New American Library, 1986.

Mill, John Stuart. *A System of Logic.* London and New York: Longmans, Green, 1948.

Miller, Delbert C. *Handbook of Research Design and Social Measurements*, 3d ed. New York: Longmans, 1977.

Moore, Nick. *How to Do Research*, 2d ed. Chicago: American Library Association, 1986.

Moran, Michael G., and Debra Journet, eds. *Research in Technical Communication: A Bibliographic Sourcebook.* Westport, NY: Greenwood Press, 1985.

Noltingh, B. E. *The Art of Research: A Guide for the Graduate.* New York: Elsevier, 1965.

Northrop, F. S. C. *The Logic of the Sciences and the Humanities.* New York: Macmillan, 1947.

Parten, Mildred. *Surveys, Polls and Samples.* New York: Harper & Brothers, 1950.

Payne, Stanley L. *The Art of Asking Questions.* Princeton, NJ: Princeton University Press, 1955.

Pearson, Karl. *Grammar of Science.* London: Everyman's Library and J.M. Dent, 1937.

Phillip, John L., Jr. *Statistical Thinking: A Structural Approach*, 2d ed. San Francisco: W.H. Freeman, 1982.

Rivers, William R. *Finding Facts: Interviewing, Observing, Using Reference Sources*. Englewood Cliffs, NJ: Prentice-Hall, 1975.

Russell, Bertrand. *An Inquiry into Meaning and Truth*. New York: W.W. Norton, 1940.

Saslow, Carol A. *Basic Research Methods*. New York: McGraw-Hill, 1982.

Sarton, George. *A History of Science*. Cambridge, MA: Harvard University Press, 1952.

Schiller, F. C. S. *Logic for Use*. New York: Harcourt, Brace, 1930.

Searles, Herbert L. *Logic and Scientific Methods*. New York: Ronald Press, 1948.

Sinclair, W. A. *An Introduction to Philosophy*. New York: Oxford University Press, 1944.

Williams, Frederick. *Reasoning With Statistics: How to Read Quantitative Research*, 3d ed. New York: Holt, Rinehart & Winston, 1986.

Wilson, E. Bright. *An Introduction to Scientific Research*. New York: McGraw-Hill, 1952.

Computers and New Information Technology

Atkinson, William. *Hypercard*. Cupertino, CA: Apple Computers, 1987.

Barrett, Edward, ed. *The Society of Text: Hypertext, Hypermedia, and the Social Structure of Information*. Cambridge, MA: MIT Press, 1989.

Barrett, Edward, ed. *Text, Context and Hypertext: Writing With and For the Computer*. Cambridge, MA: MIT Press, 1988.

Bolter, J. David. *Writing Space: The Computer, Hypertext, and the History of Writing*. New York: Lawrence Erlbaum, 1990.

Bove, Tony, Cheryl Rhodes, and Wes Thomas. *The Art of Desktop Publishing: Using Personal Computers to Publish It Yourself*, 2d ed. New York: Bantam Books, 1988.

Burns, Diane, S. Venit, and Rebecca Hausene. *The Electronic Publisher*. New York: Brady/Simon & Schuster, 1988.

Cavuoto, James, and Jesse Berst. *Inside Xerox Ventura Publisher*, 2d ed. Torrance, CA: Micro Publishing, 1989.

Chicago Guide to Preparing Electronic Manuscripts, for Authors and Publishers. Chicago: University of Chicago Press, 1987.

Consumers Should Know: How to Buy a Personal Computer. Washington, DC: Electronic Industries Association, 1986.

Dvorak, John C., and Nick Anis. *Dvorak's Guide to PC Telecommunications*. New York: Osborne McGraw-Hill, 1990.

Erickson, Fritz J., and John A. Vonk. *Easy PageMaker, A Guide to Learning PageMaker for the IBM PC, Featuring Version 3*. New York: Macmillan, 1991.

Felici, Jim, and Ted Nace. *Desktop Publishing Skills: A Primer for Typesetting with Computers and Laser Printers*. Reading, MA: Addison-Wesley, 1987.

Foerster, Scott. *The Printer Bible*. Carmel, IN: Que Corporation, 1990.

Hobart, R. Dale, Sharon Octernaud, and Sid Sytsma. *Hands-On Computing Using Lotus 1-2-3, WordPerfect 5.0, and dBase IV*. Columbus, OH: Merrill, 1990.

Horn, Robert E. *Mapping Hypertext: The Analysis, Organization, and Display of Knowledge for the Next Generation of On-Line Text and Graphics*. Lexington, MA: The Lexington Institute, 1989.

Horton, William K. *Designing & Writing Online Documentation.* New York: John Wiley, 1989.

Kleper, Michael L. *The Illustrated Handbook of Desktop Publishing and Typesetting.* Blue Ridge Summit, PA: TAB Books, 1987.

Krull, Robert, ed. *Word Processing for Technical Writing.* Amityville, NY: Baywood, 1988.

The Language of Computer Publishing. San Diego, CA: Brenner Information Group, 1990.

Mitchell, Joan P. *The New Writer: Techniques for Writing with a Computer.* Redmond, WA: Microsoft Press, 1987.

Nielson, Jakob. *Hypertext and Hypermedia.* San Diego, CA: Academic Press, 1990.

Pfaffenberger, Brian. *Que's Computer User's Dictionary.* Carmel, IN: Que Corporation, 1990.

Ralston, Anthony, ed. *Encyclopedia of Computer Science and Engineering*, 2d ed. New York: Van Nostrand Reinhold, 1983.

Rardin, Kevin. *Desktop Publishing on the MAC: A Step-by-Step Guide to the New Technology.* New York: New American Library, 1986.

Rizk, A. et al., eds. *Hypertexts: Concepts, Systems, and Applications.* New York: Cambridge University Press, 1990.

Rosenberg, Jerry M. *Dictionary of Computers, Information Processing and Telecommunications.* New York: John Wiley, 1987.

Schneiderman, Ben and Greg Kearsley. *HYPERTEXT HANDS-ON! An Introduction to a New Way of Organizing and Accessing Information.* Reading, MA: Addison-Wesley, 1989.

Shore, John. *The Sachertorte Algorithm and Other Antidotes to Computer Anxiety.* Markham, Ontario: Penguin Books, 1986.

Shushan, Ronnie, and Don Wright. *Desktop Publishing Design.* Redmond, WA: Microsoft Press, 1989.

Sitarz, Daniel. *The Desktop Publisher's Legal Handbook: A Comprehensive Guide to Computer Publishing Law.* Carbondale, IL: Nova, 1989.

The Software Encyclopedia. New York: R.R. Bowker, 1989.

Spencer, Donald D. *The Illustrated Computer Dictionary*, 3d ed. Columbus, OH: Merrill, 1986.

Spencer, Donald D. *An Introduction to Computers, Developing Computer Literacy.* Columbus, OH: Merrill, 1983.

Standera, Oldrich. *The Electronic Era of Publishing.* New York: Elsevier, 1987.

Szymanski, Robert A., et al. *Introduction to Computers and Information Systems.* Columbus, OH: Merrill, 1988.

Wang, W. E., and Joe Kraynak. *The First Book of Personal Computing.* Carmel, IN: Howard W. Sams, 1990.

Webster's New World Dictionary of Computer Terms, 3d ed. New York: Webster's New World, 1988.

Oral Communication

Adams, David. *Preparing and Delivering Technical Presentations.* Boston: Artech House, 1987.

Bryant, Donald C., et al. *Oral Communication*, 5th ed. Englewood Cliffs, NJ: Prentice-Hall, 1982.

Clevenger, T., and J. Matthews. *The Speech Communication Process*. Glenview, IL: Scott, Foresman, 1971.

Ehninger, Douglas, et al. *Principles and Types of Speech Communication*, 10th ed. Glenview, IL: Scott, Foresman, 1986.

Gronbeck, Bruce E., et al. *Principles of Speech Communication*, 10th ed. Glenview, IL: Scott, Foresman, 1988.

Jeffrey, Robert, and Owen Peterson. *Speech: A Basic Text*, 3d ed. New York: Harper & Row, 1988.

Kenny, Peter. *A Handbook for Public Speaking for Scientists & Engineers*. Bristol, England: Adam Hilger, 1982.

Lucas, Stephen E. *The Art of Public Speaking*, 2d ed. New York: Random House, 1986.

Seiler, William J. *Introduction to Speech Communication*. Glenview, IL: Scott, Foresman, 1988.

Skinner, Paul H., and Ralph L. Shelton. *Speech, Language and Hearing*, 2d ed. New York: John Wiley, 1978.

Smith, Terry C. *Making Successful Presentations, A Self-Teaching Guide*. New York: John Wiley, 1984.

Tacey, William S. *Business and Professional Speaking*, 4th ed. Dubuque, IA: Wm. C. Brown, 1983.

Turk, Christopher. *Effective Speaking, Communicating in Speech*. New York: Methuen, 1986.

Wiksell, Wesley. *Do They Understand You?* New York: Macmillan, 1960.

Glossary of Computer and Desktop Publishing Terms

Abort. The procedure for terminating a computer program when a mistake or a malfunction occurs.

Alphanumerics. Alphabetical letters A–Z, numerical digits 0–9, punctuation marks, and special characters such as: #, $, <, *, /, ¦, +, etc. that are capable of being processed by a computer.

Application. The task to be performed by a computer program or system.

Architecture. The organization and interconnection of the components of a computer sytem.

Artificial intelligence (AI). The ability of computers to perform human-like thinking and intelligence.

Assembly language. Programming language that allows a computer user to write a program with mnemonics instead of numeric instructions.

Back matter. The elements of a document that supplement its information, such as a bibliography, appendix, glossary, and index.

Backup (BK). Copy of a file or data set kept for reference in memory in case the original is lost or destroyed.

Batch. A group of records or programs that is considered as a single unit for processing on a computer.

Bar code. Code used on labels to be read by a scanner. Bar codes are used to identify retail sales items.

Baud (B). A unit for measuring data transmission speed. One baud is one bit per second.

Basic input/output system (BIOS). A computer chip that determines how the computer reacts to software commands.

Bit. The smallest unit of information a computer stores. (It takes eight *bits* to make one character and those eight bits make one byte. A megabyte is a million characters. A double-spaced page of text is about 2000 bytes, or 2K. K is the symbol for 1024 bytes. One megabyte is about 500 pages.)

Binary system. The base 2 numbering system, which uses only the digits 0 and 1. Computer data are represented by this system. 0 represents off and 1 represents on.

Block move. Process in which a block of text is moved from one part of a document or file to another place in the document.

Boilerplate. Portions of text that get used over and over again, word for word in different documents. To *boilerplate*, as a verb, means inserting such previously written text from another file into the document being worked on.

Boldface. Type with heavier, darker appearance than regular type.

Boolean algebra. A branch of symbolic logic similar to algebra, which deals in logical relationships rather than numerical relationships.

Boot. To start or restart a computer. It involves loading part of the operating system into the computer's main memory. If the computer is already turned on, it is a "warm" boot; if not, it is a "cold" boot.

Bpi. The abbreviation for bits per inch. Bytes per inch is abbreviated BPI, also.

Break. The beginning or ending point of a line, column, or page.

Bug. A mistake in a computer's program or system, or a malfunction in a computer's hardware. To *debug* means to remove mistakes and correct malfunctions.

Bullet. A bold dot or a similar symbol used to emphasize or set off items in a list.

Byte. A group of adjacent eight bits.

Callout. A label describing parts of a subject in an illustration, often with arrows pointing to the described object.

Camera-ready copy. Text and illustrations laid out for a page in proper size and position, ready to be photographed for a printing plate.

Caption. A title or explanation placed above, below, or to the side of an illustration.

Centering. A software feature that automatically places a word or a group of words in the center of a line or page.

Central processing unit (CPU). The component of a computer system that controls the function of interpreting and executing instructions to the computer.

Character. Any symbol, digit, letter, or punctuation mark, including a blank space, stored and processed by a computer.

Character enhancement. A software feature that allows the user to alter the type style for purposes of emphasis, such as boldfacing or italicizing.

Character recognition. Technology of using electronic machines to identify human-readable symbols automatically, and then to express their identities in machine-readable codes.

Chip. An electronic device, a type of complex on/off switch. The chip can be permanently rigged to perform a certain task or it can be designed to store information. By adding more chips, the user enables the computer to perform many more functions and tasks.

Clip art. Art items in public domain, which can be used free of charge in publications.

Code. To write a program or routine for a computer. Used as a noun, it is a set of rules defining the way in which data are represented.

Coding. Writing a set of instructions to cause a computer to perform specified operations.

Cold boot/cold start. To turn on a computer and load an operating program in it.

Collate. To sort multiple reproduction pages into correctly ordered sets.

Commands. Action a user takes to enact a word processing function, usually by hitting a key or combination of keys.

Command-driven. A software program that requires the operator to enter or command the program to make a desired operation.

Compatibility. The ability of software and hardware of one computer system to work with that of another computer system.

Computer. An electronic machine that can accept data, perform certain functions on that data, and present the results of those operations. A machine that enables the use of a word processing or other software program.

Computerese. The jargon or shoptalk of persons working with computers and information processing systems.

Computer graphics. A software program with the capability of converting digital information into illustrations on the display screen or hard-copy printer.

Computer literacy. A general knowledge of what computers are and how they are used.

Computer network. A system of two or more computers linked by data communication channels.

Computer program. A series of instructions that direct the activities of a computer.

Computer science. The field of knowledge covering all aspects of the design and use of computers.

Copyfitting. Editing text to fit a specific space.

Crop. To cut or trim an illustration or other graphic element.

Crop marks. Tiny marks on the corners or edges of drawings or photographs that tell the camera operator which portions of the illustration to include and which to eliminate (crop out).

Cursor. An on-screen position indicator, usually a blinking line or square.

Cut-and-paste. A feature in some word processing and graphic programs that moves text or graphics from one location in a document to another.

Daisywheel printer. A type of printer that uses a print wheel resembling a daisy. At the end of each "petal" is a fully formed character.

Data. Representations of facts or concepts usually by means of letters, characters, or graphics suitable for communication, interpretation, or processing by persons or machines. Also, the raw material of information.

Database. A collection of related files of information stored together in a logical manner.

Data processing. The process of transforming data into useful information by a computer. It is also referred to as information processing.

Debug. The activity of finding and correcting an error or malfunction in a program or computer system.

Default. A designed action that a program takes in the absence of another specified command.

Desktop publishing. A process for printing that combines the use of a computer with graphics-oriented page-composition software and a high-quality laser printer to prepare the reproduction copy of a document. The quantity printing of that document can either be done by the laser printer or the reproduction copy can serve as a master for offset printing.

Dedicated word processing system. A computer system designed for the function of word processing.

Delete. A software feature that allows the user to remove existing text, data, records, or files.

Density. The amount of data that can be stored in a given area of a storage medium, like a diskette. The higher the density, the more data that can be stored.

Digitize. To convert an image into a series of dots stored by the computer so that the image can be manipulated and placed in publications.

Directory. A listing containing the names and locations of all the files (documents) in a storage medium.

Disk. A magnetic device for storing information and programs in a computer. It can be either a rigid platter (hard disk) or a sheet of flexible plastic (floppy disk). Both types have tracks on which data are stored.

Disk drive. The device that reads data from a magnetic disk and copies it into the computer's memory so it can be used later.

Disk operating system (DOS). The set of programs that controls and supervises the computer; the system uses disks to assemble, edit, and apply programs. DOS is also the proprietary name for the machine language used to program and operate IBM and compatible computers.

Display. Physical presentation of data on the monitor's screen.

Distribution logic. A shared computer system that allows parts of memory and control circuits to be placed at individual work stations.

Document. Writing that has permanence and that can be read by people or machines.

Dot-matrix printer. An impact printer that uses a print head containing from nine to twenty-four pins that produce characters by printing patterns of dots.

Downtime. Length of time a computer is inoperative because of a malfunction.

Drum printer. A line-at-a-time impact printer that uses a rotating drum of 80–132 print positions, with each print position containing a complete set of characters.

Dummy. A mock-up of a document. Sometimes it is a rough preliminary layout; other times it is made from blank pages or photocopies of galley pages.

Electronic bulletin board. An information exchange technology that uses a computer and a modem for users to post and receive messages and information.

Electronic mail (e-mail). A communication service for computer users by which messages are sent to a central computer system ("electronic mailbox") and later retrieved by the addressee.

Ellipsis. Equally spaced periods, usually three, to indicate the omission of text.

End user. Anyone who uses a computer system or its output.

Error message. Messages that appear on the computer screen when the user makes an error. Some messages tell not only what the error is but also how to correct it.

Escape key. A control key on the computer keyboard used to take control of the computer away from a program or to stop a program.

Exchange mode. An editing software feature in which newly inserted characters take the place of (overwrite) existing material occupying the same place. Also called replace mode.

Exit key. The key to press when the user wants to leave the program in operation.

Facsimile (FAX). A form of electronic mail that copies and sends text and graphics via phone lines over long distances.

Fallback. Backup system brought into use in an emergency situation; the reserve database and programs that are switched on in the event of a detected fault in the system being operated.

Fanfold paper. One long, continuous sheet of paper perforated at regular intervals to mark page boundaries and folded fan-style into a stack.

File. A collection of related records treated as a unit.

Floppy disk. A flexible magnetic disk (diskette) for storing a computer's data and programs.

Font. All letter characters, numbers, and symbols of one size and typeface.

Footer. Text of one or more lines of type programmed to appear automatically at the bottom of every page of a document.

Format. The overall appearance of a document based on its design elements such as page size, margins, column width, spacing, and page layout.

Front matter. The pages that precede the main text of a document, including title page, copyright page, preface, acknowledgements, and table of contents.

Function keys. Special keys on a keyboard that are preprogrammed so that when the user presses one, it performs a predetermined function.

Galley. A reproduction of a column of type, usually printed on a long sheet of paper.

Garbage in/garbage out (GIGO). A term used to describe bad data inputted into a computer, resulting in bad output.

Glossary. A word processing program that allows abbreviations to be typed instead of longer words or phrases. As the user types the abbreviations, the full word or phrase appears on the screen.

Grammar checker. A software program that locates grammar and punctuation errors in a document.

Graphics. Software that produces a computer-generated picture on screen, paper, or film. The graphics range from simple line or bar graphs to colored and detailed images.

Gray scale. The number of gradations of gray between white and black that can be generated by a scanner or by software.

Greeking. Simulating text as gray bars to show position of that text on the page; used in layout design of a page.

Grid. A matrix of horizontal and vertical lines used by designers and pasteup artists to align and position the elements of a page.

Gutter. The space between two facing pages or the space between two columns.

Hacker. A term previously applied to a computer enthusiast who was experienced in using computers and enjoyed solving complex or unusual problems on a computer. Presently, it is being applied to persons who intentionally break into other computer systems, whether maliciously or not.

Halftone. The representation of a continuous-tone photograph or illustration as a series of dots that look like gray tones when printed.

Hard copy. The printed output of a computer.

Hard hyphen. The hyphen required by the spelling of the term, for example: two-level dwelling or re-enter.

Hardware. The physical components of a computer system that enable it to function.

Header. Software-programmed text that appears automatically at the top of a page, often a title, subtitle, author's name, or page number.

Help system. Messages on the screen to help users. Some software messages are always in view; others must be called up by command.

HyperCard. A hypertext system developed by Apple Computer. It is called HyperCard because the modules of information agglomerate like stacks of cards.

Hypertext. A generic term for computer software systems that enable on-line identification, callup, and linkage of text modules on specific topics in a nonlinear way.

Initial caps. Text in which the first letter of each word is capitalized.

Inputting. Process of entering data into a computer system. Also called writing.

Insert mode. An editing software feature in which all text to the right of an addition shifts to make room for new material.

Italic. A slanted typeface.

Interface. Points of meeting between a computer and an external entity.

Interactive. A system in which the user is in direct and continual communication with the computer.

Joystick. An electromechanical lever to enable a user to move the cursor.

Justify. To place spaces between words and characters in a column so that both left and right margins are even.

Kerning. The reduction of excess white space between characters to make them fit more tightly on a line.

Keyboard. The input device—similar to a typewriter's—used to enter programs and data into the computer.

Keystroke. The action of pressing a single or combination of keys on a keyboard.

Keyword. A significant and informative word in a title or document that describes the content of the document.

Kludge. A collection of mismatched components that have been assembled into a system.

Laser printer. A nonimpact printer that produces images on paper by directing a laser beam onto a drum, leaving a negative charge in the form of a character to which positively charged toner powder will stick. The toner powder is transferred to paper as it rolls by the drum and is bonded to the paper by hot rollers, creating near-typeset quality text and graphics.

Layout. The arrangement and positioning of text, white space, and graphics on a page.

Letter-quality print. Print of the highest quality, having fully formed characters, and similar to a good typewriter print.

Line art. A drawing that contains no grays or middle tones; line art is made up exclusively of lines and white space.

Light pen. A light sensitive, penlike electronic device that allows the user to communicate with the computer by touching the pen to certain positions on the screen.

Lowercase. Small letters, as opposed to capital letters.

Machine language. The basic language of computers. It is based on electronic states and operates on the binary system of 1's and 0's.

Macro. A software feature that makes use of previously recorded keystrokes or commands that can be executed with one or two keystrokes.

Makeup. The physical assembly of components of a page, usually on stiff cardboard. Sometimes called pasteup.

Mainframe. A large-scale, general-purpose computer, often having multiple users.

Mechanicals. Camera-ready pages on art boards, with text and art in position.

Memory. The computer's storage space.

Menu. A list of options in a software program that allows the user to choose which to operate.

Menu-driven. A software program that gives the user a menu (list) of choices—commands or directions on how to enter input.

Microform. A photographic image, usually of text pages, too small to be seen without magnification. Microfiche and microfilm are the names given to two types of microform display formats.

Modem. A computerese abbreviation for modulator/demodulator. A device that enables computers to send information over telephone lines.

Monitor. A television-like device for displaying data.

Mouse. A small hand-held device that allows the user to position the cursor on the screen of the monitor without using the keyboard.

Multimedia/hypermedia. Multiple forms of communication media, such as a book (text, photographs, and illustrations) and video (audio and moving visuals) that are controlled, coordinated, and integrated by a computer.

Natural language processing. The ability of a computer to understand and perform operations from commands issued in a natural language, such as English.

Near-letter-quality print. Print made from dots rather than fully formed characters, which approach the appearance of letter-quality print.

Network. A system of interconnected computer systems and terminals.

Numeric pad. The set of numbered keys to the right of the typewriter keys on the keyboard. Most of these keys are preset to perform word processing functions when the appropriate software is used.

Off-line. Equipment, devices, or persons not in direct communication—not connected to—the central processing unit of the computer.

Offset printing. A process by which a page is reproduced photographically on a metal plate attached to a revolving cylinder. Ink is transferred from the plate to a rubber blanket, from which it is transferred to paper.

On-line. Equipment, devices, and persons in direct communication—connected with—the central processing unit of the computer.

Operating system. Software that controls and supervises a computer system's hardware and provides computer services to users.

Optical character reader. An electronic device that accepts a printed document as input and processes it in the computer so that it may be edited with word processing software.

Orphan. The first line of a paragraph isolated at the bottom of a page.

Outliner. A software feature that helps the user to organize ideas by presenting them as headings or subheadings that can be arranged in a logical pattern.

Output. Information produced by the computer. In word processing, *output* is what the user keys into the computer and sees on the screen or that which is printed.

Page break. A software feature that controls the number of lines to a page. After the set number of lines are reached, the additional text is automatically placed on the next page.

Pagination. The process of numbering pages. In desktop publishing, it is the process of laying out graphics and blocks of text for the purpose of designing all pages of a document.

Paintbrush. A graphics software program that provides the user with a variety of brush shapes on the display screen by means of a mouse or joystick. As the brush moves, it leaves behind a trail in the selected shape.

Pan. A graphics software feature that moves the entire graphic content on the screen from side to side, allowing the user to create and view images that are wider than the screen.

Paperfeed. Method by which paper is pulled through a printer.

Password. A special word, code, or symbol that must be presented to the computer system to gain access to its resources.

PC. Computerese for personal computer.

Peripheral equipment. Input/output devices and auxiliary storage units of a computer system, attached by cables to the central processing unit. Sometimes called peripherals.

Pixel. Short for picture element, it is the smallest unit on a computer screen. The clarity of the screen depends on the number of pixels per inch on the monitor; the more pixels available, the clearer the image.

Preprogrammed keys (macros). Single keys or a limited number of keystroke combinations that are predefined and used to insert a word or a whole passage into a document.

Printer. Output device that produces hard copy.

Printout. Output printed on a page by a printer.

Processing. The steps a computer takes to convert data into information.

Program. A series of instructions that tells the computer what to do.

Prompt. A message the software provides to indicate it is ready to accept keyboard input. Usually, it is an on-screen question or instruction that tells the user what data to enter or what action to take.

Proof. A trial copy of a page or publication used to check the accuracy of its information. (Also, short for *proofread,* meaning to check for mistakes.)

Protocol. A set of rules and procedures for transmitting and receiving data so that different devices can communicate with one another.

Purge. To erase a file.

Qwerty. The standard keyboard arrangement of characters. It is called Qwerty after the first six letters on the top alphabetic line of the keyboard.

Random-access memory (RAM). The working but temporary memory of the computer. It is the memory into which the user enters information and instructions and from which the user can call up data. If power is shut off or lost before the user directs the computer to save it, the data are lost.

Read-only memory (ROM). A solid state storage chip, programmed at the time of manufacture to have a set of frequently used instructions. ROM does not lose its programs when the computer is shut off, but the user cannot reprogram it.

Reboot. To restart the operating system. It usually occurs by human intervention resulting from a problem.

Record. A collection of related items of data treated as a unit.

Recto. A right-hand page.

Register. Precise alignment of printing plates or negatives.

Resolution. The density of dots or pixels on a page or display, usually measured in dots per inch (dpi). The higher the resolution, the finer appearance of text or graphics. Fineness of resolution depends on the printer. In current desktop publishing technology, resolution ranges from 300 dpi in most laser printers to 2540 dpi in Linotronic imagesetters.

Retrieve. To call back from the computer's storage input that the user wants to review or edit.

Roman. A vertical-style type as opposed to italics or oblique.

Roman-style typefaces. Typefaces having serifs in their characters.

Routine. A set of computer instructions for carrying out a specific processing operation; the term is sometimes used as a synonym for program.

Running head. A line at the top of the page that helps orient the reader, containing such items of information as title, author, chapter, date, or page number. Also called header.

Sans serif. A typeface without finishing strokes at the end of the character. *Sans* is the French word meaning *without*. Sans serif typefaces are sometimes called block or gothic type.

Save. A word processing function that when keyed stores what is being written within the computer's random access memory. What is saved is stored permanently and can be retrieved for review, revision, and printing. Also called store.

Scanner. A device that electronically converts a photograph or art piece into a collection of dots that can be manipulated by a software program and placed into a page layout for reproduction.

Screen. A part of the terminal that displays what is being entered into the computer.

Scroll. A software feature that permits moving of the data into view on the screen, up, down, or across, so the viewer can review elements of the text that has been entered.

Search and replace. A software feature that allows the user to look for a word or phrase and replace the item with a corrected term wherever it occurs in the text.

Serif. A line or curve projecting from the end of a letter form.

Shared logic. Concurrent use of a single computer by multiple users.

Shareware. Free software that users can get from user groups and electronic bulletin boards.

Signature. A section of a book or document consisting of a large sheet of paper folded after printing into smaller page sizes within the final publication. Most printers use signatures of 16 or 32 ultimate pages.

Soft copy. Data presented on the display screen of the terminal.

Soft hyphen. A hyphen printed only to break a word between syllables at the end of a line.

Software. The instructions that direct the operations of a computer.

Sort. To arrange data according to a logical system.

Spelling checker. Sometimes called a dictionary, it is a software program that locates misspelled words or words not in the word processing program's dictionary.

Split screen. Display screen that can be partitioned into two or more areas (windows) so that within each screen area formats, data, or help information can be shown at the same time.

Work station. A combination of a monitor and keyboard used to enter input, view output, and edit it within shared logic systems; it does not contain a central processing unit.

Wraparound. Text that wraps around a graphic. Also called runaround.

Writing. The process of entering data into a computer. Also called inputting.

WYSIWYG (Pronounced wizzy-wig). The acronym for "What you see is what you get." It refers to the representation on a computer screen of text and graphic elements as they will look on the printed page.

Xerography. The process used by laser printers and photocopiers in which light conveys an image to an electrostatically charged surface. The image is picked up by toner and transferred to paper.

Zoom. To view an enlarged (zoom in) or reduced (zoom out) portion of a page or screen.

Index